Cardiac and Vascular Biology

Volume 3

The book series gives an overview on all aspects of state-of-the-art research on the cardiovascular system in health and disease. Basic research aspects of medically relevant topics are covered and the latest advances and methods covering diverse disciplines as epigenetics, genetics, mechanobiology, platelet research or stem cell biology are featured. The book series is intended for researchers, experts and graduates, both basic and clinically oriented, that look for a carefully selected collection of high quality review articles on their respective field of expertise.

More information about this series at http://www.springer.com/series/13128

Viacheslav Nikolaev • Manuela Zaccolo
Editors

Microdomains in the Cardiovascular System

Springer

Editors
Viacheslav Nikolaev
Institute of Experimental Cardiovascular Research
University Medical Center
Hamburg-Eppendorf
Germany

Manuela Zaccolo
Department of Physiology, Anatomy and Genetics
University of Oxford
Oxford
United Kingdom

ISSN 2509-7830 ISSN 2509-7849 (electronic)
Cardiac and Vascular Biology
ISBN 978-3-319-54578-3 ISBN 978-3-319-54579-0 (eBook)
DOI 10.1007/978-3-319-54579-0

Library of Congress Control Number: 2017952370

Printed on acid-free paper

This Springer imprint is published by Springer Nature
The registered company is Springer International Publishing AG
The registered company address is: Gewerbestrasse 11, 6330 Cham, Switzerland

Preface

Cardiovascular disease is the leading cause of death in the western world with an enormous global health care burden. Heart failure affects around 26 million individuals worldwide costing dozens of billion dollar in the USA or Europe every year. Despite intensive medication, the 5-year survival for heart failure patients remains at only ~50%, and projections indicate a 46% prevalence increase from 2012 to 2030. Available therapeutic strategies are based on general blockade of neurohormonal mechanisms using β-blockers and inhibitors of the renin-angiotensin-aldosterone system. In addition, inhibitors or natriuretic peptide degradation was recently introduced into the medical practice, while designer natriuretic peptides are in clinical trials. However, treatment remains symptomatic and results are unsatisfactory for a significant number of patients: current drugs associate with serious side effects and are ineffective for some patient groups. Present therapeutic strategies largely ignore signalling processes occurring in cardiomyocytes at the subcellular level.

Evidence accumulating over the past few decades clearly shows that cardiac disease is associated with dramatic alterations in functionally relevant subcellular microdomains where the major second messengers cAMP, cGMP and calcium come into action. To understand how such microdomains work and regulate cardiac function and disease, several new technologies, including biosensors for live cell imaging as well as sophisticated biochemical, electrophysiological and non-optical imaging techniques, have been developed and widely applied to cardiac and vascular cells. The wealth of data obtained this way has transformed the theory of second messenger compartmentation into a widely accepted paradigm. It is also becoming apparent that microdomains might be the key to development of more specific cardiovascular therapeutics which could potentially lead to a breakthrough in heart failure therapy and long desired improvement of patients' survival and care.

This volume covers recent findings in the research area of second messenger microdomains with the focus on the cardiovascular system and disease, presented by the renowned leaders in the field. The first part covers cyclic nucleotide microdomains and includes chapters on cAMP and cGMP microdomains. The second part is dedicated to calcium microdomains.

We are grateful to all authors who provided outstanding chapters for this book. The book illustrates the work done in the field over the last decades and gives a good account of how this research has significantly advanced our understanding of the role of second messenger microdomains in cardiovascular physiology and pathophysiology. With this knowledge it is now our common interest and goal to carry these exciting insights over translational paths to new therapeutic applications.

Hamburg, Germany — Viacheslav Nikolaev
Oxford, UK — Manuela Zaccolo
2017

The original version of this volume was revised as it was originally published unnumbered. The revised version now has been numbered; numbering is done following the order of appearance in the book series Cardiac and Vascular Biology. The respective correction note can be found at https://doi.org/10.1007/978-3-319-54579-0_18

Contents

Part I

Cyclic Nucleotide Microdomains

Receptor-Cyclic Nucleotide Microdomains in the Heart

1

Nadja I. Bork and Viacheslav O. Nikolaev

Contents

Abstract

Cyclic nucleotides 3′,5′-cyclic adenosine (cAMP) and 3′,5′-cyclic guanosine monophosphates (cGMP) are important second messengers which regulate cardiac function and disease by acting in spatially separated subcellular microdomains. Function of these microdomains includes but is not limited to the modulation of calcium cycling, excitation-contraction coupling, and cardiac hypertrophy. In recent years, visualization of local compartmentalized cAMP and cGMP dynamics became possible due to rapid development of optical and nonoptical imaging techniques. In this chapter, we will briefly review these state-of-the-art biophysical methods and available fluorescent biosensors which can be used to understand microdomain-specific signaling and its involvement in cardiovascular function and disease.

N.I. Bork • V.O. Nikolaev (✉)
Institute of Experimental Cardiovascular Research, University Medical Center Hamburg-Eppendorf, Martinistr. 52, 20246 Hamburg, Germany

DZHK (German Center for Cardiovascular Research) Partner Site Hamburg/Kiel/Lübeck, Hamburg, Germany
e-mail: v.nikolaev@uke.de

V. Nikolaev, M. Zaccolo (eds.), *Microdomains in the Cardiovascular System*, Cardiac and Vascular Biology 3, DOI 10.1007/978-3-319-54579-0_1

Abbreviations

cAMP	3′,5′-Cyclic adenosine monophosphate
CM	Cardiomyocyte
CFP	Cyan fluorescent protein
cGMP	3′,5′-Cyclic guanosine monophosphate
FRET	Förster resonance energy transfer
GFP	Green fluorescent protein
IBMX	3-Isobutyl-1-methylxanthine
ISO	Isoproterenol
PDE	Phosphodiesterase
PLN	Phospholamban
SICM	Scanning ion conductance microscopy
SR	Sarcoplasmic reticulum
YFP	Yellow fluorescent protein
β-AR	β-Adrenergic receptor

1.1 Introduction

Many cells of our body including cardiomyocytes (CMs) are under constant extracellular stimulation with hormones and neurotransmitters. These molecules help adapt the heart function to acute and chronic needs of the body, enabling effective and optimal oxygen supply of all tissues. Nature has selected a limited number of molecules called second messengers that convey input signals sensed by hormone and neurotransmitter receptors to transduce them into a plethora of functional responses at the intracellular side. After the discovery of 3′,5′-cyclic adenosine monophosphate (cAMP) as a second messenger by Earl Sutherland (Sutherland and Rall 1958) and isolation of 3′,5′-cyclic guanosine monophosphate (cGMP) from rat urine (Ashman et al. 1963), the question of how just a couple of second messengers can specifically trigger myriads of often opposing cellular responses had remained unanswered for almost three decades (Beavo and Brunton 2002; Fischmeister et al. 2006).

It was only in the late 1980s that classical biochemical methods including cell fractionation and enzyme activity measurements provided the first evidence that stimulation of CMs with two different receptor ligands, the β-AR agonist isoproterenol (ISO) and prostaglandin E1, leads to cAMP synthesis in distinct subcellular compartments, in this case soluble and membrane fractions, which differentially affect CM contractility (Buxton and Brunton 1983; Hayes et al. 1980). The subsequently developed theory of cyclic nucleotide compartmentation now became a widely accepted paradigm. The formation of distinct spatially segregated subcellular signaling microdomains or macromolecular signalosomes enables a plethora of physiological responses engaged by the same second messengers in one cell (Fischmeister et al. 2006; Perera and Nikolaev 2013; Zaccolo 2009). In this book, the most recent findings and the current state of the art for cAMP, cGMP, and calcium microdomains in the cardiovascular system are presented by the leaders in the field. This first chapter gives a short overview of the modern biophysical techniques and tools capable of visualizing second messengers in intact cells and tissues.

1.2 Biosensors for cAMP and cGMP

Classical antibody-based techniques such as radioimmunoassays, enzyme-linked immunosorbent assays, or Western blots have been traditionally used to measure cAMP and cGMP concentrations or to monitor phosphorylation of target proteins by downstream protein kinases. However, these methods share the disadvantage of measuring only total rather than physiologically more relevant free cyclic nucleotide concentrations and the entire lack of spatial resolution at the subcellular level (Brooker et al. 1979; Williams 2004). To study cyclic nucleotide compartmentation and dynamics in living cells within high temporal and spatial resolution, new biochemical techniques have recently been developed and widely applied to cardiovascular research (Sprenger and Nikolaev 2013).

In particular, multiple fluorescent biosensors for cAMP (see Table 1.1) and cGMP (see Table 1.2) have been developed and used in various settings. The majority of these biosensors rely on the principle of Förster resonance energy transfer (FRET). This is a quanto-mechanical phenomenon which allows a nonradiative energy transfer between two fluorescent molecules, the excited donor and a closely located (typically at a distance below 10 nm) acceptor fluorophore. Upon excitation of the donor molecule, FRET leads to a fluorescent emission of the acceptor as well (which does not need to be directly excited), with an efficiency inversely proportional to the sixth power of distance between the fluorophores. This principle is widely used in cell biology, typically with various green fluorescent protein (GFP) mutants such as cyan (CFP) and yellow fluorescent protein (YFP) as donor and acceptor fluorophores, respectively, to monitor protein-protein interactions and conformational changes (Zaccolo 2004; Zhang et al. 2002). These two options are employed to generate FRET-based biosensors for various biologically active molecules and biochemical processes.

One of the first studies which provided direct visualization of microdomains with high cAMP concentrations in neonatal rat CMs was possible due to the development of a genetically encoded FRET biosensor in which regulatory (R) and catalytic (C) subunits of the cAMP-dependent protein kinase (PKA) were fused to CFP and YFP (Zaccolo and Pozzan 2002). By monitoring a decrease of FRET resulting from R and C subunit dissociation, cAMP levels could be measured at the functionally important sites of PKA localization. Later on, many more biosensors including those with a simpler design have been developed and used in CMs and other cells. The main information about these biosensors is summarized in Table 1.1. In addition to evenly localized cytosolic probes, several targeted versions of these biosensors were generated to visualize cAMP in the subcellular microdomains and to study the mechanisms of its compartmentation.

Similar development has occurred for cGMP biosensors as well, although there is still a very limited number of targeted version available (Table 1.2). In addition,

Table 1.1 Förster resonance energy transfer (FRET)-based cAMP biosensors used to visualize global- and microdomain-specific signals in cardiomyocytes (CMs)

Name, sensitivity	Design	Properties	Targeted versions, use in myocytes	References
FlCRhR (PKA based) cAMP EC_{50} = 90 nM	Purified PKA R and C subunits chemically labeled with fluorescein and rhodamine	Not genetically encoded. Labeling, purification, and microinjection required. Relatively slow kinetics	Not available No	Adams et al. (1991)
R-CFP, C-YFP (PKA based) cAMP EC_{50} = 0.5–0.9 μM	Genetically encoded PKA R and C subunits fused to CFP and YFP	Tetrameric complex. Relatively slow kinetics	pmPKA-R-CFP/C-YFP for subsarcolemmal domain (contains CAAX box) Mostly in neonatal and rarely in adult CMs, AGT	Mongillo et al. (2004), Zaccolo et al. (2000), Zaccolo and Pozzan (2002), Dyachok et al. (2006), Warrier et al. (2005)
PKA camps (PKA based) cAMP EC_{50} = 1.9 μM	Single cAMP-binding domain B from the R subunit sandwiched between CFP and YFP	Simple single-chain design	Not available No	Nikolaev et al. (2004)
AKAR1–3	PKA activity reporter consisting of substrate sequence and 14-3-3 or FHA1 protein sandwiched between CFP and YFP variants	Measures PKA activity in real time	SR-AKAR3 (fusion to PLN transmembrane domain). AKAR3-NLS (nuclear). Neonatal and adult CMs, AGT	Allen and Zhang (2006), Zhang et al. (2005), Zhang et al. (2001), Liu et al. (2011), Haj Slimane et al. (2014)
AKAR4	As above but with other CFP and YFP mutants	Improved dynamic range	ER-AKAP4 (ER matrix, fusion to CYP450) Only mouse neonatal CMs	Depry et al. (2011), Mehta et al. (2014), Tao et al. (2015)

Epac1/2-camps (Epac based) cAMP $EC_{50} = 2.4/0.9\ \mu M$	Single cAMP-binding domain from Epac sandwiched between CFP and YFP	Single chain. Good sensitivity. Faster kinetics than for multimeric sensors. Compact design, easy for targeting, and in vivo expression	pmEpac1/2-camps (caveolar membrane targeted) AC8-Epac2-camps (fusion to adenylyl cyclase 8) Epac1-camps-HSP20 PDE3A/PDE4A-Epac1-camps (fusions with PDEs) RI_epac and RII_epac (Epac1 camps fused to N-terminal dimerization-docking domain of RI or RII PKA subunits) β_1-AR-Epac2-camps (fusion to β_1-AR) Epac1-PLN (SR targeted, fusion to PLN) Neonatal and adult CMs. Transgenic mice and flies for Epac1 camps, pmEpac1, and Epac1-PLN	Nikolaev et al. (2004), Mohamed et al. (2011), Wachten et al. (2010). Sin et al. (2011), Herg et al. (2008), Di Benedetto et al. (2008), Stangherlin et al. (2011), Richter et al. (2013), Sprenger et al. (2015) Calebiro et al. (2009), Perera et al. (2015), Shafer et al. (2008)
Epac2-camps 300 cAMP $EC_{50} = 300$ nM	Epac2-camps with K405E mutation to increase cAMP affinity	High sensitivity	Not available No	Castro et al. (2010), Norris et al. (2009)
CFP-Epac1-YFP (ΔDEP,CD) (Epac1 based) cAMP EC_{50} ~50 μM (~15 μM)	Whole or N terminally truncated Epac1 fused to CFP and YFP variants	Single chain. Relatively low sensitivity	Not available No	Ponsioen et al. (2004)
CFP-Epac2-YFP (Epac2 based)	As above	Ab above	Not available No	Herbst et al. (2011), Zhang et al. (2009)

(continued)

Table 1.1 (continued)

Name, sensitivity	Design	Properties	Targeted versions, use in myocytes	References
ICUE1–3 (Epac1 based) cAMP EC_{50} ~10–50 μM	As above	As above	pmICUE, NLS-ICUE, mitoICUE (Subsarcolemmal, nuclear, mitochondrial) Neonatal and adult CMs, AGT	DiPilato et al. (2004), Violin et al. (2008), De Arcangelis et al. (2009), Fu et al. (2014)
H171 (Epac1 based)	As above but with m Turquoise2 and cp173Venus as fluorophores	Improved brightness, affinity (by 2.5-fold with Q270E mutation) and dynamic range	Not available Adult CMs, AGT available	Klarenbeek et al. (2015)
mlCNBD-FRET cAMP EC_{50} ~70 nM	Single cAMP-binding site from bacterial MlotiK1 channel fused to Cerulean and Citrine	Single chain. Compart design. High sensitivity	Not available No	Mukherjee et al. (2016)
HCN-camps cAMP EC_{50} ~6 μM	Single cAMP-binding site from HCN2 channel	Moderate sensitivity. For cells with high basal cAMP	Adult CMs, transgenic mice	Nikolaev et al. (2006a)

AGT adenoviral gene transfer, *C* PKA catalytic subunit, *CFP* cyan fluorescent protein, *ER* endoplasmic reticulum, *NLS* nuclear localization signal, *PLN* phospholamban, *R* PKA regulatory subunit, *SR* sarcoplasmic reticulum, *YFP* yellow fluorescent protein

Table 1.2 Fluorescent cGMP biosensors and their use in cardiomyocytes (CMs)

Name, sensitivity	Design	Properties	Targeted versions, use in myocytes	References
CGY-Del1 cGMP $EC_{50} = 20$ nM	N terminally truncated cGKI fused to CFP and YFP	Single chain. Low cGMP/cAMP selectivity	Not available No	Nikolaev et al. (2006b), Sato et al. (2000)
Cygnet-1/2 cGMP $EC_{50} = 1.5/1.9$ μM	Similar but slightly shorter N-terminus	Better cGMP/cAMP selectivity. Relatively low sensitivity and temporal resolution	RI_Cygnet, RII_Cygnet (fusions to N-terminal dimerization-docking domain of RI or RII PKA subunits)	Honda et al. (2001)
			Only neonatal CMs	Stangherlin et al. (2011)
cGi-500/3000/6000 cGMP $EC_{50} = 500/3000/6000$ nM	Both cGMP-binding domains from cGKI sandwiched between CFP and YFP	Small size. Moderate selectivity and sensitivity. High dynamic range. Fast kinetics	No published fusions cGi500 transgenic mice and worms available	Russwurm et al. (2007), Thunemann et al. (2013), Couto et al. (2013)
FlincG1–3 cGMP $EC_{50} = 0.17–0.89$ μM	Non-FRET sensor. Both cGMP-binding domains from cGKI fused to circularly permuted GFP	Good dynamic range. Rapid kinetics. Moderate cGMP/cAMP selectivity	Not available No	Nausch et al. (2008), Bhargava et al. (2013)
cGES-DE2/5 (PDE based) cGMP $EC_{50} = 0.9/1.5$ μM	Single cGMP-binding (GAF) domains from PDE2 or PDE5 fused to CFP and YFP	Single chain, fast kinetics, moderate sensitivity	cGES-DE2-PDE5A (fusion to PDE5A) No	Herget et al. (2008), Nikolaev et al. (2006b)
Red cGES-DE5 cGMP EC_{50} ~40 nM	As above but with T-Sapphire and Dimer2 as fluorophores	High sensitivity	Not available Adult CMs, transgenic mice	Niino et al. (2009), Belge et al. (2014), Götz et al. (2014)

cGKI cGMP-dependent protein kinase type I, *GFP* green fluorescent protein; other abbreviations are in Table 1.1

low sensitivity prevents many of these probes to be successfully used in physiologically most relevant adult CMs which have relatively low nanomolar cGMP concentrations (Götz et al. 2014). In these cells, only a highly sensitive biosensor red cGES-DE5 could reliably record cGMP signals in response to natriuretic peptides and phosphodiesterase (PDE) inhibitors, at least from what has been reported so far (Götz et al. 2014).

Another important development which is very helpful for visualization of cAMP and cGMP in adult CMs is the generation of transgenic mice expressing various cytosolic and targeted FRET biosensors in the heart (Calebiro et al. 2009; Götz et al. 2014; Nikolaev et al. 2006a; Perera et al. 2015; Sprenger et al. 2015). Since these cells are difficult to transfect and keep in culture without loss of their highly complex membrane structure, the use of freshly isolated cells from healthy and diseased transgenic biosensor mice represents a powerful approach to study microdomain-specific cAMP and cGMP signaling (Froese and Nikolaev 2015).

1.3 Mechanisms of Compartmentation in Microdomains

cAMP and cGMP biosensors were instrumental for our understanding of second messenger microdomains and the mechanisms of cyclic nucleotide compartmentations. In the heart, such microdomains are associated, for example, with calcium-handling proteins. cAMP synthetized upon catecholamine stimulation of cell membrane β-adrenergic receptors (β-ARs) modulates excitation-contraction coupling by PKA-dependent phosphorylation of L-type calcium channels responsible for calcium influx upon depolarization, ryanodine receptors responsible for calcium release from sarcoplasmic reticulum (SR) and phospholamban (PLN), a small SR transmembrane protein which regulates diastolic calcium reuptake (Bers 2002; Lompre et al. 2010). Each of these microdomains is differentially regulated by several mechanisms of cyclic nucleotide compartmentation, which are thoroughly discussed in various chapters of this book.

Firstly, **receptor localization to various membrane microdomains** such as caveolae, transverse (T) tubules, lipid rafts, and noncaveolar domains generates specific signalosomes and a strictly localized pattern of cAMP and cGMP production (see Chap. 2).

Secondly, **A-kinase anchoring proteins** serve as scaffolds to bring together the most important molecular players which regulate synthesis, degradation, and phosphorylation of specific molecular targets for cyclic nucleotides and their respective kinases (see Chaps. 3, 4, and 5).

Thirdly, **phosphodiesterases (PDEs)** are key enzymes responsible for cAMP and cGMP degradation and shaping their intracellular gradients in various microdomains. Several PDE families, mechanisms of their regulation, and their central role in cyclic nucleotide compartmentation are extensively discussed in Chaps. 6–10.

Last but not the least, **physical barriers for cyclic nucleotide diffusion** such as mitochondria and cAMP buffering by the PKA are especially relevant for adult CMs due to their highly organized cellular structure. These issues are briefly discussed in Chap. 2.

cGMP signals are compartmentalized in various functionally important microdomains as well, where they regulate cardiac function and disease-related processes (Chaps. 11–13). Recent discoveries in the field of calcium microdomains formed around different calcium channels, and their role in cardiac arrhythmias are presented in the second part of the book (Chaps. 14–17). At the end of this chapter, we would like to briefly summarize how receptor-microdomain signaling can be precisely studied in CMs.

1.4 Receptor-cAMP/cGMP Microdomain Visualization

As mentioned above, stimulation of different receptors at the CM membrane can generate distinct pattern of second messenger responses. The difference between the effects of ISO and prostaglandin, the former inducing a strong increase in CM contractility and the latter being without effect on contraction, has been recently better clarified using FRET biosensors. Fusion of Epac1-camps biosensor to the dimerization-docking domains of either regulatory type I (RI) or type II subunit (RII) of the PKA allowed targeting of the sensor to intracellular compartments where endogenous PKA types I and II typically reside. β-AR stimulation led to an increase of cAMP in the RII microdomain, while prostaglandin stimulation increased cAMP exclusively in the RI microdomain, explaining differential PKA substrate phosphorylation after ISO and prostaglandin stimulation (Di Benedetto et al. 2008).

β_1- and β_2-adrenergic receptor (AR) subtypes are the major CM receptors for catecholamines which act predominantly by producing cAMP (Brodde et al. 2006; Lohse et al. 2003). However, it was unclear for a long time why the functional effects of these two receptors are so different in respect to cardiac contractility and remodeling in disease. By developing the first transgenic mouse expressing the HCN2-camps biosensor in CMs, it became possible to perform FRET imaging of cAMP signals diffusing from locally stimulated pools of receptors through the adult CM cytosol. This study showed that while β_1-AR/cAMP signals can rapidly diffuse throughout the whole cell, β_2-AR/cAMP is highly locally confined at the site of stimulation (Nikolaev et al. 2006a). This could better explain why β_1-AR has strong effects on contractility and induces CM apoptosis and cardiac remodeling, while β_2-AR has only a slight positive inotropic response and protects from apoptosis. However, the exact localization of these receptors to different CM membrane structures and the mechanisms involved in compartmentation of β_2-AR/cAMP response were not completely clear.

To study the localization of functional receptors at the CM membrane, which was not possible by standard techniques because of insufficient specificity of respective antibodies, we developed a new imaging method which combines scanning ion conductance microscopy (SICM) with FRET-based detection of cAMP (Nikolaev et al. 2010). SICM is a nonoptical imaging technique (see also Chap. 14) which uses a small glass nanopipette to obtain a highly resolved morphological profile of a living cell membrane based on ion current measurement (Hansma et al. 1989; Korchev et al. 1997). By applying receptor agonists to precise membrane locations from the scanning pipette and simultaneous recording of intracellular cAMP using FRET, we could show that while β_1-AR is evenly distributed across all membrane regions, β_2-AR is exclusively located in the T-tubules of healthy myocytes. β_2-AR/cAMP responses at the T-tubular membrane are highly compartmentalized by PDE4 activity and local pools of PKA which can partially buffer locally produced cAMP. In heart failure, β_2-AR can redistribute to detubulated membrane areas and induce far-reaching cAMP signals similar to those of β_1-AR, which might even worsen the clinical signs of an established heart failure (Nikolaev et al. 2010).

In the future, it will be exciting to clarify the exact localization of other membrane receptors, especially those stimulating cGMP signals, and to better understand various receptor-microdomain interactions. This should be possible to achieve using FRET and SICM/FRET imaging techniques.

Acknowledgment The work in authors' laboratory is supported by the grants from the German Research Foundation ("Deutsche Forschungsgemeinschaft" grants NI 1301/1, NI 1301/2, FOR 2060) and by the Gertraud und Heinz-Rose Stiftung.

Compliance with Ethical Standards

Conflict of Interest Statement The authors declare that they have no conflict of interest.

References

Adams SR, Harootunian AT, Buechler YJ et al (1991) Fluorescence ratio imaging of cyclic AMP in single cells. Nature 349:694–697

Allen MD, Zhang J (2006) Subcellular dynamics of protein kinase a activity visualized by FRET-based reporters. Biochem Biophys Res Commun 348:716–721

Ashman DF, Lipton R, Melicow MM et al (1963) Isolation of adenosine 3′, 5′-monophosphate and guanosine 3′, 5′-monophosphate from rat urine. Biochem Biophys Res Commun 11:330–334

Beavo JA, Brunton LL (2002) Cyclic nucleotide research—still expanding after half a century. Nat Rev Mol Cell Biol 3:710–718

Belge C, Hammond J, Dubois-Deruy E et al (2014) Enhanced expression of beta3-adrenoceptors in cardiac myocytes attenuates neurohormone-induced hypertrophic remodeling through nitric oxide synthase. Circulation 129:451–462

Bers DM (2002) Cardiac excitation-contraction coupling. Nature 415:198–205

Bhargava Y, Hampden-Smith K, Chachlaki K et al (2013) Improved genetically-encoded, FlincG-type fluorescent biosensors for neural cGMP imaging. Front Mol Neurosci 6:26

Brodde OE, Bruck H, Leineweber K (2006) Cardiac adrenoceptors: physiological and pathophysiological relevance. J Pharmacol Sci 100:323–337

Brooker G, Harper JF, Terasaki WL et al (1979) Radioimmunoassay of cyclic AMP and cyclic GMP. Adv Cyclic Nucleotide Res 10:1–33

Buxton IL, Brunton LL (1983) Compartments of cyclic AMP and protein kinase in mammalian cardiomyocytes. J Biol Chem 258:10233–10239

Calebiro D, Nikolaev VO, Gagliani MC et al (2009) Persistent cAMP-signals triggered by internalized G-protein-coupled receptors. PLoS Biol 7:e1000172

Castro LR, Gervasi N, Guiot E et al (2010) Type 4 phosphodiesterase plays different integrating roles in different cellular domains in pyramidal cortical neurons. J Neurosci 30:6143–6151

Couto A, Oda S, Nikolaev VO et al (2013) In vivo genetic dissection of O_2-evoked cGMP dynamics in a *Caenorhabditis elegans* gas sensor. Proc Natl Acad Sci U S A 110:E3301–E3310

De Arcangelis V, Liu R, Soto D et al (2009) Differential association of phosphodiesterase 4D isoforms with beta2-adrenoceptor in cardiac myocytes. J Biol Chem 284:33824–33832

Depry C, Allen MD, Zhang J (2011) Visualization of PKA activity in plasma membrane microdomains. Mol Biosyst 7:52–58

Di Benedetto G, Zoccarato A, Lissandron V et al (2008) Protein kinase a type I and type II define distinct intracellular signaling compartments. Circ Res 103:836–844
DiPilato LM, Cheng X, Zhang J (2004) Fluorescent indicators of cAMP and Epac activation reveal differential dynamics of cAMP signaling within discrete subcellular compartments. Proc Natl Acad Sci U S A 101:16513–16518
Dyachok O, Isakov Y, Sagetorp J et al (2006) Oscillations of cyclic AMP in hormone-stimulated insulin-secreting beta-cells. Nature 439:349–352
Fischmeister R, Castro LR, Abi-Gerges A et al (2006) Compartmentation of cyclic nucleotide signaling in the heart: the role of cyclic nucleotide phosphodiesterases. Circ Res 99: 816–828
Froese A, Nikolaev VO (2015) Imaging alterations of cardiomyocyte cAMP microdomains in disease. Front Pharmacol 6:172
Fu Q, Kim S, Soto D et al (2014) A long lasting beta1 adrenergic receptor stimulation of cAMP/protein kinase a (PKA) signal in cardiac myocytes. J Biol Chem 289:14771–14781
Götz KR, Sprenger JU, Perera RK et al (2014) Transgenic mice for real-time visualization of cGMP in intact adult cardiomyocytes. Circ Res 114:1235–1245
Haj Slimane Z, Bedioune I, Lechene P et al (2014) Control of cytoplasmic and nuclear protein kinase a by phosphodiesterases and phosphatases in cardiac myocytes. Cardiovasc Res 102:97–106
Hansma PK, Drake B, Marti O et al (1989) The scanning ion-conductance microscope. Science 243:641–643
Hayes JS, Brunton LL, Mayer SE (1980) Selective activation of particulate cAMP-dependent protein kinase by isoproterenol and prostaglandin E1. J Biol Chem 255:5113–5119
Herbst KJ, Coltharp C, Amzel LM et al (2011) Direct activation of Epac by sulfonylurea is isoform selective. Chem Biol 18:243–251
Herget S, Lohse MJ, Nikolaev VO (2008) Real-time monitoring of phosphodiesterase inhibition in intact cells. Cell Signal 20:1423–1431
Honda A, Adams SR, Sawyer CL et al (2001) Spatiotemporal dynamics of guanosine 3',5'-cyclic monophosphate revealed by a genetically encoded, fluorescent indicator. Proc Natl Acad Sci U S A 98:2437–2442
Klarenbeek J, Goedhart J, van Batenburg A et al (2015) Fourth-generation Epac-based FRET sensors for cAMP feature exceptional brightness, photostability and dynamic range: characterization of dedicated sensors for FLIM, for ratiometry and with high affinity. PLoS One 10:e0122513
Korchev YE, Bashford CL, Milovanovic M et al (1997) Scanning ion conductance microscopy of living cells. Biophys J 73:653–658
Liu S, Zhang J, Xiang YK (2011) FRET-based direct detection of dynamic protein kinase a activity on the sarcoplasmic reticulum in cardiomyocytes. Biochem Biophys Res Commun 404:581–586
Lohse MJ, Engelhardt S, Eschenhagen T (2003) What is the role of beta-adrenergic signaling in heart failure? Circ Res 93:896–906
Lompre AM, Hajjar RJ, Harding SE et al (2010) Ca2+ cycling and new therapeutic approaches for heart failure. Circulation 121:822–830
Mehta S, Aye-Han NN, Ganesan A et al (2014) Calmodulin-controlled spatial decoding of oscillatory Ca^{2+} signals by calcineurin. Elife 3:e03765
Mohamed TM, Oceandy D, Zi M et al (2011) Plasma membrane calcium pump (PMCA4)-neuronal nitric-oxide synthase complex regulates cardiac contractility through modulation of a compartmentalized cyclic nucleotide microdomain. J Biol Chem 286:41520–41529
Mongillo M, McSorley T, Evellin S et al (2004) Fluorescence resonance energy transfer-based analysis of cAMP dynamics in live neonatal rat cardiac myocytes reveals distinct functions of compartmentalized phosphodiesterases. Circ Res 95:67–75
Mukherjee S, Jansen V, Jikeli JF et al (2016) A novel biosensor to study cAMP dynamics in cilia and flagella. Elife 5
Nausch LW, Ledoux J, Bonev AD et al (2008) Differential patterning of cGMP in vascular smooth muscle cells revealed by single GFP-linked biosensors. Proc Natl Acad Sci U S A 105:365–370

Niino Y, Hotta K, Oka K (2009) Simultaneous live cell imaging using dual FRET sensors with a single excitation light. PLoS One 4:e6036
Nikolaev VO, Bunemann M, Hein L et al (2004) Novel single chain cAMP sensors for receptor-induced signal propagation. J Biol Chem 279:37215–37218
Nikolaev VO, Bunemann M, Schmitteckert E et al (2006a) Cyclic AMP imaging in adult cardiac myocytes reveals far-reaching beta1-adrenergic but locally confined beta2-adrenergic receptor-mediated signaling. Circ Res 99:1084–1091
Nikolaev VO, Gambaryan S, Lohse MJ (2006b) Fluorescent sensors for rapid monitoring of intracellular cGMP. Nat Methods 3:23–25
Nikolaev VO, Moshkov A, Lyon AR et al (2010) Beta2-adrenergic receptor redistribution in heart failure changes cAMP compartmentation. Science 327:1653–1657
Norris RP, Ratzan WJ, Freudzon M et al (2009) Cyclic GMP from the surrounding somatic cells regulates cyclic AMP and meiosis in the mouse oocyte. Development 136:1869–1878
Perera RK, Nikolaev VO (2013) Compartmentation of cAMP signalling in cardiomyocytes in health and disease. Acta Physiol (Oxf) 207:650–662
Perera RK, Sprenger JU, Steinbrecher JH et al (2015) Microdomain switch of cGMP-regulated phosphodiesterases leads to ANP-induced augmentation of beta-adrenoceptor-stimulated contractility in early cardiac hypertrophy. Circ Res 116:1304–1311
Ponsioen B, Zhao J, Riedl J et al (2004) Detecting cAMP-induced Epac activation by fluorescence resonance energy transfer: Epac as a novel cAMP indicator. EMBO Rep 5:1176–1180
Richter W, Mika D, Blanchard E et al (2013) beta1-adrenergic receptor antagonists signal via PDE4 translocation. EMBO Rep 14:276–283
Russwurm M, Mullershausen F, Friebe A et al (2007) Design of fluorescence resonance energy transfer (FRET)-based cGMP indicators: a systematic approach. Biochem J 407:69–77
Sato M, Hida N, Ozawa T et al (2000) Fluorescent indicators for cyclic GMP based on cyclic GMP-dependent protein kinase I alpha and green fluorescent proteins. Anal Chem 72:5918–5924
Shafer OT, Kim DJ, Dunbar-Yaffe R et al (2008) Widespread receptivity to neuropeptide PDF throughout the neuronal circadian clock network of drosophila revealed by real-time cyclic AMP imaging. Neuron 58:223–237
Sin YY, Edwards HV, Li X et al (2011) Disruption of the cyclic AMP phosphodiesterase-4 (PDE4)-HSP20 complex attenuates the beta-agonist induced hypertrophic response in cardiac myocytes. J Mol Cell Cardiol 50:872–883
Sprenger JU, Nikolaev VO (2013) Biophysical techniques for detection of cAMP and cGMP in living cells. Int J Mol Sci 14:8025–8046
Sprenger JU, Perera RK, Steinbrecher JH et al (2015) In vivo model with targeted cAMP biosensor reveals changes in receptor-microdomain communication in cardiac disease. Nat Commun 6:6965
Stangherlin A, Gesellchen F, Zoccarato A et al (2011) cGMP signals modulate cAMP levels in a compartment-specific manner to regulate catecholamine-dependent signaling in cardiac myocytes. Circ Res 108:929–939
Sutherland EW, Rall TW (1958) Fractionation and characterization of a cyclic adenine ribonucleotide formed by tissue particles. J Biol Chem 232:1077–1091
Tao W, Rubart M, Ryan J et al (2015) A practical method for monitoring FRET-based biosensors in living animals using two-photon microscopy. Am J Physiol Cell Physiol 309:C724–C735
Thunemann M, Wen L, Hillenbrand M et al (2013) Transgenic mice for cGMP imaging. Circ Res 113:365–371
Violin JD, DiPilato LM, Yildirim N et al (2008) beta2-adrenergic receptor signaling and desensitization elucidated by quantitative modeling of real time cAMP dynamics. J Biol Chem 283:2949–2961
Wachten S, Masada N, Ayling LJ et al (2010) Distinct pools of cAMP Centre on different isoforms of adenylyl cyclase in pituitary-derived GH3B6 cells. J Cell Sci 123:95–106

Warrier S, Belevych AE, Ruse M et al (2005) Beta-adrenergic- and muscarinic receptor-induced changes in cAMP activity in adult cardiac myocytes detected with FRET-based biosensor. Am J Physiol Cell Physiol 289:C455–C461
Williams C (2004) cAMP detection methods in HTS: selecting the best from the rest. Nat Rev Drug Discov 3:125–135
Zaccolo M (2004) Use of chimeric fluorescent proteins and fluorescence resonance energy transfer to monitor cellular responses. Circ Res 94:866–873
Zaccolo M (2009) cAMP signal transduction in the heart: understanding spatial control for the development of novel therapeutic strategies. Br J Pharmacol 158:50–60
Zaccolo M, De Giorgi F, Cho CY et al (2000) A genetically encoded, fluorescent indicator for cyclic AMP in living cells. Nat Cell Biol 2:25–29
Zaccolo M, Pozzan T (2002) Discrete microdomains with high concentration of cAMP in stimulated rat neonatal cardiac myocytes. Science 295:1711–1715
Zhang CL, Katoh M, Shibasaki T et al (2009) The cAMP sensor Epac2 is a direct target of antidiabetic sulfonylurea drugs. Science 325:607–610
Zhang J, Campbell RE, Ting AY et al (2002) Creating new fluorescent probes for cell biology. Nat Rev Mol Cell Biol 3:906–918
Zhang J, Hupfeld CJ, Taylor SS et al (2005) Insulin disrupts beta-adrenergic signalling to protein kinase a in adipocytes. Nature 437:569–573
Zhang J, Ma Y, Taylor SS et al (2001) Genetically encoded reporters of protein kinase a activity reveal impact of substrate tethering. Proc Natl Acad Sci U S A 98:14997–15002

Membrane Microdomains and cAMP Compartmentation in Cardiac Myocytes

2

Shailesh R. Agarwal, Rennolds S. Ostrom, and Robert D. Harvey

Contents

Abstract

Signaling through the diffusible second messenger, 3′,5′-cyclic adenosine monophosphate (cAMP) is critical to the regulation of cardiac function. Several different G-protein-coupled receptors, including β-adrenergic receptors, muscarinic receptors, and E-type prostaglandin receptors, elicit distinct responses using this ubiquitous second messenger. One critical paradigm that has emerged to explain this behavior is that cAMP signaling is compartmentalized. Spatially confining specific receptors and their downstream effector proteins to form subcellular signaling complexes has been proposed to allow for the high efficiency and fidelity in producing specific functional responses. In cardiac myocytes, lipid rafts created by cholesterol- and sphingolipid-rich membrane microdomains have

S.R. Agarwal • R.D. Harvey (✉)
Department of Pharmacology, University of Nevada School of Medicine,
Reno, NV 89557, USA
e-mail: rdharvey@med.unr.edu

R.S. Ostrom
Department of Biomedical and Pharmaceutical Sciences,
Chapman University School of Pharmacy, Irvine, CA 92618, USA

V. Nikolaev, M. Zaccolo (eds.), *Microdomains in the Cardiovascular System*,
Cardiac and Vascular Biology 3, DOI 10.1007/978-3-319-54579-0_2

been demonstrated to act as one means of sorting appropriate receptors and corresponding effectors to relevant subcellular locations. Caveolae, which represent a specific subset of lipid rafts, can dynamically attract or exclude specific signaling proteins through a variety of mechanisms to create highly localized and self-sufficient multi-molecular signaling complexes. Furthermore, disruption of this organization in disease states such as heart failure has been found to alter cAMP responses. In this review, we summarize the current understanding of the role of membrane domains in cAMP signaling in cardiac myocytes. We also highlight the insights gained from previous studies to offer new avenues of research in this expanding field of study.

Abbreviations

AC	Adenylyl cyclase
ACh	Acetylcholine
AKAP	A-kinase-anchoring protein
cAMP	3′,5′-Cyclic adenosine monophosphate
Cav3	Caveolin 3
CSD	Caveolin scaffolding domain
DRM	Detergent-resistant membrane
EC	Excitation-contraction
eNOS	Endothelial nitric oxide synthase
Epac	Exchange protein directly activated by cAMP
EPR	E-type prostaglandin receptor
GPCR	G-protein-coupled receptor
GPI	Glycosylphosphatidylinositol
M_2R	M_2 muscarinic receptor
MβCD	Methyl-β-cyclodextrin
NO	Nitric oxide
PDE	Phosphodiesterase
PKA	Protein kinase A
PLB	Phospholamban
SICM	Scanning ion conductance microscopy
SR	Sarcoplasmic reticulum
T tubule	Transverse tubule
β_1AR	β_1-Adrenergic receptor
β_2AR	β_2-Adrenergic receptor

2.1 Introduction

The diffusible second messenger cAMP is involved in mediating responses to a wide array of neurotransmitters, hormones, and physiologically active chemicals operating through a variety of G-protein-coupled receptors (GPCRs) in virtually every cell in the human body (Gancedo 2013). This ubiquitous signaling molecule regulates an extensive range of cellular functions, including gene expression, glucose and lipid metabolism, steroidogenesis, insulin secretion, fluid and electrolyte secretion, muscle contraction and relaxation, as well as nerve and muscle electrical excitability (Robison et al. 1968). What is even more remarkable is that although cAMP is regulating many of these activities in the same cell, not all receptors that stimulate cAMP production elicit the same set of responses. The classic example is in cardiac myocytes, where both β-adrenergic receptors (βARs) and E-type prostaglandin receptors (EPRs) stimulate cAMP production, yet only βARs produce acute functional responses (Corbin et al. 1977; Brunton et al. 1979; Hayes et al. 1979; Steinberg and Brunton 2001; Warrier et al. 2007; Agarwal et al. 2011). This can be explained if cAMP signaling is compartmentalized, but it is by no means unique to cardiac myocytes or to those specific receptors.

One important factor in maintaining the fidelity of receptor-mediated responses is the formation of signaling complexes that organize effectors of cAMP, such as protein kinase A (PKA), together with the target proteins they regulate. This is often through interactions with A-kinase-anchoring proteins (AKAPs) (Scott et al. 2013). Numerous studies have demonstrated that disrupting PKA-AKAP interactions can alter cAMP-mediated responses in the heart (Perino et al. 2012). However, effector signaling complexes alone are not sufficient to explain compartmentation. If stimulation of every receptor produced a uniform increase in cAMP throughout the cell, they would all elicit the same responses. This means that there must be some means of keeping cAMP from moving freely throughout the cell, and just as importantly, there must be some means of creating sites of localized cAMP production. The emphasis of this review is on what is known about the role of membrane microdomains in generating the localized production of cAMP in cardiac myocytes.

2.2 Membrane Microdomains

In the original fluid mosaic model, it was proposed that membrane proteins are able to diffuse freely throughout the lipid bilayer (Singer and Nicolson 1972). This led to the idea that signal transduction involving these proteins occurs through the random process of collision coupling (Tolkovsky and Levitzki 1978). However, the density of signaling proteins is actually too low for this idea to explain the rapid and reliable responses that occur (Ostrom et al. 2000a). Furthermore, it is now known that mechanisms exist that actually restrict the movement of membrane proteins (Bethani et al. 2010). This suggests that there must be some means of segregating the plasma membrane into distinct microdomains where different signaling proteins can be concentrated and their diffusion limited. One such mechanism involves the concentration of cholesterol and sphingolipids together with certain phospholipids to form gel-like, liquid-ordered domains called lipid rafts (Jacobson et al. 2007). Membrane fractionation studies have demonstrated that certain signaling proteins are concentrated in the buoyant, detergent-resistant membrane (DRM) fractions that represent lipid rafts, while others are specifically excluded (Brown 2006; Allen et al. 2007).

The mechanisms that result in the targeting or exclusion of proteins to or from lipid raft domains of the plasma membrane are not fully understood. A specific type of receptor found in lipid rafts in one cell type may actually be excluded from that very same membrane domain in another (Ostrom and Insel 2004). In some cases, whether or not transmembrane proteins partition into lipid rafts appears to be based on their affinity for an ordered membrane environment (Melkonian et al. 1999). In other cases, there is evidence that these proteins are actually able to bind lipid components of rafts. For example, certain GPCRs are able to bind cholesterol, which may explain their preference for these membrane domains in some systems (Hanson et al. 2008). On the other hand, agonist binding can also affect receptor affinity for cholesterol and may contribute to translocation of receptors between membrane domains (Chini and Parenti 2004).

Posttranslational modification may also play an important role in targeting proteins to lipid rafts. This can be achieved through attachment of glycosylphosphatidylinositol (GPI), as rafts are associated with high concentrations of GPI-anchored proteins (Brown and Rose 1992). Acylation involving the attachment of palmitate or myristate to neighboring amino acid residues is another common means of targeting proteins to lipid rafts, since these fatty acids prefer a liquid-ordered environment. Conversely, prenylation is often associated with membrane-targeted proteins that are excluded from lipid raft domains. This involves the attachment of highly branched farnesyl or geranyl-geranyl moieties, which do not fit well in a highly ordered environment (Melkonian et al. 1999; Resh 2006). The α subunits of many heterotrimeric G proteins are palmitoylated, which favors localization in lipid raft domains. However, βγ subunits are often prenylated, which favors association with non-lipid raft locations (Moffett et al. 2000). N-glycosylation can also affect the membrane distribution of proteins. There is evidence that this is responsible for targeting of Ca^{2+}-regulated adenylyl cyclase (AC) isoforms (AC1, 5, 6, and 8) to lipid rafts (Pagano et al. 2009). Thus, multiple factors may be involved in ultimately determining how membrane-associated signaling proteins are distributed.

Raft targeting may also involve protein-protein interactions. Caveolins in particular are commonly associated with lipid rafts due to three palmitoylation sites near the C-terminus (Williams and Lisanti 2004). These are 18–22 kDa proteins that insert asymmetrically into the intracellular face of the plasma membrane in a hairpin-like configuration. They tend to form oligomers that result in the formation of flask-shaped invaginations of the plasma membrane called caveolae, which represent a subset of lipid rafts (see Fig. 2.1). There are three caveolin isoforms, Cav1, Cav2, and Cav3. Cav3 is sometimes referred to as the muscle-specific isoform because it is found primarily in cardiac, skeletal, and smooth muscle. Although Cav3 is thought to be the predominant isoform found in cardiac myocytes, evidence for expression all three isoforms has been reported (Head et al. 2005). All caveolins possess a 20-amino acid sequence called the caveolin scaffolding domain (CSD), which can interact with a variety of proteins. In cardiac myocytes, Cav3 has been shown to interact directly with βARs, the α subunits of both stimulatory (G_s) and inhibitory (G_i) G proteins, adenylyl cyclase 5/6, and the regulatory subunit of type II PKA (Rybin et al. 2000; Xiang et al. 2002b; Balijepalli et al. 2006; Nichols

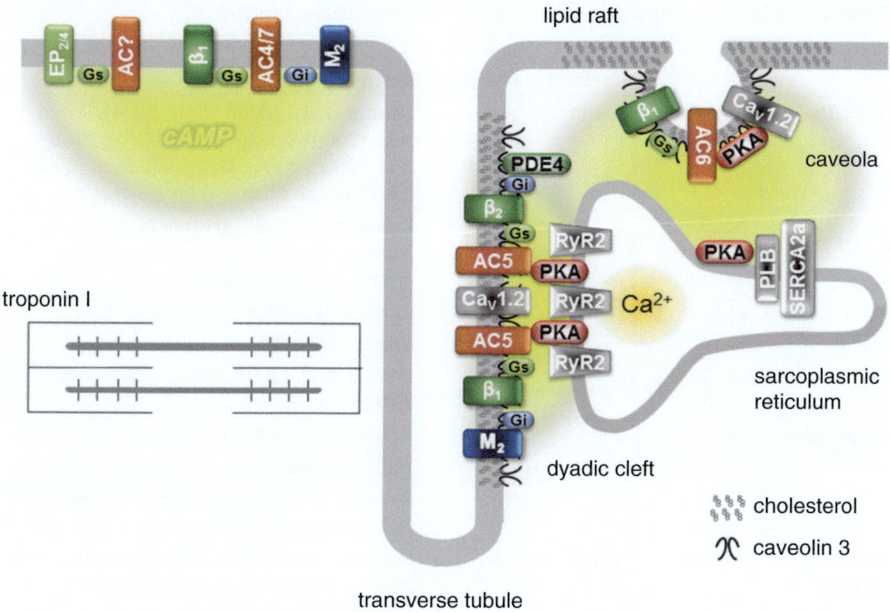

Fig. 2.1 Compartmentalized cAMP signaling in adult ventricular myocytes. Organization of membrane microdomains and cAMP-signaling proteins in adult cardiac ventricular myocytes. *β1* and *β2* β-adrenergic receptors, *M2* muscarinic receptors, *EP2/4* E-type prostaglandin receptors, *Gs* and *Gi* inhibitory and stimulatory G proteins, *AC5, AC6, AC4/7* adenylyl cyclases, *PKA* protein kinase A, *CaV1.2* L-type Ca^{2+} channels, *RyR2* ryanodine receptors, *PLB*, phospholamban, *SERCA2a* sarcoplasmic reticulum Ca^{2+} ATPase. See text for details

et al. 2010). This interaction with the CSD of caveolins can also regulate the intrinsic activity of many of these signaling proteins (Toya et al. 1998; Allen et al. 2007).

It has been reported that disrupting microtubules and actin microfilaments results in a loss of signaling proteins from caveolar membrane fractions, which suggests that the cytoskeletal interactions may play an important role in the targeting and/or stabilization of caveolar proteins, as well (Head et al. 2006).

It is important to note that the partitioning of certain signaling proteins in different membrane domains can also be affected by receptor activation. For example, in adult ventricular myocytes, M_2 muscarinic receptors (M_2Rs) have been reported to exist predominantly in non-caveolar fractions of the plasma membrane under unstimulated conditions, but then shift to caveolar fractions following agonist binding (Feron et al. 1997). β_2ARs, on the other hand, are normally found in caveolar membrane fractions, but they move out following agonist stimulation (Rybin et al. 2000; Ostrom et al. 2001; Xiang et al. 2002b). The movement of β_2ARs may be related to, at least in part, receptor internalization via an arrestin and clathrin-coated pit-mediated pathway (Rybin et al. 2000; Xiang et al. 2002b; Xiang and Kobilka 2003). β_1ARs are minimally affected by this process, because of anchoring to PSD-95 or a related protein via a PDZ-binding domain interaction (Xiang et al. 2002a).

Table 2.1 Membrane microdomain location of cAMP-signaling proteins

		Caveolar	Non-caveolar	References
GPCRs	β_1AR	Yes	Yes	Ostrom et al. (2000b, 2001, 2004), Rybin et al. (2000), Xiang et al. (2002b), Head et al. (2005, 2006), Balijepalli et al. (2006), Nichols et al. (2010), Agarwal et al. (2011)
	β_2AR	Yes	No	Rybin et al. (2000), Ostrom et al. (2001), Xiang et al. (2002b), Head et al. (2005, 2006), Balijepalli et al. (2006)
	M_2R	Yes*	Yes	Feron et al. (1997), Rybin et al. (2000), Head et al. (2005)
	EP_2R, EP_4R	No	Yes	Ostrom et al. (2001, 2004), Agarwal et al. (2011)
G proteins	G_s	Yes	Yes	Rybin et al. (2000), Head et al. (2005, 2006), Balijepalli et al. (2006)
	G_i	Yes	No	Rybin et al. (2000), Head et al. (2005), Balijepalli et al. (2006)
Adenylyl cyclase	AC5/6	Yes	Yes	Ostrom et al. (2000b, 2001, 2004), Rybin et al. (2000), Head et al. (2005, 2006), Balijepalli et al. (2006)
	AC4, AC7	No	Yes	Ostrom and Insel (2004), Willoughby and Cooper (2007)

*At least some fraction of these receptors translocate to this domain following agonist stimulation

In neonatal myocytes, receptor internalization has been shown to be necessary for activation of G_i-dependent signaling (Xiang et al. 2002a, b). Following agonist stimulation, β_2AR phosphorylation by a G-protein-coupled receptor kinase (GRK) leads to binding of the adapter protein arrestin. This results in receptor internalization, which has been shown to recruit the binding of phosphodiesterases (PDEs) as well as activation of other signaling pathways (Shenoy and Lefkowitz 2011).

Table 2.1 lists the membrane microdomains commonly associated with signaling proteins involved in regulating cAMP production in cardiac myocytes.

2.3 Receptor-Mediated Responses

2.3.1 β_1-Adrenergic Receptors (β_1ARs)

β_1ARs are the predominant βAR subtype found in the heart, making up ~80% of the total βAR population in most species (Steinberg 1999). Selective stimulation of these receptors produces a positive inotropic as well as a positive lusitropic effect on cardiac muscle contraction (Bers 2002). Both involve cAMP production and subsequent activation of protein kinase A (PKA). The positive inotropic effect is associated with an increase in cytosolic Ca^{2+} due to phosphorylation-dependent regulation of L-type Ca^{2+} channels in the plasma membrane, ryanodine receptors in the junctional sarcoplasmic reticulum (SR), and phospholamban (PLB) in the non-junctional SR. The positive lusitropic effect is associated with an increase in the rate of Ca^{2+} uptake by the SR during relaxation as a consequence of PLB phosphorylation, as well as a decrease in Ca^{2+}-binding affinity of myofilaments caused by phosphorylation of the regulatory protein troponin-I. In addition to these effects, β_1AR stimulation also regulates the activity of a number of different ion channels, not just L-type Ca^{2+} channels (Hartzell 1988). In the adult heart, this limits changes in action potential duration and ensures normal impulse propagation. In neonatal ventricular myocytes, it contributes to an increase in the spontaneous beating rate.

β_1ARs have been found in all fractions of the plasma membrane of both neonatal and adult cardiac myocyte. This includes caveolar and non-caveolar lipid raft fractions as well as non-lipid raft heavy fractions (Ostrom et al. 2000b, 2001, 2004; Rybin et al. 2000; Xiang et al. 2002b; Head et al. 2005, 2006; Balijepalli et al. 2006; Nichols et al. 2010; Agarwal et al. 2011). Many of the same studies found that the stimulatory protein G_s, which couples the β_1AR to adenylyl cyclase (AC), exhibits a similar distribution pattern. There are multiple AC isoforms expressed in cardiac tissue, the most abundant being AC5 and AC6, which are expressed in caveolar as well as non-caveolar membrane fractions (Ostrom et al. 2000b; Rybin et al. 2000; Head et al. 2005; Balijepalli et al. 2006). Most evidence suggests that the association of AC5/6 with caveolar fractions of the plasma membrane does so at least in part because of direct interactions with Cav3 (Rybin et al. 2000; Xiang et al. 2002a; Head et al. 2005; Balijepalli et al. 2006). However, at least one study has suggested that the interaction between Cav3 and AC5/6 involves the A-kinase-anchoring protein, AKAP5 (Nichols et al. 2010).

The widespread distribution of β_1ARs raises the question of whether cAMP produced by receptors found in different membrane domains plays an equal role in regulating the various targets phosphorylated by PKA. As a way of addressing this question, investigators have looked at the effect that disrupting caveolae and/or lipid rafts have on different responses. One way of doing this is by depleting membrane cholesterol using agents such as methyl-β-cyclodextrin (MβCD), which removes cholesterol directly from the lipid bilayer (Harvey and Calaghan 2012). This has been shown to cause a shift of caveolin from buoyant to heavy membrane fractions. It has also been found to cause a significant reduction in caveolar density as determined by electron microscopy. It has been argued that this technique preferentially affects caveolar fractions. However, it is most likely affecting non-caveolar lipid raft fractions of the membrane as well. Caveolar signaling can also be affected by knocking down expression of Cav3 using molecular approaches (Balijepalli et al. 2006), by using Cav3 antibodies (Calaghan and White 2006) or peptides that mimic the CSD to block caveolin function (Feron et al. 1998; Macdougall et al. 2012), or by inhibiting cholesterol biosynthesis (Pugh et al. 2014).

It has been reported that cholesterol depletion or Cav3 knockdown with siRNA does not alter the effect that maximal β_1AR activation has on the spontaneous beating rate (Xiang et al. 2002b) or L-type Ca^{2+} channel function (Balijepalli et al. 2006) in neonatal cardiac myocytes. It has also been reported that MβCD treatment has no effect on the maximal response that β_1AR stimulation has on contraction or L-type Ca^{2+} channel function in adult ventricular myocytes (Calaghan and White 2006). However, MβCD treatment can dramatically increase the sensitivity of contraction and L-type Ca^{2+} channel activity to β_1AR stimulation without altering the maximal response (Agarwal et al. 2011). This is consistent with studies demonstrating that cholesterol depletion enhances the sensitivity of cardiac myocytes to β_1AR stimulation of cAMP production (Rybin et al. 2000; Head et al. 2005). This can be explained by the loss of cholesterol disrupting the inhibitory effect that direct interaction with Cav3 has on AC activity (Toya et al. 1998; Rybin et al. 2000).

This question of whether or not cAMP produced by β_1ARs associated with caveolae is restricted to localized responses or contributes to more global changes has been addressed using genetically encoded FRET-based biosensors expressed in different subcellular locations. One such biosensor, constructed using type II PKA, is targeted to the same locations as endogenous type II PKA through its interactions with A-kinase-anchoring proteins (AKAPs), which includes caveolae (Zaccolo and Pozzan 2002). Changes in cAMP activity detected by this probe have been shown to correlate with PKA-dependent functional responses (Warrier et al. 2005, 2007; Agarwal et al. 2011). Another biosensor, constructed using the cAMP-binding domain of the type 2 exchange protein activated by cAMP (Epac2-camps), lacks any targeting sequences allowing it to freely diffuse throughout the cytoplasm of the cell (Nikolaev et al. 2004). This biosensor responds to global changes in cAMP (Warrier et al. 2007; Iancu et al. 2008; Agarwal et al. 2011). Using these probes, it was found that MβCD treatment increases the sensitivity of the cAMP response in PKA signaling domains, but not global cAMP responses (Agarwal et al. 2011).

The increase in sensitivity of cAMP responses detected by the type II PKA probe was found to correlate with changes in the sensitivity of L-type Ca^{2+} channel function as well as changes in both inotropic and lusitropic responses (Agarwal et al. 2011). Although these results suggest that β_1ARs found in caveolae contribute to the regulation of targets found in various locations throughout the cell, the cAMP produced there does not contribute to global cytosolic changes like those detected by the Epac2-based probe. These data support the conclusion that β_1ARs found in caveolae are capable of producing a compartmentalized cAMP response, while receptors found in other locations are likely contributing to the global cytosolic response. Consistent with this idea, Nichols et al. (2010) demonstrated that AKAP5 anchors AC5/6 to a Cav3 signaling complex that includes the β_1AR and L-type Ca^{2+} channels. It was shown that activation of β_1ARs associated with this complex selectively regulates a subpopulation of L-type Ca^{2+} channels that are part of the dyadic complex involved in excitation-contraction (EC) coupling. Knocking out AKAP5 expression resulted in the translocation of AC5/6 to non-lipid raft domains, where it then regulated Ca^{2+} channels not involved in EC coupling.

Although globally expressed, cytosolic biosensors have been shown to detect cAMP responses that are propagated throughout the cell following β_1AR stimulation (Nikolaev et al. 2006, 2010; Wright et al. 2014), at least one study suggest that this is not the case (Richards et al. 2016). This raises the possibility that the movement of cAMP produced outside of caveolar membrane domains is restricted as well. It has recently been demonstrated that PKA buffering may be an important factor limiting cAMP movement throughout the cell (Agarwal et al. 2016).

The AC antibodies used in most studies are not able to distinguish between AC5 and AC6 (although see Nichols et al. 2010), often leading to the assumption that these two isoforms exhibit a common distribution pattern. However, experiments using transgenic animals in which AC5 or AC6 are selectively knocked out suggest that this is not the case (Timofeyev et al. 2013). Using these mice, together with methods that separate the T tubules from the peripheral sarcolemma, Timofeyev et al. (Timofeyev et al. 2013) demonstrated that β_1AR responses occur in both the T tubules and the peripheral sarcolemma of adult ventricular myocytes. However, T tubule responses were mediated by AC5, while the responses associated with the peripheral sarcolemma involved AC6 (see Fig. 2.1).

2.3.2 β_2-Adrenergic Receptors (β_2ARs)

β_2ARs make up ~20% of the total βAR population in cardiac myocytes, and although they also couple to G_s-dependent stimulation of AC and cAMP production, the functional responses they produce can be quite different from those elicited by β_1ARs. In adult ventricular myocytes of certain species, it has been reported that β_2AR stimulation is capable of producing a positive inotropic effect, but not the positive lusitropic effect elicited by β_1ARs (Xiao 2001). The explanation for this behavior is that β_2ARs produce a localized cAMP response that is restricted to regulating L-type Ca^{2+} channel activity (Chen-Izu et al. 2000). There is little or no PKA-dependent phosphorylation of phospholamban or troponin-I, which is necessary to speed relaxation (Xiao et al. 1994; Kuschel et al. 1999a, b; Macdougall et al. 2012). The ability of β_2ARs to produce a localized cAMP response has been attributed to the fact that they can couple to a G_i-dependent mechanism (Xiao et al. 1995, 1999; Kuschel et al. 1999a), although the exact nature of the G_i signaling pathway involved is a point of debate.

In neonatal ventricular myocytes, β_2AR stimulation produces both a positive inotropic and positive lusitropic effect (Kuznetsov et al. 1995). β_2ARs also produce an increase in the rate of spontaneous beating of these cells (Xiang et al. 2002a, b). However, this is followed by a subsequent decrease. While the increase in beating rate is associated with an increase in cAMP production, the subsequent decrease in beating rate is due to activation of a G_i-dependent mechanism (Devic et al. 2001; Xiang et al. 2002b).

As discussed above, β_2ARs are found primarily in caveolin-enriched buoyant fractions of the plasma membrane of both neonatal and adult cardiac ventricular myocytes, along with G_s, G_i, AC5/6, and type II PKA (Ostrom et al. 2000b, 2001, 2004; Rybin et al. 2000; Xiang et al. 2002b; Head et al. 2005, 2006; Balijepalli et al. 2006). Because β_2AR activation elicits localized functional responses associated with the selective regulation of L-type Ca^{2+} channels, one might predict that cAMP production would be highly restricted. Consistent with this hypothesis, at least one group has found that β_2AR stimulation does not produce a detectible change in cAMP measured using more traditional biochemical techniques

(Kuznetsov et al. 1995). However, studies using more sensitive FRET-based biosensor approaches have shown that β_2AR stimulation produces cAMP responses that can be detected by the type II PKA-based probe, as might be expected for a caveolar-targeted receptor (Macdougall et al. 2012). β_2AR stimulation also produces a response detected by more globally distributed cytosolic biosensors (Nikolaev et al. 2006, 2010; Macdougall et al. 2012; Wright et al. 2014). Although these observations might suggest that β_2AR production of cAMP is not necessarily restricted, it has been shown that cAMP produced by β_2AR stimulation does not readily diffuse throughout the cell (Nikolaev et al. 2006, 2010; Wright et al. 2014).

Disrupting caveolae by depleting cholesterol with MβCD was found to enhance cAMP production detected in type II PKA signaling domains, without affecting the size of global responses (Macdougall et al. 2012). This could be explained by the disruption of inhibitory effect that Cav3 has on AC activity. However, it has also been reported that disrupting caveolar function by either MβCD treatment or overexpression of a dominant negative form of Cav3 that disrupts CBD interactions actually causes a redistribution of β_2ARs, resulting in their ability to produce a cAMP response that then propagates throughout the cell. These observations paralleled those seen in myocytes following pressure overload-induced heart failure (Nikolaev et al. 2010; Wright et al. 2014).

An explanation for the difference between the β_1AR- and β_2AR-mediated responses may lie in the precise location of these receptors. Although caveolae are found in the peripheral plasma membrane as well as the T tubules of adult cardiac myocytes (Levin and Page 1980), Nikolaev et al. (2010) used scanning ion conductance microscopy (SICM) in combination with FRET-based biosensors to elegantly demonstrate that β_2AR production of cAMP is associated specifically with the T tubules, while β_1AR stimulation produces cAMP in the T tubules as well as the peripheral sarcolemma. The idea that β_2AR signaling is confined to the T tubules is consistent with the evidence that β_2AR regulation of L-type Ca^{2+} channels only occurs there as well. Furthermore, this response was found to be mediated specifically by AC5 (Timofeyev et al. 2013) (see Fig. 2.1).

A large proportion of the L-type Ca^{2+} channels found in the T tubules of adult myocytes are associated with dyadic clefts, where they come in close proximity to ryanodine receptors in the junctional SR (Scriven et al. 2000). Although caveolae are believed to be excluded from dyadic clefts (Levin and Page 1980), Cav3 has been shown to bind L-type Ca^{2+} channels as well as β_2ARs (Balijepalli et al. 2006), and it is postulated that Cav3 may form signaling complexes that are not necessarily associated with caveolae (Insel et al. 2005; Nichols et al. 2010; Wright et al. 2014).

The formation of a Cav3 signaling complex in T tubules may explain how β_2AR activation leads to selective regulation of L-type Ca^{2+} channels. The ability of these receptors to activate a G_i signaling mechanism is also clearly involved, since blocking the inhibitory G-protein pathway with pertussis toxin results in the ability of β_2ARs to stimulate phosphorylation of the non-junctional SR protein phospholamban and generate a positive lusitropic effect (Xiao et al. 1995; Kuschel et al. 1999b). One potential mechanism by which β_2AR stimulation produces a localized cAMP response is through recruitment of PDEs (Perry et al. 2002; Baillie et al. 2003), which have been proposed to limit cAMP diffusion to more distant targets. The recruitment of PDEs has also been reported to explain the transient nature of β_2AR effects on the rate of spontaneous beating in neonatal ventricular myocytes (Xiang et al. 2005). Alternatively, it has been suggested that G_i signaling is regulating phosphatase activity (Kuschel et al. 1999a). Consistent with this idea, it has been shown that disrupting caveolae not only allows β_2AR stimulation to produce a positive lusitropic effect, it also results in a loss of phosphatase-dependent regulation of phospholamban phosphorylation (Macdougall et al. 2012).

2.3.3 M_2 Muscarinic Receptors (M_2Rs)

M_2Rs are the primary type of muscarinic receptor found in cardiac myocytes (Dhein et al. 2001). These receptors are associated with a variety of functional responses throughout the heart (Löffelholz and Pappano 1985). M_2R activation causes a decrease in the rate of spontaneous firing of sinoatrial node cells, a decrease in action potential duration of atrial myocytes, a slowing of action potential propagation in atrioventricular node cells, and antagonism of β-adrenergic responses in all cardiac myocytes, including ventricular myocytes.

These responses are mediated by two general mechanisms. The first involves M_2R regulation of acetylcholine (ACh)-activated K^+ channels through a direct, G-protein-dependent mechanism (Wickman and Clapham 1995). This occurs primarily in myocytes found in the sinoatrial node, the atria, and the atrioventricular node. The second mechanism involves regulation of cAMP production (Harvey and Belevych 2003). This occurs in cells throughout the heart, including ventricular myocytes. The canonical mechanism associated with this later effect involves M_2R regulation of cAMP production through direct G_i-dependent inhibition of AC by the α subunit of the activated G protein. In ventricular myocytes, this response is most pronounced when AC activity is first enhanced by βAR stimulation. This is referred to as accentuated antagonism (Levy 1971).

There is also evidence that in some cells M_2R activation may inhibit cAMP production by stimulating endothelial nitric oxide synthase (eNOS) activity, resulting in the generation of nitric oxide (NO). NO-dependent nitrosylation can directly inhibit AC5/6 (Ostrom et al. 2004). NO may also indirectly affect cAMP by stimulating guanylyl cyclase activity (Balligand 1999). The resulting production of cGMP can then either decrease cAMP levels by stimulating type 2 PDE activity or increase cAMP levels by inhibiting the type 3 PDE activity. However, knocking out eNOS activity does not affect M_2R regulation of cAMP-dependent responses in adult ventricular myocytes (Vandecasteele et al. 1999; Belevych and Harvey 2000; Gödecke et al. 2001). This suggests that NO production is not directly involved in mediating M_2R responses in these cells.

M_2Rs are excluded from caveolar fractions of the plasma membrane in adult myocytes (Feron et al. 1997; Head et al. 2005). However, following agonist stimulation, at least some of these receptors have been reported to translocate to caveolar fractions over a time course of several minutes. Experiments using the type II PKA FRET-based biosensor support the idea that M_2R inhibition of cAMP production occurs in caveolar domains (Warrier et al. 2005), but whether M_2R inhibition of AC activity involves receptor translocation or perhaps activation of a residual population of caveolar receptors is unclear.

In neonatal ventricular myocytes, it has been reported that M_2R stimulation inhibits cAMP production by β_1ARs but not β_2ARs (Aprigliano et al. 1997). Because β_2ARs are thought to exist exclusively in caveolae, it is tempting to suggest that M_2R inhibition of cAMP production occurs outside of caveolae. However, β_1ARs are found in both caveolar and extra-caveolar domains. Furthermore, M_2Rs are found in caveolar as well as non-caveolar membrane fractions in neonatal myocytes (Rybin et al. 2000). It is interesting to note that while M_2R stimulation does not antagonize the β_2AR-induced positive inotropic effect observed in these cells, it does inhibit the positive lusitropic effect (Aprigliano et al. 1997). It is unclear whether M_2R activation antagonizes β_2AR responses in adult myocytes.

In adult myocytes, the response to M_2R activation is also more complex than simple inhibition of cAMP production (Harvey and Belevych 2003). These receptors can also stimulate cAMP production (Warrier et al. 2005; Iancu et al. 2008). The consequence is a complex biphasic response, whereby exposure to ACh produces a rapid inhibition of cAMP levels followed by a rebound increase upon washout. This behavior can be explained by M_2R activation producing both inhibitory and stimulatory responses simultaneously. The inhibitory effect turns on and off quickly, while the stimulatory effect turns on and off slowly. In the presence of agonist, the inhibitory effect is dominant. However, upon termination of receptor activation, the inhibitory effect turns off rapidly, revealing the stimulatory response, which turns off more slowly. In atrial cardiac myocytes, the stimulatory effect has been attributed to eNOS-dependent generation of NO, stimulation of cGMP production, and subsequent inhibition of PDE3 activity (Wang et al. 1998). However, the inhibitory as well as the stimulatory responses to M_2R activation are intact in myocytes isolated from eNOS-knockout mice (Belevych and Harvey 2000). In ventricular myocytes, this complex behavior can be explained by the ability of M_2R activation of G_i to regulate different isoforms of AC activity. In addition to AC5/6, cardiac myocytes are also known to express AC4 and AC7 (Ishikawa and Homcy 1997; Defer et al. 2000). While G_i signaling inhibits AC5/6 activity via a G_α-dependent mechanism, it actually stimulates AC4 and AC7 via $G_{\beta\gamma}$ (Sunahara et al. 1996). Therefore, the inhibitory effect can be explained by M_2R inhibition of AC5/6 while at the same time stimulating AC4/7 (Belevych et al. 2001; Harvey and Belevych 2003).

It has been hypothesized that the complex time course of the M_2R-mediated cAMP responses can be explained by the fact that different AC isoforms are expressed in different membrane domains (Iancu et al. 2007). While AC5/6 is found in caveolar membrane fractions of most cells, including cardiac myocytes, AC4 and AC7 are consistently excluded from caveolar fractions wherever it has been examined (Ostrom and Insel 2004; Willoughby and Cooper 2007) (see Fig. 2.1). Computational modeling indicates that inhibition of cAMP production in caveolar domains and stimulation of cAMP production in extra-caveolar domains are a feasible explanation for the biphasic response to M_2R activation in adult ventricular myocytes (Iancu et al. 2007, 2008).

2.3.4 Prostaglandin Receptors (EPRs)

The first evidence that cAMP signaling is compartmentalized came from studies comparing βAR and EPR responses in cardiac tissue. Early studies demonstrated that activation of either type of receptor is able to elevate intracellular cAMP, yet only βAR stimulation causes functional changes as well as PKA-dependent phosphorylation of the associated regulatory proteins (Keely 1977; Brunton et al. 1979; Hayes et al. 1979). Furthermore, βAR stimulation was found to activate PKA (primarily type II) in the particulate fraction of cardiac tissue homogenates, while EPR stimulation activated PKA (primarily type I) in the soluble fraction (Hayes et al. 1980). These observations lead to the original hypothesis that cAMP signaling is compartmentalized and that βARs and EPRs are capable of stimulating cAMP production in different subcellular locations (Corbin et al. 1977; Hayes et al. 1980). Although it was demonstrated early on that this phenomenon could be observed in isolated adult ventricular myocytes (Buxton and Brunton 1983), it wasn't until the development of genetically encoded FRET-based biosensors that it was possible to examine this behavior in intact, living cells.

One of the first studies using FRET-based biosensors compared the responses detected by the targeted type II PKA-based probe and the diffusible Epac2-camps biosensor in adult guinea pig ventricular myocytes (Warrier et al. 2007). It was found that the PKA-based probe was unable to detect any cAMP produced by EPR stimulation under normal circumstances. However, this probe was able to detect a response if targeting of the probe was first disrupted by blocking its interaction with AKAPs, allowing it to distribute throughout the cell (Warrier et al. 2007). This supported the idea that EPRs stimulate cAMP production in subcellular locations where particulate PKA signaling normally does not occur. Consistent with this idea, Epac2-camps, which is expressed throughout the cytoplasmic compartment, was able to detect cAMP produced by EPR stimulation.

Cardiac myocytes actually express multiple prostaglandin receptors subtypes. This includes EP_2 and EP_4Rs, which couple to G_s and stimulate cAMP production, as well as EP_3Rs, which couple to G_i. Although G_i signaling has been associated with signaling mechanisms that can lead to compartmentalized cAMP responses, it was found that blocking this pathway did not alter the compartmentalized behavior in these cells (Warrier et al. 2007).

Unlike β_1ARs and β_2ARs, which can be found in caveolar membrane fractions, cardiac EP_2 and EP_4Rs are excluded from this domain (Ostrom et al. 2001, 2004; Agarwal et al. 2011) (see Fig. 2.1). To evaluate whether this might contribute to compartmentation of cAMP responses to EPR activation, the effect of disrupting caveolae by cholesterol depletion with MβCD was examined. Unlike the guinea pig, in adult rat ventricular myocytes, EPR stimulation produces cAMP responses that can be detected by both the type II PKA-based probe and Epac2-camps. However, there is still no detectible change in myocyte contraction or L-type Ca^{2+} channel function (Agarwal et al. 2011). Furthermore, disrupting caveolae with MβCD treatment enhanced β_1AR responses, yet it had no effect on the EPR-mediated responses detected by either probe. This supports the idea that EPRs are stimulating cAMP production in non-caveolar membrane domains and that this contributes to cAMP compartmentation. It also suggests that, at least in some species, production of cAMP in non-caveolar membrane domains is able to activate type II PKA in locations not associated with the regulation of myocyte contraction. Finally, the fact that disrupting caveolae did not change the inability of EPR stimulation to regulate myocyte contraction and L-type Ca^{2+} channel function suggests that these receptors are not simply excluded from caveolae/lipid rafts and that other factors regulate their distribution.

A similar study was subsequently conducted in neonatal ventricular myocytes, where it was found that EPR stimulation could produce cAMP responses detected by an Epac1-based probe targeted to both type I and type II PKA signaling domains (Di Benedetto et al. 2008). Unlike the results in adult myocytes, inhibition of G_i signaling was found to enhance the response detected in type II but not type I signaling domains.

2.4 Conclusions and Future Directions

Compartmentation of cAMP signaling is central to our understanding of the varied responses elicited by activation of different G-protein-coupled receptors. This hypothesis was originally supported by studies showing differences in cAMP production in cytosolic and membrane fractions of homogenized preparations following specific receptor activation. While providing important information, such approaches are limited in their ability to evaluate the role that inhomogeneities in the distribution of membrane receptors and other signaling proteins play in compartmentation of cAMP signaling. The development of genetically encoded probes that respond to changes in cAMP levels or PKA activity now makes it possible to study the role that the localized production of cAMP plays in generating compartmentalized cAMP responses in live cells (Sprenger and Nikolaev 2013).

Most studies in cardiac myocytes have focused on the use of biosensors that are either expressed in the cytoplasmic compartment of the cell or targeted to PKA signaling domains. Using these probes in combination with other methods (i.e., lipid raft/caveolar disruption or targeted receptor activation by SICM), it has been possible to demonstrate that the nonuniform distribution of receptors plays an essential role in compartmentation of cAMP signaling. Future studies using cAMP-sensing probes targeted to different membrane domains should provide even more detailed information. For example, using probes targeted specifically to lipid raft and non-lipid raft domains by attaching acylation and prenylation sequences, respectively, it has been possible to show that EPR and βAR stimulations produce distinctly different cAMP responses in HEK293 cells (Agarwal et al. 2014). It should be possible to use these probes to investigate the specificity of EPR and βAR responses in cardiac myocytes. They may also be critical in testing the hypothesis that M_2Rs inhibit and stimulate cAMP production in different subcellular locations of these cells as well.

Localized cAMP production is just one factor contributing to cAMP compartmentation. PDE activity has long been known to be essential to generate such responses (Fischmeister et al. 2006; Mika et al. 2012). Yet, other critical factors are also becoming apparent. One such factor is slowed diffusion of cAMP due to molecular crowding and PKA buffering (Agarwal et al. 2016). However, modeling studies have revealed that even when taking into account slow diffusion, cAMP levels would still be expected to disperse uniformly throughout most of the cell. Only in when cAMP production is localized to anatomically restricted spaces, such dyadic clefts created by apposition of the junctional SR and T tubule, is PDE activity in conjunction with slow diffusion able to produce compartmentalized responses (Yang et al. 2016). It is still not entirely clear what the relationship is between different membrane microdomains and these restricted spaces. Advanced techniques, such as super-resolution microscopy (Eggeling et al. 2009), in conjunction with the use of genetically encoded biosensors, SICM, and computational modeling, are poised to offer a more detailed picture of the role that caveolar/lipid raft and non-lipid raft microdomains play in compartmentation of cAMP in cardiac myocytes.

Acknowledgements This work was supported by the National Institutes of Health grants GM101928 and GM107094.

Compliance with Ethical Standards

Conflict of Interest Statement The authors declare that they have no conflict of interest.

References

Agarwal SR, Clancy CE, Harvey RD (2016) Mechanisms restricting diffusion of intracellular cAMP. Sci Rep 6:19577. doi:10.1038/srep19577

Agarwal SR, Macdougall DA, Tyser R et al (2011) Effects of cholesterol depletion on compartmentalized cAMP responses in adult cardiac myocytes. J Mol Cell Cardiol 50:500–509

Agarwal SR, Yang PC, Rice M et al (2014) Role of membrane microdomains in compartmentation of cAMP signaling. PLoS One 9:e95835

Allen JA, Halverson-Tamboli RA, Rasenick MM (2007) Lipid raft microdomains and neurotransmitter signalling. Nat Rev Neurosci 8:128–140

Aprigliano O, Rybin VO, Pak E et al (1997) β_1- and β_2-adrenergic receptors exhibit differing susceptibility to muscarinic accentuated antagonism. Am J Physiol 272:H2726–H2735

Baillie GS, Sood A, McPhee I et al (2003) beta-Arrestin-mediated PDE4 cAMP phosphodiesterase recruitment regulates beta-adrenoceptor switching from Gs to Gi. Proc Natl Acad Sci U S A 100:940–945

Balijepalli RC, Foell JD, Hall DD et al (2006) Localization of cardiac L-type Ca(2+) channels to a caveolar macromolecular signaling complex is required for beta(2)-adrenergic regulation. Proc Natl Acad Sci U S A 103:7500–7505

Balligand JL (1999) Regulation of cardiac beta-adrenergic response by nitric oxide. Cardiovasc Res 43:607–620

Belevych AE, Harvey RD (2000) Muscarinic inhibitory and stimulatory regulation of the L-type Ca^{2+} current is not altered in cardiac ventricular myocytes from mice lacking endothelial nitric oxide synthase. J Physiol (Lond) 528:279–289

Belevych AE, Sims C, Harvey RD (2001) ACh-induced rebound stimulation of L-type Ca(2+) current in guinea-pig ventricular myocytes, mediated by Gbetagamma-dependent activation of adenylyl cyclase. J Physiol (Lond) 536:677–692

Bers DM (2002) Cardiac excitation-contraction coupling. Nature 415:198–205

Bethani I, Skanland SS, Dikic I et al (2010) Spatial organization of transmembrane receptor signalling. EMBO J 29:2677–2688

Brown DA (2006) Lipid rafts, detergent-resistant membranes, and raft targeting signals. Physiology 21:430–439
Brown DA, Rose JK (1992) Sorting of GPI-anchored proteins to glycolipid-enriched membrane subdomains during transport to the apical cell surface. Cell 68:533–544
Brunton LL, Hayes JS, Mayer SE (1979) Hormonally specific phosphorylation of cardiac troponin I and activation of glycogen phosphorylase. Nature 280:78–80
Buxton IL, Brunton LL (1983) Compartments of cyclic AMP and protein kinase in mammalian cardiomyocytes. J Biol Chem 258:10233–10239
Calaghan S, White E (2006) Caveolae modulate excitation-contraction coupling and beta2-adrenergic signalling in adult rat ventricular myocytes. Cardiovasc Res 69:816–824
Chen-Izu Y, Xiao RP, Izu LT et al (2000) G(i)-dependent localization of beta(2)-adrenergic receptor signaling to L-type Ca(2+) channels. Biophys J 79:2547–2556
Chini B, Parenti M (2004) G-protein coupled receptors in lipid rafts and caveolae: how, when and why do they go there? J Mol Endocrinol 32:325–338
Corbin JD, Sugden PH, Lincoln TM et al (1977) Compartmentalization of adenosine 3′:5′-monophosphate and adenosine 3′:5′-monophosphate-dependent protein kinase in heart tissue. J Biol Chem 252:3854–3861
Defer N, Best-Belpomme M, Hanoune J (2000) Tissue specificity and physiological relevance of various isoforms of adenylyl cyclase. Am J Physiol 279:F400–F416
Devic E, Xiang Y, Gould D et al (2001) Beta-adrenergic receptor subtype-specific signaling in cardiac myocytes from beta(1) and beta(2) adrenoceptor knockout mice. Mol Pharmacol 60:577–583
Dhein S, Van Koppen CJ, Brodde OE (2001) Muscarinic receptors in the mammalian heart. Pharmacol Res 44:161–182
Di Benedetto G, Zoccarato A, Lissandron V et al (2008) Protein kinase A type I and type II define distinct intracellular signaling compartments. Circ Res 103:836–844
Eggeling C, Ringemann C, Medda R et al (2009) Direct observation of the nanoscale dynamics of membrane lipids in a living cell. Nature 457:1159–1162
Feron O, Dessy C, Opel DJ et al (1998) Modulation of the endothelial nitric-oxide synthase-caveolin interaction in cardiac myocytes. Implications for the autonomic regulation of heart rate. J Biol Chem 273:30249–30254
Feron O, Smith TW, Michel T et al (1997) Dynamic targeting of the agonist-stimulated m2 muscarinic acetylcholine receptor to caveolae in cardiac myocytes. J Biol Chem 272:17744–17748
Fischmeister R, Castro LR, Abi-Gerges A et al (2006) Compartmentation of cyclic nucleotide signaling in the heart: the role of cyclic nucleotide phosphodiesterases. Circ Res 99:816–828
Gancedo JM (2013) Biological roles of cAMP: variations on a theme in the different kingdoms of life. Biol Rev Camb Philos Soc 88:645–668
Gödecke A, Heinicke T, Kamkin A et al (2001) Inotropic response to beta-adrenergic receptor stimulation and anti- adrenergic effect of ACh in endothelial NO synthase-deficient mouse hearts. J Physiol (Lond) 532:195–204
Hanson MA, Cherezov V, Griffith MT et al (2008) A specific cholesterol binding site is established by the 2.8 A structure of the human beta2-adrenergic receptor. Structure 16:897–905
Hartzell HC (1988) Regulation of cardiac ion channels by catecholamines, acetylcholine and second messenger systems. Prog Biophys Mol Biol 52:165–247
Harvey RD, Belevych AE (2003) Muscarinic regulation of cardiac ion channels. Br J Pharmacol 139:1074–1084
Harvey RD, Calaghan SC (2012) Caveolae create local signalling domains through their distinct protein content, lipid profile and morphology. J Mol Cell Cardiol 52:366–375
Hayes JS, Brunton LL, Brown JH et al (1979) Hormonally specific expression of cardiac protein kinase activity. Proc Natl Acad Sci U S A 76:1570–1574
Hayes JS, Brunton LL, Mayer SE (1980) Selective activation of particulate cAMP-dependent protein kinase by isoproterenol and prostaglandin E1. J Biol Chem 255:5113–5119

Head BP, Patel HH, Roth DM et al (2005) G-protein-coupled receptor signaling components localize in both sarcolemmal and intracellular caveolin-3-associated microdomains in adult cardiac myocytes. J Biol Chem 280:31036–31044

Head BP, Patel HH, Roth DM et al (2006) Microtubules and actin microfilaments regulate lipid raft/caveolae localization of adenylyl cyclase signaling components. J Biol Chem 281: 26391–26399

Iancu RV, Jones SW, Harvey RD (2007) Compartmentation of cAMP signaling in cardiac myocytes: a computational study. Biophys J 92:3317–3331

Iancu RV, Ramamurthy G, Warrier S et al (2008) Cytoplasmic cAMP concentrations in intact cardiac myocytes. Am J Physiol Cell Physiol 295:C414–C422

Insel PA, Head BP, Patel HH et al (2005) Compartmentation of G-protein-coupled receptors and their signalling components in lipid rafts and caveolae. Biochem Soc Trans 33:1131–1134

Ishikawa Y, Homcy CJ (1997) The adenylyl cyclases as integrators of transmembrane signal transduction. Circ Res 80:297–304

Jacobson K, Mouritsen OG, Anderson RGW (2007) Lipid rafts: at a crossroad between cell biology and physics. Nat Cell Biol 9:7–14

Keely SL (1977) Activation of cAMP-dependent protein kinase without a corresponding increase in phosphorylase activity. Res Commun Chem Pathol Pharmacol 18:283–290

Kuschel M, Zhou YY, Cheng H et al (1999a) G(i) protein-mediated functional compartmentalization of cardiac beta(2)-adrenergic signaling. J Biol Chem 274:22048–22052

Kuschel M, Zhou YY, Spurgeon HA et al (1999b) Beta2-adrenergic cAMP signaling is uncoupled from phosphorylation of cytoplasmic proteins in canine heart. Circulation 99:2458–2465

Kuznetsov V, Pak E, Robinson RB et al (1995) β_2-adrenergic receptor actions in neonatal and adult rat ventricular myocytes. Circ Res 76:40–52

Levin KR, Page E (1980) Quantitative studies on plasmalemmal folds and caveolae of rabbit ventricular myocardial cells. Circ Res 46:244–255

Levy MN (1971) Sympathetic-parasympathetic interactions in the heart. Circ Res:437–445

Löffelholz K, Pappano AJ (1985) The parasympathetic neuroeffector junction of the heart. Pharmacol Rev 37:1–24

Macdougall DA, Agarwal SR, Stopford EA et al (2012) Caveolae compartmentalise beta2-adrenoceptor signals by curtailing cAMP production and maintaining phosphatase activity in the sarcoplasmic reticulum of the adult ventricular myocyte. J Mol Cell Cardiol 52:388–400

Melkonian KA, Ostermeyer AG, Chen JZ et al (1999) Role of lipid modifications in targeting proteins to detergent-resistant membrane rafts. Many raft proteins are acylated, while few are prenylated. J Biol Chem 274:3910–3917

Mika D, Leroy J, Vandecasteele G et al (2012) PDEs create local domains of cAMP signaling. J Mol Cell Cardiol 52:323–329

Moffett S, Brown DA, Linder ME (2000) Lipid-dependent targeting of G proteins into rafts. J Biol Chem 275:2191–2198

Nichols CB, Rossow CF, Navedo MF et al (2010) Sympathetic stimulation of adult cardiomyocytes requires association of AKAP5 with a subpopulation of L-type calcium channels. Circ Res 107:747–756

Nikolaev VO, Bunemann M, Hein L et al (2004) Novel single chain cAMP sensors for receptor-induced signal propagation. J Biol Chem 279:37215–37218

Nikolaev VO, Bunemann M, Schmitteckert E et al (2006) Cyclic AMP imaging in adult cardiac myocytes reveals far-reaching beta-1 adrenergic but locally confined beta-2 adrenergic receptor-mediated signaling. Circ Res 99:1084–1091

Nikolaev VO, Moshkov A, Lyon AR et al (2010) Beta2-adrenergic receptor redistribution in heart failure changes cAMP compartmentation. Science 327:1653–1657

Ostrom RS, Bundey RA, Insel PA (2004) Nitric oxide inhibition of adenylyl cyclase type 6 activity is dependent upon lipid rafts and caveolin signaling complexes. J Biol Chem 279: 19846–19853

Ostrom RS, Gregorian C, Drenan RM et al (2001) Receptor number and caveolar co-localization determine receptor coupling efficiency to adenylyl cyclase. J Biol Chem 276:42063–42069
Ostrom RS, Insel PA (2004) The evolving role of lipid rafts and caveolae in G protein-coupled receptor signaling: implications for molecular pharmacology. Br J Pharmacol 143:235–245
Ostrom RS, Post SR, Pa I (2000a) Stoichiometry and compartmentation in G protein-coupled receptor signaling: implications for therapeutic interventions involving G(s). J Pharmacol Exp Ther 294:407–412
Ostrom RS, Violin JD, Coleman S et al (2000b) Selective enhancement of beta-adrenergic receptor signaling by overexpression of adenylyl cyclase type 6: colocalization of receptor and adenylyl cyclase in caveolae of cardiac myocytes. Mol Pharmacol 57:1075–1079
Pagano M, Clynes MA, Masada N et al (2009) Insights into the residence in lipid rafts of adenylyl cyclase AC8 and its regulation by capacitative calcium entry. Am J Physiol Cell Physiol 296:C607–C619
Perino A, Ghigo A, Scott JD et al (2012) Anchoring proteins as regulators of signaling pathways. Circ Res 111:482–492
Perry SJ, Baillie GS, Kohout TA et al (2002) Targeting of cyclic AMP degradation to beta 2-adrenergic receptors by beta-arrestins. Science 298:834–836
Pugh SD, MacDougall DA, Agarwal SR et al (2014) Caveolin contributes to the modulation of basal and beta-adrenoceptor stimulated function of the adult rat ventricular myocyte by simvastatin: a novel pleiotropic effect. PLoS One 9:e106905
Resh MD (2006) Trafficking and signaling by fatty-acylated and prenylated proteins. Nat Chem Biol 2:584–590
Richards M, Lomas O, Jalink K et al (2016) Intracellular tortuosity underlies slow cAMP diffusion in adult ventricular myocytes. Cardiovasc Res 110:395–407
Robison GA, Butcher RW, Sutherland EW (1968) Cyclic AMP. Annu Rev Biochem 37:149–174
Rybin VO, Xu X, Lisanti MP et al (2000) Differential targeting of beta -adrenergic receptor subtypes and adenylyl cyclase to cardiomyocyte caveolae. A mechanism to functionally regulate the cAMP signaling pathway. J Biol Chem 275:41447–41457
Scott JD, Dessauer CW, Tasken K (2013) Creating Order from Chaos: Cellular Regulation by Kinase Anchoring. Annu Rev Pharmacol Toxicol 53:187–210
Scriven DR, Dan P, Moore ED (2000) Distribution of proteins implicated in excitation-contraction coupling in rat ventricular myocytes. Biophys J 79:2682–2691
Shenoy SK, Lefkowitz RJ (2011) beta-Arrestin-mediated receptor trafficking and signal transduction. Trends Pharmacol Sci 32:521–533
Singer SJ, Nicolson GL (1972) The fluid mosaic model of the structure of cell membranes. Science 175:720–731
Sprenger JU, Nikolaev VO (2013) Biophysical techniques for detection of cAMP and cGMP in living cells. Int J Mol Sci 14:8025–8046
Steinberg SF (1999) The molecular basis for distinct beta-adrenergic receptor subtype actions in cardiomyocytes. Circ Res 85:1101–1111
Steinberg SF, Brunton LL (2001) Compartmentation of G protein-coupled signaling pathways in cardiac myocytes. Annu Rev Pharmacol Toxicol 41:751–773
Sunahara RK, Dessauer CW, Gilman AG (1996) Complexity and diversity of mammalian adenylyl cyclases. Annu Rev Pharmacol Toxicol 36:461–480
Timofeyev V, Myers RE, Kim HJ et al (2013) Adenylyl cyclase subtype-specific compartmentalization: differential regulation of L-type Ca^{2+} current in ventricular myocytes. Circ Res 112:1567–1576
Tolkovsky AM, Levitzki A (1978) Mode of coupling between the beta-adrenergic receptor and adenylate cyclase in turkey erythrocytes. Biochemistry 17:3795
Toya Y, Schwencke C, Couet J et al (1998) Inhibition of adenylyl cyclase by caveolin peptides. Endocrinology 139:2025–2031
Vandecasteele G, Eschenhagen T, Scholz H et al (1999) Muscarinic and beta-adrenergic regulation of heart rate, force of contraction and calcium current is preserved in mice lacking endothelial nitric oxide synthase. Nat Med 5:331–334

Wang YG, Rechenmacher CE, Lipsius SL (1998) Nitric oxide signaling mediates stimulation of L-type Ca^{2+} current elicited by withdrawal of acetylcholine in cat atrial myocytes. J Gen Physiol 111:113–125

Warrier S, Belevych AE, Ruse M et al (2005) Beta-adrenergic and muscarinic receptor induced changes in cAMP activity in adult cardiac myocytes detected using a FRET based biosensor. Am J Physiol Cell Physiol 289:C455–C461

Warrier S, Ramamurthy G, Eckert RL et al (2007) cAMP microdomains and L-type Ca^{2+} channel regulation in guinea-pig ventricular myocytes. J Physiol (Lond) 580:765–776

Wickman K, Clapham DE (1995) Ion channel regulation by G proteins. Physiol Rev 75:865–885

Williams TM, Lisanti MP (2004) The caveolin proteins. Genome Biol 5:214

Willoughby D, Cooper DM (2007) Organization and Ca^{2+} regulation of adenylyl cyclases in cAMP microdomains. Physiol Rev 87:965–1010

Wright PT, Nikolaev VO, O'Hara T et al (2014) Caveolin-3 regulates compartmentation of cardiomyocyte beta2-adrenergic receptor-mediated cAMP signaling. J Mol Cell Cardiol 67:38–48

Xiang Y, Devic E, Kobilka B (2002a) The PDZ binding motif of the beta 1 adrenergic receptor modulates receptor trafficking and signaling in cardiac myocytes. J Biol Chem 277: 33783–33790

Xiang Y, Kobilka B (2003) The PDZ-binding motif of the beta2-adrenoceptor is essential for physiologic signaling and trafficking in cardiac myocytes. Proc Natl Acad Sci U S A 100:10776–10781

Xiang Y, Naro F, Zoudilova M et al (2005) Phosphodiesterase 4D is required for beta2 adrenoceptor subtype-specific signaling in cardiac myocytes. Proc Natl Acad Sci U S A 102:909–914

Xiang Y, Rybin VO, Steinberg SF et al (2002b) Caveolar localization dictates physiologic signaling of beta 2-adrenoceptors in neonatal cardiac myocytes. J Biol Chem 277:34280–34286

Xiao RP (2001) Beta-adrenergic signaling in the heart: dual coupling of the beta2-adrenergic receptor to G(s) and G(i) proteins. Science STKE 2001:RE15

Xiao RP, Avdonin P, Zhou YY et al (1999) Coupling of beta2-adrenoceptor to Gi proteins and its physiological relevance in murine cardiac myocytes. Circ Res 84:43–52

Xiao RP, Hohl C, Altschuld R et al (1994) Beta 2-adrenergic receptor-stimulated increase in cAMP in rat heart cells is not coupled to changes in Ca^{2+} dynamics, contractility, or phospholamban phosphorylation. J Biol Chem 269:19151–19156

Xiao RP, Ji X, Lakatta EG (1995) Functional coupling of the beta 2-adrenoceptor to a pertussis toxin-sensitive G protein in cardiac myocytes. Mol Pharmacol 47:322–329

Yang PC, Boras BW, Jeng MT et al (2016) A computational modeling and simulation approach to investigate mechanisms of subcellular cAMP compartmentation. PLoS Comput Biol 12:e1005005

Zaccolo M, Pozzan T (2002) Discrete microdomains with high concentration of cAMP in stimulated rat neonatal cardiac myocytes. Science 295:1711–1715

Function to Failure: Compartmentalization of Cardiomyocyte Signaling by A-Kinase-Anchoring Proteins

3

John M. Redden, Kimberly L. Dodge-Kafka, and Michael S. Kapiloff

Contents

J. M. Redden
Department of Physiology and Neurobiology, University of Connecticut, Storrs, CT, USA
e-mail: john.redden@uconn.edu

K.L. Dodge-Kafka
Calhoun Center for Cardiology, University of Connecticut Health Center, Farmington, CT, USA
e-mail: dodge@uchc.edu

M.S. Kapiloff (✉)
Cardiac Signal Transduction and Cellular Biology Laboratory, Departments of Pediatrics and Medicine, Leonard M. Miller School of Medicine, Interdisciplinary Stem Cell Institute, University of Miami, Miami, FL, USA
e-mail: mkapiloff@med.miami.edu

V. Nikolaev, M. Zaccolo (eds.), *Microdomains in the Cardiovascular System*, Cardiac and Vascular Biology 3, DOI 10.1007/978-3-319-54579-0_3

Abstract

Compartmentalization of signaling enzymes allows cardiomyocytes to make contextually specific decisions using a common set of second messengers. Though first identified by their role in localizing the pleiotropic cAMP-dependent protein kinase A (**PKA**) to specific intracellular organelles and compartments, A-kinase-anchoring proteins (**AKAPs**) are a structurally and functionally diverse family of multivalent scaffolds that organize "signalosomes" constituting critical nodes in the cell-type-specific network of intracellular signaling pathways. This chapter summarizes the role of AKAPs in cardiomyocytes, with a focus on the intersection of compartmentalized signaling and cardiac pathophysiology.

3.1 Introduction

The heart is a dynamic organ, central to human life beyond 6 weeks of gestation. The adult human heart is tasked with circulating over 7000 L of blood each day to meet the energetic demands of the body—the end result of 100,000 individual beats. As the needs of the tissues change in response to internal and external stimuli, the heart rises to the challenge. During exercise, for example, a typical adult may experience an increase in cardiac output of fivefold or more in order to compensate for the increased oxygen demands of the body's skeletal muscle (Grimby et al. 1966). At the opposite extreme is heart failure, a clinical syndrome defined as pump dysfunction accompanied by shortness of breath, fatigue, and/or fluid retention (Remme and Swedberg 2001). Although the heart as an organ has been studied for thousands of years, many pioneering studies on heart failure were only conducted in the early twentieth century (Katz 2008). Changes in cardiac function are a hallmark of cardiovascular disease, which is currently responsible for one third of all deaths worldwide (World Health Organization 2005). The high morbidity and mortality associated with cardiac dysfunction, therefore, provides a sobering call to action for more innovative and precise treatments, stemming from a more thorough understanding of cardiac pathophysiology.

From a histological perspective, the heart is heterogeneous in its composition. In fact, the cardiomyocytes that produce the characteristic pumping of the heart are outnumbered 2:1 by other smaller cell types (Nag 1980). Though nonmyocyte cell populations such as fibroblasts and endothelial cells play crucial roles in the normal maintenance of cardiac structure and function, it is ultimately the contractile cardiomyocytes that define heart function at the organ level. Each beat requires a spatially and temporally synchronized contraction of nearly three billion individual cardiomyocytes. These cells, at the molecular level, must precisely relay electrochemical messages to the intracellular excitation-contraction machinery on a millisecond timescale (Bers 2008). This intricate process is initiated by the cardiac action potential propagating across the sarcolemma. As depolarization spreads through the t-tubule system, voltage-dependent, L-type Ca^{2+} channels (**Cav1.2**) are brought to threshold, allowing the entry of extracellular Ca^{2+}. This Ca^{2+} entry elicits the passive release of Ca^{2+} ions from the sarcoplasmic reticulum through the ryanodine receptor (**RyR**) in a process termed Ca^{2+}-induced Ca^{2+} release. Ca^{2+} binding to troponin C liberates tropomyosin from the myosin binding site on actin, allowing the myofilaments to slide and the myocyte to contract (Bers 2002). These events characterize the *inotropic* phase of the cycle. The *lusitropic* phase, during which the cardiomyocyte relaxes, requires the release of Ca^{2+} ions from troponin C, Ca^{2+} efflux, and Ca^{2+} reuptake into the sarcoplasmic reticulum via the sarco-/endoplasmic reticulum Ca^{2+} ATPase 2a (**SERCA2a**). The cell repolarizes back to its resting potential through the efflux of potassium ions and awaits the next depolarizing stimulus to begin the cycle anew (Bartos et al. 2015).

The heart has a limited repertoire of responses to acute and chronic changes in demands for cardiac output. Acutely, heart rate (chronotropy), the strength of myocyte contraction (inotropy), and the rate of myocyte relaxation (lusitropy) can all be increased in response to sympathetic stimulation (Katz and Lorell 2000). Canonical experiments in the mid-twentieth century documented a positive lusitropic effect in response to sympathetic stimulation and exercise (Levy 1971). Activation of the β-adrenergic receptor, for example, was shown to increase Ca^{2+} reuptake via SERCA2a. Positive inotropic and chronotropic changes were later documented. These effects were ultimately characterized as an effect of the second messenger 3′5′cyclic adenosine monophosphate (**cAMP**) and its effector PKA (Keely and Corbin 1977). In contrast to short-term, acute stimulation, long-term stress will induce cardiac hypertrophy, i.e., non-mitotic growth of the cardiac myocytes (Grossman et al. 1975; Hill and Olson 2008).

Physiologic stress, such as in pregnancy or extreme exercise, results in a relatively symmetric growth in length and width of the stereotypically cylindrical myocyte and the heart overall. In disease, pathological hypertrophy is often asymmetrical, with concentric growth in width of the myocyte and the ventricular wall in response to pressure overload and an eccentric growth in length of the myocyte and ventricular dilation in response to volume overload (Hill and Olson 2008). In contrast to physiological hypertrophy, pathological hypertrophy is associated with changes in myocyte contractility and metabolism, myocyte apoptosis, and the induction of interstitial myocardial fibrosis. This pathological remodeling ultimately results in heart failure, a clinical syndrome with an average 5-year mortality of 50% (Mozaffarian et al. 2015). Because PKA signaling plays a central role in the response of the heart to both acute and chronic stress, in both health and disease, a major effort in the field of cardiac signal transduction has been to understand how specificity and efficacy are conferred for this pleiotropic kinase (Lymperopoulos et al. 2013) .

In the late 1970s, it was observed that although cAMP was unquestionably the second messenger responsible for enhancing cardiac myocyte contractility, only the cAMP produced from sympathetic stimulation produced these acute effects. Activation of the prostaglandin E1 receptor, for example, had little effect on contractility despite leading to a global increase in cAMP levels (Hayes et al. 1980). This paradox led to a compartmentalized model for cAMP signaling within the cardiomyocyte. This model was substantiated by the observation that regulatory subunits of PKA were bound directly by scaffold proteins called A-kinase-anchoring proteins (**AKAPs**) targeting distinct pools of PKA holoenzyme to diverse intracellular organelles and compartments. In the decades following this discovery, the anchoring hypothesis has expanded significantly. It is now evident that the intracellular space within the cardiomyocyte is carved into heterogeneous microenvironments, in which receptor-mediated signals are locally generated, transduced to target proteins, and locally terminated (Wong and Scott 2004). AKAP scaffolds can organize multiprotein complexes consisting of an adenylyl cyclase (**AC**) to generate cAMP, the PKA holoenzyme (activated by cAMP), a PKA substrate, and a phosphodiesterase (**PDE**) to hydrolyze cAMP and end the signaling cycle. These enzymes are localized at a specific cellular region by the AKAP through protein-protein or protein-lipid interactions. Of particular interest, many AKAPs now have documented interactions with enzymes regulated by other second messengers and upstream signals, making them important communication hubs within the cell.

PKA is a heterotetramer composed of two regulatory R-subunits and two catalytic C-subunits in a C-R-R-C configuration. The R-subunit N-terminal docking and dimerization domains form an X-type, antiparallel four-helix bundle with a hydrophobic groove that can bind the hydrophobic face of the amphipathic helix that typically comprises an AKAP PKA-binding domain (Newlon et al. 1999). Over 50 AKAPs have been identified that have varying selectively for type I and type II PKA holoenzyme (as defined by the presence of RIα and RIβ or RIIα and RIIβ subunits). At least 20 different AKAPs are known to be expressed in the heart, each creating a unique signaling compartment at a discrete subcellular location (Table 3.1). As the role of AKAPs in regulating a diverse subset of cardiac functions continues to broaden, links between these scaffolds and the various etiologies of heart failure have also emerged. As there have been many excellent reviews on cardiac myocyte AKAPs, in this chapter we limit the discussion below to select examples for which there have been recent important findings (Kritzer et al. 2012; Diviani et al. 2013).

Table 3.1 Cardiac AKAPs

Gene	Aliases	Known interactions	Localization	Associated functions
AKAP1	D-AKAP, s-AKAP84, AKAP121, AKAP149	PKAs I and II, protein tyrosine phosphatase PTPD1, Src, PKCα, Lfc, PDE7A, RSK1, PP1, PP2A, CaN, mRNA, AMY-1, lamin-B, HIV-1 RT	Mitochondrion, endoplasmic reticulum, nuclear envelope	Cardiac remodeling, ischemia
AKAP5	AKAP79, AKAP75, AKAP150	PKAII, PKC, CaN, KCNQ2, CaV1, β1-AR, AC, SAP-97, PP1, AC5/AC6, PP2B, TRPV1	Plasmalemma and T-tubules	Calcium cycle/ contractility?
AKAP6	mAKAPβ	PKAII, AC5, PDE4D3, PP2A B56δ, RyR2, CaNAβ, NFATc, HIF-1a, VHL, Siah2, Epac1, Rap1, ERK5, MEK5, RSK3, PDK1, NCX1, Nesprin-1α, myopodin	Nuclear envelope	Cardiac remodeling, hypertrophy, ischemia
AKAP7	AKAP15, AKAP18 (α,β,γ,δ)	PKAII, CaV1, NaV1.2a channel, phospholamban, inhibitor-1, PP1, PKC, AQP2, 5′-AMP, AKAP7	Plasmalemma and sarcoplasmic reticulum, nuclear, plasma membrane	Calcium cycle/ contractility?
AKAP8	AKAP95	PKAII, PDE7A, MCM2, p68 RNA helicase, HDAC3, AMY-1, cyclin D/cyclin E, condensin	Nucleus	None
AKAP9	Yotiao, AKAP450, AKAP350, CG-NAP, hyperion	PKAII, AC, PP1, PP2A, PDE4D3, KCNQ1, IP3R, PKCε, PKN, casein kinase 1, chloride intracellular channel (CLIC), NMDAR, GCP2/3	Plasmalemma, centrosomes, Golgi	Arrhythmia (LQTS)
AKAP10	D-AKAP2	PKA I and II, Rab11, Rab4, PDZK1, PP1, RSK1	Outer mitochondrial membrane	Arrhythmia
AKAP11	AKAP220	PKAII, PP1, GSK3β, GABACR, AQP2	Peroxisomes	Unknown
AKAP12	Gravin, SSeCKS	PKAII, PDE4D3, PKC Src, CaN, β2-AR, calmodulin, Cyclin D	Actin cytoskeleton, plasma membrane (myristoylation)	Adrenergic receptor regulation, hypertrophy
AKAP13	AKAP-Lbc, Ht31, BRX-1	PKAII, Gα12, Rho, PKNα, MLTK, MKK3, p38α, KSR1, Raf, MEK1/MEK2, ERK1/ERK2, 14-3-3, PKCη, PKD, SHp2, HSP20, α-catulin, LC3	Cytoskeleton	Cardiac remodeling, hypertrophy
ARGEF2	BIG2	PKA I and II, formin-binding protein 3, PDE3A, TNFR1,MLC, β-catenin	Cytoplasm, internal membranes including Golgi	Unknown
EZR	Ezrin/AKAP78	PKAI and II, CFTR, EBP50/NHERF, NHE3, calmodulin, Rho kinase, actin, α1AR, E-cadherin, β-catenin, EGFR, Fas-R, PKCα, S100, FAK, SAP-97, moesin, radixin, FAK, merlin	Cytoskeleton	Ischemia, hypertrophy

(continued)

Table 3.1 (continued)

MAP2	MAP2B	PKAII, tubulin, CaV1, myosin VIIa	Microtubules	Cardiac development
CMYA5	Myospryn	PKAII, dysbindin, titin, calpain-3, desmin, dystrophin	Sarcomere	Cardiac development
SPHKAP	SKIP	PKAI, sphingosine kinase type 1	Cytosol	Cardiac development
SYNM	Synemin	PKAII, desmin, vimentin, dystrobrevin, desmuslin, zyxin, talin, vinculin	Intercalated disks, Z-lines, intermediate filaments	Heart failure
TNNT2	Troponin T	Troponin I, troponin C, actin, PKA	Sarcomere thin filaments	Contractility
LDB3	Cypher, ZASP	PKAII, CaN, L-Type calcium channel	Sarcomere Z-lines	Unknown
C2orf88	smAKAP	PKAI	Cell junctions, filopodia	Unknown
PCNT	Pericentrin	Calmodulin, y-tubulin, PKA	Centrosome	Cardiac development
WASF-1	Wave-1	Actin nucleation core Arp2/3 complex, BAIAP2, profilin 1	Actin cytoskeleton	Cardiac development
ACBD3	PAP7	Giantin, PPM1L, PKAI	Mitochondria	Unknown
NBEA	Neurobeachin	SAP102, PKA	Golgi	Unknown
AKAP14	Akap28	PKA	Ciliary axonemes	Unknown
CBFA2T3	Myeloid translocation gene (MTG) 8 and 16b	Plexin, PKA1 and PKA2	Golgi	Unknown
RAB32	Rab32	PKA	Mitochondria	Unknown
MYRIP	Myosin VIIa- and Rab-interacting protein	PKA, myosin VA	Perinuclear	Contractility
NF2	Merlin	PKA, HGS, ezrin, cullin-4A, syntenin-1, VPRBP, RIT1, SPTBN1, MED28, DDB1	Unknown	Ischemia, cardiac development
AKAP17A	SFRS17A	PKA	Nucleus	Cardiac development
PIK3CG	P110g	PI3K, PKA	Sarcolemma	Adrenergic receptor
PDE4DIP	CMYA2, MMGL, myomegalin	cMyBPC, PKA1, PKA2	Sarcomere (myosin)	Contractility

AKAPs that are at least potentially expressed in the heart (not only myocytes) are included in the table. Other AKAPs whose expression has not been detected in the heart include AKAP2, AKAP3, AKAP4, T-AKAP80, AKAP140, AKAP85, and α4-integrins. We regret not being able to cite the original work for the AKAPs listed in this table due to space limitations

3.2 AKAPs and Regulation of the Calcium Cycle

3.2.1 AKAP7 (Small Isoforms)

As a result of alternative splicing, the AKAP7 gene produces at least four unique isoforms of the AKAP7 protein, named by increasing molecular weight as AKAP7α, AKAP7**β**, AKAP7**γ**, and AKAP7**δ** (Trotter et al. 1999). The first to be characterized were the lower molecular weight isoforms AKAP7α and AKAP7**β**, which are palmitoylated and myristoylated at the N-terminus and targeted to the plasma membrane (Fraser et al. 1998). In polarized cells, the additional residues on AKAP7**β** direct the complex to the basolateral membrane, though the significance of this in cardiomyocytes, if any, is unclear (Fraser et al. 1998). These small isoforms of AKAP7 possess a modified leucine zipper motif that can mediate protein-protein interactions, including binding to α1.2 subunit of the $Ca_v1.2$ L-type channel that is critical for initiating excitation-contraction coupling (Hulme et al. 2002). Numerous studies have now documented physical and functional interactions between low molecular weight isoforms of AKAP7 and Ca^{2+} channels in muscle tissue spanning several species including rabbit, rat, and mouse, as well as in heterologous expression systems. Early pharmacological approaches using cell-permeable and virally expressed peptides, as well as more recent studies using small molecule inhibitors, all reported a blunted response to adrenergic upon stimulation when the AKAP7/$Ca_v1.2$ binding was disrupted (Gray et al. 1998). Based upon this evidence, it has been proposed that the small isoforms of AKAP7 direct the phosphorylation of $Ca_v1.2$ and play a role in orchestrating the sympathetic control of inotropy.

In the context of heart failure, β-adrenergic regulation of excitation-contraction coupling is downregulated due to changes in the expression and function of multiple proteins. This has detrimental effects on Ca^{2+} cycling and contractility. Conflicting reports exist on the expression levels of $Ca_v1.2$ in the context of heart failure. Some investigators report downregulation of the channels, others report increases, and others still report no change. However, changes in channel activity may not always correlate with expression. For example, it has been suggested that even if channel expression is unchanged, the assembly and makeup of channel subunits may change. New auxiliary subunits may also associate with the channel (Bodi et al. 2005). The role of AKAP7 in regulating the L-type Ca^{2+} channel may become even more evident if studied under these conditions.

AKAP7α also interacts with enzymes outside of the cAMP pathway, including all classes of protein kinase C (**PKC**). Immunoprecipitates of PKC from rat heart were reported to contain AKAP7, and a direct interaction between the scaffold and kinase was later shown using purified proteins (Redden et al. 2012). The various isoforms of PKC play diverse and occasionally antagonistic roles in regulating cardiac function (Satin 2013). The tonic and evoked activity of $Ca_v1.2$ channel clusters, for example, has been reported to be both positively and negatively regulated by protein kinase C in all muscle types, though this phenomenon has been more extensively characterized in smooth muscle tissue (Kamp and Hell 2000). While it is known that AKAP7 is important for controlling the localization of protein kinase C at the plasma membrane, the functional consequences of this interaction on the $Ca_v1.2$ conductance have yet to be investigated.

3.2.2 AKAP7 (Large)

The larger products of the AKAP7 gene, AKAP7γ and AKAP7δ, are localized primarily through protein-protein interactions. In cardiomyocytes, binding to the small integral membrane protein phospholamban (**PLN**) recruits the large isoforms of the scaffold to the sarcoplasmic reticulum (Lygren et al. 2007). Phospholamban is a negative regulator of SERCA2a function and plays a key role in regulating the lusitropic phase of the cardiac cycle. The underlying mechanism of inhibition is unresolved but is thought to involve either direct competition between PLN and Ca^{2+} or allosteric stabilization of a SERCA2a conformation with a lower affinity for Ca^{2+} (Simmerman and Jones 1998; Bidwell et al. 2011). Despite being a small protein, PLN contains several phosphorylation sites. When PLN is phosphorylated by PKA at Ser16, its inhibition of SERCA is alleviated, expediting the transport of Ca^{2+} back into the sarcoplasmic reticulum. In rat cardiomyocytes, disrupting the binding between AKAP7 and PLN reduced the phosphorylation of PLN by 50% following β-adrenergic stimulation, in accord with reports that the Ht31 PKA anchoring disruptor peptide inhibited β-adrenergic-induced PLN phosphorylation (Fink et al. 2001; Lygren et al. 2007; McConnell et al. 2009).

Recent biochemical and computational studies on AKAP7 have further advanced our understanding of this process by giving insight into the kinetics and stoichiometry. Prior to these studies, it was unclear how an AKAP with a relatively low expression could impact the function of a protein like PLN that is expressed several orders of magnitude higher. Surface plasmon resonance was used to demonstrate that the interaction between PLN and AKAP7 is dependent on the phosphorylation state of PLN. When PLN becomes phosphorylated, the affinity for AKAP7 decreases dramatically, while the affinity of dephosphorylated PLN remains higher (Rigatti et al. 2015). In a physiological context, this difference in affinity would allow a single AKAP/PKA complex to cycle between multiple molecules of PLN, amplifying its effect beyond what would be expected if binding to the complex was state independent and static. Secondly, the stoichiometry of AKAP7-PLN binding is unlikely to be 1:1, based upon the finding that the large isoforms of AKAP7 are capable of forming homo and heterooligomers in vitro and in transfected cells (Singh et al. 2015). The unstructured domains found in many AKAPs are thought to accommodate simultaneous interactions with many binding partners, and AKAP7 contains at least two binding sites for other AKAP7 molecules. Models generated to analyze the impact of AKAP7 oligomerization have predicted that AKAP7 multimers may create a positive feedback loop following adrenergic stimulation. In other words, the release of a single PKA catalytic subunit leads to the release of more catalytic subunits when oligomers are present in comparison to conditions where the signaling occurs through single complexes [Singh et al. 2015]. Thus, dimerization enables more catalytic subunits of PKA to be activated, faster release of those subunits from their regulatory subunits, and further propagation of the signal its point of initiation. This creates a high concentration of active PKA around substrates like PLN to ensure its efficient phosphorylation in response to elevated local cAMP levels.

The evidence described above provides strong support for the idea that the large isoforms of AKAP7 are important for regulating the control of SR Ca^{2+} reuptake by

PKA-dependent PLN phosphorylation. Recent studies suggest that AKAP7 also regulates phosphatase activity in this compartment. The dephosphorylation of PLN, which restores its inhibition of SERCA2a, is due to the serine/threonine phosphatase protein phosphatase 1 (**PP1**) (Champion 2005). While the catalytic subunits of phosphatases are notoriously promiscuous, they are tightly regulated by a diverse complement of regulatory subunits. SR-associated PP1 is thought to be regulated by protein phosphatase inhibitor 1 (**I-1**) (Nicolaou et al. 2009). Several years ago, I-1 was shown to bind directly to the large isoforms of AKAP7, with PP1 associated through an indirect, unknown mechanism (Singh et al. 2011). Interestingly, I-1 itself is a PKA substrate. When phosphorylated by PKA, I-1 becomes a potent inhibitor of PP1 activity. When I-1 is in a protein complex with AKAP7, I-1 is a more potent inhibitor of PP1 than when I-1 is unbound. This is because AKAP7 enhances the phosphorylation of I-1 by PKA. By decreasing phosphatase activity while simultaneously increasing kinase activity, AKAP7 potentiates PLN phosphorylation by multiple mechanisms. To add an additional layer of complexity to this regulation, the potency of I-1 inhibition decreases when it is phosphorylated by PKC, another AKAP7-associated enzyme. While it has yet to be explicitly demonstrated that AKAP7 regulates the phosphorylation of I-1 by PKC, there are no other PKC substrates known to be associated with the complex. Additionally, a recent study has shown that AKAP7 increases the speed and magnitude of phosphorylation for PKC substrates in the complex (Greenwald et al. 2013). Thus, the scaffold may fine-tune phosphatase activity at the SR to a greater extent than is currently appreciated.

The large isoforms of AKAP7 are also likely to play an important role in the context of heart failure, where Ca^{2+} transients are impaired. The changes in excitation-contraction coupling in heart failure are complex, but include an increase in the ratio of PLN:SERCA2a, downregulation of β-adrenergic receptors, increased phosphatase activity, PLN hypophosphorylation, and inhibition of SERCA2a (Spragg et al. 2003). Each of these events is a contributing factor to the change in intracellular Ca^{2+} gradients that delays the relaxation phase of contraction, depletes SR Ca^{2+} stores, and interferes with the heart's ability to pump blood. The current body of research on AKAP7 all supports the idea that many signals including cAMP, lipids, and Ca^{2+} are likely to converge on AKAP7 with the effect of altering SR Ca^{2+} reuptake and influencing these gradients (Redden and Dodge-Kafka 2011). Modulating these signals through AKAP7 may ultimately create new therapeutic options. In support of this idea are clinical data from two patients with PLN mutations (PLN R9C and PLNΔ14). Both mutations lead to hypophosphorylation and heart failure. It has recently been demonstrated that these mutated forms of PLN do not bind AKAP7 (Rigatti et al. 2015).

Recently, however, the role of AKAP7 in the cardiac myocyte has been called into question. Conflicting data resulted from the characterization of a new AKAP7 knockout mouse (Jones et al. 2012). The AKAP7 knockout allele lacked exon 7 that is conserved among all AKAP7 isoforms and contains both the leucine zipper motif and the PKA-binding domain of the scaffold. In this study, no difference in the Ca^{2+} transients of myocytes from knockout and wild-type animals was detected either at baseline or in response to acute β-adrenergic stimulation, leading the authors to suggest that AKAP7 is not relevant to in vivo to Ca^{2+} handling. However, it is possible that constitutive, global knockout of AKAP7 resulted in compensation by other myocyte AKAPs or by altered expression of PKA and other cAMP signaling enzymes. In addition, these mice were not studied upon chronically stressed conditions, in which a functional role for the AKAP might be elicited. Further investigation of this model is required to understand how deletion of this region of AKAP7 perturbs signaling enzymes that interact with the smaller AKAP7 fragment that is still expressed. This study also called into question the expression of the various AKAP7 isoforms in murine tissue, warranting further characterization of the species-specific differences between humans and the model organisms employed in the study of cardiac physiology and disease.

3.2.3 AKAP5 (AKAP75/AKAP79/AKAP150)

The AKAP5 gene encodes the rodent homolog AKAP150 and, like AKAP7α, is proposed to regulate the L-type Ca^{2+} channel (Hall et al. 2007; Nichols et al. 2010). This signaling complex has been well characterized in neurons, but is expressed in the cardiomyocyte as well. AKAP5 binds numerous enzymes including PKA, PKC, the phosphatase calcineurin (CaN), and AC5/AC6. In the heart, AKAP5 has been found in a protein complex with Cav1.2 and Caveolin 3 (Nichols et al. 2010). Biochemical and electrophysiological characterization of an AKAP5 knockout mouse led to the conclusion that AKAP5 is an important regulator of a subset of $Ca_v1.2$ channels under conditions of β-adrenergic stimulation. AKAP5 knockout animals also display hypophosphorylated PLN and RyR due to the loss of AC localization. However, a comprehensive assessment of PKA, PKC, and CaN activity in the knockout animal has not yet been conducted. Other functions of AKAP5 have been characterized in other myocyte populations, including arterial smooth muscle (Scott and Santana 2010).

It is important to note that the regulation of the L-type Ca^{2+} channel by AKAP complexes remains controversial. The overlapping functions of AKAP5 and AKAP7 may help to explain the negative results obtained for the AKAP7 knockout mouse (Jones et al. 2012). Additionally, *Cypher/ZASP* is a CaN-binding AKAP that has also been reported to bind and regulate $Ca_v1.2$ in cardiac myocytes (Lin et al. 2013). Cypher/ZASP null myocytes display reduced phosphorylation of $Ca_v1.2$. The individual contributions of these three (and potentially other) AKAPS to the overall process of excitation-contraction coupling may depend upon conditions within the cell and, under physiological conditions, allow for contextual regulation of the Ca^{2+} transient. This is supported by the observation that AKAP5 only regulates a subpopulation of $Ca_v1.2$ channels associated with caveolin (Nichols et al. 2010).

3.3 AKAPs and Cardiac Remodeling

Damage and stress on the heart set in motion the process of cardiac remodeling, a complex series of events characterized by altered cardiac structure and function at the molecular, cellular, histological, and organ levels. Although the underlying causes of cardiac remodeling in disease are diverse (e.g., pressure/volume overload, myocardial infarction, inflammation), the remodeling process is characterized by several factors, including hypertrophic cardiomyocyte growth. Concentric hypertrophy can decrease wall stress in response to increased afterload (Law of LaPlace), while eccentric hypertrophy that increases ventricular volumes accommodates increased preload without overstretching the sarcomere. With persistent stress, however, hypertrophied tissue underperforms. During this decompensated phase, deterioration of cardiac systolic and/or diastolic function is induced by maladaptive changes in gene transcription, signaling fidelity, metabolism, and excitation-contraction coupling and apoptosis (Cohn et al. 2000). An extensive network has been defined that regulates pathological myocyte hypertrophy. Notably, there are several AKAPs with recognized roles in coordinating these changes (Diviani et al. 2011; Passariello et al. 2015).

3.3.1 AKAP6

Named most commonly in the literature as "muscle A-kinase-anchoring protein (mAKAP)," AKAP6 is the product of the AKAP6 gene. Alternative 5′ splicing of AKAP6 mRNAs results in expression of AKAP6α in neurons and AKAP6β in striated muscle, such that AKAP6β is identical to AKAP6α residues 245–2314 (rat numbering). The originally identified "AKAP100" is an AKAP6 fragment that does not appear to be expressed in vivo (McCartney et al. 1995). At 227 kD, the relatively large AKAP6β scaffold interacts with a large cohort of diverse signaling enzymes and, despite its relatively low expression level in the heart, has established the paradigm for compartmentalized signaling. As an AKAP, one of the first characterized functions of the scaffold was the coordination of cAMP-dependent signaling at the outer membrane of the nuclear envelope, where the scaffold is exclusively localized by interactions between its spectrin-like repeat domains and the integral membrane protein nesprin-1α (Kapiloff et al. 1999). In this compartment, AKAP6 is poised to signal changes in gene transcription.

The current understanding of how AKAP6 regulates cAMP is the result of numerous genetic, biochemical, and in vivo studies. It is unparalleled in its complexity on account of the fact that both cAMP generation and degradation are orchestrated by the complex (Dodge-Kafka et al. 2005, 2006). In addition to PKA, AKAP6 also interacts with phosphodiesterase-4D3 (**PDE4D3**), creating a classic negative feedback loop. When PKA is activated following a rise in cAMP, AKAP6 coordinates PDE4D3 phosphorylation at Ser residues 13 and 54. This increases its catalytic activity, promotes cAMP breakdown, and terminates the signal. However, the catalytic activity of PDE4D3 is inhibited when it is phosphorylated by extracellular regulated kinases (**ERK**), including ERK5, such that cAMP signaling is potentiated by ERK activation (Hoffmann et al. 1999). ERK5 is localized to the scaffold through its association with PDE4D3 (Dodge-Kafka et al. 2005). A further layer of complexity comes from the association of a guanine nucleotide exchange protein activated by cAMP (**EPAC**). When cAMP levels rise, EPAC inhibits ERK5 activity via a Rap1-dependent mechanism and prevents ERK phosphorylation of PDE4D3. This synergizes with the phosphorylation by PKA to bring phosphodiesterase activity to its peak and allows the cardiomyocyte to maintain tight spatial and temporal control over cAMP fluctuations at the nuclear envelope despite sustained adrenergic stimulation. Interestingly, AKAP6 also coordinates the local synthesis of cAMP through its binding of adenylyl cyclase 5 (**AC5**), which is also regulated through a negative feedback mechanism (Kapiloff et al. 2009). AC5 catalytic activity is inhibited by PKA phosphorylation.

The ability to maintain cAMP gradients within a narrow range is directly relevant to the onset and progression of heart failure. Prolonged adrenergic stimulation is sufficient to induce hypertrophy in isolated cardiomyocytes and heart failure in animal models. In humans, levels of circulating epinephrine are negatively correlated with overall cardiovascular health (Viquerat et al. 1985). Given the central role of AKAP6 in propagating pro-hypertrophic signals to the nuclear envelope, modulating these intricately choreographed interactions may influence outcomes. Most recently, it was observed that loss of AKAP6 protects knockout animals from catecholamine-induced hypertrophy and pressure overload-induced heart failure, while not interfering with normal cardiac development or function (Kritzer et al. 2014).

AKAP6 orchestrates hypertrophic signaling by other pathways as well. Recent studies have defined a role in concentric hypertrophy for ribosomal S6 kinase type 3 (**RSK3**) that also binds directly to AKAP6 (Li et al. 2013a). Although the downstream targets of activated RSK3 are still poorly characterized, RSKs have been linked to changes in transcriptional activity, metabolic activity, and protein expression in multiple tissues and are associated with several other disease states including cancer (Romeo et al. 2012). RSK3 activity is stimulated by multiple upstream kinases in the MAPK family. Several, including ERK5 (discussed previously) and mitogen-activated protein kinase 5 (**MEK5**), are co-localized to AKAP6 (Dodge-Kafka et al. 2005). Given the association of AKAP6 with the nuclear envelope, it is likely that the most significant effect of activated RSK3 in cardiomyocytes is altered gene expression. RSK3 knockout had a cardioprotective effect, attenuating the concentric hypertrophy that is typically observed following pressure overload or catecholamine stimulation (Li et al. 2013a). Loss of activity through this pathway is likely to contribute to the cardioprotective phenotype observed in AKAP6 knockout animals (described above).

Transcriptional activity in the heart is regulated by transcription factors that bind AKAP6, including nuclear factor of activated T cells (**NFAT**) and myocyte enhancer factor (**MEF2**), as well as the type IIa histone deacetylase **HDAC4** (Molkentin 2004; Potthoff and Olson 2007). Activity of these gene regulatory proteins has been linked to changes in the expression levels of signaling enzymes, proteases, components of the cytoskeleton, and other transcriptional regulators. These changes in protein expression ultimately lead to mechanical dysfunction, ventricular dilation, and cardiomyopathy. NFAT activity is regulated by its localization to the cytosol or nucleus, which varies depending on its phosphorylation state. When dephosphorylated by the Ca^{2+}/calmodulin-dependent protein phosphatase type 2B (PP2B, calcineurin, **CaN**), NFAT translocates to the nucleus and induces gene expression. In cardiomyocytes, AKAP6 expression is required for NFAT dephosphorylation, nuclear translocation, and hypertrophy (Pare et al. 2005; Li et al. 2010). Similarly, activation of NFAT is also contingent upon CaN being co-localized in the complex. Disrupting the association using anchoring disruptor peptides diminishes its hypertrophic effects. In light of CaN's role in disease onset and progression, it has been identified as a promising therapeutic target for quite some time (Frey et al. 2004). Displacing it from the scaffold may provide one avenue of curtailing its detrimental effects while minimizing the impact on other substrates that reside outside of the complex.

MEF2 is also regulated by CaN. When dephosphorylated, MEF2 initiates a gene transcription program involving genes required for myogenic differentiation and cardiac hypertrophy (Vargas et al. 2012; Li et al. 2013b). MEF2-CaN binding is dependent upon AKAP6 expression in the myocyte. AKAP6 may also promote MEF2 activity in other ways, for example, by regulating its DNA binding affinity through PKA phosphorylation or by promoting an activating phosphorylation by complex-associated ERK5. Given the importance of this enzyme in muscle development and disease, the mechanisms of MEF2 regulation by AKAP6 is an area worthy of future study.

Lastly, AKAP6 is also known to be involved in transducing phosphatidylinositol 4-phosphate-dependent signals to the nucleus through its association with phospholipase C (**PLCε**) and protein kinase D (**PKD**) (Zhang et al. 2011, 2013). Activation of PLCε induces hypertrophy by activating PKD, which in turn phosphorylates HDAC4 and derepresses MEF2-dependent gene transcription (Kritzer et al. 2014). In isolated myocytes, preventing the association of AKAP6 with either PLCε or PKD precludes HDAC4 activation and pressure overload-induced hypertrophy. Despite the extensive characterization of AKAP6's role in cardiac hypertrophy, the role this multivalent scaffold plays in coordinating the actions of its many binding partners continues to be an area of active research.

3.3.2 AKAP13

Also known as AKAP-lbc or brx-1, AKAP13 is a large, 2817 amino acid protein expressed in cardiomyocytes. In contrast to AKAP6 that has a key role in the onset of pathological cardiac hypertrophy, AKAP13 has a cardioprotective role. By transducing early hypertrophic signals from the plasma membrane to the nucleus, AKAP13 has emerged as an important regulator of compensated cardiac hypertrophy. AKAP-lbc was the first AKAP identified that is a guanine nucleotide exchange factor (**GEF**) for the Ras homolog gene family member A (**RhoA**), thereby potentiating hypertrophic signaling through MAPK pathways upon stimulation of α-adrenergic and stretch receptors (Diviani et al. 2001; Appert-Collin et al. 2007). RhoA effectors are numerous, but two major targets in cardiomyocytes are protein kinase N1 (**PKNα**) and inhibitor of nuclear factor-κB kinase subunit B (**IKKB**). A series of elegant biochemical experiments demonstrated a direct association between AKAP13 and PKNα, and through this association, three more kinases are recruited and sequentially activated: mitogen-activated protein kinase kinase (**MLTK**), mitogen-activated protein kinase kinase 3 (**MKK3**), and p38alpha map kinases (**p38alpha**) (Cariolato et al. 2011). This phosphorylation cascade ultimately activates the mammalian target of rapamycin complex (**mTOR**), which is a master regulator of protein synthesis. Downstream signaling from mTOR is complex and nuanced but is generally accepted to promote ribosome and mitochondrial biogenesis and changes in metabolism through its effectors including p70S6K and 4E-BP1(Pérez López et al. 2013). Thus, though it was originally characterized as a generalized stress-activated pathway regulating inflammation, growth, and proliferation, it is now known to produce some of the cellular changes in cell size, sarcomere organization, and contractility that take place during the early stages of cardiac remodeling (Sciarretta et al. 2014). The central role of AKAP13 in facilitating this process is demonstrated by studies in which the silencing of AKAP13 expression attenuated hypertrophic changes and induced premature apoptosis in response to various GPCR-mediated stimuli. Interestingly, signaling through this pathway likely complements other functions of AKAP13. Specifically, AKAP13 maintains chaperone activity of complex-associated heat shock protein 20 (**HSP20**), which is thought to protect the cardiomyocyte from ischemic injury, necrosis, and apoptosis when phosphorylated by PKA (Edwards et al. 2012).

AKAP13 is also involved in compensatory changes resulting from activation of kinases in other pathways, such as PKCn and PKD (Carnegie et al. 2008). PKD becomes active in response to phosphorylation by PKC at Ser744 and 748. Early studies demonstrated the PKC required for this activation is provided by AKAP13. Active PKD is released from the complex when AKAP13 is phosphorylated by PKA. Thus, signals downstream of both Gs and Gq GPCRs converge on AKAP13 to regulate PKD activity. Like AKAP6, AKAP13-mediated signaling results in phosphorylation and nuclear export of type IIa HDACs such as HDAC5 and leads to an increase in the transcriptional activity of MEF2. Ablation of binding between PKD and AKAP13 reduces hypertrophic growth of the cardiomyocytes, though apoptotic cell death is increased, and morbidity increases (Taglieri et al. 2014). This supports the idea that the early stages of cardiac remodeling mediated by AKAP13 are cardioprotective against hormonally and surgically (TAC) induced heart failure. The functional overlap in signaling enzymes between AKAP13 and AKAP6 highlights a gap in the current knowledge of compartmentalized signaling, since it is still unclear if there is redundancy in the pathway or if there is still another layer of compartmentalization that is not yet understood. However, the available data does fit a "modular" model of scaffold-mediated signaling.

3.4 Other AKAPs with Characterized Functions

AKAP9 (Yotiao, AKAP350, AKAP450) is a large scaffold with characterized functions in the brain and heart. In cardiomyocytes, the leucine zipper motif facilitates an interaction with the KCNQ1 channel (Chen et al. 2005; Chen and Kass 2006). During the normal action potential, repolarization of the cardiomyocyte membrane potential precedes diastole and is due to efflux of potassium ions through delayed rectifier potassium channels including KCNQ1. While there are several channels that contribute to the potassium current, KCNQ1 is of particular interest since mutations of this channel in humans prolong the action potential duration and lead to a cardiac arrhythmia known as long QT syndrome (**LQTS**) (Chen et al. 2007). Delaying the repolarization of individual myocytes produces an overall delay in the repolarization of the ventricle and can result in fatal ventricular arrhythmias. On the electrocardiogram, this is seen as an increase in time between the start of the QRS wave, which corresponds to ventricular contraction, and the T wave, which corresponds to ventricular relaxation. Thus, ventricular diastole is prolonged. The ability to exert chronotropic effects on the heart are inextricably linked to the speed at which the cardiomyocyte action potential can terminate. Increases in heart rate following adrenergic stimulation, for example, increase the kinetics of depolarization *and* repolarization (Terrenoire et al. 2005). The phosphorylation of KCNQ1 at Ser27 by PKA provides the requisite change in channel kinetics (Chen et al. 2007). Importantly, this phosphorylation requires scaffolding by AKAP9, which not only co-localizes PKA but also regulates the amount of cAMP available by recruiting AC9 and PDE4D3 to the complex (Terrenoire et al. 2009; Li et al. 2012). AKAP9 also recruits PP1 to the channel in order to regulate its dephosphorylation (Chen and Kass 2006). Loss of AKAP9 results in a concomitant loss of both kinase and phosphatase regulation. Studies in humans have demonstrated that mutations in AKAP9 that prevent association with KCNQ1 result in LQTS, as do mutations in KCNQ1 that ablate binding to AKAP9 (Chen et al. 2007).

Other aspects of the cardiac rhythm are thought to be regulated by other AKAP signaling complexes. Mutations in the PKA-RI binding site on *AKAP10* (D-AKAP2), for example, have been shown to decrease the P-R internal on EKG recordings, though the underlying mechanism responsible for this change is currently unknown (Tingley et al. 2007).

Signaling at the β_2-adrenergic receptor is orchestrated by *AKAP12*, also known as Gravin. This scaffold anchors PKA, PKC, and CaN in proximity to the receptor and plays an important role in receptor internalization and desensitization (Shih et al. 1999). Mice expressing a truncated form of AKAP12 exhibit potentiation of adrenergic signaling and increased phosphorylation of the cardiac excitation-contraction machinery (Guillory et al. 2013). Emerging data suggests that AKAP12 may also stabilize adrenergic receptor density in cardiac myocytes in response to hypertrophic stimuli, establishing a new cardioprotective role for this scaffold (McConnell et al. 2016). β-adrenergic receptor density is also regulated through the lesser characterized AKAP *p110g*, which is reported to promote cAMP degradation via anchored PDE3B and to inhibit the local production of phosphatidylinositol (3,4,5)-trisphosphate in this compartment (Perino et al. 2011).

Other AKAPs are important for localizing PKA to myofilaments and regulating contractility. *Myospryn*, for example, is found at the costamere as a component of the dystrophin-glycoprotein complex, which couples the sarcolemma to the sarcomere. Although its function in the heart is still unclear, myospryn is transcriptionally regulated by MEF2 and may be involved in myofibrillogenesis during development (Durham et al. 2006; Reynolds et al. 2007). Interestingly, in skeletal muscle models of Duchenne muscular dystrophy, myospryn and PKA are mislocalized (Reynolds et al. 2008). *Dystrobrevin*, another AKAP, is also localized to this complex, though its targets are also uncharacterized. However, dystrobrevin knockout mice have better outcomes than WT mice in catecholamine-induced cardiac hypertrophy (Strakova et al. 2014). The sarcomeric AKAP *myomegalin* coordinates the phosphorylation of myosin binding protein c and the subsequent rearrangement of thick filaments, ultimately leading to enhanced contractility (Uys et al. 2011). *Synemin*, whose expression is upregulated in human heart failure, anchors PKA to the Z-lines (Russell et al. 2006). *Cardiac troponin T*, a dual-specificity AKAP, anchors PKA to the thin filaments (Sumandea et al. 2011).

PKA localized to the mitochondria and endoplasmic reticulum by *AKAP1* (D-AKAP1) may play a role in regulating some of the metabolic and proapoptotic signals that underlie heart failure (Perrino et al. 2010). Interestingly, the AKAP121 isoform of AKAP1 negatively regulates cardiac hypertrophy through a mechanism that is thought to involve NFAT and CaN (Abrenica et al. 2009). The nuclear-localized AKAP chromodomain helicase binding protein 8 (Chd8) may play a role in cardiac development by regulating chromatin dynamics and the β-catenin pathway (Shanks et al. 2012), though its function is still poorly characterized in cardiac tissue.

Though many unanswered questions remain, AKAPs are intricately involved in a growing number of processes within the cardiomyocyte—some appear to have unilateral control over a single process, whereas others appear to have overlapping functions. Further characterization of compartmentalized signaling within the cardiomyocyte will provide a more thorough understanding of normal cardiac function and will yield novel therapeutic targets for the treatment of cardiovascular disease.

Compliance with Ethical Standards

Funding This work was funded by the State of Connecticut Department of Public Health Grant 2014-0133 (K.D.K.) and the US National Institutes of Health Grants HL126825 (K.D.K. and M.S.K.) and HL075398 (M.S.K.)

Conflict of Interest Statement Drs. Kapiloff and Dodge-Kafka are coinventors of patented intellectual property concerning the use of RSK3 and AKAP6 inhibitors for the treatment of heart failure and by which they and the University of Miami may gain royalties from future commercialization. Dr. Kapiloff is the manager of Anchored RSK3 Inhibitors, LLC, and president of Cardiac RSK3 Inhibitors, LLC, companies interested in developing RSK3-targeted therapies and in which Dr. Kapiloff holds equity.

References

Abrenica B, AlShaaban M, Czubryt MP (2009) The A-kinase anchor protein AKAP121 is a negative regulator of cardiomyocyte hypertrophy. J Mol Cell Cardiol 46:674–681

Appert-Collin A, Cotecchia S, Nenniger-Tosato M et al (2007) The A-kinase anchoring protein (AKAP)-Lbc-signaling complex mediates α1 adrenergic receptor-induced cardiomyocyte hypertrophy. Proc Natl Acad Sci 104:10140–10145. doi:10.1073/pnas.0701099104

Bartos DC, Grandi E, Ripplinger CM (2015) Ion channels in the heart. Compr Physiol 5: 1423–1464. doi:10.1002/cphy.c140069

Bers DM (2002) Cardiac excitation-contraction coupling. Nature 415:198–205. doi:10.1038/415198a

Bers DM (2008) Calcium cycling and signaling in cardiac myocytes. Annu Rev Physiol 70:23–49. doi:10.1146/annurev.physiol.70.113006.100455

Bidwell P, Blackwell DJ, Hou Z et al (2011) Phospholamban binds with differential affinity to calcium pump conformers. J Biol Chem 286:35044–35050. doi:10.1074/jbc.M111.266759

Bodi I, Mikala G, Koch SE et al (2005) The L-type calcium channel in the heart: the beat goes on. J Clin Invest 115(12):3306–3317. doi:10.1172/JCI27167

Cariolato L, Cavin S, Diviani D (2011) A-kinase anchoring protein (AKAP)-Lbc anchors a PKN-based signaling complex involved in α1-adrenergic receptor-induced p38 activation. J Biol Chem 286:7925–7937. doi:10.1074/jbc.M110.185645

Carnegie GK, Soughayer J, Smith FD et al (2008) AKAP-Lbc mobilizes a cardiac hypertrophy signaling pathway. Mol Cell 32:169–179. doi:10.1016/j.molcel.2008.08.030
Champion HC (2005) Targeting protein phosphatase 1 in heart failure. Circ Res 96:708–710. doi:10.1161/01.RES.0000164359.95588.25
Chen L, Kass RS (2006) Dual roles of the A kinase-anchoring protein Yotiao in the modulation of a cardiac potassium channel: a passive adaptor versus an active regulator. Eur J Cell Biol 85:623–626. doi:10.1016/j.ejcb.2006.03.002
Chen L, Kurokawa J, Kass RS (2005) Phosphorylation of the A-kinase-anchoring protein Yotiao contributes to protein kinase A regulation of a heart potassium channel. J Biol Chem 280:31347–31352. doi:10.1074/jbc.M505191200
Chen L, Marquardt ML, Tester DJ et al (2007) Mutation of an A-kinase-anchoring protein causes long-QT syndrome. Proc Natl Acad Sci U S A 104:20990–20995. doi:10.1073/pnas.0710527105
Cohn JN, Ferrari R, Sharpe N, Forum I (2000) Cardiac remodeling—concepts and clinical implications: a consensus paper from an international forum on cardiac remodeling. J Am Coll Cardiol 35(3):569–582
Diviani D, Soderling J, Scott JD (2001) AKAP-Lbc anchors protein kinase A and nucleates Galpha 12-selective Rho-mediated stress fiber formation. J Biol Chem 276:44247–44257. doi:10.1074/jbc.M106629200
Diviani D, Dodge-Kafka KL, Li J, Kapiloff MS (2011) A-kinase anchoring proteins: scaffolding proteins in the heart. AJP Hear Circ Physiol 301:H1742–H1753. doi:10.1152/ajpheart.00569.2011
Diviani D, Maric D, Perez Lopez I et al (2013) A-kinase anchoring proteins: molecular regulators of the cardiac stress response. Biochim Biophys Acta 1833:901–908. doi:10.1016/j.bbamcr.2012.07.014
Dodge-Kafka KL, Soughayer J, Pare GC et al (2005) The protein kinase a anchoring protein mAKAP coordinates two integrated cAMP effector pathways. Nature 437:574–578. doi:10.1038/nature03966
Dodge-Kafka KL, Langeberg L, Scott JD (2006) Compartmentation of cyclic nucleotide signaling in the heart: the role of A-kinase anchoring proteins. Circ Res 98:993–1001. doi:10.1161/01.RES.0000218273.91741.30
Durham JT, Brand OM, Arnold M et al (2006) Myospryn is a direct transcriptional target for MEF2A that encodes a striated muscle, alpha-actinin-interacting, costamere-localized protein. J Biol Chem 281:6841–6849. doi:10.1074/jbc.M510499200
Edwards HV, Scott JD, Baillie GS (2012) The A-kinase-anchoring protein AKAP-lbc facilitates cardioprotective PKA phosphorylation of Hsp20 on Ser(16). Biochem J 446:437–443. doi:10.1042/BJ20120570
Fink MA, Zakhary DR, Mackey JA et al (2001) AKAP-mediated targeting of protein kinase a regulates contractility in cardiac myocytes. Circ Res 88:291–297
Fraser LDC, Tavalin SJ, Lester LB et al (1998) A novel lipid-anchored A-kinase anchoring protein facilitates cAMP-responsive membrane events. EMBO J 17:2261–2272. doi:10.1093/emboj/17.8.2261
Frey N, Katus HA, Olson EN, Hill JA (2004) Hypertrophy of the heart: a new therapeutic target? Circulation 109:1580–1589. doi:10.1161/01.CIR.0000120390.68287.BB
Gray PC, Johnson BD, Westenbroek RE et al (1998) Primary structure and function of an a kinase anchoring protein associated with calcium channels. Neuron 20:1017–1026. doi:10.1016/S0896-6273(00)80482-1
Greenwald EC, Redden JM, Dodge-Kafka KL, Saucerman JJ (2013) Scaffold state-switching amplifies, accelerates and insulates protein kinase C signaling. J Biol Chem 289(4):2353–2360. doi:10.1074/jbc.M113.497941
Grimby G, Nilsson NJ, Saltin B (1966) Cardiac output during submaximal and maximal exercise in active middle-aged athletes. J Appl Physiol 21:1150–1156
Grossman W, Jones D, McLaurin LP (1975) Wall stress and patterns of hypertrophy in the human left ventricle. J Clin Invest 56:56–64. doi:10.1172/JCI108079
Guillory AN, Yin X, Wijaya CS et al (2013) Enhanced cardiac function in Gravin mutant mice involves alterations in the β-adrenergic receptor signaling cascade. PLoS One 8:e74784. doi:10.1371/journal.pone.0074784

Hall DD, Davare MA, Shi M et al (2007) Critical role of cAMP-dependent protein kinase anchoring to the L-type calcium channel Cav1.2 via A-kinase anchor protein 150 in neurons. Biochemistry 46:1635–1646. doi:10.1021/bi062217x

Hayes JS, Brunton LL, Mayer SE (1980) Selective activation of particulate cAMP-dependent protein kinase by isoproterenol and prostaglandin E1. J Biol Chem 255:5113–5119

Hill JA, Olson EN (2008) Cardiac plasticity. N Engl J Med 358:1370–1380. doi:10.1056/NEJMra072139

Hoffmann R, Baillie GS, MacKenzie SJ et al (1999) The MAP kinase ERK2 inhibits the cyclic AMP-specific phosphodiesterase HSPDE4D3 by phosphorylating it at Ser579. EMBO J 18:893–903

Hulme JT, Ahn M, Hauschka SD et al (2002) A novel leucine zipper targets AKAP15 and cyclic AMP-dependent protein kinase to the C terminus of the skeletal muscle Ca2+ channel and modulates its function. J Biol Chem 277:4079–4087. doi:10.1074/jbc.M109814200

Jones BW, Brunet S, Gilbert ML et al (2012) Cardiomyocytes from AKAP7 knockout mice respond normally to adrenergic stimulation. Proc Natl Acad Sci U S A 109:17099–17104. doi:10.1073/pnas.1215219109

Kamp TJ, Hell JW (2000) Regulation of cardiac L-type calcium channels by protein kinase A and protein kinase C. Circ Res 87:1095–1102. doi:10.1161/01.RES.87.12.1095

Kapiloff MS, Schillace RV, Westphal AM, Scott JD (1999) mAKAP: an A-kinase anchoring protein targeted to the nuclear membrane of differentiated myocytes. J Cell Sci 112(Pt 1):2725–2736

Kapiloff MS, Piggott LA, Sadana R et al (2009) An adenylyl cyclase-mAKAPbeta signaling complex regulates cAMP levels in cardiac myocytes. J Biol Chem 284:23540–23546. doi:10.1074/jbc.M109.030072

Katz AM (2008) The "modern" view of heart failure: how did we get here? Circ Heart Fail 1:63–71. doi:10.1161/CIRCHEARTFAILURE.108.772756

Katz AM, Lorell BH (2000) Regulation of cardiac contraction and relaxation. Circulation 102:Iv-69–Iv-74

Keely SL, Corbin JD (1977) Involvement of cAMP-dependent protein kinase in the regulation of heart contractile force. Am J Physiol Heart Circ Physiol 233:H269–H275

Kritzer MD, Li J, Dodge-Kafka K, Kapiloff MS (2012) AKAPs: the architectural underpinnings of local cAMP signaling. J Mol Cell Cardiol 52:351–358

Kritzer MD, Li J, Passariello CL et al (2014) The scaffold protein muscle A-kinase anchoring protein beta orchestrates cardiac myocyte hypertrophic signaling required for the development of heart failure. Circ Heart Fail 7:663–672. doi:10.1161/CIRCHEARTFAILURE.114.001266

Levy MN (1971) Brief reviews: sympathetic-parasympathetic interactions in the heart. Circ Res 29:437–445. doi:10.1161/01.RES.29.5.437

Li J, Negro A, Lopez J et al (2010) The mAKAPbeta scaffold regulates cardiac myocyte hypertrophy via recruitment of activated calcineurin. J Mol Cell Cardiol 48:387–394. doi:10.1016/j.yjmcc.2009.10.023

Li Y, Chen L, Kass RS, Dessauer CW (2012) The A-kinase anchoring protein Yotiao facilitates complex formation between adenylyl cyclase type 9 and the IKs potassium channel in heart. J Biol Chem 287:29815–29824. doi:10.1074/jbc.M112.380568

Li J, Kritzer MD, Michel JJC et al (2013a) Anchored p90 ribosomal S6 kinase 3 is required for cardiac myocyte hypertrophy. Circ Res 112:128–139. doi:10.1161/CIRCRESAHA.112.276162

Li J, Vargas MAX, Kapiloff MS, Dodge-Kafka KL (2013b) Regulation of MEF2 transcriptional activity by calcineurin/mAKAP complexes. Exp Cell Res 319:447–454. doi:10.1016/j.yexcr.2012.12.016

Lin C, Guo X, Lange S et al (2013) Cypher/ZASP is a novel A-kinase anchoring protein. J Biol Chem 288:29403–29413. doi:10.1074/jbc.M113.470708

Lygren B, Carlson CR, Santamaria K et al (2007) AKAP complex regulates Ca2+ re-uptake into heart sarcoplasmic reticulum. EMBO Rep 8:1061–1067. doi:10.1038/sj.embor.7401081

Lymperopoulos A, Rengo G, Koch WJ (2013) Adrenergic nervous system in heart failure: pathophysiology and therapy. Circ Res 113:739–753. doi:10.1161/CIRCRESAHA.113.300308

McCartney S, Little BM, Langeberg LK, Scott JD (1995) Cloning and characterization of A-kinase anchor protein 100 (AKAP100). A protein that targets A-kinase to the sarcoplasmic reticulum. J Biol Chem 270:9327–9333

McConnell BK, Popovic Z, Mai N et al (2009) Disruption of protein kinase a interaction with A-kinase-anchoring proteins in the heart in vivo effects on cardiac contractility, protein kinase a phosphorylation, and troponin i proteolysis. J Biol Chem 284:1583–1592. doi:10.1074/jbc.M806321200

McConnell B, Suryavanshi S, Fa'ak F et al (2016) Disruption of Gravin's Scaffolding protects against isoproterenol induced heart failure. FASEB J 30:718.6–718.6

Molkentin JD (2004) Calcineurin-NFAT signaling regulates the cardiac hypertrophic response in coordination with the MAPKs. Cardiovasc Res 63:467–475. doi:10.1016/j.cardiores.2004.01.021

Mozaffarian D, Benjamin EJ, Go AS et al (2015) Heart disease and stroke statistics-2016 update: a report from the American Heart Association. Circulation 133(4):e38–360

Nag AC (1980) Study of non-muscle cells of the adult mammalian heart: a fine structural analysis and distribution. Cytobios 28:41–61

Newlon MG, Roy M, Morikis D et al (1999) The molecular basis for protein kinase A anchoring revealed by solution NMR. Nat Struct Biol 6:222–227. doi:10.1038/6663

Nichols CB, Rossow CF, Navedo MF et al (2010) Sympathetic stimulation of adult cardiomyocytes requires association of AKAP5 with a subpopulation of L-type calcium channels. Circ Res 107:747–756. doi:10.1161/CIRCRESAHA.109.216127

Nicolaou P, Hajjar RJ, Kranias EG (2009) Role of protein phosphatase-1 inhibitor-1 in cardiac physiology and pathophysiology. J Mol Cell Cardiol 47:365–371

Pare GC, Bauman AL, McHenry M et al (2005) The mAKAP complex participates in the induction of cardiac myocyte hypertrophy by adrenergic receptor signaling. J Cell Sci 118:5637–5646. doi:10.1242/jcs.02675

Passariello CL, Li J, Dodge-Kafka K, Kapiloff MS (2015) mAKAP-a master scaffold for cardiac remodeling. J Cardiovasc Pharmacol 65:218–225. doi:10.1097/FJC.0000000000000206

Pérez López I, Cariolato L, Maric D et al (2013) A-kinase anchoring protein Lbc coordinates a p38 activating signaling complex controlling compensatory cardiac hypertrophy. Mol Cell Biol 33:2903–2917. doi:10.1128/MCB.00031-13

Perino A, Ghigo A, Ferrero E et al (2011) Integrating cardiac PIP3 and cAMP signaling through a PKA anchoring function of p110γ. Mol Cell 42:84–95. doi:10.1016/j.molcel.2011.01.030

Perrino C, Feliciello A, Schiattarella GG et al (2010) AKAP121 downregulation impairs protective cAMP signals, promotes mitochondrial dysfunction, and increases oxidative stress. Cardiovasc Res 88:101–110. doi:10.1093/cvr/cvq155

Potthoff MJ, Olson EN (2007) MEF2: a central regulator of diverse developmental programs. Development 134:4131–4140. doi:10.1242/dev.008367

Redden JM, Dodge-Kafka KL (2011) AKAP phosphatase complexes in the heart. J Cardiovasc Pharmacol 58:354–362

Redden JM, Le AV, Singh A et al (2012) Spatiotemporal regulation of PKC via interactions with AKAP7 isoforms. Biochem J 446:301–309

Remme WJ, Swedberg K (2001) Guidelines for the diagnosis and treatment of chronic heart failure. Eur Heart J 22:1527–1560. doi:10.1053/euhj.2001.2783

Reynolds JG, McCalmon SA, Tomczyk T, Naya FJ (2007) Identification and mapping of protein kinase A binding sites in the costameric protein myospryn. Biochim Biophys Acta 1773:891–902. doi:10.1016/j.bbamcr.2007.04.004

Reynolds JG, McCalmon SA, Donaghey JA, Naya FJ (2008) Deregulated protein kinase A signaling and myospryn expression in muscular dystrophy. J Biol Chem 283:8070–8074. doi:10.1074/jbc.C700221200

Rigatti M, Le AV, Gerber C et al (2015) Phosphorylation state-dependent interaction between AKAP7δ/γ and phospholamban increases phospholamban phosphorylation. Cell Signal 27:1807–1815. doi:10.1016/j.cellsig.2015.05.016

Romeo Y, Zhang X, Roux PP (2012) Regulation and function of the RSK family of protein kinases. Biochem J 441:553–569. doi:10.1042/BJ20110289

Russell MA, Lund LM, Haber R et al (2006) The intermediate filament protein, synemin, is an AKAP in the heart. Arch Biochem Biophys 456:204–215. doi:10.1016/j.abb.2006.06.010

Satin J (2013) The long and short of PKC modulation of the L-type calcium channel. Channels 7:57–58. doi:10.4161/chan.24147

Sciarretta S, Volpe M, Sadoshima J (2014) Mammalian target of rapamycin signaling in cardiac physiology and disease. Circ Res 114:549–564. doi:10.1161/CIRCRESAHA.114.302022

Scott JD, Santana LF (2010) A-kinase anchoring proteins: getting to the heart of the matter. Circulation 121:1264–1271. doi:10.1161/CIRCULATIONAHA.109.896357

Shanks MO, Lund LM, Manni S et al (2012) Chromodomain helicase binding protein 8 (Chd8) is a novel A-kinase anchoring protein expressed during rat cardiac development. PLoS One 7:e46316. doi:10.1371/journal.pone.0046316

Shih M, Lin F, Scott JD et al (1999) Dynamic complexes of beta2-adrenergic receptors with protein kinases and phosphatases and the role of gravin. J Biol Chem 274:1588–1595

Simmerman HK, Jones LR (1998) Phospholamban: protein structure, mechanism of action, and role in cardiac function. Physiol Rev 78:921–947

Singh A, Redden JMJ, Kapiloff MSM, Dodge-Kafka KL (2011) The large isoforms of AKAP18 mediate the phosphorylation of inhibitor-1 by PKA and the inhibition of PP1 activity. Mol Pharmacol 79:533–540. doi:10.1124/mol.110.065425

Singh A, Rigatti M, Le AV et al (2015) Analysis of AKAP7γ dimerization. J Signal Transduct 2015:371626. doi:10.1155/2015/371626

Spragg DD, Leclercq C, Loghmani M et al (2003) Regional alterations in protein expression in the dyssynchronous failing heart. Circulation 108:929–932. doi:10.1161/01.CIR.0000088782.99568.CA

Strakova J, Dean JD, Sharpe KM et al (2014) Dystrobrevin increases dystrophin's binding to the dystrophin-glycoprotein complex and provides protection during cardiac stress. J Mol Cell Cardiol 76:106–115. doi:10.1016/j.yjmcc.2014.08.013

Sumandea CA, Garcia-Cazarin ML, Bozio CH et al (2011) Cardiac troponin T, a sarcomeric AKAP, tethers protein kinase A at the myofilaments. J Biol Chem 286:530–541. doi:10.1074/jbc.M110.148684

Taglieri DM, Johnson KR, Burmeister BT et al (2014) The C-terminus of the long AKAP13 isoform (AKAP-Lbc) is critical for development of compensatory cardiac hypertrophy. J Mol Cell Cardiol 66:27–40. doi:10.1016/j.yjmcc.2013.10.010

Terrenoire C, Clancy CE, Cormier JW et al (2005) Autonomic control of cardiac action potentials: role of potassium channel kinetics in response to sympathetic stimulation. Circ Res 96:e25–e34. doi:10.1161/01.RES.0000160555.58046.9a

Terrenoire C, Houslay MD, Baillie GS, Kass RS (2009) The cardiac IKs potassium channel macromolecular complex includes the phosphodiesterase PDE4D3. J Biol Chem 284:9140–9146. doi:10.1074/jbc.M805366200

Tingley WG, Pawlikowska L, Zaroff JG et al (2007) Gene-trapped mouse embryonic stem cell-derived cardiac myocytes and human genetics implicate AKAP10 in heart rhythm regulation. Proc Natl Acad Sci U S A 104:8461–8466. doi:10.1073/pnas.0610393104

Trotter KW, Fraser IDC, Scott GK et al (1999) Alternative splicing regulates the subcellular localization of A-kinase anchoring protein 18 isoforms. J Cell Biol 147:1481–1492. doi:10.1083/jcb.147.7.1481

Uys GM, Ramburan A, Loos B et al (2011) Myomegalin is a novel A-kinase anchoring protein involved in the phosphorylation of cardiac myosin binding protein C. BMC Cell Biol 12:18. doi:10.1186/1471-2121-12-18

Vargas MAX, Tirnauer JS, Glidden N et al (2012) Myocyte enhancer factor 2 (MEF2) tethering to muscle selective A-kinase anchoring protein (mAKAP) is necessary for myogenic differentiation. Cell Signal 24:1496–1503. doi:10.1016/j.cellsig.2012.03.017

Viquerat CE, Daly P, Swedberg K et al (1985) Endogenous catecholamine levels in chronic heart failure. Relation to the severity of hemodynamic abnormalities. Am J Med 78:455–460. doi:10.1016/0002-9343(85)90338-9

Wong W, Scott JD (2004) AKAP signalling complexes: focal points in space and time. Nat Rev Mol Cell Biol 5:959–970. doi:10.1038/nrm1527

World Health Organization (2005) Preventing chronic diseases. A vital investment: WHO global report. World Health Organization, Geneva

Zhang L, Malik S, Kelley GG et al (2011) Phospholipase C epsilon scaffolds to muscle-specific A kinase anchoring protein (mAKAPbeta) and integrates multiple hypertrophic stimuli in cardiac myocytes. J Biol Chem 286:23012–23021. doi:10.1074/jbc.M111.231993

Zhang L, Malik S, Pang J et al (2013) Phospholipase Cε hydrolyzes perinuclear phosphatidylinositol 4-phosphate to regulate cardiac hypertrophy. Cell 153:216–227. doi:10.1016/j.cell.2013.02.047

Pharmacological Approaches for Delineating Functions of AKAP-Based Signalling Complexes and Finding Therapeutic Targets

4

Katharina Schrade and Enno Klussmann

Contents

K. Schrade
Max Delbrueck Center for Molecular Medicine Berlin (MDC) in the Helmholtz Association, Robert-Rössle-Str. 10, 13125 Berlin, Germany

E. Klussmann (✉)
Max Delbrueck Center for Molecular Medicine Berlin (MDC) in the Helmholtz Association, Robert-Rössle-Str. 10, 13125 Berlin, Germany

DZHK (German Centre for Cardiovascular Research), Partner Site Berlin, Oudenarder Strasse 16, 13347 Berlin, Germany
e-mail: enno.klussmann@mdc-berlin.de

V. Nikolaev, M. Zaccolo (eds.), *Microdomains in the Cardiovascular System*, Cardiac and Vascular Biology 3, DOI 10.1007/978-3-319-54579-0_4

Abstract
A-kinase anchoring proteins (AKAPs) comprise a family of scaffolding proteins that direct their interacting partners to defined cellular compartments. The interacting partners can comprise all proteins of canonical cAMP signalling: protein kinase A (PKA), PKA substrates, adenylyl cyclases, phosphodiesterases (PDEs) and protein phosphatases. AKAPs are central for compartmentalising these components and thus for achieving specificity of cAMP signalling cascades. Since AKAPs can additionally bind proteins of other signalling cascades, they constitute nodes for the integration of cellular signalling. Although general functions have been ascribed to several AKAPs, a detailed understanding of the roles of most of their individual protein-protein interactions is lacking. In particular, knowledge of the functions of individual AKAP-PKA interactions is scarce, as they are mediated by conserved domains and difficult to disrupt selectively. In this article, we will discuss pharmacological agents for interference with individual protein-protein interactions of AKAPs. We will mainly focus on recent progress in targeting AKAP-PKA interactions. Since AKAP-directed signalling is dysregulated in some diseases, such agents may be suitable for validating AKAPs as potential drug targets.

Abbreviations

AC	Adenylyl cyclase
AKAP	A-kinase anchoring protein
AKB	A-kinase-binding domain
D/D domain	Docking/dimerisation domain of PKA
Epac	Exchange proteins directly activated by cAMP
PDE	Phosphodiesterase
PKA	cAMP-dependent protein kinase A
STAD	Stapled anchoring disruptor

4.1 Introduction

4.1.1 The cAMP Signalling Cascade

First messengers are defined as extracellular signalling molecules, e.g. neurotransmitters, growth factors or hormones, which cannot enter cells due to their hydrophilic properties but which elicit specific cellular responses. In order to translate first messenger signals into cellular responses, first messengers activate cellular receptors and induce second messenger systems. Second messengers such as cyclic adenosine-3′,5′-monophosphate (cAMP), cyclic guanosine-3′,5′-monophosphate (cGMP), diacylglycerol, phosphatidylinositol, Ca^{2+} or nitric oxide (NO) are molecules which are generated in response to first messengers or which are preformed intracellular molecules (Lodish et al. 2000; Sutherland and Rall 1958).

A plethora of first messengers activates G protein-coupled receptors (GPCR), which in turn stimulate heterotrimeric G proteins. A subclass of GPCRs couples to the stimulatory G protein, G_S. Activation of these G_S-coupled GPCRs causes the release of the α-subunit of the G protein. This activates adenylyl cyclase (AC), which converts adenosine triphosphate (ATP) into cAMP.

4.1.2 cAMP Effectors

There are three main effectors of cAMP (Fig. 4.1): exchange proteins directly activated by cAMP (Epacs), cyclic nucleotide-gated ion channels (CNGCs) and protein kinase A (PKA, cAMP-dependent protein kinase).

Epacs comprise two isoforms, Epac1 and Epac2, which are products of two genes, *Epac1* and *Epac2*. Epac1 is ubiquitously expressed with enhanced levels found in the thyroid, kidney, ovary, skeletal muscle and specific brain regions, whereas Epac2 is mainly expressed in the brain and adrenal gland (Kawasaki et al. 1998; de Rooij et al. 1998). Both Epac1 and 2 function as guanine-nucleotide

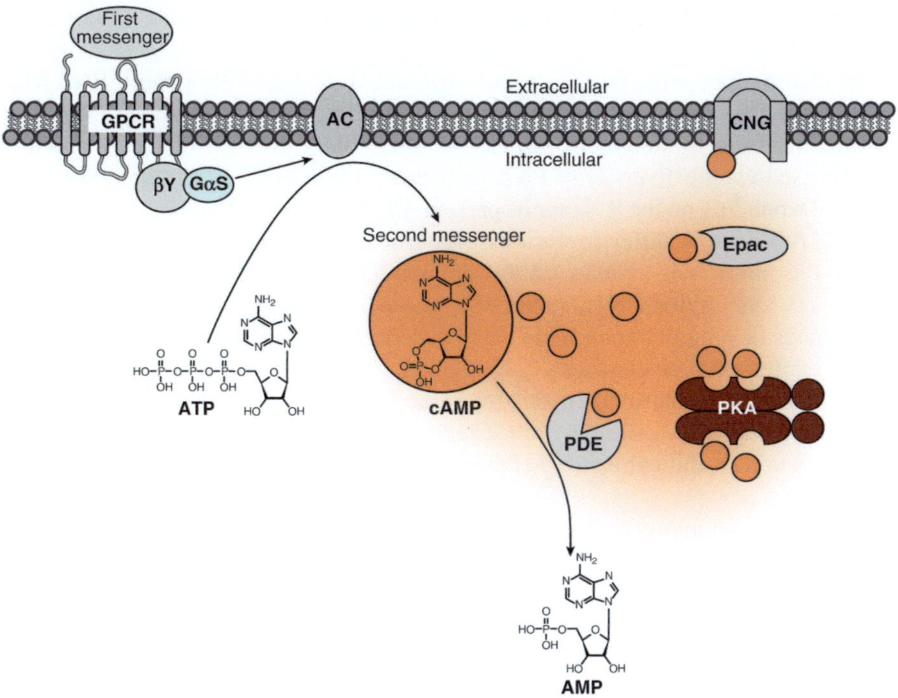

Fig. 4.1 General scheme of cAMP signalling. The stimulation of G_s-coupled seven-transmembrane receptors by a first messenger activates adenylyl cyclase (AC). AC catalyses the conversion of ATP to the second messenger cAMP, which in turn activates its effectors: cyclic nucleotide-gated channels (CNG), exchange proteins directly activated by cAMP (Epac) and protein kinase A (PKA). Phosphodiesterases (PDEs) hydrolyse cAMP to AMP and thereby silence the signalling cascade

exchange factors (GEFs) for the two small Ras-related GTPases Rap1 and Rap2 (Gloerich and Bos 2010; Rehmann 2006). They catalyse the exchange of guanosine diphosphate (GDP) for guanosine triphosphate (GTP) on Rap1 and 2 and thereby promote their transition from the inactive into the active state. Rap signalling is involved in cell adhesion and cell-cell junction formation (Pannekoek et al. 2013, 2009).

CNGCs are nonselective cation channels. They do not distinguish between alkali (Na^+) and alkaline earth ions (Ca^{2+}). CNGCs are found in the plasma membranes of various tissues and cell types, predominantly in sensory cells like retinal cells and olfactory sensory neurons (Biel and Michalakis 2009; Kaupp and Seifert 2002).

The third and major cAMP effector is PKA. PKA was discovered around 50 years ago (Walsh et al. 1968). It is a serine/threonine kinase and phosphorylates serines and threonines within the consensus sequences R-R-X-S/T, R-K-X-S/T, K-R-X-S/T or K-K-X-S/T (Shabb 2001). PKA holoenzyme assembles as an inactive tetramer composed of two regulatory (R) and two catalytic subunits (C). The R-subunits form a dimer and each of the R-subunits interacts with one catalytic subunits. Each regulatory subunit can bind two molecules of cAMP. The binding leads to a conformational change; the catalytic subunits dissociate from the holoenzyme and phosphorylate their targets (Skalhegg and Tasken 2000; Taylor et al. 2012, 2013).

In mammalian cells, there are four different isoforms of R-subunit, RIα, RIβ, RIIα and RIIβ, and three isoforms of the catalytic subunit, Cα, Cβ and Cγ. The R-subunits form homodimers, which can assemble with each of the C-subunits giving rise to some diversity of PKA isoforms (Taylor et al. 2012, 2013; Wu et al. 2007). Depending on the R-subunits present, PKA is defined PKA-I or PKA-II. The two PKA isoforms differ in various properties, e.g. in terms of cAMP affinity (PKA-I is more sensitive to cAMP compared to PKA-II) (Cadd et al. 1990; Dostmann and Taylor 1991; Gamm et al. 1996)). Their expression pattern due to differences in subunit expression is also different: while RIα and RIIα are ubiquitously expressed (Lee et al. 1983; Scott et al. 1987), RIβ is primarily found in the brain and RIIβ is elevated in fat, reproductive and endocrine tissue and brain (Cadd and Stanley McKnight 1989; Clegg et al. 1988; Jahnsen et al. 1986). RI is mainly found in the cytosol, whereas RII is predominantly found in particulate fractions.

PKA plays an important role in many cellular processes such as cell differentiation, proliferation and metabolism. For example, the kinase is involved in long-term memory formation by modulating synaptic plasticity through affecting late-phase long-term potentiation (Huang et al. 2006). PKA regulates several ion channels, e.g. L-type Ca^{2+} channels in skeletal and cardiac myocytes, as well as the Ca^{2+}-activated K^+ channel in neurons (Gray et al. 1998). In olfactory receptor neurons, PKA phosphorylates voltage-gated Na^+ and Ca^{2+} channels, leading to Ca^{2+} influx and thereby to membrane depolarisation and generation of action potentials (Wetzel et al. 2001). PKA is also involved in transcriptional control of many genes through the transcription factors of the cAMP response element-binding (CREB) family (Gonzalez and Montminy 1989; Siu and Jin 2007).

4.1.3 cAMP Signalling Compartments

Since cAMP and its effectors are ubiquitous and regulate a plethora of distinct cellular processes, the question arises how specificity of cAMP signalling can be achieved. Fluorescence resonance energy transfer (FRET) and other techniques revealed that cAMP is not distributed uniformly throughout cells (Berrera et al. 2008; Jurevicius and Fischmeister 1996; Tsien et al. 1993; Willoughby and Cooper 2008; Zaccolo et al. 2000; Zaccolo and Pozzan 2002). It rather occurs in gradients. They are shaped by ACs synthesising cAMP and phosphodiesterases (PDEs) hydrolysing it.

ACs are organised into six classes (I–VI). All known eukaryotic ACs belong exclusively to Class III, whereas prokaryotic ACs are found in all classes (Cann 2004; Steegborn 2014). In mammals, ten AC isoforms (1–10) are expressed; AC1–9 contain transmembrane domains and are mainly regulated by GPCRs. AC10 is soluble and therefore termed soluble AC (sAC). It can be activated by bicarbonate and acts as a bicarbonate and pH sensor and modulator within the cell (Cann 2004; Pastor-Soler et al. 2003). ACs are expressed in all tissues and are involved in a whole variety of cellular processes, e.g. AC1 contributes to memory formation (Wu et al. 1995), AC5 and AC6 are involved in controlling cardiac myocyte contractility (Ishikawa et al. 2005) and sAC regulates mitochondrial oxidative phosphorylation (Acin-Perez et al. 2009).

Hydrolysis of cAMP by PDEs is the major way for termination of cAMP signalling. PDEs degrade cAMP and/or cGMP by hydrolysing the 3′-cyclic phosphate bond resulting in either adenosine monophosphate (AMP) or guanosine monophosphate (GMP). Among the 12 known PDE families, substrate specificity is different. While PDE4, 7 and 8 are specific for cAMP, PDE5, 6 and 9 are cGMP-selective, and PDE1, 2, 3, 10 and 11 hydrolyse both cAMP and cGMP (Bender and Beavo 2006). PDE12 degrades 2′-5′ oligoadenylates (Poulsen et al. 2012). PDEs are constitutively active and positioned at specific subcellular locations. The resulting local reduction of cAMP establishes gradients and cAMP microdomains, which are sensed by the cAMP effectors. Several PDEs associate directly with cellular compartments through their own domains. For example, a transmembrane domain directs PDE3A1 to the membrane of the sarcoplasmic reticulum in cardiac myocytes (Wechsler et al. 2002). The tryptophan anchoring phosphatidic acid selective-binding domain 1 (TAPAS-1 domain) of PDE4A1 mediates an interaction with phosphatidic acid and insertion into lipid bilayers (Baillie et al. 2002; Huston et al. 2006). Other PDEs are directed to defined cellular compartments through protein-protein interactions. In some instances, the interacting proteins are AKAPs (Klussmann 2016). PDE3A1, for example, interacts with an AKAP18γ-based complex in human cardiac myocytes that tethers the PDE to the sarcoplasmic reticulum (Ahmad et al. 2015). In rat renal collecting duct principal cells, PDE4D directly binds AKAP18δ and the complex is located on intracellular exocytic vesicles (Stefan et al. 2007). Further means for terminating cAMP signalling include, for example, desensitisation of seven-transmembrane receptors that couple to G_s or the extrusion of cAMP by ATP-binding cassette (ABC) transporters (also termed multidrug resistance proteins, MRP). The export of cAMP by ABCC4 from cardiac myocytes, for instance, was cardioprotective in a murine pressure overload model (Sassi et al. 2014).

4.2 A-Kinase Anchoring Proteins (AKAPs)

AKAPs comprise a family of around 50 heterogeneous proteins, which share the common feature of binding R-subunits of PKA. The binding of AKAPs to PKA is mediated by structurally conserved A-kinase-binding (AKB) domains of 14–18 amino acid residues in length. AKB domains form an amphipathic helix whose hydrophobic side binds the N-terminal dimerisation/docking (D/D) domain of the regulatory subunits of PKA. By virtue of their anchoring domains, AKAPs tether the AKAP-PKA complex to defined cellular compartments including mitochondria, the endoplasmic reticulum or the plasma membrane bringing PKA into close proximity of its substrates (Burgers et al. 2012; Carr et al. 1992; Chen et al. 1997; Furusawa et al. 2001; Huang et al. 1999).

AKAPs cannot only bind PKA, PKA substrates, phosphatases and PDEs. Several AKAPs can bind further signalling proteins such as kinases, which are activated by second messengers other than cAMP, e.g. PKC by Ca^{2+}. These AKAPs facilitate cross talk between signalling systems. Therefore, AKAPs are crucial in the spatiotemporal coordination of multiple signalling pathways (Skroblin et al. 2010) (Fig. 4.1).

The ability to facilitate cross talk between signalling systems is illustrated by the AKAP GSK3β interaction protein (GSKIP). GSKIP is required for a proper craniofacial development and postnatal life in mice (Deák et al. 2016), and its gene duplication in humans is associated with familial myeloid malignancy (Saliba et al. 2015). GSKIP binds PKA and glycogen synthase kinase 3β (GSK3β) (Hundsrucker et al. 2010). The interactions with both of the kinases contribute to the control of β-catenin and thus Wnt signalling. Wnt signals lead to β-catenin accumulation and translocation into the nucleus, where it initiates Wnt target gene expression (Lin et al. 2009; Dema et al. 2016). GSKIP-bound PKA is involved in the control of the β-catenin phosphorylation at Ser-675, which stabilises it, whereas GSKIP-bound GSK3β facilitates control of the phosphorylation of β-catenin at Ser-33/Ser-37/Thr-41, which destabilises it. Wnt signalling regulates processes such as embryonic development, cell cycle progression, glycogen metabolism and immune responses. Dysregulation of Wnt signalling is associated with widespread diseases, e.g. cancer, type 2 diabetes and inflammatory, Alzheimer's and Parkinson's diseases (Fancy et al. 2009; White et al. 2012).

4.2.1 AKAP-PKA Interactions

AKAPs differ in terms of specificity of their AKB domains for the different R-subunits of PKA. The majority of AKAPs binds RII with higher affinity than RI; several preferentially or specifically bind RI. For instance, sphingosine kinase interacting protein (SKIP) and small membrane AKAP (smAKAP) are both specific RI

Table 4.1 Structural information on selected AKAP-PKA interactions

AKAP	RIα (1–44)	RIIα (1–44)	References
GSKIP		28–52	Hundsrucker et al. (2010)
AKAP18β		43–83	Götz et al. (2016)
smAKAP	61–74		Burgers et al. (2012, 2016)
D-AKAP1	285–387	285–387	Carlson et al. (2003)
AKAP-Lbc		493–515	Newlon et al. (1999)
AKAP79		392–408	Newlon et al. (1999, 2001)

The numbers indicate amino acid residues in the respective AKAP, i.e. the AKB domain. In all structural analyses, the D/D domains (amino acid residues 1–44) of RIα or RIIα were used as indicated

binders (Burgers et al. 2012; Kovanich et al. 2010). Some AKAPs binding both isoforms are termed 'dual-specific (D)-AKAPs'; these include D-AKAP1, D-AKAP2 and Opa1 (Huang et al. 1997a, b; Pidoux et al. 2011).

Using the information provided by NMR and crystal structures of AKB domains in complex with D/D domains (Table 4.1), a deeper understanding of structural organisation and isoform selectivity was achieved. In general, an AKB-D/D domain interaction requires an antiparallel X-type four-helix bundle formed by the D/D domain that builds a hydrophobic groove into which the hydrophobic side of the amphipathic helix of the AKB domain docks (Newlon et al. 1999, 2001; Banky et al. 2003; Skroblin et al. 2010) (Fig. 4.2). The crystal structures of D-AKAP2 with the D/D domains of either RIα or RIIα and of smAKAP with the D/D domain of RIα provided instructive insights into the interactions and isoform selectivity (Burgers et al. 2016; Kinderman et al. 2006; Sarma et al. 2010): whereas RII-specific AKB domains show only two, RI-specific AKB domains contain four residue pairs interacting with hydrophobic pockets of D/D domains. Burgers et al. revealed that the two disulphide bridges between Cys16-Cys37′ and Cys16′-Cys37 of RI form two additional hydrophobic pockets which are involved in the binding of smAKAP.

We elucidated the 3D structure of an extended AKB domain of AKAP18β bound to the D/D domain of RIIα-subunits and identified three hydrophilic anchor points in AKAP18β outside the core hydrophobic residues that dock into the hydrophobic pocket of the D/D domain. Such anchor points are conserved within AKAPs that bind RII-subunits, and a different set of anchor points is present in AKAPs binding RI-subunits. The loss of the anchor points abolishes or attenuates the interactions of AKAP18γ with RIIα and of smAKAP with RIα. We had previously shown that the affinity of the binding of AKAP18δ is increased by hydrophilic interactions (Hundsrucker and Klussmann 2008). Thus, AKAP-PKA interactions do not exclusively rely on hydrophobic interactions between the AKB and D/D domains. Anchor points apparently modulate the affinity of the interactions.

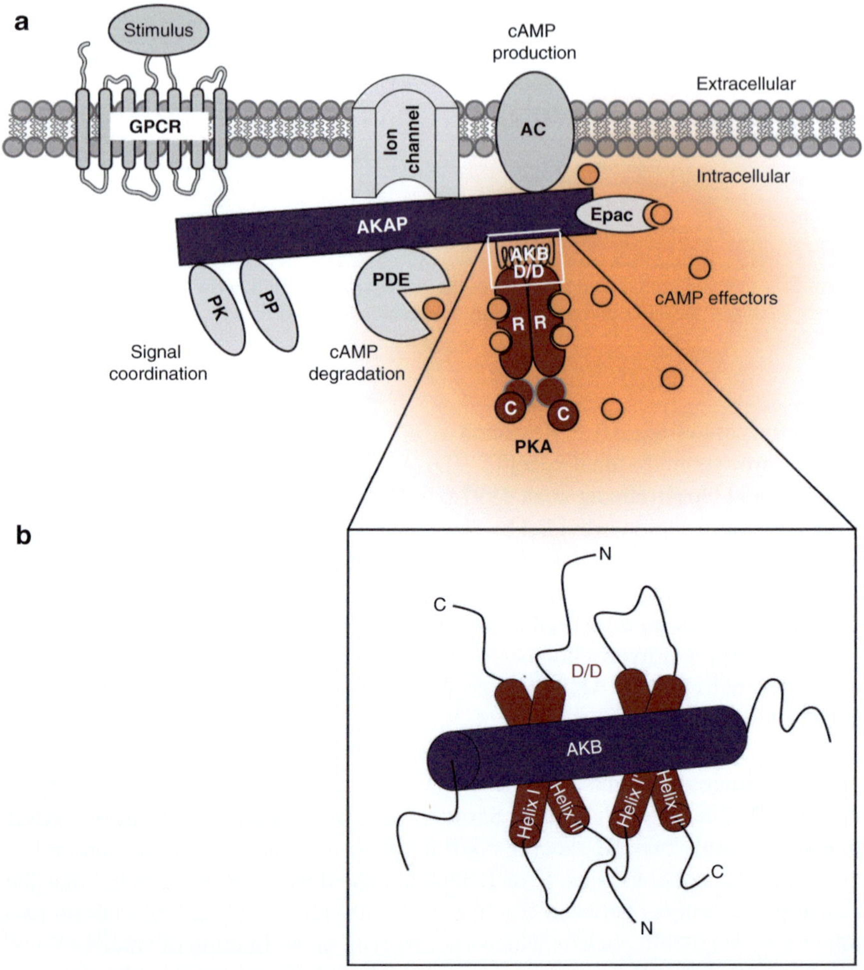

Fig. 4.2 (**a**) The role of AKAPs in integrating multiple signalling pathways. (**b**) Schematic representation of AKAP-PKA interactions. D/D domains of R-subunits of PKA form X-type helix bundles with a hydrophobic groove. The hydrophobic sides of the α-helices of AKB domains of AKAPs dock into these pockets. *AC* adenylyl cyclase; *AKB* A-kinase-binding domain; *D/D* dimerisation and docking domain; *Epac* exchange protein directly activated by cAMP; *GPCR* G protein-coupled receptor; *PDE* phosphodiesterase; *PKA* protein kinase A; *PP* protein phosphatase; *PK* protein kinase

4.2.2 The Role of AKAPs in the Heart and Cardiovascular Diseases Exemplifies Their Potential Value as Drug Targets

In the heart, PKA phosphorylates various targets, including the two sarcomeric proteins myosin-binding protein C (Colson et al. 2012) and troponin I (Bodor et al. 1997), phospholamban (PLN), a negative regulator of the sarcoplasmic reticulum Ca^{2+}-ATPase (SERCA2) (Hasenfuss 1998), the ryanodine receptor (RyR) (Kapiloff et al. 2001; Wehrens et al. 2006) and $Ca_V1.2$ channels (Hulme et al. 2006). Together, the PKA-mediated phosphorylations regulate cardiac myocyte contractility by controlling Ca^{2+} cycling (Deak and Klussmann 2015; Kranias and Hajjar 2012).

Several AKAPs tether PKA to these substrates. One of them is AKAP18α (AKAP15). It tethers PKA to the $Ca_V1.2$ channel (Hulme et al. 2003). In response to β-adrenergic stimulation, PKA phosphorylates the channel leading to an increased Ca^{2+} influx into the cytosol, which contributes to the plateau phase of the cardiac action potential (Weiss et al. 2013). AKAP79 (AKAP150/AKAP5) forms the basis of a complex encompassing PKA, PKC (Tavalin 2008), the calmodulin-activated phosphatase calcineurin (PP2B) (Coghlan et al. 1995; Oliveria et al. 2007), and AC5 and AC6 (Zhang et al. 2013). AKAP79 like AKAP18α regulates phosphorylation of the $Ca_V1.2$ channel (Nichols et al. 2010). Muscle-selective AKAP (mAKAP) facilitates phosphorylation of RyR receptors by PKA upon adrenergic stimuli (Soni et al. 2014). In addition, mAKAP anchors PDE4D3 which locally hydrolyses cAMP and thereby resets the signalling complex to the resting state (Carlisle Michel et al. 2004). Another isoform of AKAP18, AKAP18γ, forms the basis of a multi-protein complex composed of SERCA2, PDE3A1, PLN and PKA (Ahmad et al. 2015). PKA-mediated phosphorylation of PLN attenuates its inhibitory effect on SERCA2 and leads to enhanced Ca^{2+} reuptake into the sarcoplasmic reticulum and thus contributes to the relaxation of the myocytes (Haghighi et al. 2014; Kranias and Bers 2007; Kranias and Hajjar 2012; MacLennan et al. 2003; MacLennan and Kranias 2003). The AKAP Yotiao forms a complex with the K^+ channel $I(K_S)$ which mediates the slow outward K^+ current. Inhibition of Yotiao-mediated K^+ channel phosphorylation provokes long QT syndrome and may lead to sudden death (Kurokawa et al. 2004; Li et al. 2012).

Dysregulation of AKAP complexes in the heart is involved in the development of cardiac hypertrophy and heart failure (Carnegie and Burmeister 2011; Diviani et al. 2011). The expression level and interactions with PKA of several AKAPs are altered in hypertrophic and failing hearts (Aye et al. 2012; Soni et al. 2014). Prominent examples of AKAPs involved in the development of these diseases are AKAP-Lbc (AKAP13) and mAKAP (Appert-Collin et al. 2007; Carnegie et al. 2008; Li et al. 2010; Pare et al. 2005; Taglieri et al. 2014). For example, mAKAPβ forms a complex with the Ca^{2+}/calmodulin-dependent phosphatase calcineurin. Phosphorylation of RyR by PKA leads to an increased cytoplasmic Ca^{2+} level, which activates calcineurin. Activated calcineurin, in turn, dephosphorylates the mAKAPβ-bound transcription factor NFATc3. The dephosphorylated NFATc3 translocates into the nucleus where it induces the expression of hypertrophic genes, which eventually causes cardiac remodelling (Li et al. 2010). mAKAP additionally scaffolds phospholipase C (PLC). Disruption of the interaction results in a decrease of endothelin1-dependent hypertrophy in neonatal cardiac myocytes (Zhang et al. 2011).

The involvement of AKAPs in the regulation and modulation of heart diseases such as cardiac hypertrophy and heart failure suggests AKAPs as potential targets for treating these diseases. In addition, it appears that deregulation of AKAPs and their protein-protein interactions plays roles in other diseases, too, e.g. obesity, sickle cell disease, kidney and immune diseases, cancer and neurological disorders (for recent reviews, see (Deak and Klussmann 2015; Dema et al. 2015)).

4.3 AKAP-PKA Disruptors

The conservation of the interactions of AKB and D/D domains extremely complicates the elucidation of functions of specific AKAP-PKA interactions. In particular, the generation of agents for the specific disruption of individual AKAP-PKA interactions is a difficult task. Such agents would allow elucidation of the functions of individual AKAP-PKA interactions and may provide compounds as starting points for the development of drugs. Several routes have been taken for disruption of these interactions. The established disruptors can be classified into three main groups: peptides, peptidomimetics and small molecules.

4.3.1 Peptides

4.3.1.1 Binding to R-Subunits of PKA

Peptides disrupting AKAP-PKA interactions typically consist of 14–25 amino acid residues, which form an amphipathic helix mimicking AKB domains. The characteristic of amphipathic helices is the ordered separation of hydrophobic and hydrophilic amino acids on the opposing faces of the helix. The peptides bind via their hydrophobic residues into the hydrophobic pocket formed by the D/D domains of R-subunits and thereby uncouple PKA from AKAPs.

The first peptide disrupting AKAP-PKA interactions through this mechanism was discovered 25 years ago (Carr et al. 1991). This peptide, Ht31, was derived from the AKB domain of AKAP13 (AKAP-Lbc) and binds to R-subunits of PKA with a K_D value in the low nanomolar range (4 nM). However, Ht31 does not distinguish between RI- and RII-subunits (Alto et al. 2003; Herberg et al. 2000). Using Ht31, involvements of AKAP-PKA interactions in diverse physiological functions were elucidated, e.g. insulin secretion, renal water reabsorption and regulation of synaptic functions in neurons (Deak and Klussmann 2015; Dema et al. 2015; Szaszak et al. 2008). Fink et al. showed an increased rate and amplitude of cell shortening and relaxation upon treatment of cardiac myocytes with this peptide (Fink et al. 2001). Additionally, Ht31-treated cardiac myocytes exhibit a reduced PKA-dependent phosphorylation of troponin I and myosin-binding protein C in response to β-adrenergic stimulation compared to controls (Fink et al. 2001).

Ht31 was followed by R-subunit isoform preferring disruptor peptides. The use of bioinformatics approaches and peptide arrays discovered peptides such as AKAP-in silico (AKAP-*IS*), which binds to RIIα-subunits with a $K_D = 0.45$ nM and to RIβ with a $K_D = 227$ nM (Alto et al. 2003). Further improved disruptor peptides include super-AKAP-*IS* with a fourfold elevated affinity for RIIα (Gold et al. 2006), RIAD (RI-anchoring disruptor) as RI-preferring (K_D(RIα) = 1.0 nM; K_D(RIIα) = 1760 nM) (Carlson et al. 2006), AKB(RI) as RI-preferring (K_D(RIα) = 5.2 nM; K_D(RIIα) = 456 nM) and AKB(RII) as RII-preferring (K_D(RIα) = 2493 nM; K_D(RIIα) = 2.7 nM)) (Burns-Hamuro et al. 2003).

Until recently, all disruptor peptides developed bind to R-subunits of PKA. By this, it was not possible to distinguish any cellular functions of individual AKAP-PKA interactions except differences between RI- and RII-specific AKAPs.

4.3.1.2 Binding to AKB Domains of AKAPs

Gold et al. chose another approach. Through phage display technology, they identified PKA-derived peptides (Gold et al. 2013), which preferentially bind to AKAP2 or AKAP18, R_{select}AKAP2 and R_{select}AKAP18 (both RII specific). Using these peptides and potentially additional ones developed through a structure-based phage selection assay, the role of individual AKAP-PKA interactions may be elucidated.

4.3.1.3 Modified Peptides

Peptides suffer of several drawbacks. They are characterised by low membrane permeation abilities, short half-life and, in terms of drug development and experiments in whole organisms, low oral bioavailability. To increase their membrane permeation abilities, peptides can be modified by the incorporation of cell-penetrating sequences such as poly-arginine, penetratin or HIV-1 TAT sequences or by chemical modifications such as the addition of stearate or myristoyl moieties (Derossi et al. 1994; Futaki et al. 2001; Hundsrucker 2008; Vivès et al. 1997). Successful examples for these modifications are the disruptor peptide TAT-AKAP-IS and TAT-conjugated A-kinase anchoring disruptor (TAT-AKAD). TAT belongs to the class of cell-penetrating peptides (CPP) (Bechara and Sagan 2013). The TAT sequence is derived from the trans-activator of transcription (Tat) protein of HIV-1, which enhances efficiency of viral transcription in the host cell. The Tat protein efficiently enters cells due to its transduction domain. The transduction domain sequence, YGRKKRRQRRR, is highly basic, and its positive charges interact with negative charges of the phospholipid bilayer of the plasma membrane leading to internalisation of the peptide (Brooks et al. 2005; Subrizi et al. 2012). TAT-AKAP-IS represents a TAT-conjugated derivative of AKAP-*IS* with improved cell permeation properties. It dose dependently prevented glucagon-induced potentiation of insulin release from pancreatic β-cells by disruption of endogenous AKAP-PKA interactions and thereby changing PKA localisation (Faruque et al. 2009). Another example is TAT-AKAD, which was derived from the PKA-binding region of AKAP10, a dual-specific AKAP. In cultured cardiac myocytes, TAT-AKAD decreases beating frequency, cell contraction and relaxation (Patel et al. 2010).

The modifications can partially change physicochemical parameters of peptide such as lipophilicity (stearate) or charge (TAT sequence). This might lead to preferred association with membranes or organelles distinct from their anticipated targets (Wang et al. 2014). Additionally, the peptides are still susceptible to proteolysis. In order to overcome these drawbacks, carbon-stapled peptides were devised. In stapled peptides, carbon braces lock the secondary structure of the peptide. Carbon staples are introduced chemically into the peptide sequence: Non-natural olefinic amino acids are inserted into the peptide sequence followed by ruthenium-catalysed ring-closing metathesis resulting in macrocyclic ring formation (Schafmeister et al. 2000; Wang et al. 2014). This stabilised peptide structure is favourable over unmodified peptides in terms of resistance against proteolytic degradation and cell permeation. All carbon-stapled peptide disruptors can permeate plasma membranes. One reason for this is the hydrophobic character of the carbon bridge shielding hydrophilic-binding regions of the peptide. This enhances the interaction of the peptide with the plasma membrane. Another reason is that a fixed α-helical structure seems generally beneficial for cellular uptake. In addition, the net charge of the peptide and the staple type (staples can differ in length of alkenyl substituents or stereochemistry) affect permeation properties. For more details, the reader is referred to Chu et al. (2015).

Stapled peptides were originally engineered for cancer targets, namely, for targeting activated BH3 domains of the BCL-2 protein family. BH3 domains trigger activation of the pro-apoptotic proteins BAX and BAK, which leads to release of cytochrome C from mitochondria and activation of the mitochondrial programme of apoptosis. Mimicking BH3 with a hydrocarbon-stapled BH3, the helix resulted in increased apoptosis of leukaemia cells (Walensky et al. 2004). Chang et al. developed a carbon-stapled peptide inhibiting the interaction of MDM2 and MDMX. This peptide mediated activation of the p53 tumour suppressor pathway and thus apoptosis in cancer cells (Chang et al. 2013).

In the development of stapled peptides disrupting AKAP-PKA interactions, the effort was focused on three candidates, namely, RIAD, AKAP220 and smAKAP, which all engage in hydrophilic interactions with D/D domains in addition to the core hydrophobic interactions. Three carbon-stapled peptides were developed, the so-called stapled anchoring disruptors (STAD)1–3. In vitro, all of the three peptides preferentially bind RII- compared to RI-subunits (STAD-1: RIα = 93 nM, RIIα = 50 nM; STAD-2: RIα > 1 μM, RIIα = 31 nM; STAD-3: RIα = 144 nM, RIIα = 3 nM). This RII preference is even more prominent in a cellular context for STAD-2 and STAD-3. STAD-1 binds only weakly to RII within cells (Wang et al. 2014). STAD-2 has an antimalarial effect (Flaherty et al. 2015). However, this seems mediated through a PKA-independent mechanism. Besides carbon-stapled peptides with preference for RII, an RI-preferring candidate was also developed, RI-STAD (Wang et al. 2015).

Although peptides have been improved with regard to their half-life, their ability to cross plasma membranes and their bioavailability, they are in general still considered difficult for drug development. On the other hand, peptides often outperform classical drugs, i.e. small molecules in particular in terms of selectivity. Due to this advantage, peptide drugs are supposed to have fewer side effects. Examples for peptide drugs on the market are insulin, somatostatin, cyclosporin A or salmon calcitonin (Craik et al. 2013; Otvos and Wade 2014).

4.3.2 Peptidomimetics

A different class of molecules for the disruption of AKAP-PKA interactions are peptidomimetics. Peptidomimetics combine target selectivity and potency of peptides with good pharmacokinetic properties. Peptidomimetics are composed of a combination of none-natural, partially natural and natural amino acids.

The first peptidomimetics to disrupt AKAP-PKA interactions was a follow-up of the above-mentioned disruptor peptide RIAD. RIAD was modified by incorporation of non-natural amino acids into the peptide sequence resulting in an enhanced resistance against proteolysis (Torheim et al. 2009). One of the RIAD mimetics, RIAD-P3, reduced HIV-1 viral load and increased CD4 cell number in human peripheral blood mononuclear cells and humanised NOD/SCID/IL2γnull (NSG) mice, which represent a murine model for AIDS (Singh et al. 2014).

A different approach yielded peptidomimetics without containing any natural amino acid (Schäfer et al. 2013). These terpyridine-based peptidomimetics contain three pyridine moieties as the basis. In order to obtain these large molecules with a protein-like backbone (molecular weight ≈ 1000), defined parts of the final product were synthesised in a first reaction and then combined in a second one. In order to mimic a protein backbone structure in terms of three-dimensional orientation, the single moieties needed to be coupled by carbon-carbon (C-C) bonds. Chemically, this is considered difficult. One reaction to overcome this problem is the Suzuki-Miyaura reaction. In the Suzuki-Miyaura reaction, a boronic acid reacts with an organohalide catalysed by a palladium(0) complex resulting in C-C coupling (Burzicki et al. 2009). In the chemical synthesis of terpyridines for the inhibition of AKAP-PKA interactions, two Suzuki-Miyaura reactions combined three educt moieties (Schäfer et al. 2013). In order to achieve a region-selective reaction, the middle component carried its Bromo substituents in positions 2 and 5. Optimisation of Pd^0 catalyst and solvent yielded in $[Pd_2(dba)_3]$ as palladium source and dioxane as solvent. The terpyridine backbone mimics the α-helix of the AKB domain of AKAP18δ. It forms an α-helical structure on which the amino acid-derived side chains are arranged in similar positions and angles as in the original AKB domain. The terpyridines disrupt the AKAP18δ-RIIα interaction in vitro. To elucidate the effect of terpyridines on AKAP-PKA interactions in cells, a negative feedback mechanism involving AKAP-PKA interactions in a HEK293 cell system was studied. The cells transiently expressed CNGCs, which are opened by cAMP and allow Ca^{2+} influx. The cells were loaded with Fura-2 and the Ca^{2+} influx detected by binding of Ca^{2+} to Fura-2, which emits fluorescence. PGE_1 stimulation was used to elevate cAMP levels. The AKAP gravin anchors PDE4D, whose activity is enhanced approximately twofold by PKA phosphorylation. PDE4D terminates cAMP signalling. In addition, AKAP150 forms a complex with AC and PKA and facilitates the phosphorylation and thereby inactivation of AC by PKA. One of the terpyridine derivatives in a concentration of 20 μM increased cAMP levels, confirming that the peptidomimetics enters cells, disrupts AKAP-PKA interactions and thereby inhibits the negative feedback loop. Despite the high K_D and IC_{50} values (K_D (ITC) = 31–148 μM, IC_{50} (HTRF) = 38 μM), terpyridines represent promising starting points for the development of highly selective agents for the disruption of AKAP-PKA interactions (Schäfer et al. 2013).

4.3.3 Small Molecules

Potentially, the most versatile and cheapest alternative to peptides and peptidomimetics are small molecules. Unlike peptides, small molecules possess good pharmacokinetic properties such as good bioavailability, membrane permeation ability and stability. Moreover, their synthesis is faster and cheaper. However, developing small molecules for disruption of protein-protein interactions is challenging. Protein-protein interactions are mediated through large and shallow surfaces of 1500–3000 Å^2, while deep pockets such as catalytic domains of enzymes that could be effectively targeted are rarely present. The contact area between a small molecule and its protein target is much smaller: 300–1000 Å^2 (Smith and Gestwicki 2012). A major challenge is to find a small molecule still in accordance with the Rule of Five (Lipinski et al. 2001). The molecular weight should not exceed 500 g/mol but should still be large enough to overcome the distributed free energy of the protein-protein interaction. An important step towards finding effective small molecules was the development of the hot spot theory. According to this, the affinity of two interaction partners is not distributed equally across the interaction surface. In contrast, a small subset of individual amino acids or regions of amino acids, so-called 'hot spots', contribute most of the free energy of binding (Clackson and Wells 1995). To determine which amino acids are important for a binding event, an alanine scan can be performed. By substituting single amino acids with alanine and measuring changes in binding affinity, hot spots can be identified (Klussmann 2016; Scott et al. 2016). On the other hand, alanine scanning can lead to false-positive results since also amino acids are identified as hot spots, which only stabilise the unbound state or lead to a conformational change of the protein. Computational methods are often used as a complementary method to alanine scans (Guo et al. 2014).

Small molecules disrupting a protein-protein interaction can be classified into either orthosteric or allosteric disruptors. Orthosteric disruptors interact directly with the interface of one of the two proteins, whereas allosteric disruptors are binding outside the interface and thereby often provoke a conformational change of the protein which prevents binding its cognate protein-binding partner (Fischer et al. 2015).

Two small molecules disrupting protein-protein interactions have reached the market: maraviroc and tirofiban. Maraviroc is an allosteric inhibitor of the CCR5 receptor on the surface of cells. Upon binding of maraviroc to CCR5, CCR5 acquires a conformation in which it is unable to interact with the Gp120 envelope protein of HIV-1 (Lagane et al. 2013). Maraviroc prevents entry of the virus into the host cells. It was approved by the FDA in 2007. Tirofiban was approved by the FDA in 1999. It mimics the RGD motif of fibrinogen and is thus binding to glycoprotein IIa/IIIb of platelets (Clackson and Wells 1995). It blocks the coagulation cascade and has antithrombotic activity. No inhibitor of intracellular protein-protein interactions has reached the market yet although several are in clinical trials.

For the disruption of AKAP-PKA interactions, so far only one small molecule, FMP-API-1, was discovered. It was identified in an enzyme-linked immunosorbent assay (ELISA)-based screen of 20,000 molecules. The AKAP-PKA disruptor peptide Ht31 bound additionally to FMP-API-1 to RIIα. This indicated a binding site of FMP-API-1 distinct from the interacting AKB and D/D domains. STD NMR experiments with RIIα deletion constructs showed that FMP-API-1 and its derivatives bind in a region C-terminally from the D/D domain. Through this allosteric mode of action, FMP-API-1 disrupted the interaction of AKAP18δ with both RIIα and RIIβ in vitro and in cultured cardiac myocytes. The region C-terminal from the D/D domain (amino acids 92–102 in human RIIα) is the autoinhibitory site of PKA. Binding of FMP-API-1 within this domain could provoke loss of the inhibitory effect leading to a dissociation of the C-subunits and thereby to activation of PKA. Indeed, FMP-API-1 activated PKA. FMP-API-1 increased contractility of neonatal rat cardiac myocytes and of isolated whole rat hearts. However, due its dual mode of action, i.e. disruption of AKAP-PKA interactions and simultaneous activation of PKA, severe side effects can be expected which exclude the molecule from further development towards a drug candidate (Christian et al. 2011).

For identification of new small molecule disruptors, ELISA assays are widely used (Schächterle et al. 2015). However, they often led to false positives, e.g. due to the unspecific binding of proteins to plastic surfaces (Terato et al. 2014). More reliable are, for instance, homogenous time-resolved fluorescence (HTRF) assays. Here, protein-protein interactions are investigated in solution. Thereby, all parts of the protein are available for a potential interaction; background signals are eliminated by using time-resolved measurements of fluorescence. The introduction of a time delay between excitation of the sample and measuring its fluorescence, short-lived background fluorescence is easily eliminated and specificity is increased (Degorce et al. 2009). HTRF assays are suitable for screening of small molecule libraries for disruptors of AKAP-PKA interactions (Schächterle et al. 2015). HTRF assays were used for characterising the potency of terpyridine-based α-helix mimetics to interfere with the AKAP18-RII interaction (Schäfer et al. 2013).

4.3.4 Disruptors of Interaction Between AKAPs and Proteins Other Than PKA

In addition to AKAP-PKA interactions, other AKAP-dependent protein-protein interactions have also been targeted. Lygren et al. developed a peptide based on the AKAP18δ-binding site of PLN to disrupt the interaction of PLN with this AKAP. The peptide decreased Ca^{2+} reuptake into the sarcoplasmic reticulum upon adrenergic stimulation and revealed the crucial role of this interaction in the control of cardiac myocyte contractility (Lygren et al. 2007).

Another AKAP18 splice variant, AKAP18α, was the basis for the development of a peptide disruptor. An AKAP18α-derived peptide interfered with the interaction of AKAP18α and the Ca^{2+} channel $Ca_V1.2$. This resulted in a decreased Ca^{2+} influx into cardiac myocytes upon β-adrenergic stimuli, again revealing a function in cardiac myocyte contractility (Hulme et al. 2003).

4.4 Conclusions and Outlook

AKAPs play key roles in the coordination of cAMP signalling and cross talk with other signalling systems. They are involved in fundamental processes such as cardiac myocyte contractility discussed here but also in many others such as water reabsorption in the kidney, blood pressure control or memory formation (Nystoriak et al. 2014; Poppinga et al. 2014; Sonkusare et al. 2014; Vukićević et al. 2016). Dysregulation of AKAPs and their protein-protein interactions is associated with diseases, some of which, such as heart failure, have an unmet medical need and thus a requirement for innovative targets and drugs. A new route may be disruption (or strengthening) of the disease-relevant AKAP-dependent protein-protein interactions. In particular, AKAP-PKA interactions seem a relevant target. For example, there are indications that their disruption in the heart beneficially influences heart failure (Christian et al. 2011; Deak and Klussmann 2015; Dema et al. 2015; Fink et al. 2001; Mauban et al. 2009; McConnell et al. 2009; Patel et al. 2010; Vukićević et al. 2016). Several agents for inhibition of AKAP-PKA interactions have been/or are being developed: peptides, peptidomimetics and small molecules, whereby the main focus is on peptides. Peptide disruptors for preferentially inhibiting AKAP-RI or AKAP-RII interactions have been identified. However, no agents are available that allow disruption of the interaction of a specific AKAP with PKA. The development of such agents in the future appears feasible on the basis of the recently identified anchor points, i.e. hydrophilic amino acid residues in AKAPs that reside outside the hydrophobic core interaction sites which bind the D/D domains of R-subunits (Götz et al. 2016). The sequence specificity of each AKAP around the anchor points and the requirement of these points for the tight binding of PKA may allow the development of selective inhibitors or modulators of these interactions. Such specific pharmacological interference would not only allow unequivocally ascribing functions to specific AKAP-PKA interactions, it may also pave the way to new concepts for treatment of human diseases.

Thus, a focus in AKAP research will continue to be on the development of approaches for selective inhibition of specific AKAP-dependent protein-protein interactions that effectively define functions and are employable for validation of such interactions as drug targets. The development of novel chemical entities may be pushed by the recent progress in elucidating 3D structures of AKAPs. The interactions of the AKB domains of AKAP18 and of smAKAP with the D/D domains of RIIα and RIα, respectively, have recently been solved (Burgers et al. 2016; Götz et al. 2016). They offer unique opportunities for in silico design of novel and specific disruptors, in particular in combination with the anchor points. Prospectively, a new option for assessing functions of specific AKAP-dependent protein-protein interactions and validating them as drug targets may be abolishing the interactions through CRISPR/Cas-based gene editing (Makarova et al. 2015). This approach may lead quicker than conventional gene targeting approaches to new cell and animal models for probing functions and validation of interactions as drug targets.

Acknowledgements This work was supported by grants from the Else Kröner-Fresenius-Stiftung (2013_A145), the German-Israeli Foundation (G.I.F. I-1210-286.13/2012), the German Centre for Cardiovascular Research (DZHK 81X210012) and the Deutsche Forschungsgemeinschaft (DFG KL1415/7-1) to E. K.

Compliance with Ethical Standards

Conflict of Interest Statement The authors declare that they have no conflict of interest.

References

Acin-Perez R, Salazar E, Kamenetsky M, Buck J, Levin LR, Manfredi G (2009) Cyclic AMP produced inside mitochondria regulates oxidative phosphorylation. Cell Metab 9:265–276. doi:10.1016/j.cmet.2009.01.012

Ahmad F, Shen W, Vandeput F, Szabo-Fresnais N, Krall J, Degerman E, Goetz F, Klussmann E, Movsesian M, Manganiello V (2015) Regulation of sarcoplasmic reticulum Ca^{2+} ATPase 2

(SERCA2) activity by phosphodiesterase 3A (PDE3A) in human myocardium: phosphorylation-dependent interaction of PDE3A1 with SERCA2. J Biol Chem 290:6763–6776. doi:10.1074/jbc.M115.638585

Alto NM, Soderling SH, Hoshi N, Langeberg LK, Fayos R, Jennings PA, Scott JD (2003) Bioinformatic design of A-kinase anchoring protein-in silico: a potent and selective peptide antagonist of type II protein kinase A anchoring. Proc Natl Acad Sci U S A 100:4445. doi:10.1073/pnas.0330734100

Appert-Collin A, Cotecchia S, Nenniger-Tosato M, Pedrazzini T, Diviani D (2007) The A-kinase anchoring protein (AKAP)-Lbc-signaling complex mediates α1 adrenergic receptor-induced cardiomyocyte hypertrophy. Proc Natl Acad Sci U S A 104:10140–10145. doi:10.1073/pnas.0701099104

Aye TT, Soni S, van Veen TAB, van der Heyden MAG, Cappadona S, Varro A, de Weger RA, de Jonge N, Vos MA, Heck AJR, Scholten A (2012) Reorganized PKA-AKAP associations in the failing human heart. J Mol Cell Cardiol 52:511–518. doi:10.1016/j.yjmcc.2011.06.003

Baillie GS, Huston E, Scotland G, Hodgkin M, Gall I, Peden AH, MacKenzie C, Houslay ES, Currie R, Pettitt TR, Walmsley AR, Wakelam MJO, Warwicker J, Houslay MD (2002) TAPAS-1, a novel microdomain within the unique N-terminal region of the PDE4A1 cAMP-specific phosphodiesterase that allows rapid, Ca^{2+}-triggered membrane association with selectivity for interaction with phosphatidic acid. J Biol Chem 277:28298–28309. doi:10.1074/jbc.M108353200

Banky P, Roy M, Newlon MG, Morikis D, Haste NM, Taylor SS, Jennings PA (2003) Related protein–protein interaction modules present drastically different surface topographies despite a conserved helical platform. J Mol Biol 330:1117–1129. doi:10.1016/S0022-2836(03)00552-7

Bechara C, Sagan S (2013) Cell-penetrating peptides: 20 years later, where do we stand? FEBS Lett 587:1693–1702. doi:10.1016/j.febslet.2013.04.031

Bender AT, Beavo JA (2006) Cyclic nucleotide phosphodiesterases: molecular regulation to clinical use. Pharmacol Rev 58:488–520. doi:10.1124/pr.58.3.5

Berrera M, Dodoni G, Monterisi S, Pertegato V, Zamparo I, Zaccolo M (2008) A toolkit for real-time detection of cAMP: insights into compartmentalized signaling. Handb Exp Pharmacol 186:285–298. doi:10.1007/978-3-540-72843-6_12

Biel M, Michalakis S (2009) Cyclic nucleotide-gated channels. Handb Exp Pharmacol 191:111–136. doi:10.1007/978-3-540-68964-5_7

Bodor GS, Oakeley AE, Allen PD, Crimmins DL, Ladenson JH, Anderson PA (1997) Troponin I phosphorylation in the normal and failing adult human heart. Circulation 96:1495–1500

Brooks H, Lebleu B, Vivès E (2005) Tat peptide-mediated cellular delivery: back to basics. Adv Drug Deliv Rev 57:559–577. doi:10.1016/j.addr.2004.12.001

Burgers PP, Ma Y, Margarucci L, Mackey M, van der Heyden MAG, Ellisman M, Scholten A, Taylor SS, Heck AJR (2012) A small novel A-kinase anchoring protein (AKAP) that localizes specifically protein kinase A-regulatory subunit I (PKA-RI) to the plasma membrane. J Biol Chem 287:43789–43797. doi:10.1074/jbc.M112.395970

Burgers PP, Bruystens J, Burnley RJ, Nikolaev VO, Keshwani M, Wu J, Janssen BJC, Taylor SS, Heck AJR, Scholten A (2016) Structure of smAKAP and its regulation by PKA-mediated phosphorylation. FEBS J 283(11):2132–2148. doi:10.1111/febs.13726

Burns-Hamuro LL, Ma Y, Kammerer S, Reineke U, Self C, Cook C, Olson GL, Cantor CR, Braun A, Taylor SS (2003) Designing isoform-specific peptide disruptors of protein kinase A localization. Proc Natl Acad Sci 100:4072–4077. doi:10.1073/pnas.2628038100

Burzicki G, Voisin-Chiret AS, Sopkovà-de Oliveira Santos J, Rault S (2009) Synthesis of dihalo bi- and terpyridines by regioselective Suzuki–Miyaura cross-coupling reactions. Tetrahedron 65:5413–5417. doi:10.1016/j.tet.2009.04.049

Cadd G, Stanley McKnight G (1989) Distinct patterns of cAMP-dependent protein kinase gene expression in mouse brain. Neuron 3:71–79. doi:10.1016/0896-6273(89)90116-5

Cadd GG, Uhler MD, McKnight GS (1990) Holoenzymes of cAMP-dependent protein kinase containing the neural form of type I regulatory subunit have an increased sensitivity to cyclic nucleotides. J Biol Chem 265:19502–19506

Cann M (2004) Bicarbonate stimulated adenylyl cyclases. IUBMB Life 56:529–534. doi:10.1080/15216540400013861

Carlisle Michel JJ, Dodge KL, Wong W, Mayer NC, Langeberg LK, Scott JD (2004) PKA-phosphorylation of PDE4D3 facilitates recruitment of the mAKAP signalling complex. Biochem J 381:587–592. doi:10.1042/BJ20040846

Carlson CR, Ruppelt A, Taskén K (2003) A kinase anchoring protein (AKAP) interaction and dimerization of the RIα and RIβ regulatory subunits of protein kinase A in vivo by the yeast two hybrid system. J Mol Biol 327:609–618. doi:10.1016/S0022-2836(03)00093-7

Carlson CR, Lygren B, Berge T, Hoshi N, Wong W, Taskén K, Scott JD (2006) Delineation of type I protein kinase A-selective signaling events using an RI anchoring disruptor. J Biol Chem 281:21535–21545. doi:10.1074/jbc.M603223200

Carnegie GK, Burmeister BT (2011) A-kinase anchoring proteins that regulate cardiac remodeling. J Cardiovasc Pharmacol 58:451–458. doi:10.1097/FJC.0b013e31821c0220

Carnegie GK, Soughayer J, Smith FD, Pedroja BS, Zhang F, Diviani D, Bristow MR, Kunkel MT, Newton AC, Langeberg LK, Scott JD (2008) AKAP-Lbc mobilizes a cardiac hypertrophy signaling pathway. Mol Cell 32:169–179. doi:10.1016/j.molcel.2008.08.030

Carr DW, Stofko-Hahn RE, Fraser ID, Bishop SM, Acott TS, Brennan RG, Scott JD (1991) Interaction of the regulatory subunit (RII) of cAMP-dependent protein kinase with RII-anchoring proteins occurs through an amphipathic helix binding motif. J Biol Chem 266: 14188–14192

Carr DW, Stofko-Hahn RE, Fraser ID, Cone RD, Scott JD (1992) Localization of the cAMP-dependent protein kinase to the postsynaptic densities by A-kinase anchoring proteins. Characterization of AKAP 79. J Biol Chem 267:16816–16823

Chang YS, Graves B, Guerlavais V, Tovar C, Packman K, To KH, Olson KA, Kesavan K, Gangurde P, Mukherjee A, Baker T, Darlak K, Elkin C, Filipovic Z, Qureshi FZ, Cai H, Berry P, Feyfant E, Shi XE, Horstick J, Annis DA, Manning AM, Fotouhi N, Nash H, Vassilev LT, Sawyer TK (2013) Stapled α–helical peptide drug development: a potent dual inhibitor of MDM2 and MDMX for p53-dependent cancer therapy. Proc Natl Acad Sci U S A 110:E3445. doi:10.1073/pnas.1303002110

Chen Q, Lin RY, Rubin CS (1997) Organelle-specific targeting of protein kinase AII (PKAII). Molecular and in situ characterization of murine A kinase anchor proteins that recruit regulatory subunits of PKAII to the cytoplasmic surface of mitochondria. J Biol Chem 272: 15247–15257

Christian F, Szaszák M, Friedl S, Drewianka S, Lorenz D, Goncalves A, Furkert J, Vargas C, Schmieder P, Götz F, Zühlke K, Moutty M, Göttert H, Joshi M, Reif B, Haase H, Morano I, Grossmann S, Klukovits A, Verli J, Gáspár R, Noack C, Bergmann M, Kass R, Hampel K, Kashin D, Genieser H-G, Herberg FW, Willoughby D, Cooper DMF, Baillie GS, Houslay MD, von Kries JP, Zimmermann B, Rosenthal W, Klussmann E (2011) Small molecule AKAP-protein kinase A (PKA) interaction disruptors that activate PKA interfere with compartmentalized cAMP signaling in cardiac myocytes. J Biol Chem 286:9079. doi:10.1074/jbc.M110.160614

Chu Q, Moellering RE, Hilinski GJ, Kim YW, Grossmann TN, Yeh JTH, Verdine GL (2015) Towards understanding cell penetration by stapled peptides. Med Chem Comm 6:111–119. doi:10.1039/C4MD00131A

Clackson T, Wells JA (1995) A hot spot of binding energy in a hormone-receptor interface. Science 267:383–386. doi:10.1126/science.7529940

Clegg CH, Cadd GG, McKnight GS (1988) Genetic characterization of a brain-specific form of the type I regulatory subunit of cAMP-dependent protein kinase. Proc Natl Acad Sci U S A 85:3703

Coghlan VM, Perrino BA, Howard M, Langeberg LK, Hicks JB, Gallatin WM, Scott JD (1995) Association of protein kinase A and protein phosphatase 2B with a common anchoring protein. Science 267:108–111

Colson BA, Patel JR, Chen PP, Bekyarova T, Abdalla MI, Tong CW, Fitzsimons DP, Irving TC, Moss RL (2012) Myosin binding protein-C phosphorylation is the principal mediator of protein kinase A effects on thick filament structure in myocardium. J Mol Cell Cardiol 53:609–616. doi:10.1016/j.yjmcc.2012.07.012

Craik DJ, Fairlie DP, Liras S, Price D (2013) The future of peptide-based drugs. Chem Biol Drug Des 81:136–147. doi:10.1111/cbdd.12055

Deak VA, Klussmann E (2015) Pharmacological interference with protein-protein interactions of A-kinase anchoring proteins as a strategy for the treatment of disease. Curr Drug Targets 17(10):1147–1171

Deák VA, Skroblin P, Dittmayer C, Knobeloch KP, Bachmann S, Klussmann E (2016) The A-kinase anchoring protein GSKIP regulates GSK3β activity and controls palatal shelf fusion in mice. J Biol Chem 291:681–690. doi:10.1074/jbc.M115.701177

Degorce F, Card A, Soh S, Trinquet E, Knapik GP, Xie B (2009) HTRF: a technology tailored for drug discovery—a review of theoretical aspects and recent applications. Curr Chem Genomics 3:22. doi:10.2174/1875397300903010022

Dema A, Perets E, Schulz MS, Deák VA, Klussmann E (2015) Pharmacological targeting of AKAP-directed compartmentalized cAMP signalling. Cell Signal 27:2474–2487. doi:10.1016/j.cellsig.2015.09.008

Dema A, Schröter MF, Perets E, Skroblin P, Moutty MC, Deàk VA, Birchmeier W, Klussmann E (2016) The A-kinase anchoring protein (AKAP) glycogen synthase kinase 3β interaction protein (GSKIP) regulates β-catenin through its interactions with both protein kinase A (PKA) and GSK3β. J Biol Chem 291:19618–19630. doi:10.1074/jbc.M116.738047

Derossi D, Joliot AH, Chassaing G, Prochiantz A (1994) The third helix of the Antennapedia homeodomain translocates through biological membranes. J Biol Chem 269:10444–10450

Diviani D, Dodge-Kafka KL, Li J, Kapiloff MS (2011) A-kinase anchoring proteins: scaffolding proteins in the heart. Am J Physiol Heart Circ Physiol 301:H1742–H1753. doi:10.1152/ajpheart.00569.2011

Dostmann WRG, Taylor SS (1991) Identifying the molecular switches that determine whether (Rp)-cAMPS functions as an antagonist or an agonist in the activation of cAMP-dependent protein kinase I. Biochemistry (Mosc) 30:8710–8716. doi:10.1021/bi00099a032

Fancy SPJ, Baranzini SE, Zhao C, Yuk DI, Irvine KA, Kaing S, Sanai N, Franklin RJM, Rowitch DH (2009) Dysregulation of the Wnt pathway inhibits timely myelination and remyelination in the mammalian CNS. Genes Dev 23:1571–1585. doi:10.1101/gad.1806309

Faruque OM, Le-Nguyen D, Lajoix A-D, Vives E, Petit P, Bataille D, Hani E-H (2009) Cell-permeable peptide-based disruption of endogenous PKA-AKAP complexes: a tool for studying the molecular roles of AKAP-mediated PKA subcellular anchoring. Am J Physiol Cell Physiol 296:C306–C316. doi:10.1152/ajpcell.00216.2008

Fink MA, Zakhary DR, Mackey JA, Desnoyer RW, Apperson-Hansen C, Damron DS, Bond M (2001) AKAP-mediated targeting of protein kinase A regulates contractility in cardiac myocytes. Circ Res 88:291–297. doi:10.1161/01.RES.88.3.291

Fischer G, Rossmann M, Hyvönen M (2015) Alternative modulation of protein–protein interactions by small molecules. Curr Opin Biotechnol 35:78–85. doi:10.1016/j.copbio.2015.04.006

Flaherty BR, Wang Y, Trope EC, Ho TG, Muralidharan V, Kennedy EJ, Peterson DS (2015) The stapled AKAP disruptor peptide STAD-2 displays antimalarial activity through a PKA-independent mechanism. PLoS One 10:e0129239. doi:10.1371/journal.pone.0129239

Furusawa M, Ohnishi T, Taira T, Iguchi-Ariga SM, Ariga H (2001) AMY-1, a c-Myc-binding protein, is localized in the mitochondria of sperm by association with S-AKAP84, an anchor protein of cAMP-dependent protein kinase. J Biol Chem 276:36647–36651. doi:10.1074/jbc.M103885200

Futaki S, Suzuki T, Ohashi W, Yagami T, Tanaka S, Ueda K, Sugiura Y (2001) Arginine-rich peptides an abundant source of membrane-permeable peptides having potential as carriers for intracellular protein delivery. J Biol Chem 276:5836–5840. doi:10.1074/jbc.M007540200

Gamm DM, Baude EJ, Uhler MD (1996) The major catalytic subunit isoforms of cAMP-dependent protein kinase have distinct biochemical properties in vitro and in vivo. J Biol Chem 271:15736–15742. doi:10.1074/jbc.271.26.15736

Gloerich M, Bos JL (2010) Epac: defining a new mechanism for cAMP action. Annu Rev Pharmacol Toxicol 50:355–375. doi:10.1146/annurev.pharmtox.010909.105714

Gold MG, Lygren B, Dokurno P, Hoshi N, McConnachie G, Taskén K, Carlson CR, Scott JD, Barford D (2006) Molecular basis of AKAP specificity for PKA regulatory subunits. Mol Cell 24:383–395. doi:10.1016/j.molcel.2006.09.006
Gold MG, Fowler DM, Means CK, Pawson CT, Stephany JJ, Langeberg LK, Fields S, Scott JD (2013) Engineering A-kinase anchoring protein (AKAP)-selective regulatory subunits of protein kinase A (PKA) through structure-based phage selection. J Biol Chem 288:17111. doi:10.1074/jbc.M112.447326
Gonzalez GA, Montminy MR (1989) Cyclic AMP stimulates somatostatin gene transcription by phosphorylation of CREB at serine 133. Cell 59:675–680. doi:10.1016/0092-8674(89)90013-5
Götz F, Roske Y, Schulz MS, Autenrieth K, Bertinetti D, Faelber K, Zühlke K, Kreuchwig A, Kennedy E, Krause G, Daumke O, Herberg FW, Heinemann U, Klussmann E (2016) AKAP18:PKA-RIIα structure reveals crucial anchor points for recognition of regulatory subunits of PKA. Biochem J 473(13):1881–1894. doi:10.1042/BCJ20160242
Gray PC, Scott JD, Catterall WA (1998) Regulation of ion channels by cAMP-dependent protein kinase and A-kinase anchoring proteins. Curr Opin Neurobiol 8:330–334. doi:10.1016/S0959-4388(98)80057-3
Guo W, Wisniewski JA, Ji H (2014) Hot spot-based design of small-molecule inhibitors for protein–protein interactions. Bioorg Med Chem Lett 24:2546–2554. doi:10.1016/j.bmcl.2014.03.095
Haghighi K, Bidwell P, Kranias EG (2014) Phospholamban interactome in cardiac contractility and survival: a new vision of an old friend. J Mol Cell Cardiol 77:160–167. doi:10.1016/j.yjmcc.2014.10.005
Hasenfuss G (1998) Alterations of calcium-regulatory proteins in heart failure. Cardiovasc Res 37:279–289. doi:10.1016/S0008-6363(97)00277-0
Herberg FW, Maleszka A, Eide T, Vossebein L, Tasken K (2000) Analysis of A-kinase anchoring protein (AKAP) interaction with protein kinase A (PKA) regulatory subunits: PKA isoform specificity in AKAP binding. J Mol Biol 298:329–339. doi:10.1006/jmbi.2000.3662
Huang LJ, Durick K, Weiner JA, Chun J, Taylor SS (1997a) Identification of a novel protein kinase A anchoring protein that binds both type I and type II regulatory subunits. J Biol Chem 272:8057–8064. doi:10.1074/jbc.272.12.8057
Huang LJ, Durick K, Weiner JA, Chun J, Taylor SS (1997b) D-AKAP2, a novel protein kinase A anchoring protein with a putative RGS domain. Proc Natl Acad Sci U S A 94:11184
Huang LJ, Wang L, Ma Y, Durick K, Perkins G, Deerinck TJ, Ellisman MH, Taylor SS (1999) NH2-terminal targeting motifs direct dual specificity A-kinase-anchoring protein 1 (D-AKAP1) to either mitochondria or endoplasmic reticulum. J Cell Biol 145:951–959
Huang T, McDonough CB, Abel T (2006) Compartmentalized PKA signaling events are required for synaptic tagging and capture during hippocampal late-phase long-term potentiation. Eur J Cell Biol 85:635. doi:10.1016/j.ejcb.2006.02.005
Hulme JT, Lin TWC, Westenbroek RE, Scheuer T, Catterall WA (2003) β-adrenergic regulation requires direct anchoring of PKA to cardiac CaV1.2 channels via a leucine zipper interaction with A kinase-anchoring protein 15. Proc Natl Acad Sci 100:13093–13098. doi:10.1073/pnas.2135335100
Hulme JT, Westenbroek RE, Scheuer T, Catterall WA (2006) Phosphorylation of serine 1928 in the distal C-terminal domain of cardiac CaV1.2 channels during beta1-adrenergic regulation. Proc Natl Acad Sci U S A 103:16574–16579. doi:10.1073/pnas.0607294103
Hundsrucker C, Klussmann E (2008) Direct AKAP-mediated protein-protein interactions as potential drug targets. Handb Exp Pharmacol 186:483–503
Hundsrucker C, Skroblin P, Christian F, Zenn H-M, Popara V, Joshi M, Eichhorst J, Wiesner B, Herberg FW, Reif B, Rosenthal W, Klussmann E (2010) Glycogen synthase kinase 3β interaction protein functions as an A-kinase anchoring protein. J Biol Chem 285:5507–5521. doi:10.1074/jbc.M109.047944
Huston E, Gall I, Houslay TM, Houslay MD (2006) Helix-1 of the cAMP-specific phosphodiesterase PDE4A1 regulates its phospholipase-D-dependent redistribution in response to release of Ca^{2+}. J Cell Sci 119:3799–3810. doi:10.1242/jcs.03106

Ishikawa Y, Iwatsubo K, Tsunematsu T, Okumura S (2005) Genetic manipulation and functional analysis of cAMP signalling in cardiac muscle: implications for a new target of pharmacotherapy. Biochem Soc Trans 33:1337–1340. doi:10.1042/BST0331337

Jahnsen T, Hedin L, Lohmann SM, Walter U, Richards JS (1986) The neural type II regulatory subunit of cAMP-dependent protein kinase is present and regulated by hormones in the rat ovary. J Biol Chem 261:6637–6639

Jurevicius J, Fischmeister R (1996) cAMP compartmentation is responsible for a local activation of cardiac Ca^{2+} channels by beta-adrenergic agonists. Proc Natl Acad Sci U S A 93:295–299

Kapiloff MS, Jackson N, Airhart N (2001) mAKAP and the ryanodine receptor are part of a multi-component signaling complex on the cardiomyocyte nuclear envelope. J Cell Sci 114:3167–3176

Kaupp UB, Seifert R (2002) Cyclic nucleotide-gated ion channels. Physiol Rev 82:769–824. doi:10.1152/physrev.00008.2002

Kawasaki H, Springett GM, Mochizuki N, Toki S, Nakaya M, Matsuda M, Housman DE, Graybiel AM (1998) A family of cAMP-binding proteins that directly activate Rap1. Science 282:2275–2279. doi:10.1126/science.282.5397.2275

Kinderman FS, Kim C, von Daake S, Ma Y, Pham BQ, Spraggon G, Xuong N-H, Jennings PA, Taylor SS (2006) A novel and dynamic mechanism for AKAP binding to RII isoforms of cAMP-dependent protein kinase. Mol Cell 24:397–408. doi:10.1016/j.molcel.2006.09.015

Klussmann E (2016) Protein–protein interactions of PDE4 family members—functions, interactions and therapeutic value. Cell Signal 28:713–718. doi:10.1016/j.cellsig.2015.10.005

Kovanich D, van der Heyden MAG, Aye TT, van Veen TAB, Heck AJR, Scholten A (2010) Sphingosine kinase interacting protein is an A-kinase anchoring protein specific for type I cAMP-dependent protein kinase. Chembiochem Eur J Chem Biol 11:963–971. doi:10.1002/cbic.201000058

Kranias EG, Bers DM (2007) Calcium and cardiomyopathies. Subcell Biochem 45:523–537

Kranias EG, Hajjar RJ (2012) Modulation of cardiac contractility by the phospholamban/SERCA2a regulatome. Circ Res 110:1646–1660. doi:10.1161/CIRCRESAHA.111.259754

Kurokawa J, Motoike HK, Rao J, Kass RS (2004) Regulatory actions of the A-kinase anchoring protein Yotiao on a heart potassium channel downstream of PKA phosphorylation. Proc Natl Acad Sci U S A 101:16374–16378. doi:10.1073/pnas.0405583101

Lagane B, Garcia-Perez J, Kellenberger E (2013) Modeling the allosteric modulation of CCR5 function by Maraviroc. Drug Discov Today Technol 10:e297–e305. doi:10.1016/j.ddtec.2012.07.011

Lee DC, Carmichael DF, Krebs EG, McKnight GS (1983) Isolation of a cDNA clone for the type I regulatory subunit of bovine cAMP-dependent protein kinase. Proc Natl Acad Sci U S A 80:3608–3612

Li J, Negro A, Lopez J, Bauman AL, Henson E, Dodge-Kafka K, Kapiloff MS (2010) The mAKAPβ scaffold regulates cardiac myocyte hypertrophy via recruitment of activated calcineurin. J Mol Cell Cardiol 48:387. doi:10.1016/j.yjmcc.2009.10.023

Li Y, Chen L, Kass RS, Dessauer CW (2012) The A-kinase anchoring protein Yotiao facilitates complex formation between adenylyl cyclase type 9 and the IKs potassium channel in heart. J Biol Chem 287:29815–29824. doi:10.1074/jbc.M112.380568

Lin CC, Chou CH, Howng SL, Hsu CY, Hwang CC, Wang C, Hsu CM, Hong YR (2009) GSKIP, an inhibitor of GSK3beta, mediates the N-cadherin/beta-catenin pool in the differentiation of SH-SY5Y cells. J Cell Biochem 108:1325–1336. doi:10.1002/jcb.22362

Lipinski CA, Lombardo F, Dominy BW, Feeney PJ (2001) Experimental and computational approaches to estimate solubility and permeability in drug discovery and development settings. Adv Drug Deliv Rev 46:3–26

Lodish H, Berk A, Zipursky SL, Matsudaira P, Baltimore D, Darnell J (2000) Molecular Cell Biology, 4th edition. Chapter 20.6 Second Messengers. New York: W.H. Freeman. ISBN-10: 0-7167-3136-3

Lygren B, Carlson CR, Santamaria K, Lissandron V, McSorley T, Litzenberg J, Lorenz D, Wiesner B, Rosenthal W, Zaccolo M, Taskén K, Klussmann E (2007) AKAP complex regulates Ca^{2+} re-uptake into heart sarcoplasmic reticulum. EMBO Rep 8:1061. doi:10.1038/sj.embor.7401081

MacLennan DH, Kranias EG (2003) Phospholamban: a crucial regulator of cardiac contractility. Nat Rev Mol Cell Biol 4:566–577. doi:10.1038/nrm1151

MacLennan DH, Asahi M, Tupling AR (2003) The regulation of SERCA-type pumps by phospholamban and sarcolipin. Ann N Y Acad Sci 986:472–480

Makarova KS, Wolf YI, Alkhnbashi OS, Costa F, Shah SA, Saunders SJ, Barrangou R, Brouns SJJ, Charpentier E, Haft DH, Horvath P, Moineau S, Mojica FJM, Terns RM, Terns MP, White MF, Yakunin AF, Garrett RA, van der Oost J, Backofen R, Koonin EV (2015) An updated evolutionary classification of CRISPR-Cas systems. Nat Rev Microbiol 13:722–736. doi:10.1038/nrmicro3569

Mauban JRH, O'Donnell M, Warrier S, Manni S, Bond M (2009) AKAP-scaffolding proteins and regulation of cardiac physiology. Physiology (Bethesda) 24:78–87. doi:10.1152/physiol.00041.2008

McConnell BK, Popovic Z, Mal N, Lee K, Bautista J, Forudi F, Schwartzman R, Jin JP, Penn M, Bond M (2009) Disruption of protein kinase A interaction with A-kinase-anchoring proteins in the heart in vivo: effects on cardiac contractility, protein kinase A phosphorylation, and troponin I proteolysis. J Biol Chem 284:1583–1592. doi:10.1074/jbc.M806321200

Newlon MG, Roy M, Morikis D, Hausken ZE, Coghlan V, Scott JD, Jennings PA (1999) The molecular basis for protein kinase A anchoring revealed by solution NMR. Nat Struct Mol Biol 6:222–227. doi:10.1038/6663

Newlon MG, Roy M, Morikis D, Carr DW, Westphal R, Scott JD, Jennings PA (2001) A novel mechanism of PKA anchoring revealed by solution structures of anchoring complexes. EMBO J 20:1651. doi:10.1093/emboj/20.7.1651

Nichols CB, Rossow CF, Navedo MF, Westenbroek RE, Catterall WA, Santana LF, McKnight GS (2010) Sympathetic stimulation of adult cardiomyocytes requires association of AKAP5 with a subpopulation of L-type calcium channels. Circ Res 107:747–756. doi:10.1161/CIRCRESAHA.109.216127

Nystoriak MA, Nieves-Cintrón M, Nygren PJ, Hinke SA, Nichols CB, Chen C-Y, Puglisi JL, Izu LT, Bers DM, Dell'acqua ML, Scott JD, Santana LF, Navedo MF (2014) AKAP150 contributes to enhanced vascular tone by facilitating large-conductance Ca^{2+}-activated K^{+} channel remodeling in hyperglycemia and diabetes mellitus. Circ Res 114:607–615. doi:10.1161/CIRCRESAHA.114.302168

Oliveria SF, Dell'Acqua ML, Sather WA (2007) AKAP79/150 anchoring of calcineurin controls neuronal L-type Ca^{2+} channel activity and nuclear signaling. Neuron 55:261–275. doi:10.1016/j.neuron.2007.06.032

Otvos L, Wade JD (2014) Current challenges in peptide-based drug discovery. Front Chem 2:62. doi:10.3389/fchem.2014.00062

Pannekoek W-J, Kooistra MRH, Zwartkruis FJT, Bos JL (2009) Cell–cell junction formation: the role of Rap1 and Rap1 guanine nucleotide exchange factors. Biochim Biophys Acta 1788:790–796. doi:10.1016/j.bbamem.2008.12.010

Pannekoek WJ, Linnemann JR, Brouwer PM, Bos JL, Rehmann H (2013) Rap1 and Rap2 antagonistically control endothelial barrier resistance. PLoS One 8:e57903. doi:10.1371/journal.pone.0057903

Pare GC, Bauman AL, McHenry M, Michel JJC, Dodge-Kafka KL, Kapiloff MS (2005) The mAKAP complex participates in the induction of cardiac myocyte hypertrophy by adrenergic receptor signaling. J Cell Sci 118:5637–5646. doi:10.1242/jcs.02675

Pastor-Soler N, Beaulieu V, Litvin TN, Silva ND, Chen Y, Brown D, Buck J, Levin LR, Breton S (2003) Bicarbonate-regulated adenylyl cyclase (sAC) is a sensor that regulates pH-dependent V-ATPase recycling. J Biol Chem 278:49523–49529. doi:10.1074/jbc.M309543200

Patel HH, Hamuro LL, Chun BJ, Kawaraguchi Y, Quick A, Rebolledo B, Pennypacker J, Thurston J, Rodriguez-Pinto N, Self C, Olson G, Insel PA, Giles WR, Taylor SS, Roth DM (2010) Disruption of protein kinase A localization using a trans-activator of transcription (TAT)-conjugated A-kinase-anchoring peptide reduces cardiac function. J Biol Chem 285:27632. doi:10.1074/jbc.M110.146589

Pidoux G, Witczak O, Jarnæss E, Myrvold L, Urlaub H, Stokka AJ, Küntziger T, Taskén K (2011) Optic atrophy 1 is an A-kinase anchoring protein on lipid droplets that mediates adrenergic control of lipolysis. EMBO J 30:4371. doi:10.1038/emboj.2011.365

Poppinga WJ, Muñoz-Llancao P, González-Billault C, Schmidt M (2014) A-kinase anchoring proteins: cAMP compartmentalization in neurodegenerative and obstructive pulmonary diseases. Br J Pharmacol 171:5603–5623. doi:10.1111/bph.12882

Poulsen JB, Andersen KR, Kjær KH, Vestergaard AL, Justesen J, Martensen PM (2012) Characterization of human phosphodiesterase 12 and identification of a novel 2′-5′ oligoadenylate nuclease—the ectonucleotide pyrophosphatase/phosphodiesterase 1. Biochimie 94:1098–1107. doi:10.1016/j.biochi.2012.01.012

Rehmann H (2006) Characterization of the activation of the Rap-specific exchange factor Epac by cyclic nucleotides. Methods Enzymol 407:159–173.

de Rooij J, Zwartkruis FJT, Verheijen MHG, Cool RH, Nijman SMB, Wittinghofer A, Bos JL (1998) Epac is a Rap1 guanine-nucleotide-exchange factor directly activated by cyclic AMP. Nature 396:474–477. doi:10.1038/24884

Saliba J, Saint-Martin C, Di Stefano A, Lenglet G, Marty C, Keren B, Pasquier F, Valle VD, Secardin L, Leroy G, Mahfoudhi E, Grosjean S, Droin N, Diop M, Dessen P, Charrier S, Palazzo A, Merlevede J, Meniane JC, Delaunay-Darivon C, Fuseau P, Isnard F, Casadevall N, Solary E, Debili N, Bernard OA, Raslova H, Najman A, Vainchenker W, Bellanné-Chantelot C, Plo I (2015) Germline duplication of ATG2B and GSKIP predisposes to familial myeloid malignancies. Nat Genet 47:1131–1140. doi:10.1038/ng.3380

Sarma GN, Kinderman FS, Kim C, von Daake S, Chen L, Wang B-C, Taylor SS (2010) Structure of D-AKAP2:PKA RI complex: insights into AKAP specificity and selectivity. Struct Lond Engl 1993(18):155–166. doi:10.1016/j.str.2009.12.012

Sassi Y, Ahles A, Truong DJ, Baqi Y, Lee SY, Husse B, Hulot JS, Foinquinos A, Thum T, Müller CE, Dendorfer A, Laggerbauer B, Engelhardt S (2014) Cardiac myocyte-secreted cAMP exerts paracrine action via adenosine receptor activation. J Clin Invest 124:5385–5397. doi:10.1172/JCI74349

Schächterle C, Christian F, Fernandes J, Klussmann E (2015) Screening for small molecule disruptors of AKAP–PKA interactions. In: Zaccolo M (ed) cAMP signaling, methods in molecular biology. Springer, New York, pp 151–166

Schäfer G, Milić J, Eldahshan A, Götz F, Zühlke K, Schillinger C, Kreuchwig A, Elkins JM, Abdul Azeez KR, Oder A, Moutty MC, Masada N, Beerbaum M, Schlegel B, Niquet S, Schmieder P, Krause G, von Kries JP, Cooper DMF, Knapp S, Rademann J, Rosenthal W, Klussmann E (2013) Highly functionalized terpyridines as competitive inhibitors of AKAP–PKA interactions. Angew Chem Int Ed 52:12187–12191. doi:10.1002/anie.201304686

Schafmeister CE, Po J, Verdine GL (2000) An all-hydrocarbon cross-linking system for enhancing the Helicity and metabolic stability of peptides. J Am Chem Soc 122:5891–5892. doi:10.1021/ja000563a

Scott JD, Glaccum MB, Zoller MJ, Uhler MD, Helfman DM, McKnight GS, Krebs EG (1987) The molecular cloning of a type II regulatory subunit of the cAMP-dependent protein kinase from rat skeletal muscle and mouse brain. Proc Natl Acad Sci U S A 84:5192–5196

Scott DE, Bayly AR, Abell C, Skidmore J (2016) Small molecules, big targets: drug discovery faces the protein-protein interaction challenge. Nat Rev Drug Discov. doi:10.1038/nrd.2016.29

Shabb JB (2001) Physiological substrates of cAMP-dependent protein kinase. Chem Rev 101:2381–2412. doi:10.1021/cr000236l

Singh M, Singh P, Vaira D, Torheim EA, Rahmouni S, Taskén K, Moutschen M (2014) The RIAD peptidomimetic inhibits HIV-1 replication in humanized NSG mice. Eur J Clin Investig 44:146–152. doi:10.1111/eci.12200

Siu YT, Jin DY (2007) CREB—a real culprit in oncogenesis. FEBS J 274:3224–3232. doi:10.1111/j.1742-4658.2007.05884.x

Skalhegg BS, Tasken K (2000) Specificity in the cAMP/PKA signaling pathway. Differential expression, regulation, and subcellular localization of subunits of PKA. Front Biosci J Virtual Libr 5:D678–D693

Skroblin P, Grossmann S, Schäfer G, Rosenthal W, Klussmann E (2010) Chapter five—mechanisms of protein kinase A anchoring. In: Jeon K (ed) International review of cell and molecular biology. Academic Press, New York, pp 235–330

Smith MC, Gestwicki JE (2012) Features of protein–protein interactions that translate into potent inhibitors: topology, surface area and affinity. Expert Rev Mol Med 14:e16. doi:10.1017/erm.2012.10

Soni S, Scholten A, Vos MA, Veen TAB (2014) Anchored protein kinase A signalling in cardiac cellular electrophysiology. J Cell Mol Med 18:2135–2146. doi:10.1111/jcmm.12365

Sonkusare SK, Dalsgaard T, Bonev AD, Hill-Eubanks DC, Kotlikoff MI, Scott JD, Santana LF, Nelson MT (2014) AKAP150-dependent cooperative TRPV4 channel gating is central to endothelium-dependent vasodilation and is disrupted in hypertension. Sci Signal 7(333):ra66. doi:10.1126/scisignal.2005052

Steegborn C (2014) Structure, mechanism, and regulation of soluble adenylyl cyclases—similarities and differences to transmembrane adenylyl cyclases. Biochim Biophys Acta 1842(12 Pt B): 2535–2547. doi:10.1016/j.bbadis.2014.08.012

Stefan E, Wiesner B, Baillie GS, Mollajew R, Henn V, Lorenz D, Furkert J, Santamaria K, Nedvetsky P, Hundsrucker C, Beyermann M, Krause E, Pohl P, Gall I, MacIntyre AN, Bachmann S, Houslay MD, Rosenthal W, Klussmann E (2007) Compartmentalization of cAMP-dependent signaling by phosphodiesterase-4D is involved in the regulation of vasopressin-mediated water reabsorption in renal principal cells. J Am Soc Nephrol 18:199–212. doi:10.1681/ASN.2006020132

Subrizi A, Tuominen E, Bunker A, Róg T, Antopolsky M, Urtti A (2012) Tat(48–60) peptide amino acid sequence is not unique in its cell penetrating properties and cell-surface glycosaminoglycans inhibit its cellular uptake. J Control Release Off J Control Release Soc 158:277–285. doi:10.1016/j.jconrel.2011.11.007

Sutherland EW, Rall TW (1958) Fractionation and characterization of a cyclic adenine ribonucleotide formed by tissue particles. J Biol Chem 232:1077–1092

Szaszak M, Christian F, Rosenthal W, Klussmann E (2008) Compartmentalized cAMP signalling in regulated exocytic processes in non-neuronal cells. Cell Signal 20:590–601

Taglieri DM, Johnson KR, Burmeister BT, Monasky MM, Spindler MJ, DeSantiago J, Banach K, Conklin BR, Carnegie GK (2014) The C-terminus of the long AKAP13 isoform (AKAP-Lbc) is critical for development of compensatory cardiac hypertrophy. J Mol Cell Cardiol 66:27–40. doi:10.1016/j.yjmcc.2013.10.010

Tavalin SJ (2008) AKAP79 selectively enhances protein kinase C regulation of GluR1 at a Ca^{2+}-calmodulin-dependent protein kinase II/protein kinase C site. J Biol Chem 283:11445–11452. doi:10.1074/jbc.M709253200

Taylor SS, Ilouz R, Zhang P, Kornev AP (2012) Assembly of allosteric macromolecular switches: lessons from PKA. Nat Rev Mol Cell Biol 13:646. doi:10.1038/nrm3432

Taylor SS, Zhang P, Steichen JM, Keshwani MM, Kornev AP (2013) PKA: lessons learned after twenty years. Biochim Biophys Acta 1834:1271. doi:10.1016/j.bbapap.2013.03.007

Terato K, Do CT, Cutler D, Waritani T, Shionoya H (2014) Preventing intense false positive and negative reactions attributed to the principle of ELISA to re-investigate antibody studies in autoimmune diseases. J Immunol Methods 407:15–25. doi:10.1016/j.jim.2014.03.013

Torheim EA, Jarnæss E, Lygren B, Taskén K (2009) Design of proteolytically stable RI-anchoring disruptor peptidomimetics for in vivo studies of anchored type I protein kinase A-mediated signalling. Biochem J 424:69–78. doi:10.1042/BJ20090933

Tsien RY, Bacskai BJ, Adams SR (1993) FRET for studying intracellular signalling. Trends Cell Biol 3:242–245. doi:10.1016/0962-8924(93)90124-J

Vivès E, Brodin P, Lebleu B (1997) A truncated HIV-1 tat protein basic domain rapidly translocates through the plasma membrane and accumulates in the cell nucleus. J Biol Chem 272:16010–16017. doi:10.1074/jbc.272.25.16010

Vukićević T, Schulz M, Faust D, Klussmann E (2016) The trafficking of the water channel aquaporin-2 in renal principal cells-a potential target for pharmacological intervention in cardiovascular diseases. Front Pharmacol 7:23. doi:10.3389/fphar.2016.00023

Walensky LD, Kung AL, Escher I, Malia TJ, Barbuto S, Wright RD, Wagner G, Verdine GL, Korsmeyer SJ (2004) Activation of apoptosis in vivo by a hydrocarbon-stapled BH3 helix. Science 305:1466. doi:10.1126/science.1099191

Walsh DA, Perkins JP, Krebs EG (1968) An adenosine 3′,5′-monophosphate-dependant protein kinase from rabbit skeletal muscle. J Biol Chem 243:3763–3765

Wang Y, Ho TG, Bertinetti D, Neddermann M, Franz E, Mo GCH, Schendowich LP, Sukhu A, Spelts RC, Zhang J, Herberg FW, Kennedy EJ (2014) Isoform-selective disruption of AKAP-localized PKA using hydrocarbon stapled peptides. ACS Chem Biol 9:635–642. doi:10.1021/cb400900r

Wang Y, Ho TG, Franz E, Hermann JS, Smith FD, Hehnly H, Esseltine JL, Hanold LE, Murph MM, Bertinetti D, Scott JD, Herberg FW, Kennedy EJ (2015) PKA-type I selective constrained peptide disruptors of AKAP complexes. ACS Chem Biol 10:1502–1510. doi:10.1021/acschembio.5b00009

Wechsler J, Choi YH, Krall J, Ahmad F, Manganiello VC, Movsesian MA (2002) Isoforms of cyclic nucleotide phosphodiesterase PDE3A in cardiac myocytes. J Biol Chem 277:38072–38078. doi:10.1074/jbc.M203647200

Wehrens XHT, Lehnart SE, Reiken S, Vest JA, Wronska A, Marks AR (2006) Ryanodine receptor/calcium release channel PKA phosphorylation: a critical mediator of heart failure progression. Proc Natl Acad Sci U S A 103:511–518. doi:10.1073/pnas.0510113103

Weiss S, Oz S, Benmocha A, Dascal N (2013) Regulation of cardiac L-type Ca^{2+} channel CaV1.2 via the β-adrenergic-cAMP-protein kinase A pathway: old dogmas, advances, and new uncertainties. Circ Res 113:617–631. doi:10.1161/CIRCRESAHA.113.301781

Wetzel CH, Spehr M, Hatt H (2001) Phosphorylation of voltage-gated ion channels in rat olfactory receptor neurons. Eur J Neurosci 14:1056–1064

White BD, Chien AJ, Dawson DW (2012) Dysregulation of Wnt/β-catenin signaling in gastrointestinal cancers. Gastroenterology 142:219–232. doi:10.1053/j.gastro.2011.12.001

Willoughby D, Cooper DMF (2008) Live-cell imaging of cAMP dynamics. Nat Methods 5:29–36. doi:10.1038/nmeth1135

Wu ZL, Thomas SA, Villacres EC, Xia Z, Simmons ML, Chavkin C, Palmiter RD, Storm DR (1995) Altered behavior and long-term potentiation in type I adenylyl cyclase mutant mice. Proc Natl Acad Sci 92:220–224

Wu J, Brown SHJ, von Daake S, Taylor SS (2007) PKA type IIα holoenzyme reveals a combinatorial strategy for isoform diversity. Science 318:274. doi:10.1126/science.1146447

Zaccolo M, Pozzan T (2002) Discrete microdomains with high concentration of cAMP in stimulated rat neonatal cardiac myocytes. Science 295:1711–1715. doi:10.1126/science.1069982

Zaccolo M, De Giorgi F, Cho CY, Feng L, Knapp T, Negulescu PA, Taylor SS, Tsien RY, Pozzan T (2000) A genetically encoded, fluorescent indicator for cyclic AMP in living cells. Nat Cell Biol 2:25–29. doi:10.1038/71345

Zhang L, Malik S, Kelley GG, Kapiloff MS, Smrcka AV (2011) Phospholipase C epsilon scaffolds to muscle-specific A kinase anchoring protein (mAKAPbeta) and integrates multiple hypertrophic stimuli in cardiac myocytes. J Biol Chem 286:23012–23021. doi:10.1074/jbc.M111.231993

Zhang M, Patriarchi T, Stein IS, Qian H, Matt L, Nguyen M, Xiang YK, Hell JW (2013) Adenylyl cyclase anchoring by a kinase anchor protein AKAP5 (AKAP79/150) is important for postsynaptic β-adrenergic signaling. J Biol Chem 288:17918–17931. doi:10.1074/jbc.M112.449462

Chatting Second Messengers: PIP3 and cAMP

5

Alessandra Ghigo, Flora Pirozzi, Mingchuan Li, and Emilio Hirsch

Contents

Abstract

3′-5′-cyclic adenosine monophosphate (cAMP) and phosphatidylinositol 3,4,5 trisphosphate (PIP3) are pleiotropic second messengers generated in response to activation of G protein-coupled receptors (GPCRs) by a wide array of hormones and neurotransmitters. Although these small molecules engage distinct and seemingly unrelated downstream signal transducers, a growing body of evidence points to a strict cooperation of cAMP and PIP3 cascades in the control of cardiomyocyte functions. Dynamic macromolecular complexes of cAMP and PIP3 molecular switches assemble into spatially and temporally restricted microdomains. Deciphering how these compartmentalized complexes form and

A. Ghigo • M. Li • E. Hirsch (✉)
Department of Molecular Biotechnology and Health Sciences, University of Torino, Via Nizza 52, 10126 Torino, Italy
e-mail: Emilio.hirsch@unito.it

F. Pirozzi
Division of Internal Medicine, Department of Translational Medical Sciences, Federico II University, Naples, Italy

V. Nikolaev, M. Zaccolo (eds.), *Microdomains in the Cardiovascular System*, Cardiac and Vascular Biology 3, DOI 10.1007/978-3-319-54579-0_5

affect the interactions between the two signaling systems is of crucial importance, since both pathways are severely deregulated in major cardiac diseases, such as heart failure. This chapter summarizes recently described mechanisms governing the bidirectional cross talk between cAMP and PIP3 signaling pathways in the pathophysiological control of cardiovascular function. In particular, we will describe how membrane-located PIP3 affects both initiation and termination of cAMP signaling as well as the negative feedback loop whereby the small and diffusible intracellular messenger, cAMP, influences PIP3 production.

5.1 Introduction

The ability of the heart to monitor and respond appropriately to changes in the extracellular environment is governed by the complex interaction between plasma membrane receptors and their cognate signaling pathways. The most prominent regulators of cardiovascular pathophysiology are the receptors that signal via the activation of heterotrimeric G proteins (G protein-coupled receptors, GPCRs) that also represent the largest group of cell surface receptors encoded by the mammalian genome (>1% of human genes).

Receptor engagement by GPCR agonists promotes the interaction with a heterotrimeric G protein, comprising α, β, and γ subunits, the exchange of GTP for GDP on the Gα subunit, and, ultimately, the dissociation between Gα and Gβγ. Free Gα and Gβγ proteins, in turn, trigger distinct effector molecules which initiate the generation of second messengers that, by regulating the degree of phosphorylation of intracellular proteins, eventually drive the appropriate cellular response.

Gα primarily affects the activity of adenylyl cyclases (ACs) that are responsible for the generation of 3′-5′-cyclic adenosine monophosphate (cAMP), a small and diffusible intracellular second messenger. cAMP activates a limited number of intracellular targets including the most common protein kinase A (PKA), but also exchange proteins activated by cAMP (Epac) and cyclic nucleotide-gated channels (CNG). Accordingly, even at the single cell level, the cAMP/PKA signaling cascade controls a bewildering number of cellular functions, ranging from cell growth and differentiation to cell movement and migration and from learning and memory formation to control of hormone secretion, metabolism, and gene transcription (Zaccolo 2009).

On the other hand, the Gβγ branch may simultaneously engage phosphoinositide 3-kinases (PI3Ks) and the ensuing production of a plasma membrane-located second messenger, phosphatidylinositol 3,4,5 trisphosphate (PIP3). This phospholipid, in turn, acts by recruiting downstream effectors that contain specific lipid-binding domains and, in some cases, by activating them in an allosteric fashion. Among these, a well characterized is the protein kinase Akt/PKB, which contains a PIP3-specific binding module, the pleckstrin homology domain (PH domain), and plays major roles in the regulation of metabolism, survival, cell cycle, inflammation, and cardiovascular homeostasis (Salamon and Backer 2013).

Therefore, both cAMP and PIP3 are responsible for the functional response of cardiomyocytes to a wide array of hormones and neurotransmitters, and this also raises the question on how the cell can decode these signals and generate the appropriate functional outputs to distinct extracellular cues. The most accepted model is that both cAMP and PIP3 signaling machineries are organized in distinct spatially and temporally restricted domains (Salamon and Backer 2013; Zaccolo 2009). Despite their compartmentalized nature that guarantees specificity of response, cAMP and PIP3 signaling are tightly interconnected downstream of GPCR activation and cooperate in the pathophysiological control of heart rate, contractility, and hypertrophy. The importance of these connections is highlighted by the evidence that the breaking of these "liaisons" underpins major pathological conditions, including heart failure and arrhythmias.

In this chapter, we will summarize recently elucidated cross-talk mechanisms underlying the reciprocal regulation of cAMP/PKA and PIP3/PI3K signaling pathways in the control of cardiovascular function in health and disease.

5.2 PIP3: A Signaling Lipid on Membranes

PIP3 represents the exclusive product of class I PI3Ks. Depending on the mechanism of activation and the preferred lipid substrate, the PI3K superfamily, including eight different isoenzymes, can be organized into three different classes. Among these, class I PI3Ks are peculiarly activated downstream of membrane receptors, including GPCRs, and preferentially phosphorylate phosphatidylinositol 4,5 bisphosphate (PIP2) on the D3 position of the inositol ring to generate PIP3. These enzymes are obligate heterodimers composed of a p110 catalytic subunit, coupled to a regulatory subunit that favors the interaction of the catalytic module with the activating receptor. Class IA p110α, β, and δ are linked to p85/55 adaptors and are primarily triggered by growth factor receptor tyrosine kinases (RTKs), although recent evidence highlights the ability of p110β to be also activated downstream of GPCRs by direct binding to Gβγ (Dbouk et al. 2012). Conversely, the unique member of class IB PI3K, p110γ, can couple to either p101 or p84/87 regulatory subunits and is exclusively engaged downstream of GPCRs (Vanhaesebroeck et al. 2010).

Despite the fact that all class I PI3Ks produce the same signaling lipid, PIP3, extensive studies have demonstrated that the distinct PI3K isoforms produce a variety of different downstream responses, likely originating from a combination of specific expression patterns, responsiveness to upstream inputs, and sites of action (Salamon and Backer 2013). The best characterized PI3K isoforms in the cardiovascular system are PI3Kα and PI3Kγ, wherein PI3Kα is ubiquitously expressed and enriched in cardiomyocytes, while PI3Kγ functions in both leukocytes and cardiac cells. Consequently, these two enzymes differentially contribute to the maintenance of cardiovascular homeostasis. PI3Kα is primarily implicated in growth factor/RTK-dependent cardiomyocyte postnatal growth and coronary angiogenesis, thereby favoring physiological cardiac hypertrophy as well as cardioprotection in contexts of heart disease (Ghigo and Li 2015). On the other hand, because of its peculiar ability to bind Gβγ subunits, PI3Kγ specifically affects GPCR-mediated responses. In particular, PI3Kγ negatively controls the activity of the predominant subset of GPCRs in cardiomyocytes, namely, β-adrenergic receptors (β-ARs), that are well-established regulators of heart rate, contractility, and hypertrophy. Hence, the PI3Kγ/PIP3 signaling has the potential to directly interfere with the cAMP/PKA cascade. Compelling studies suggest that this cross talk may occur at different levels, including not only receptor-activated cAMP production but also cAMP degradation by specialized hydrolases, named phosphodiesterases (PDEs) (Ghigo et al. 2012; Patrucco et al. 2004; Perino et al. 2011). On the other hand, recent studies demonstrate the existence of a feedback loop whereby cAMP may affect PI3K activity and the ensuing PIP3 production (Perino et al. 2011).

5.3 Mechanisms of Cross-Talk between PIP3 and cAMP Signaling

In the following sections, we will describe the molecular mechanisms underlying the reciprocal regulation of cAMP/PKA and PI3K/PIP3 signaling pathways in the control of cardiovascular function, in health and disease. In particular, we will focus on the subcellular compartments involved and we will describe how PIP3 controls cAMP signaling activation via β-ARs at the plasma membrane (Sect. 5.3.1) or cAMP signaling termination via intracellular PDEs (Sect. 5.3.2) and, finally, how cAMP controls PI3K/PIP3 signaling (Sect. 5.3.3).

5.3.1 PIP3-Mediated Downregulation of β-AR/cAMP Signaling in Heart Failure

A major site for the cross talk between cAMP and PIP3 is the plasma membrane where PIP3 determines the abundance of GPCRs that initiate cAMP production. Compelling studies highlight a key role for PIP3 in the downregulation of both GPCRs and RTKs, which stems from the ability of this signaling lipid to promote clathrin-mediated endocytosis of membrane receptors (Laketa et al. 2014; Perino et al. 2011). In the heart, PIP3 primarily contributes to the pathological decrease of β-AR density during the natural history of heart failure, through the recruitment of PH domain-containing proteins, such as the clathrin adaptor protein AP-2, required for the assembly of β-AR downregulation machinery (Naga Prasad et al. 2001). Class IB PI3Kγ is likely the major responsible for the generation of this pool of PIP3 (Fig. 5.1). Perino and colleagues demonstrate that β-AR downregulation is completely prevented in mice expressing a knock-in kinase-inactive PI3Kγ (PI3Kγ kinase-dead, PI3Kγ KD), but not in PI3Kβ KD animals, in a model of pressure overload-induced heart failure. The involvement of a PI3Kγ-dependent pool of PIP3 in this process is further corroborated by the finding that PIP3 levels are significantly upregulated in pressure-overloaded wild-type, but not PI3Kγ KD, hearts. Intriguingly, pharmacological inhibition of PI3Kγ with AS-605240 recapitulates the effects of genetic PI3Kγ blockade. β-AR density is fully recovered in pressure-overloaded mice treated with the compound, and this increase in receptor density is accompanied by a significant rescue of systolic function (Perino et al. 2011).

PI3Kγ further contributes to β-AR downregulation at the plasma membrane by interacting with β-AR kinases, also known as G protein-coupled receptor kinases

(GRKs), which phosphorylate agonist-occupied receptors and promote their desensitization and internalization. This process involves the recruitment of β-arrestin to the phosphorylated receptor which, in turn, prevents the coupling with the G protein and concomitantly promotes the internalization via clathrin-coated vesicles (Madamanchi 2007; Rockman et al. 2002). PI3Kγ participates to this process by directly interacting with β-AR kinase 1 (βARK1). This was first uncovered by Perrino and coworkers showing that overexpression of an inactive form of PI3K reduces βARK1 activity and restores β-AR density as well as AC activity (Perrino et al. 2005). Furthermore, βARK1-associated PI3K activity is significantly upregulated in a mouse model of heart failure induced by a mutation of calcium-binding protein calsequestrin (CSQ). In these animals, the displacement of endogenous PI3K from βARK1 preserves β-AR density, ameliorates cardiac dysfunction, and improves survival. This regulation is likely peculiar of PI3Kγ and does not involve the other major cardiac isoform PI3Kα, since PI3Kγ activity is exclusively increased in a mouse model of heart failure and significantly reduced by overexpression of a PI3K-inactive transgene (Perrino et al. 2005). In human patients, a significant recovery of plasma membrane β-AR density occurs after implantation of left ventricular assist device (LVAD) and correlates with improved clinical conditions when the loading of the left ventricle is reduced. Similar to the preclinical model, the activity of the PI3Kγ/βARK1 complex is increased in heart failure patients before the implantation of LVAD, but is decreased to control levels after mechanical unloading (Perrino et al. 2007).

Besides these PIP3-related regulations, PI3Kγ may control β-AR density via mechanisms that involve its protein kinase rather than lipid kinase function. A cellular substrate for PI3Kγ protein kinase activity in cardiomyocytes is the cytoskeletal protein non-muscle tropomyosin, which is involved in the rearrangement of the actin cytoskeleton during endocytosis. Naga Prasad and colleagues show that PI3Kγ-mediated phosphorylation on Ser-61 of tropomyosin is an essential step for agonist-dependent β-AR internalization (Naga Prasad et al. 2005). Furthermore, the protein kinase activity of PI3Kγ can contribute to the resensitization phase of β-AR recycling, a process that is regulated by protein phosphatase 2A (PP2A), an enzyme that dephosphorylates the receptor in early endosomes. Vasudevan et al. show that β-AR-associated phosphatase activity is significantly increased in transgenic mice overexpressing an inactive PI3Kγ, leading to receptor dephosphorylation and resensitization and, ultimately, preserved cardiac function. This protection relies on the ability of PI3Kγ to inhibit PP2A at the β-AR complex (Fig. 5.1). In particular, PI3Kγ phosphorylates an intracellular inhibitor of PP2A (I2PP2A) on serine residues 9 and 93, resulting in enhanced binding to PP2A (Vasudevan et al. 2011).

Altogether, compelling studies identify a central role for both the lipid and the protein kinase activities of PI3Kγ in the downregulation of β-AR/cAMP signaling and envisage the possibility of manipulating this cross talk pharmacologically in heart failure treatment. However, a recent study suggests that PI3K lipid kinase activity, and the ensuing PIP3 generation, may affect cAMP signaling initiation not only at receptor level but also through the modulation of major cAMP-generating enzymes, such as AC5 and AC6, in an Akt-independent manner (Reddy et al. 2015). Future studies are expected to elucidate the mechanisms underlying this further cross talk between PI3K/PIP3 and cAMP/PKA signaling.

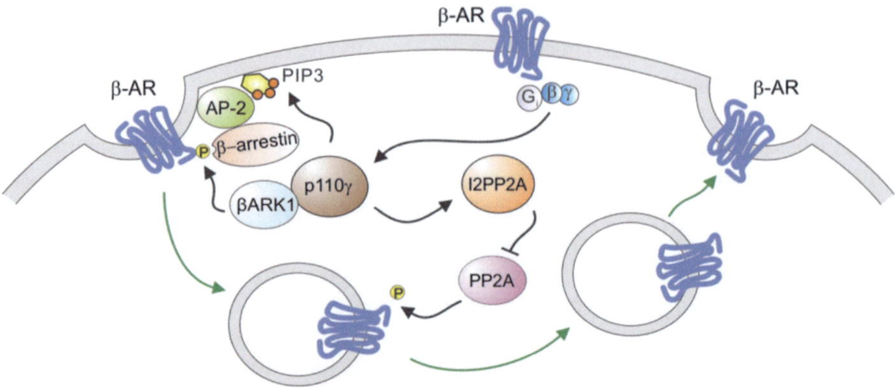

Fig. 5.1 PI3K/PIP3 mediates downregulation of β-AR/cAMP signaling in heart failure. PI3Kγ can contribute to the internalization of plasma membrane β-ARs by directly interacting with βARK1, which in turn phosphorylates the receptor and promotes the recruitment of β-arrestin. PI3Kγ-dependent production of PIP3 further contributes to β-AR downregulation by favoring the recruitment of the clathrin adaptor protein AP-2 and the ensuing assembly of the β-AR downregulation machinery at the plasma membrane. Finally, PI3Kγ controls β-AR resensitization via PP2A, an enzyme responsible for the dephosphorylation of the receptor in early endosomes. Notably, this process is unrelated to PI3Kγ lipid kinase activity and depends on the ability of the protein to phosphorylate the intracellular inhibitor of PP2A (I2PP2A) that ultimately inhibits PP2A activity and dampens β-AR resensitization. Abbreviations: *βARK1* β-AR kinase 1, *β-AR* β-adrenergic receptor, *PP2A* protein phosphatase 2A, *I2PP2A* intracellular inhibitor of PP2A

5.3.2 PIP3- and PI3K-Dependent Control of cAMP Hydrolysis

Besides affecting β-AR signaling and the ensuing cAMP production at the plasma membrane, PIP3 is also critically implicated in cAMP signaling termination, a process involving the action of intracellular PDEs (Fig. 5.2). In the heart, 90% of the total cAMP-hydrolyzing activity is provided by PDE3 and PDE4 isoforms (Mika et al. 2012). The catalytic function of these enzymes is finely tuned by phosphorylation events, occurring within the N-terminal regulatory regions and involving major kinases, including also the cAMP-dependent PKA and the major PIP3 effector, PKB (Omori and Kotera 2007).

PKA-mediated phosphorylation represents the major mechanism of PDE activation and results in a two- to three-fold increase of the catalytic activity of both PDE3 and PDE4. Furthermore, the PIP3-dependent kinase PKB can cooperate with PKA and further activates PDE3 (Omori and Kotera 2007). The first evidence of PDE phosphorylation by PKB traces back to 1999, when Kitamura et al. found that, in adipocytes, insulin treatment or adenovirus-mediated expression of a constitutively active p110α (myr-p110α) triggers PDE3B phosphorylation on Ser-273 and the ensuing catalytic activation (Kitamura et al. 1999). Subcellular fractionation studies reveal that insulin-dependent phosphorylation and activation of PDE3B mainly occurs within ER/Golgi membrane compartments and relies on the formation of macromolecular complexes containing PDE3B, together with other insulin-related

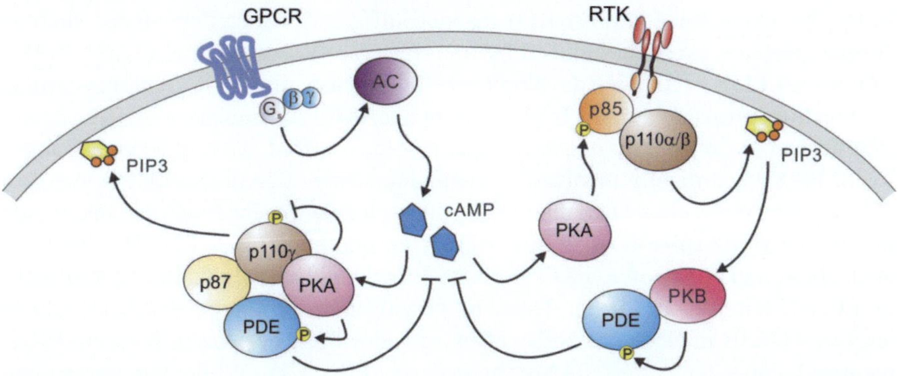

Fig. 5.2 cAMP/PIP3 cross talk affects cAMP signaling termination via PDEs. In cardiomyocytes, PI3Kγ serves as an AKAP that anchors cAMP-degrading enzymes, PDEs, in close proximity of their activator PKA. This ultimately favors PKA-mediated phosphorylation and activation of PDEs and compartmentalized cAMP-PKA responses. Furthermore, the close proximity of p110γ and PKA within the same macromolecular complex allows PKA-dependent phosphorylation of p110γ and dampens PI3Kγ lipid kinase activity and the ensuing PIP3 production. On the other hand, PKA phosphorylates p85, the regulatory subunit of PI3Kα/β, and synergizes with insulin receptor-dependent activation of PIP3/Akt signaling. As a feedback mechanism, Akt can directly bind and phosphorylate PDEs, which in turn attenuate cAMP/PKA activity. Abbreviations: *AKAP* A-kinase anchoring protein, *PDE* phosphodiesterase, *PKA* protein kinase A

molecules, including p85 and PKB. Under insulin stimulation, PDE3B and PKB interact directly, through the PH domain of PKB and N-terminal regulatory region of PDE3B, and this spatial vicinity ultimately favors efficient PDE3B activation by insulin/PI3K/PKB signaling (Ahmad et al. 2007). Additional evidence shows that the other PDE3 isoform, PDE3A, is similarly activated by PKB, via phosphorylation of Ser-290-292 (Han et al. 2006). Although PKB-dependent activation has long been considered a unique feature of PDE3 isoforms, a recent study suggests the existence of consensus sequences for PKB phosphorylation, such as RxRxxS and RRxS, on other PDE isoenzymes (Fang et al. 2015). In particular, a PKB phosphorylation site has been identified on Ser-190 of PDE4D, the major PDE subtype in airway smooth muscle (ASM) cells. Accordingly, inhibition of the PKB activator PDK1 with arctigenin prevents PDE4D activation and synergizes with β2-AR agonists in increasing intracellular cAMP levels and enhancing tracheal muscle relaxation (Fang et al. 2015). Overall, the signaling lipid PIP3 can directly affect the cAMP/PKA cascade by reducing intracellular cAMP.

In addition to PIP3/Akt-mediated control of PDE3, PI3K(γ) can indirectly activate PDE3 and PDE4 through a PIP3-unrelated mechanism. This process does not require an active kinase domain and only relies on the ability of the enzyme to serve as an A-kinase anchoring protein (AKAP), a family of scaffold proteins that allow spatial and temporal compartmentalization of cAMP signaling components within macromolecular complexes (Perino et al. 2012). The finding that PI3Kγ has a biological activity not directly related to PIP3 production first came from the observation that animals expressing a knock-in kinase-inactive PI3Kγ (PI3Kγ KD) and

PI3Kγ knockout mice (PI3Kγ KO) respond differently to cardiac stress, such as chronic pressure overload induced by transverse aortic constriction (TAC). PI3Kγ KO but not PI3Kγ KD mice develop heart failure early after TAC, and this cardiac dysfunction primarily stems from abnormal cAMP accumulation in KO hearts (Patrucco et al. 2004). These results imply that the scaffold, but not the kinase activity, of PI3Kγ is critically involved in the maintenance of cardiac cAMP homeostasis. Further studies elucidated the underlying molecular mechanisms and uncovered that the catalytic subunit of PI3Kγ, p110γ, on one hand binds PDE3B (Patrucco et al. 2004) and on the other hand interacts directly with the regulatory subunit RIIα of PKA (Perino et al. 2011). Thus, by serving as a prototypical AKAP, p110γ anchors PDE3B in close proximity of its activator PKA eventually favoring PKA-mediated enhancement of cAMP hydrolysis (Fig. 5.2). While the association between p110γ and PKA is direct, the binding with PDE3B is likely indirect and mediated by the PI3Kγ regulatory subunit p84/87, as p84/87, but not p101, co-purifies with PDE3B (Voigt et al. 2006) and PKA (Perino et al. 2011).

Notably, PI3Kγ-dependent anchoring of PKA represents a generalized mechanism of PDE activation as PI3Kγ also orchestrates the activity of other PDEs, including those with a major impact on cardiac function such as PDE4A, PDE4B, and PDE3A. These regulations occur in spatially defined compartments as PI3Kγ-dependent PDE4 primarily limits subsarcolemmal cAMP, while PDE3 is likely controlling a pool of cAMP at the sarcoplasmic reticulum. This compartmentalized signaling eventually ensures localized activation of PKA targets and homeostatic Ca^{2+} regulation. Accordingly, PI3Kγ deficiency results in hyperphosphorylation of L-type Ca^{2+} channels (LTCC) and phospholamban (PLN), but not of ryanodine receptors (RyR), upon β_2-AR activation. The physiological importance of this mechanism is further demonstrated by the finding that disruption of PI3Kγ scaffolding function exacerbates catecholamine-induced ventricular arrhythmia by amplifying cAMP/PKA-dependent Ca^{2+} spark occurrence and amplitude (Ghigo et al. 2012). PI3Kγ scaffolding plays a crucial role also in other pathological contexts, such as in sepsis-induced myocardial depression (SIMD), an early and frequent event of infection-induced systemic inflammatory response syndrome (SIRS). In a model of LPS-induced SIRS, PI3Kγ KO mice display early myocardial hyper-contractility, followed by delayed myocardial depression. This was paralleled by enhanced accumulation of intracellular cAMP, myocardial Ca^{2+} trafficking and upregulation of iNOS in LPS-treated PI3Kγ KO, but not in WT and PI3Kγ KD (Ndongson-Dongmo et al. 2015).

In summary, PI3Kγ coordinates PIP3-dependent and PIP3-independent signals converging on cAMP/PKA signaling termination in defined subcellular compartments.

5.3.3 cAMP-Mediated Regulation of PIP3 Signaling

Another aspect of the communication between PIP3 and cAMP is cAMP-mediated contro of PI3K/PIP3 signaling. This kind of regulation primarily occurs at sites of p110γ/PKA interaction. Besides favoring PKA-mediated activation of PDEs and the ensuing termination of cAMP signaling, the close proximity of p110γ and PKA within the same macromolecular complex allows PKA-dependent phosphorylation of p110γ on Thr-1024. This is an inhibitory phosphorylation that dampens PI3Kγ lipid kinase activity, and the ensuing PIP3 production, thus ensuring that p110γ activity is almost negligible in physiological conditions (Fig. 5.2) (Perino et al. 2011). Conversely, in pathological contexts, such as pressure overload-induced heart failure, p110γ and its adaptor subunit p101, but not p84/87, are upregulated. This in turn results in disruption of p84/87-dependent p110γ/PKA complexes and relieves p110γ inhibition by PKA. Accordingly, PI3Kγ-mediated PIP3 production raises and further exacerbates the β-AR desensitization and downregulation occurring in failing hearts (Perino et al. 2011).

Notably, cAMP can affect PI3Kγ activity not only via PKA but also through the other cAMP effector, EPAC1. In human arterial endothelial cells (HAECs), p110γ is recruited by its adaptor subunit p84/87 to a membrane-associated macromolecular complex, including PDE3B and EPAC1, where EPAC1 activates PI3Kγ. Peptide-based disruption of the interaction between PDE3B and EPAC1 increases cAMP binding to EPAC1, promotes ERK and Akt phosphorylation, and ultimately enhances HAEC adhesion. Intriguingly, this effect is prevented by the PI3Kγ pharmacological inhibitor AS-604850, thus indicating that PDE3B-mediated control of EPAC1 mediates cAMP-dependent regulation of PI3Kγ. Nevertheless, further investigation is required to clarify the molecular details of this regulation (Wilson et al. 2011).

In addition to PI3Kγ, other PI3K isoenzymes can be modulated by cAMP. In cardiomyocytes, β-AR agonists, including isoprenaline, salbutamol, and dobutamine, potentiate insulin-stimulated Akt activation, while they do not affect Akt phosphorylation in the absence of insulin, implying the existence of a cross talk between β-AR and insulin-triggered PI3K signaling cascades (Fig. 5.2). The finding that cAMP analogues (dibutyryl-cAMP and N6-benzoyl-cAMP) similarly increase insulin-stimulated Akt activation further corroborates a model wherein β-ARs cross talk with insulin/PI3K signaling through cAMP/PKA (Stuenaes et al. 2010).

Nevertheless, how cAMP/PKA regulate Akt is still undefined and requires further investigation. A clue comes from the finding that PKA phosphorylates p85α, one of the regulatory subunits of class IA PI3Ks, on Ser-83, in vascular smooth muscle cells (VSMCs) and endothelial cells (ECs). The phosphorylation of p85α in turn affects two different signaling pathways. On one hand, p85α binds $p21^{ras}$, reduces ERK1/2 activation, and inhibits VSMC proliferation, while on the other hand, p85α activates Akt signaling and reduces VSMC and EC apoptosis. Because VSMC proliferation is the major cause of maladaptive neointimal hyperplasia after arterial injury, local delivery of a plasmid encoding a phosphomimetic mutant p85α reduces restenosis in an experimental balloon injury model, by preventing VSMC proliferation and ensuring EC survival in rats (Torella et al. 2009).

In summary, cAMP can affect PIP3 signaling mediated by different PI3K isoforms.

Conclusion

Deciphering the molecular mechanisms that underpin the cross talk between cAMP/PKA and PI3K/PIP3 signaling appears of crucial importance as it may have severe clinical implications. Alterations in both signaling systems are a key trait of heart failure, and drugs targeting major enzymes of these cascades have been developed. β-AR downregulation, along with a decrease in the functional coupling of the remaining receptors with the downstream signaling effectors, represents the first biochemical event in the natural history of heart failure (Feldman et al. 2005). Intriguingly, the concomitant upregulation of PI3Kγ/PIP3 cascade further contributes to the derangement of the β-AR/cAMP/PKA axis, by impinging on both β-AR density and PDE-dependent compartmentalization of cAMP/PKA responses (Ghigo et al. 2012; Perino et al. 2011). Pharmacological blockade of β-ARs, although counterintuitive, is beneficial in patients with left ventricular dysfunction and heart failure, thus representing the current standard heart failure pharmacotherapy. Conversely, PI3K inhibitors are becoming clinically available, primarily as anticancer and anti-inflammatory drugs, but their cardiovascular use, both alone or in combination with conventional β-blockers, is still unexplored and requires further investigation.

Compliance with Ethical Standards

Conflict of Interest Statement Emilio Hirsch is cofounder of Kither Biotech, a company involved in the development of PI3K inhibitors. The other authors declare no conflict of interest.

References

Ahmad F, Lindh R, Tang Y, Weston M, Degerman E, Manganiello VC (2007) Insulin-induced formation of macromolecular complexes involved in activation of cyclic nucleotide phosphodiesterase 3B (PDE3B) and its interaction with PKB. Biochem J 404(2):257–268

Dbouk HA, Vadas O, Shymanets A, Burke JE, Salamon RS, Khalil BD, Barrett MO, Waldo GL, Surve C, Hsueh C, Perisic O, Harteneck C, Shepherd PR, Harden TK, Smrcka AV, Taussig R, Bresnick AR, Nurnberg B, Williams RL, Backer JM (2012) G protein-coupled receptor-mediated activation of p110beta by Gbetagamma is required for cellular transformation and invasiveness. Sci Signal 5(253):ra89

Fang R, Cui Q, Sun J, Duan X, Ma X, Wang W, Cheng B, Liu Y, Hou Y, Bai G (2015) PDK1/Akt/PDE4D axis identified as a target for asthma remedy synergistic with beta2 AR agonists by a natural agent arctigenin. Allergy 70(12):1622–1632

Feldman DS, Carnes CA, Abraham WT, Bristow MR (2005) Mechanisms of disease: beta-adrenergic receptors--alterations in signal transduction and pharmacogenomics in heart failure. Nat Clin Pract Cardiovasc Med 2(9):475–483

Ghigo A, Li M (2015) Phosphoinositide 3-kinase: friend and foe in cardiovascular disease. Front Pharmacol 6:169

Ghigo A, Perino A, Mehel H, Zahradnikova A Jr, Morello F, Leroy J, Nikolaev VO, Damilano F, Cimino J, De Luca E, Richter W, Westenbroek R, Catterall WA, Zhang J, Yan C, Conti M, Gomez AM, Vandecasteele G, Hirsch E, Fischmeister R (2012) Phosphoinositide 3-kinase gamma protects against catecholamine-induced ventricular arrhythmia through protein kinase A-mediated regulation of distinct phosphodiesterases. Circulation 126(17):2073–2083

Han SJ, Vaccari S, Nedachi T, Andersen CB, Kovacina KS, Roth RA, Conti M (2006) Protein kinase B/Akt phosphorylation of PDE3A and its role in mammalian oocyte maturation. EMBO J 25(24):5716–5725

Kitamura T, Kitamura Y, Kuroda S, Hino Y, Ando M, Kotani K, Konishi H, Matsuzaki H, Kikkawa U, Ogawa W, Kasuga M (1999) Insulin-induced phosphorylation and activation of cyclic nucleotide phosphodiesterase 3B by the serine-threonine kinase Akt. Mol Cell Biol 19(9): 6286–6296

Laketa V, Zarbakhsh S, Traynor-Kaplan A, Macnamara A, Subramanian D, Putyrski M, Mueller R, Nadler A, Mentel M, Saez-Rodriguez J, Pepperkok R, Schultz C (2014) PIP(3) induces the recycling of receptor tyrosine kinases. Sci Signal 7(308):ra5

Madamanchi A (2007) Beta-adrenergic receptor signaling in cardiac function and heart failure. Mcgill J Med 10(2):99–104

Mika D, Leroy J, Vandecasteele G, Fischmeister R (2012) PDEs create local domains of cAMP signaling. J Mol Cell Cardiol 52(2):323–329
Naga Prasad SV, Barak LS, Rapacciuolo A, Caron MG, Rockman HA (2001) Agonist-dependent recruitment of phosphoinositide 3-kinase to the membrane by beta-adrenergic receptor kinase 1. A role in receptor sequestration. J Biol Chem 276(22):18953–18959
Naga Prasad SV, Jayatilleke A, Madamanchi A, Rockman HA (2005) Protein kinase activity of phosphoinositide 3-kinase regulates β-adrenergic receptor endocytosis. Nat Cell Biol. 7(8):785–796
Ndongson-Dongmo B, Heller R, Hoyer D, Brodhun M, Bauer M, Winning J, Hirsch E, Wetzker R, Schlattmann P, Bauer R (2015) Phosphoinositide 3-kinase gamma controls inflammation-induced myocardial depression via sequential cAMP and iNOS signalling. Cardiovasc Res 108(2):243–253
Omori K, Kotera J (2007) Overview of PDEs and their regulation. Circ Res 100(3):309–327
Patrucco E, Notte A, Barberis L, Selvetella G, Maffei A, Brancaccio M, Marengo S, Russo G, Azzolino O, Rybalkin SD, Silengo L, Altruda F, Wetzker R, Wymann MP, Lembo G, Hirsch E (2004) PI3Kgamma modulates the cardiac response to chronic pressure overload by distinct kinase-dependent and -independent effects. Cell 118(3):375–387
Perino A, Ghigo A, Ferrero E, Morello F, Santulli G, Baillie GS, Damilano F, Dunlop AJ, Pawson C, Walser R, Levi R, Altruda F, Silengo L, Langeberg LK, Neubauer G, Heymans S, Lembo G, Wymann MP, Wetzker R, Houslay MD, Iaccarino G, Scott JD, Hirsch E (2011) Integrating cardiac PIP3 and cAMP signaling through a PKA anchoring function of p110gamma. Mol Cell 42(1):84–95
Perino A, Ghigo A, Scott JD, Hirsch E (2012) Anchoring proteins as regulators of signaling pathways. Circ Res 111(4):482–492
Perrino C, Naga Prasad SV, Patel M, Wolf MJ, Rockman HA (2005) Targeted inhibition of beta-adrenergic receptor kinase-1-associated phosphoinositide-3 kinase activity preserves beta-adrenergic receptor signaling and prolongs survival in heart failure induced by calsequestrin overexpression. J Am Coll Cardiol 45(11):1862–1870
Perrino C, Schroder JN, Lima B, Villamizar N, Nienaber JJ, Milano CA, Naga Prasad SV (2007) Dynamic regulation of phosphoinositide 3-kinase-gamma activity and beta-adrenergic receptor trafficking in end-stage human heart failure. Circulation 116(22):2571–2579
Reddy GR, Subramanian H, Birk A, Milde M, Nikolaev VO, Bunemann M (2015) Adenylyl cyclases 5 and 6 underlie PIP3-dependent regulation. FASEB J 29(8):3458–3471
Rockman HA, Koch WJ, Lefkowitz RJ (2002) Seven-transmembrane-spanning receptors and heart function. Nature 415(6868):206–212
Salamon RS, Backer JM (2013) Phosphatidylinositol-3,4,5-trisphosphate: tool of choice for class I PI 3-kinases. BioEssays 35(7):602–611
Stuenaes JT, Bolling A, Ingvaldsen A, Rommundstad C, Sudar E, Lin FC, Lai YC, Jensen J (2010) Beta-adrenoceptor stimulation potentiates insulin-stimulated PKB phosphorylation in rat cardiomyocytes via cAMP and PKA. Br J Pharmacol 160(1):116–129
Torella D, Gasparri C, Ellison GM, Curcio A, Leone A, Vicinanza C, Galuppo V, Mendicino I, Sacco W, Aquila I, Surace FC, Luposella M, Stillo G, Agosti V, Cosentino C, Avvedimento EV, Indolfi C (2009) Differential regulation of vascular smooth muscle and endothelial cell proliferation in vitro and in vivo by cAMP/PKA-activated p85alphaPI3K. Am J Physiol Heart Circ Physiol 297(6):H2015–H2025
Vanhaesebroeck B, Guillermet-Guibert J, Graupera M, Bilanges B (2010) The emerging mechanisms of isoform-specific PI3K signalling. Nat Rev Mol Cell Biol 11(5):329–341
Vasudevan NT, Mohan ML, Gupta MK, Hussain AK, Naga Prasad SV (2011) Inhibition of protein phosphatase 2A activity by PI3Kgamma regulates beta-adrenergic receptor function. Mol Cell 41(6):636–648
Voigt P, Dorner MB, Schaefer M (2006) Characterization of p87PIKAP, a novel regulatory subunit of phosphoinositide 3-kinase gamma that is highly expressed in heart and interacts with PDE3B. J Biol Chem 281(15):9977–9986
Wilson LS, Baillie GS, Pritchard LM, Umana B, Terrin A, Zaccolo M, Houslay MD, Maurice DH (2011) A phosphodiesterase 3B-based signaling complex integrates exchange protein activated by cAMP 1 and phosphatidylinositol 3-kinase signals in human arterial endothelial cells. J Biol Chem 286(18):16285–16296
Zaccolo M (2009) cAMP signal transduction in the heart: understanding spatial control for the development of novel therapeutic strategies. Br J Pharmacol 158(1):50–60

Cyclic Nucleotide Phosphodiesterases and Compartmentation in Normal and Diseased Heart

6

Ibrahim Bedioune, Pierre Bobin, Jérôme Leroy, Rodolphe Fischmeister, and Grégoire Vandecasteele

Contents

Abstract

Cyclic nucleotide phosphodiesterases (PDEs) degrade the second messengers cAMP and cGMP, thereby regulating multiple aspects of cardiac function. This highly diverse class of enzymes encoded by 21 genes encompasses 11 families which are not only responsible for the termination of cyclic nucleotide signalling, but are also involved in the generation of dynamic microdomains of cAMP and cGMP controlling specific cell functions in response to various neurohormonal stimuli. In myocardium, the PDE3 and PDE4 families are predominant to degrade cAMP and thereby regulate cardiac excitation-contraction coupling. PDE3 inhibitors are positive inotropes and vasodilators in human, but their use is limited to acute heart failure and intermittent claudication. PDE5 is particularly important to degrade cGMP in vascular smooth muscle, and PDE5 inhibitors are used to treat erectile dysfunction and pulmonary hypertension. However, these drugs do not seem efficient in heart failure with preserved ejection fraction.

I. Bedioune • P. Bobin • J. Leroy • R. Fischmeister (✉) • G. Vandecasteele (✉)
Faculté de Pharmacie, Inserm, UMR-S 1180, Univ. Paris-Sud, Université Paris-Saclay, F-92296 Châtenay-Malabry, France
e-mail: rodolphe.fischmeister@inserm.fr; gregoire.vandecasteele@u-psud.fr

V. Nikolaev, M. Zaccolo (eds.), *Microdomains in the Cardiovascular System*, Cardiac and Vascular Biology 3, DOI 10.1007/978-3-319-54579-0_6

There is experimental evidence that these PDEs as well as other PDE families including PDE1, PDE2 and PDE9 may play important roles in cardiac diseases such as hypertrophy and heart failure. After a brief presentation of the cyclic nucleotide pathways in cardiac cells and the major characteristics of the PDE superfamily, this chapter will present their role in cyclic nucleotide compartmentation and the current use of PDE inhibitors in cardiac diseases together with the recent research progresses that could lead to a better exploitation of the therapeutic potential of these enzymes in the future.

Abbreviations

AC	Adenylyl cyclases
AKAP	A-kinase anchoring protein
ANP	Atrial natriuretic peptide
BNP	Brain natriuretic peptide
CaM	Calmodulin
CaMKII	Ca^{2+}/calmodulin-dependent kinase II
cAMP	Cyclic adenosine monophosphate
cGMP	Cyclic guanosine monophosphate
CN	Cyclic nucleotides
CNP	C-type natriuretic peptide
ECC	Excitation-contraction coupling
Epac	Exchange protein directly activated by cAMP
ERK	Extracellular signal-regulated kinase
FRET	Förster resonance energy transfer
GAF	cGMP-stimulated phosphodiesterases, Anabaena adenylyl cyclases, Fhla transcription factor
GC	Guanylyl cyclase
HF	Heart failure
ICER	Inducible-cAMP early repressor
I/R	Ischemia/reperfusion
KO	Knockout
LTCC	L-type Ca^{2+} channels
mAKAP	Muscle AKAP
NO	Nitric oxide
NOS	NO synthase
PDE	Cyclic nucleotide phosphodiesterase
pGC	Particulate guanylyl cyclase
PGE	Prostaglandin
PI3Kγ	Phosphoinositide 3-kinase, γ isoform
PKA	cAMP-dependent protein kinase
PKG	cGMP-dependent protein kinase

PLB	Phospholamban
RyR2	Ryanodine receptor type 2
SERCA	Sarco-endoplasmic reticulum Ca^{2+}-ATPase
sGC	Soluble guanylyl cyclase
SR	Sarcoplasmic reticulum
TnI	Troponin I
β-ARs	β-adrenergic receptors

6.1 Introduction

The cyclic nucleotides (CN) cAMP and cGMP participate in the main regulations of cardiac function. They act as second messengers for sympathetic and parasympathetic systems, nitric oxide (NO) and natriuretic peptides. CN may exert beneficial or deleterious effects on the heart, depending on the strength and duration of the stimulation. Acute elevation of CN regulates cardiac excitation-contraction coupling (ECC). However, chronic elevation of cAMP contributes to the development of cardiac hypertrophy and progression to heart failure (HF), while cGMP possesses anti-hypertrophic properties. The amplitude, duration and localization of CN responses are determined by the balance between synthesis of cAMP and cGMP by adenylyl and guanylyl cyclases, respectively, and degradation by cyclic nucleotide phosphodiesterases (PDEs). PDEs represent the main route to rapidly lower CN levels inside the cells and constitute a highly diverse superfamily of enzymes. Different enzymatic properties and localization of multiple PDE isoforms within the cell participate in CN compartmentation, which is critical to determine specific physiological responses (Conti et al. 2014; Steinberg and Brunton 2001). In addition, modification in the expression and activity of specific PDEs are observed in several cardiovascular diseases. Thus, the members of the PDE superfamily are well placed to be the targets for pharmacological interventions in cardiovascular diseases. This is actually the case for a few of them, with PDE inhibitors being approved for the treatment of acute heart failure, erectile dysfunction, pulmonary hypertension and intermittent claudication. In the following, we will present an overview of the roles of PDEs in cardiac muscle, the current indication of PDE inhibitors in heart diseases and the recent research advances holding promises for future therapeutic developments in cardiovascular diseases.

6.2 CN Signalling in Cardiac Myocytes

In response to activation of G_s-coupled receptors, cAMP is produced by transmembrane adenylyl cyclases, which constitute the main source of cAMP in cardiac cells. Two types of guanylyl cyclases (GC) produce cGMP, the soluble GC (sGC) which is activated by NO, and particulate GCs (pGC) which constitute the receptors for natriuretic peptides (ANP, BNP and CNP). Once synthesized, CN exert their effects by acting through a limited number of cellular effectors: for cAMP, these include the cAMP-dependent protein kinase (PKA), the cyclic nucleotide-gated (CNG) ion channels, the exchange proteins directly activated by cAMP (Epac) and the recently discovered Popeye domain-containing proteins (Schindler and Brand 2016). For cGMP, the main effector in heart is the cGMP-dependent protein kinase (PKG). Both nucleotides also bind directly to PDEs, not only at the catalytic site but also at allosteric sites, thereby modulating their activity.

During the fight or flight response, epinephrine and norepinephrine bind to β-adrenoceptors (β-ARs) in cardiomyocytes, leading to cAMP elevation and PKA activation. PKA phosphorylation of sarcolemmal L-type Ca^{2+} channels (LTCC), ryanodine receptors type 2 (RyR2), phospholamban (PLB, which controls the activity of the Ca^{2+}-ATPase from the sarcoplasmic reticulum, SERCA2) and troponin I (TnI) enhances the amplitude and kinetics of Ca^{2+} transients in cardiomyocytes (Fig. 6.1), underlying the classical positive inotropic and lusitropic effects of acute sympathetic stimulation. However, sustained stimulation of β-ARs, as what occurs during hypertension or chronic heart diseases, is detrimental to the heart as it favours maladaptive hypertrophic remodelling, apoptosis and arrhythmias. Along with PKA, Epac is activated by cAMP and may play an important role in this context. Epac activation triggers a signalling pathway involving the phosphatase calcineurin and Ca^{2+}/calmodulin-dependent kinase II (CaMKII) to stimulate hypertrophic growth (Lezoualc'h et al. 2016). CaMKII activation, which can also result from PKA-dependent increases in Ca^{2+}, also phosphorylates RyR2 and promotes a pro-arrhythmogenic sarcoplasmic reticulum (SR) Ca^{2+} leak which may ultimately lead to chamber dilatation and HF (Ruiz-Hurtado et al. 2012) (Fig. 6.1).

In the heart, cGMP is often viewed as the mirror of cAMP, opposing its effects on cardiac function. Indeed, cGMP can exert negative inotropic effects via PKG-mediated inhibition of the L-type Ca^{2+} current (Méry et al. 1991; Yang et al. 2007) and phosphorylation of TnI to decrease myofilament sensitivity to Ca^{2+} (Layland et al. 2005). In addition, cGMP can modulate cAMP levels through regulation of distinct PDEs (see below). One proposed mechanism by which cGMP-PKG signalling exerts its anti-hypertrophic action is by inhibiting the calcineurin pathway (Tsai and Kass 2009) (Fig. 6.1).

6.3 Overview of the PDE Superfamily

Eleven PDE families that differ in their primary structure, catalytic properties and affinities for cAMP and/or cGMP, as well as in their mechanisms of regulation, are known in mammals (Fig. 6.2). Most PDE families are encoded by several genes, which together generate close to 100 different PDE isoforms by the use of different translation initiation sites and alternative mRNA splicing. Specific isoforms are designated according to a common nomenclature: PDE is followed by a family number (1–11), a capital letter indicating the gene (A, B, C or D) and a final number corresponding to the splice variant. Some PDE families selectively hydrolyze cAMP

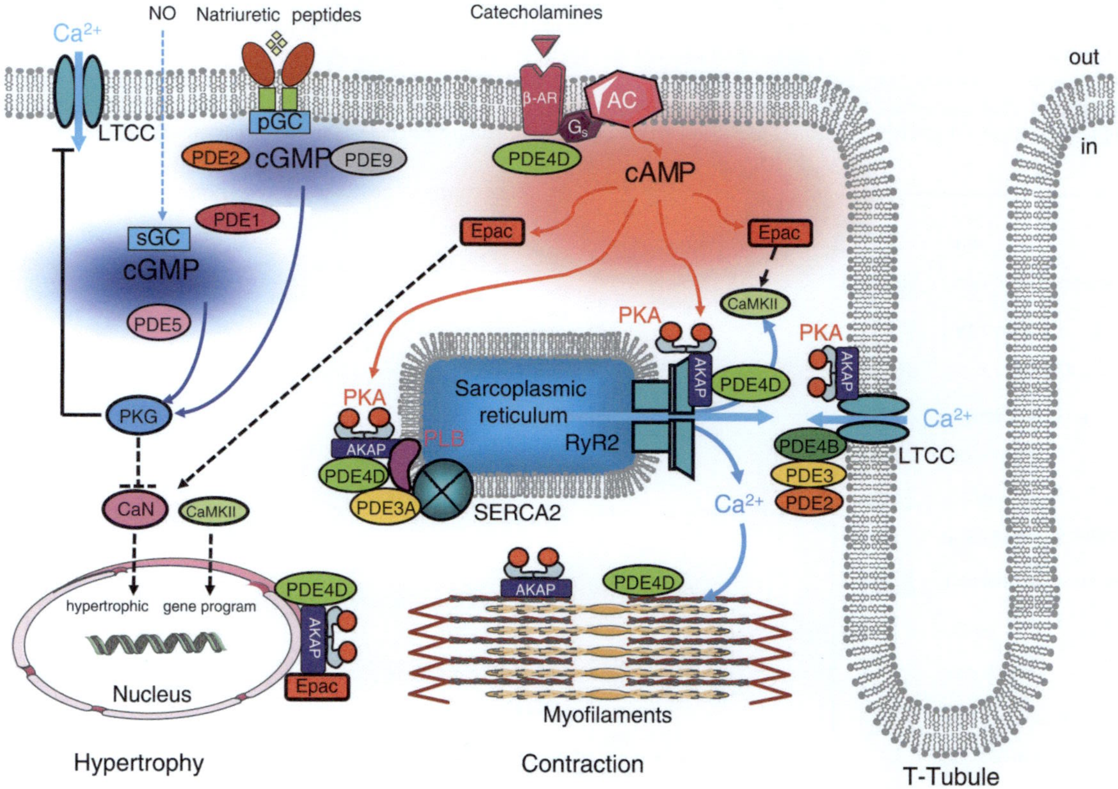

Fig. 6.1 Cyclic nucleotides metabolism in cardiac myocytes. The major PDEs expressed in cardiac myocytes are indicated together with their subcellular localization in relation to their role in regulating hypertrophic growth and excitation-contraction coupling. For simplicity, the PDE family and—when relevant—the gene are indicated but not the specific splice variants. AC, adenylyl cyclase; AKAP, A-kinase anchoring protein; CaMKII, Ca^{2+}-calmodulin kinase II; CaN, calcineurin; GPCR, G protein-coupled receptor; pGC, particulate guanylyl cyclase; sGC, soluble guanylyl cyclase

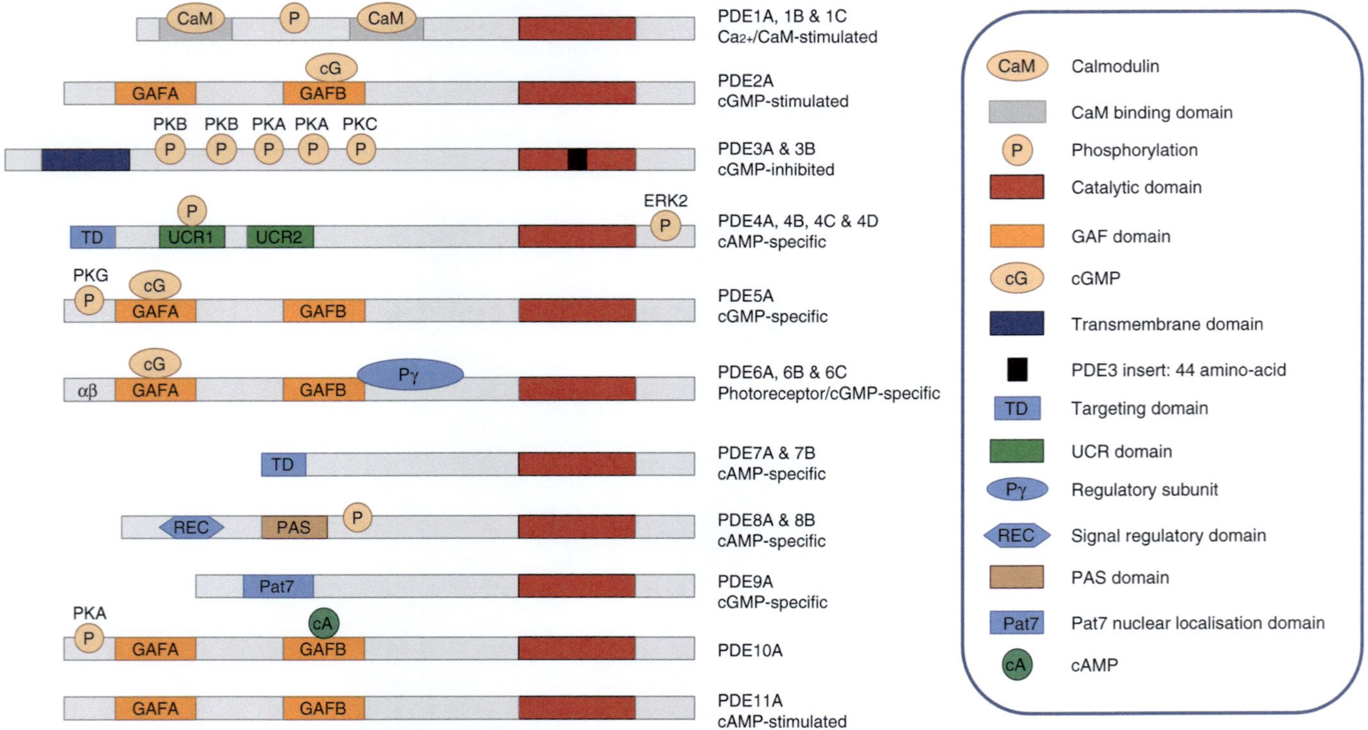

Fig. 6.2 The 11 mammalian PDE families. The conserved catalytic domain (shown in *red*) is located in the carboxy-terminal portion of the PDEs. The catalytic domain of PDE3 contains a unique 44-amino-acid insert (shown in *black*). Many of the PDE families contain amino-terminal subdomains (such as GAF domains, transmembrane domains, targeting domains, upstream conserved regions (UCRs), PAS domains and REC domains) and N-terminal hydrophobic regions that are important in subcellular localization, in the incorporation of PDEs into compartmentalized signalosomes, in interactions with signalling molecules and molecular scaffolds and in the regulation of PDE activity. GAF domains regulate the allosteric binding of cGMP (to PDE2, PDE5, PDE6 and PDE11), the allosteric binding of cAMP (to PDE10) and the regulation of catalytic activity (in PDE2, PDE5 and PDE6) (Modified from ref. Maurice et al. 2014)

(PDE4, 7, 8), while others are specific for cGMP (PDE5, 6, 9). A third category, so-called dual PDEs, hydrolyzes both cAMP and cGMP (PDE1, 2, 3, 10, 11).

PDEs share a conserved catalytic domain (C-domain) showing approximately 25–52% AA sequence identity but differ markedly in their regulatory N-domain (Fig. 6.2). N-domains contain diverse elements involved in enzyme dimerization, binding of regulatory small molecules, phosphorylation and localization. They are characteristic for each family and their variants. For instance, the unique distinguishing feature of the PDE1 family is the existence of two binding sites for the Ca^{2+}-binding protein calmodulin (CaM) in the N-domain which are responsible for enzyme stimulation by Ca^{2+}. Other important domains found in several PDE families (PDE2, PDE5, PDE6, PDE10 and PDE11) are the so-called GAF domains (this acronym is based on the first letters of the three proteins in which they were first identified: G: cGMP-stimulated phosphodiesterases; A: Anabaena adenylyl cyclase; F: Fhla transcription factor) which are involved in enzyme dimerization and allosteric regulation by cyclic nucleotides. In particular, in PDE2 and PDE5, cGMP-binding to their respective GAF domain stimulates enzymatic activity (Martins et al. 1982; Rybalkin et al. 2003). In contrast to PDE2, PDE3 is inhibited by cGMP, by direct competition at the catalytic site. The N-domains of various PDEs contain phosphorylation sites for distinct kinases which modulate enzymatic activity. For instance, long isoforms of PDE4 are phosphorylated by PKA, leading to an increase in cAMP-hydrolytic activity (Sette and Conti 1996), whereas PKG phosphorylation of PDE5 increases cGMP-hydrolytic activity (Francis et al. 2011). Long isoforms of PDE4D have been largely studied in that respect and were shown to be phosphorylated by multiple other kinases, including in particular ERK2, ERK5 and CaMKII (Mika et al. 2015). N-domains are also important for intracellular localization, through specific regions that provide membrane association or protein-protein interaction. PDEs can associate with multiple protein partners including scaffold proteins such as A-kinase anchoring proteins (AKAPs) or β-arrestin strategically located within the cells. Recently, an interaction between PDE8 and the regulatory subunit RIα of PKA was described which brings together regions spanning the phosphodiesterase active site and cAMP-binding sites of RIα to facilitate cAMP hydrolysis (Krishnamurthy et al. 2014). More detailed presentations of PDEs including their structure, regulation, physiological roles and pharmacology are available in several recent reviews (Conti and Beavo 2007; Francis et al. 2011; Keravis and Lugnier 2012; Maurice et al. 2014).

6.4 Role of PDEs in Cyclic Nucleotide Compartmentation

The notion of compartmentation arose from studies of cAMP signalling in the heart. Almost 40 years ago, Corbin and co-workers provided evidence that the two subtypes of PKA, designated type I and type II, are differentially distributed in the heart: while PKA type I was found mostly in the soluble fraction, PKA type II was predominant in the particulate fraction (Corbin et al. 1977). It was shown subsequently by Brunton and colleagues that the classical positive inotropic effect of β-AR stimulation was accompanied by an increase in cAMP and PKA in both fractions, whereas prostaglandin E_1 (PGE_1) stimulation increased cAMP and activated PKA only in the soluble fraction and was devoid of a positive inotropic effect (Buxton and Brunton 1983; Hayes et al. 1980). Two main conclusions were drawn from these experiments; the first was that cAMP generated by a given receptor does not activate all possible PKA molecules in the cell, and the second was that active PKA cannot phosphorylate all possible substrates (Steinberg and Brunton 2001). The discovery of AKAPs as scaffold proteins able to direct PKA to specific subcellular compartments provided a mean to spatially restrict PKA activity (Dodge-Kafka et al. 2006; Wong and Scott 2004). A large number of PKA type II-specific AKAPs were shown to be associated with critical components of β-AR signalling, ECC or hypertrophic signalling in the heart (for review see (Diviani et al. 2011)). The role of AKAPs in defining the range of PKA action is further supported by recent provocative experiments in HEK293 cells suggesting that when PKA is anchored to an AKAP, the catalytic subunits of the kinase may not dissociate from the regulatory subunits upon hormonal stimulation (Smith et al. 2013). Yet, these sophisticated assemblies of cAMP signalling components would not be sufficient to ensure specificity if cAMP can diffuse uniformly and indistinctly activate all PKAs present in a given cell. The observation that a local β-AR stimulation is much more efficient at stimulating local than remote Ca^{2+} channels in frog cardiomyocytes provided evidence that cAMP diffusion must be hindered. Importantly, global PDE inhibition with IBMX increased β-AR stimulation of the remote Ca^{2+} channels, implying that PDEs can act as a diffusion barrier preventing cAMP diffusion (Jurevicius and Fischmeister 1996). More direct evidence of the existence of cAMP gradients and extended cAMP spreading upon PDE inhibition were subsequently obtained by expressing cyclic nucleotide-gated channels or Förster resonance energy transfer (FRET) biosensors to measure cAMP in real time in mammalian cardiomyocytes (Leroy et al. 2008; Molina et al. 2014; Mongillo et al. 2004; Nikolaev et al. 2006a, b; Nikolaev et al. 2010; Zaccolo and Pozzan 2002). Based on experiments performed in HEK293 cells, it was proposed that differentially localized PDEs may function as local sinks that drain cAMP concentration in defined domains by locally degrading the second messenger (Terrin et al. 2006). The interaction of PDE4D isoforms with several AKAPs facilitating PKA-mediated activation of the PDE by PKA also suggest that cAMP can be controlled locally, at the level of one macromolecular complex (Dodge et al. 2001; Terrenoire et al. 2009). Additional evidence that PDEs control local cAMP signals in cardiomyocytes come from studies showing that the contribution of the different PDEs to cAMP degradation depends on the type of G_s-coupled receptors generating the signal (Nikolaev et al. 2006a, b; Rochais et al. 2006; Xiang et al. 2005). These studies highlighted the predominant role of the PDE4 family in this control, and in neonatal cardiac myocytes, distinct PDE4 variants were shown to associate with and regulate the cAMP signals generated by $β_1$-ARs versus $β_2$-ARs (Baillie et al. 2003; De Arcangelis et al. 2009; Mika et al. 2014; Richter et al. 2008; Richter et al. 2013). The use of cAMP biosensors specifically targeted to distinct subcellular compartments also suggests local cAMP degradation by distinct PDEs. For instance, targeting of FRET-based cAMP biosensors to the plasma membrane and the cytosol indicates a spatially restricted function of a specific PDE4 variant (PDE4B) to regulation of subsarcolemmal cAMP generated by β_1-ARs in neonatal cardiomyocytes (Mika et al. 2014). Moreover, cAMP biosensors targeted to PKA type I and II compartments are differentially activated by distinct receptors and PDE inhibitors: while cell stimulation with PGE_1 or PDE2 inhibition preferentially increase cAMP in the PKA type I compartment, stimulation with Iso or PDE4 inhibition preferentially increase cAMP in the PKA type II compartment (Di Benedetto et al. 2008). Activation of PKA type II correlates with phosphorylation of SR proteins such as PLB by Iso, whereas PGE_1/E_2 fails to phosphorylate PLB (Di Benedetto et al. 2008; Liu et al. 2012). The use of a FRET-based PKA biosensor (AKAR) localized to the SR confirmed a lack of PKA activation in the SR upon PGE_2 stimulation but also showed that PGE_2 is able to blunt β-AR inotropic response by activating PDE4D and preventing cAMP generated by β-ARs at the membrane to diffuse and activate PKA at the SR (Liu et al. 2012). Inhibition of PDE4 or ablation of the PDE4D gene were also shown to enhance PKA activation in the nucleus following a short stimulation of β-ARs, suggesting that PDE4D also impedes the diffusion of cAMP from the plasma membrane to the nucleus (Haj Slimane et al. 2014).

While the works related above argue for a critical role of PDEs in cAMP compartmentation, a number of modelling studies suggest that PDEs are not sufficient to establish cAMP gradients if cAMP diffusion in the cell cytoplasm is as fast as in water (Feinstein et al. 2012; Iancu et al. 2007; Rich et al. 2001; Saucerman et al. 2014; Saucerman et al. 2006). Indeed, given their in vitro kinetics and estimated concentration in cells, PDEs may not be able to metabolize cAMP fast enough to avoid its replacement by diffusion from surrounding areas. Yet, at least for PDE2, there is convincing evidence that upon activation of the enzyme by cGMP, hydrolysis of cAMP is fast enough to balance the hormonal stimulation of cAMP synthesis (Fischmeister and Hartzell 1987; Hartzell and Fischmeister 1986; Nikolaev et al. 2005). Also, it should be kept in mind that PDE activity can be enhanced during hormonal stimulation (Conti et al. 2014). Whereas several studies indicated fast diffusion rates for cAMP in neurons (Bacskai et al. 1993; Chen et al. 1999; Nikolaev et al. 2004), recent investigations suggest that cAMP diffusion is markedly slower in adult cardiomyocytes (Agarwal et al. 2016; Richards et al. 2016). The reasons for this are not entirely clear, but the above-mentioned recent studies propose an important role of the mitochondrial network, acting either as a physical barrier (Richards et al. 2016) or as a support for anchored PKA acting as a cAMP buffer system (Agarwal et al. 2016).

6.5 PDEs and Cyclic Nucleotide Compartmentation in Heart Failure

In the 1970s and 1980s, PDE3 inhibitors were discovered to exhibit cardiotonic, inotropic, bronchodilatory and vasodilatory activities in several species and were initially developed as cardiotonic agents to replace or add to cardiac glycosides in the treatment of HF (Movsesian et al. 2011). However, despite beneficial hemodynamic effects on the short term, chronic use of PDE3 inhibitors were associated with increased cardiac arrhythmias and sudden death (Packer et al. 1991). Thus, the use of PDE3 inhibitors is now limited to acute decompensated HF. Nevertheless, PDE3 inhibitors are targeting several functionally distinct isoforms which are co-expressed in the heart, raising the hope that more selective targeting might provide some benefits. PDE3 is encoded by two genes, PDE3A and PDE3B. Evidence from global PDE3A and PDE3B knockout (KO) mice indicates that PDE3A but not PDE3B is responsible for the inotropic and chronotropic effects of PDE3 inhibitors (Sun et al. 2007). Three isoforms of PDE3A are expressed in cardiomyocytes, which differ only in their N-terminal domain, giving rise to different intracellular localization (Wechsler et al. 2002). In mice and humans, PDE3A1 controls PLB-SERCA2 activity and Ca^{2+} re-uptake in the SR (Ahmad et al. 2015; Beca et al. 2013) (Fig. 6.1). Because dephosphorylated PLB and depressed SERCA2 activity are a hallmark of HF, PDE3 inhibitors targeting specifically the PDE3A1 associated with PLB-SERCA2 may improve contractile performance and provide therapy for HF (Movsesian 2015). However, currently available PDE3 inhibitors have little selectivity for PDE3A versus PDE3B isoforms, whose catalytic domains are similar, and no selectivity for individual PDE3A isoforms, whose catalytic domains are identical. Phosphorylation of PDE3A1 was recently shown to regulate its interaction with SERCA2 (Ahmad et al. 2015). Targeting this mechanism may offer an alternative to selectively enhance contractility without the harmful effects of global inhibition of PDE3 activity.

The second major PDE involved in cAMP hydrolysis in the heart is the cAMP-specific PDE4. The PDE4 family is encoded by four genes (PDE4A-D). Most of our knowledge on the role of individual PDE4 subtypes in the heart is limited to PDE4D. Deletion of this gene in mice leads to PKA hyperphosphorylation of RyR2, increased sensitivity to exercise-induced arrhythmias and a late onset dilated cardiomyopathy (Lehnart et al. 2005). PDE4D isoforms are localized in multiple compartments of the cardiomyocyte. For instance, PDE4D3 is localized at the perinuclear region, where it is part of a macromolecular complex organized by the scaffold protein mAKAP and comprising Epac1 and the kinase ERK5 to regulate cardiomyocyte hypertrophy (Dodge-Kafka et al. 2005). This isoform is also present at the sarcolemma, where it associates through another AKAP with slowly activating delayed rectifier potassium channels controlling cardiac repolarization (Terrenoire et al. 2009) and at myofilaments, in association with another scaffold protein, myomegalin (Verde et al. 2001). In addition, as indicated above, distinct PDE4D isoforms were shown to interact with β_1-ARs and β_2-ARs, either directly or indirectly through β-arrestin, and to shape specific physiological or pathophysiological responses (Baillie et al. 2003; Berthouze-Duquesnes et al. 2013; De Arcangelis et al. 2009; Richter et al. 2008; Richter et al. 2013). Finally, similarly to PDE3A, PDE4D also associates with the PLB/SERCA2 complex and regulates the SERCA pump activity in the mouse heart (Beca et al. 2011) (Fig. 6.1).

A role for PDE4B in the heart emerged recently when it was identified as an integral component of the LTCC complex and the main PDE regulating the LTCC current during β-AR stimulation (Fig. 6.1). PDE4B KO mice, as PDE4D KO, have an increased susceptibility to ventricular arrhythmias during catecholamine stimulation which may be due to enhanced Ca^{2+} influx through LTCC (Leroy et al. 2011). Although RyR2 phosphorylation by PKA did not seem to be affected in adult hearts from PDE4B KO mice (Leroy et al. 2011), a recent study indicates that it was increased in neonatal myocytes lacking PDE4B (Mika et al. 2014) suggesting that altered RyR2 regulation may also contribute to this arrhythmic phenotype. In a recent study in rat ventricular myocytes, we showed that under β-AR stimulation, inhibition of PDE4 (as well as inhibition of PDE3) exerted inotropic effects via PKA but led to spontaneous diastolic Ca^{2+} waves via both PKA and CaMKII, suggesting the potential use of CaMKII inhibitors as adjuncts to PDE inhibition to limit their pro-arrhythmic effects (Bobin et al. 2016).

As stated above, phosphorylation of certain PDE3 and PDE4 isoforms by PKA activates these enzymes, and this constitutes a powerful negative feedback for cAMP signals in cardiomyocytes (Leroy et al. 2008; Rochais et al. 2004). This regulation has been shown to be facilitated by spatial proximity of PKA and PDEs assembled by the perinuclear mAKAP (Dodge-Kafka et al. 2005) or by PI3Kγ, which in addition to its lipid kinase function also acts as an AKAP facilitating phosphorylation of PDE3B, PDE4A and PDE4B by PKA (Ghigo et al. 2012).

Although these studies underline the critical role of PDE4 in controlling β-AR stimulation in rodents, this family contributes less to the regulation of cardiac contractility in humans, where PDE3 predominates (Molenaar et al. 2013). However, in human atrial strips, inhibition of PDE3, but also of PDE4, potentiates the arrhythmogenic effect of β-AR stimulation, and PDE4 activity tends to decrease in the atria of patients with atrial fibrillation (Molina et al. 2012). A further understanding of the role of PDE4 in humans may also be important for the pro-arrhythmic effect of PDE3 inhibitors since PDE3 inhibitors such as milrinone and enoximone may also inhibit PDE4 in cardiac preparations (Bethke et al. 1992; Shakur et al. 2002).

In cardiac hypertrophy and HF, there are profound modifications in the major components of the cAMP pathway. These include a decreased density of β_1-ARs, an uncoupling of β_2-ARs from G_s, an increase in G_i and in the G protein-coupled receptor kinase GRK2 and, in certain models, a decrease in adenylyl cyclase activity (Lohse et al. 2003). In a model of pathological hypertrophy induced by pressure overload in rats, we found that the expression and activity of PDE3A, PDE4A and PDE4B were decreased, and this was associated with a blunted regulation of subsarcolemmal cAMP generated by β-ARs by PDE3 and PDE4 (Abi-Gerges et al. 2009). In contrast, in a model of cardiac hypertrophy induced by angiotensin II, an increased PDE4 activity was observed, accompanied by an increase in the 69-kDa-PDE4A isoform and a decrease in expression of 52- and 76-kDa PDE4D isoforms (Mokni et al. 2010). These results suggest that the level of expression of the isoforms of PDE3 and PDE4 is specifically regulated by the type of stimulus used to induce cardiac hypertrophy and the stage of the disease. Whereas an increase in cAMP-PDE can participate in desensitization of the β-AR pathway, a decrease could represent a compensatory mechanism to restore cAMP levels and inotropism. However, lower PDE activity also alters the degree of cAMP confinement, which could lead to illegitimate or excessive activation of certain pools of PKA or Epac, hence promoting maladaptive remodelling and rhythmic disturbances. This is supported by the results of a recent study showing that the specific PDE4D5 isoform regulates activation of hypertrophic programme by Epac1 upon stimulation of β_2-AR (Berthouze-Duquesnes et al. 2013) receptors.

HF is also accompanied by modifications of the microarchitecture of the myocytes, including that of the T-tubular system, manifested as T-tubule loss or as reorganization depending on the studies (Louch et al. 2010). These alterations in the T-tubular network not only affect the synchrony of Ca^{2+} release (Heinzel et al. 2002), but are also associated with modifications in the distribution of CN signalling components and, therefore, CN compartmentation. In a rat model of HF, it was shown that β_2-ARs redistribute from their normal T-tubular localization to the peripheral sarcolemma, leading to diffusive as opposed to constrained cAMP signals generated by these receptors (Nikolaev et al. 2010). In a more recent study, the local regulation of cAMP by PDEs in the vicinity of SERCA2 was compared in transgenic mice with cardiac-specific expression of a PLB-targeted cAMP biosensor, and modifications induced by transverse aortic constriction were analysed (Sprenger et al. 2015). In agreement with their known localization within the SERCA2 complex (Beca et al. 2013; Beca et al. 2011), both PDE3 and PDE4 were found to regulate cAMP in this microdomain. Interestingly, during hypertrophy and early HF, there was a specific rearrangement of the PDEs regulating this specific cAMP pool, with a decreased contribution of PDE4 and an increased contribution of PDE2 (Sprenger et al. 2015). These results indicate that PDE alterations in cardiac disease include redistribution of PDE variants in discrete microcompartments of cardiomyocytes, as shown for PDE2 and PDE3 at the membrane (Perera et al. 2015).

The dual specific PDE2 represents a minor part of cAMP-hydrolytic activity in the normal heart, but the cAMP-hydrolytic activity of this PDE is stimulated 5 to 30-fold by cGMP, and this was shown to inhibit cardiac LTCC in various species including humans (Fischmeister et al. 2005). Subsequently, measurements with FRET-based sensors in neonatal rat cardiomyocytes showed that by decreasing the level of cAMP, PDE2 counteracts the effects of a β-AR stimulation downstream of β_3-ARs (Mongillo et al. 2005). In contrast to PDE3 and PDE4, which expression and activity are generally decreased in pathological hypertrophy and HF (Abi-Gerges et al. 2009; Ding et al. 2005; Osadchii 2007), we found recently that PDE2 is increased in animal models as well as in human HF (Mehel et al. 2013). PDE2 inhibition partially restores β-AR responsiveness in diseased cardiomyocytes, suggesting that PDE2 enhancement in HF constitutes a protective mechanism against excessive β-AR stimulation. This hypothesis was confirmed in a more recent study using a transgenic mouse model with a cardiac-specific overexpression of PDE2 (Vettel et al. 2016). Transgenic increase in PDE2 abundance in mice lowers heart rate but preserves cardiac output due to greater cardiac force. Increased PDE2 abundance was found to be cardioprotective in vivo in acute catecholaminergic stress and after myocardial infarction without compromising contractile performance (Vettel et al. 2016). However, according to another recent study, PDE2 could exert a pro-hypertrophic effect by blunting PKA-mediated phosphorylation of NFAT (Zoccarato et al. 2015). Further studies are needed to fully understand the role of PDE2 in HF (Wagner et al. 2016; Zoccarato et al. 2016).

Similarly to PDE2, PDE1 and PDE5 were reported to be upregulated in pathological hypertrophy and HF (Miller et al. 2009; Pokreisz et al. 2009; Vandeput et al. 2007). Because PDE1 and PDE5 preferentially (PDE1A) or specifically (PDE5) degrade cGMP, their increase in HF can clearly be seen as maladaptive. Accordingly, transgenic mice with cardiac-specific overexpression of PDE5 are predisposed to adverse remodelling after myocardial infarction (Pokreisz et al. 2009), whereas on the contrary, pharmacological inhibition of PDE1 (Miller et al. 2009) or PDE5 (Takimoto et al. 2005) reduces hypertrophy and improves cardiac pressure and volume overload. Numerous animal studies have shown that PDE5 inhibitors protect against ischemia/reperfusion (I/R) injury, doxorubicin cardiotoxicity, ischemic and diabetic cardiomyopathy, and Duchenne muscular dystrophy (Das et al. 2015). However, it remains controversial whether significant levels of PDE5 are expressed in the myocardium, raising the possibility that the beneficial effects of PDE5 inhibitors involve other mechanisms including inhibition of PDE1 (Degen et al. 2015; Lukowski et al. 2014). In patients with systolic HF, sildenafil decreased pulmonary vascular pressure and increased peak oxygen consumption and cardiac index (Lewis et al. 2007). Sildenafil also improved left ventricular diastolic function, cardiac geometry and clinical status in patients with systolic HF (Guazzi et al. 2011) and improved diabetic cardiomyopathy (Giannetta et al. 2012). However, despite encouraging results in an initial mono-centre study (Guazzi et al. 2011), chronic therapy with sildenafil was not associated with clinical benefit in patients with diastolic HF in a larger, multicentre study (Redfield et al. 2013). Ongoing trials with PDE5 inhibitors include testing for the gender response to tadalafil in left ventricular hypertrophy associated to diabetic cardiomyopathy (NCT01803828).

Two other PDEs were recently proposed to participate in cGMP degradation in the heart. Experiments performed in isolated cardiomyocytes from transgenic mice expressing a FRET-based cGMP biosensor have suggested that PDE3, which is classically known to degrade preferentially cAMP, may also be involved in the control of cGMP levels (Gotz et al. 2014). In addition, the cGMP-specific PDE9 was found to be expressed in rodent and human heart and to be upregulated in hypertrophy and HF (Lee et al. 2015). PDE9 genetic ablation or pharmacological inhibition appears to protect the heart against pathological remodelling during pressure overload. Moreover, PDE9 inhibition reverses pre-established heart disease in a NO synthase (NOS) activity-independent manner, whereas PDE5 inhibition requires active NOS, which is decreased in HF. This is because PDE9 seems to hydrolyze specifically cGMP generated by natriuretic peptides, whereas PDE5 controls cGMP generated by NO (Castro et al. 2006; Takimoto et al. 2007). We had shown previously that PDE2 is critical to regulate subsarcolemmal cGMP levels in response to pGC activation in adult cardiomyocytes (Castro et al. 2006), thus raising the question of whether PDE2 and PDE9 exert a redundant or distinct regulation of natriuretic peptide signalling.

6.6 PDEs as Therapeutic Targets in Ischemia/Reperfusion Injury?

Manipulation of PDE activity may also prove protective in the context of ischemia/reperfusion (I/R) injury. Indeed, PDE5 inhibitors were shown to reduce infarct size in rabbits and mice. They also decreased cell death in isolated cardiomyocytes, suggesting that part of this effect is independent of vasodilation. Several mechanisms appear to be involved in these effects, including increased NO synthase expression, cGMP elevation, PKG activation and opening of mitochondrial K_{ATP} and Ca^{2+}-activated K^+ channels (Das et al. 2015). PDE3 inhibitors have also been reported to reduce infarct size when applied before sustained ischemia, thus mimicking the cardioprotection conferred by ischemic preconditioning (Fukasawa et al. 2008; Sanada et al. 2001; Tosaka et al. 2007). A recent study using KO mice for either PDE3A or PDE3B strongly suggests that PDE3B is the isoform mediating the cardioprotective effect of PDE3 inhibitors in this context. Indeed, PDE3B KO mice, but not PDE3A KO mice, were protected against ischemia/reperfusion injury. This protective effect appears to involve cAMP/PKA-mediated opening of mitochondrial Ca^{2+}-activated K^+ channels and assembly of ischemia-induced caveolin-3-enriched fractions (Chung et al. 2015). Somehow at odds with the above-mentioned cardioprotective effect of PDE3 inhibitors, mice with cardiac-specific overexpression of PDE3A1 were protected during ischemia/reperfusion injury (Oikawa et al. 2013). In addition to regulating SERCA2, PDE3A1 also acts as a negative regulator of cardiomyocyte apoptosis, by controlling the expression of the transcriptional repressor and pro-apoptotic factor, ICER (inducible- cAMP early repressor) (Yan et al. 2007). Inhibition of this mechanism in mice with cardiac-specific overexpression of PDE3A1 was associated with protection during ischemia/reperfusion (Oikawa et al. 2013). Collectively, these studies suggest that PDE3A and PDE3B may play an opposite role during ischemia/reperfusion, which may be linked to their differential localization and the control of discrete cAMP pools in cardiomyocytes (Chung et al. 2015).

6.7 Concluding Remarks

Soon after the discovery of cAMP and cGMP by Sutherland and colleagues more than 50 years ago, an enzymatic activity that could degrade these second messengers was described (Sutherland and Rall 1958). Since then, the large diversity and complexity of the PDE superfamily have been unveiled, and the critical role of these enzymes in the cardiovascular system demonstrated. After several disappointments in the development of PDE3 and PDE4 inhibitors due to their adverse effects (mortality and emesis, respectively), the success in PDE5 inhibitors to treat erectile dysfunction associated with an increasing understanding of PDE biology has raised new hopes that manipulating PDE activity with greater specificity is possible and should yield therapeutic benefits. However, the targeting of individual PDE isoforms located in distinct subcellular compartments to regulate local cAMP and/or cGMP concentrations and a specific cell function is challenging. Knowledge of the protein partners and of the molecular mechanisms that govern this specific localization allows to envisage the displacement of specific PDE pools by small molecules or peptides that disrupt protein-protein interaction, with potentially fewer adverse effects than global inhibition of an entire PDE family. Recent examples of such strategy include the disruption of PDE4 from heat shock protein 20 (HSP20) in cardiomyocytes, resulting in hyperphosphorylation of HSP20 and attenuated cardiac hypertrophic growth (Martin et al. 2014; Sin et al. 2011), and the disruption of Epac1-PDE3B and Epac1-PDE4D complexes involved in the control of endothelial cell adhesion, spreading and permeability (Rampersad et al. 2010; Wilson et al. 2011). Another interesting strategy is the development of allosteric inhibitors of PDEs, for instance, targeting the GAF domain of PDE5 to block enzyme activation but not its basal activity (Schultz et al. 2011) or exploiting sequence differences outside the active site to reach isoform selectivity in the case of PDE4B and PDE4D (Burgin et al. 2010; Fox et al. 2014). The allosteric mode of regulation could also allow the discovery of small molecules acting as activators of specific PDEs, which may actually be useful in correcting the detrimental effects of excessive β-AR stimulation observed in HF.

Acknowledgements This work was supported by the Fondation de France (to GV) and the Agence Nationale de la Recherche 2010 BLAN 1139-01 (to GV). PB and IB were supported by PhD fellowships from the region Ile-de-France (CORDDIM) and Fondation pour la Recherche Médicale.

Compliance with Ethical Standards

Conflict of Interest None.

References

Abi-Gerges A, Richter W, Lefebvre F et al (2009) Decreased expression and activity of cAMP phosphodiesterases in cardiac hypertrophy and its impact on ß-adrenergic cAMP signals. Circ Res 105:784–792

Agarwal SR, Clancy CE, Harvey RD (2016) Mechanisms restricting diffusion of intracellular cAMP. Sci Rep 6:19577

Ahmad F, Shen W, Vandeput F et al (2015) Regulation of sarcoplasmic reticulum Ca^{2+} ATPase 2 (SERCA2) activity by phosphodiesterase 3A (PDE3A) in human myocardium: phosphorylation-dependent interaction of PDE3A1 with SERCA2. J Biol Chem 290:6763–6776

Bacskai BJ, Hochner B, Mahaut-Smith M et al (1993) Spatially resolved dynamics of cAMP and protein kinase A subunits in Aplysia sensory neurons. Science 260:222–226

Baillie GS, Sood A, McPhee I et al (2003) β-Arrestin-mediated PDE4 cAMP phosphodiesterase recruitment regulates β-adrenoceptor switching from G_s to G_i. Proc Natl Acad Sci U S A 100:941–945

Beca S, Ahmad F, Shen W et al (2013) PDE3A regulates basal myocardial contractility through interacting with SERCA2a-signaling complexes in mouse heart. Circ Res 112:289–297

Beca S, Helli PB, Simpson JA et al (2011) Phosphodiesterase 4D regulates baseline sarcoplasmic reticulum Ca^{2+} release and cardiac contractility, independently of L-type Ca^{2+} current. Circ Res 109:1024–1030

Berthouze-Duquesnes M, Lucas A, Sauliere A et al (2013) Specific interactions between Epac1, beta-arrestin2 and PDE4D5 regulate beta-adrenergic receptor subtypes differential effects on cardiac hypertrophic signaling. Cell Signal 25:970–980

Bethke T, Eschenhagen T, Klimkiewicz A et al (1992) Phosphodiesterase inhibition by enoximone in preparations from nonfailing and failing human hearts. Arzneimittelforschung 42:437–445

Bobin P, Varin A, Lefebvre F, Fischmeister R, Vandecasteele G, Leroy J (2016) Calmodulin kinase II inhibition limits the pro-arrhythmic Ca^{2+} waves induced by cAMP-phosphodiesterase inhibitors. Cardiovasc Res 110:151–161

Burgin AB, Magnusson OT, Singh J et al (2010) Design of phosphodiesterase 4D (PDE4D) allosteric modulators for enhancing cognition with improved safety. Nat Biotechnol 28:63–70

Buxton ILO, Brunton LL (1983) Compartments of cyclic AMP and protein kinase in mammalian cardiomyocytes. J Biol Chem 258:10233–10239

Castro LR, Verde I, Cooper DMF, Fischmeister R (2006) Cyclic guanosine monophosphate compartmentation in rat cardiac myocytes. Circulation 113:2221–2228
Chen CH, Nakamura T, Koutalos Y (1999) Cyclic AMP diffusion coefficient in frog olfactory cilia. Biophys J 76:2861–2867
Chung YW, Lagranha C, Chen Y et al (2015) Targeted disruption of PDE3B, but not PDE3A, protects murine heart from ischemia/reperfusion injury. Proc Natl Acad Sci U S A 112:E2253–E2262
Conti M, Beavo JA (2007) Biochemistry and physiology of cyclic nucleotide phosphodiesterases: essential components in cyclic nucleotide signaling. Annu Rev Biochem 76:481–511
Conti M, Mika D, Richter W (2014) Perspectives on: cyclic nucleotide microdomains and signaling specificity: cyclic AMP compartments and signaling specificity: role of cyclic nucleotide phosphodiesterases. J Gen Physiol 143:29–38
Corbin JD, Sugden PH, Lincoln TM, Keely SL (1977) Compartmentalization of adenosine 3′:5′-monophosphate and adenosine 3′:5′-monophosphate-dependent protein kinase in heart tissue. J Biol Chem 252:3854–3861
Das A, Durrant D, Salloum FN, Xi L, Kukreja RC (2015) PDE5 inhibitors as therapeutics for heart disease, diabetes and cancer. Pharmacol Ther 147:12–21
De Arcangelis V, Liu R, Soto D, Xiang Y (2009) Differential association of phosphodiesterase 4D isoforms with β_2-adrenoceptor in cardiac myocytes. J Biol Chem 284:33824–33832
Degen CV, Bishu K, Zakeri R, Ogut O, Redfield MM, Brozovich FV (2015) The Emperor's new clothes: PDE5 and the heart. PLoS One 10:e0118664
Di Benedetto G, Zoccarato A, Lissandron V et al (2008) Protein kinase A type I and type II define distinct intracellular signaling compartments. Circ Res 103:836–844
Ding B, Abe J, Wei H et al (2005) A positive feedback loop of phosphodiesterase 3 (PDE3) and inducible cAMP early repressor (ICER) leads to cardiomyocyte apoptosis. Proc Natl Acad Sci U S A 102:14771–14776
Diviani D, Dodge-Kafka KL, Li J, Kapiloff MS (2011) A-kinase anchoring proteins: scaffolding proteins in the heart. Am J Physiol Heart Circ Physiol 301:H1742–H1753
Dodge-Kafka KL, Langeberg L, Scott JD (2006) Compartmentation of cyclic nucleotide signaling in the heart: the role of A-kinase anchoring proteins. Circ Res 98:993–1001
Dodge-Kafka KL, Soughayer J, Pare GC et al (2005) The protein kinase A anchoring protein mAKAP co-ordinates two integrated cAMP effector pathways. Nature 437:574–578
Dodge KL, Khouangsathiene S, Kapiloff MS et al (2001) mAKAP assembles a protein kinase A/PDE4 phosphodiesterase cAMP signaling module. EMBO J 20:1921–1930
Feinstein WP, Zhu B, Leavesley SJ, Sayner SL, Rich TC (2012) Assessment of cellular mechanisms contributing to cAMP compartmentalization in pulmonary microvascular endothelial cells. Am J Physiol Cell Physiol 302:C839–C852
Fischmeister R, Castro L, Abi-Gerges A, Rochais F, Vandecasteele G (2005) Species-and tissue-dependent effects of NO and cyclic GMP on cardiac ion channels. Comp Biochem Physiol A Mol Integr Physiol 142:136–143
Fischmeister R, Hartzell HC (1987) Cyclic guanosine 3′,5′-monophosphate regulates the calcium current in single cells from frog ventricle. J Physiol 387:453–472
Fox D 3rd, Burgin AB, Gurney ME (2014) Structural basis for the design of selective phosphodiesterase 4B inhibitors. Cell Signal 26:657–663
Francis SH, Blount MA, Corbin JD (2011) Mammalian cyclic nucleotide phosphodiesterases: molecular mechanisms and physiological functions. Physiol Rev 91:651–690
Fukasawa M, Nishida H, Sato T, Miyazaki M, Nakaya H (2008) 6-[4-(1-Cyclohexyl-1H-tetrazol-5-yl)butoxy]-3,4-dihydro-2-(1H)quinolinone (cilostazol), a phosphodiesterase type 3 inhibitor, reduces infarct size via activation of mitochondrial Ca^{2+}-activated K^+ channels in rabbit hearts. J Pharmacol Exp Ther 326:100–104
Ghigo A, Perino A, Mehel H et al (2012) PI3K$_\gamma$ protects against catecholamine-induced ventricular arrhythmia through PKA-mediated regulation of distinct phosphodiesterases. Circulation 126:2073–2083
Giannetta E, Isidori AM, Galea N et al (2012) Chronic inhibition of cGMP phosphodiesterase 5A improves diabetic cardiomyopathy: a randomized, controlled clinical trial using magnetic resonance imaging with myocardial tagging. Circulation 125:2323–2333

Gotz K, Sprenger J, Perera RK et al (2014) Transgenic mice for real time visualization of cGMP in intact adult cardiomyocytes. Circ Res 114:1235–1245

Guazzi M, Vicenzi M, Arena R, Guazzi MD (2011) PDE5 inhibition with sildenafil improves left ventricular diastolic function, cardiac geometry, and clinical status in patients with stable systolic heart failure: results of a 1-year, prospective, randomized, placebo-controlled study. Circ Heart Fail 4:8–17

Haj Slimane Z, Bedioune I, Lechêne P et al (2014) Control of cytoplasmic and nuclear protein kinase A activity by phosphodiesterases and phosphatases in cardiac myocytes. Cardiovasc Res 102:97–106

Hartzell HC, Fischmeister R (1986) Opposite effects of cyclic GMP and cyclic AMP on Ca^{2+} current in single heart cells. Nature 323:273–275

Hayes JS, Brunton LL, Mayer SE (1980) Selective activation of particulate cAMP-dependent protein kinase by isoproterenol and prostaglandin E1. J Biol Chem 255:5113–5119

Heinzel FR, Bito V, Volders PG, Antoons G, Mubagwa K, Sipido KR (2002) Spatial and temporal inhomogeneities during Ca^{2+} release from the sarcoplasmic reticulum in pig ventricular myocytes. Circ Res 91:1023–1030

Iancu RV, Jones SW, Harvey RD (2007) Compartmentation of cAMP signaling in cardiac myocytes: a computational study. Biophys J 92:3317–3331

Jurevicius J, Fischmeister R (1996) cAMP compartmentation is responsible for a local activation of cardiac Ca^{2+} channels by ß-adrenergic agonists. Proc Natl Acad Sci U S A 93:295–299

Keravis T, Lugnier C (2012) Cyclic nucleotide phosphodiesterase (PDE) isozymes as targets of the intracellular signalling network: benefits of PDE inhibitors in various diseases and perspectives for future therapeutic developments. Br J Pharmacol 165:1288–1305

Krishnamurthy S, Moorthy BS, Xin Xiang L et al (2014) Active site coupling in PDE:PkA complexes promotes resetting of mammalian cAMP signaling. Biophys J 107:1426–1440

Layland J, Solaro RJ, Shah AM (2005) Regulation of cardiac contractile function by troponin I phosphorylation. Cardiovasc Res 66:12–21

Lee DI, Zhu G, Sasaki T et al (2015) Phosphodiesterase 9A controls nitric-oxide-independent cGMP and hypertrophic heart disease. Nature 519:472–476

Lehnart SE, Wehrens XHT, Reiken S et al (2005) Phosphodiesterase 4D deficiency in the ryanodine receptor complex promotes heart failure and arrhythmias. Cell 123:23–35

Leroy J, Abi-Gerges A, Nikolaev VO et al (2008) Spatiotemporal dynamics of ß-adrenergic cAMP signals and L-type Ca^{2+} channel regulation in adult rat ventricular myocytes: role of phosphodiesterases. Circ Res 102:1091–1100

Leroy J, Richter W, Mika D et al (2011) Phosphodiesterase 4B in the cardiac L-type Ca^{2+} channel complex regulates Ca^{2+} current and protects against ventricular arrhythmias. J Clin Investig 121:2651–2661

Lewis GD, Lachmann J, Camuso J et al (2007) Sildenafil improves exercise hemodynamics and oxygen uptake in patients with systolic heart failure. Circulation 115:59–66

Lezoualc'h F, Fazal L, Laudette M, Conte C (2016) Cyclic AMP sensor EPAC proteins and their role in cardiovascular function and disease. Circ Res 118:881–897

Liu S, Li Y, Kim S et al (2012) Phosphodiesterases coordinate cAMP propagation induced by two stimulatory G protein-coupled receptors in hearts. Proc Natl Acad Sci U S A 109:6578–6583

Lohse MJ, Engelhardt S, Eschenhagen T (2003) What is the role of ß-adrenergic signaling in heart failure? Circ Res 93:896–906

Louch WE, Hake J, Jolle GF et al (2010) Control of Ca^{2+} release by action potential configuration in normal and failing murine cardiomyocytes. Biophys J 99:1377–1386

Lukowski R, Krieg T, Rybalkin SD, Beavo J, Hofmann F (2014) Turning on cGMP-dependent pathways to treat cardiac dysfunctions: boom, bust, and beyond. Trends Pharmacol Sci 35:404–413

Martin TP, Hortigon-Vinagre MP, Findlay JE, Elliott C, Currie S, Baillie GS (2014) Targeted disruption of the heat shock protein 20-phosphodiesterase 4D (PDE4D) interaction protects against pathological cardiac remodelling in a mouse model of hypertrophy. FEBS Open Bio 4:923–927

Martins TJ, Mimby MC, Beavo JA (1982) Purification and characterization of a cyclic GMP-stimulated cyclic nucleotide phosphodiesterase from bovine tissues. J Biol Chem 257:1973–1979

Maurice DH, Ke H, Ahmad F, Wang Y, Chung J, Manganiello VC (2014) Advances in targeting cyclic nucleotide phosphodiesterases. Nat Rev Drug Discov 13:290–314

Mehel H, Emons J, Vettel C et al (2013) Phoshodiesterase-2 is upregulated in human failing hearts and blunts ß-adrenergic responses in cardiomyocytes. J Am Coll Cardiol 62:1596–1606

Méry P-F, Lohmann SM, Walter U, Fischmeister R (1991) Ca^{2+} current is regulated by cyclic GMP-dependent protein kinase in mammalian cardiac myocytes. Proc Natl Acad Sci U S A 88:1197–1201

Mika D, Richter W, Conti M (2015) A CaMKII/PDE4D negative feedback regulates cAMP signaling. Proc Natl Acad Sci U S A 112:2023–2028

Mika D, Richter W, Westenbroek RE, Catterall WA, Conti M (2014) PDE4B mediates local feedback regulation of β_1-adrenergic cAMP signaling in a sarcolemmal compartment of cardiac myocytes. J Cell Sci 127:1033–1042

Miller CL, Oikawa M, Cai Y et al (2009) Role of Ca^{2+}/calmodulin-stimulated cyclic nucleotide phosphodiesterase 1 in mediating cardiomyocyte hypertrophy. Circ Res 105:956–964

Mokni W, Keravis T, Etienne-Selloum N et al (2010) Concerted regulation of cGMP and cAMP phosphodiesterases in early cardiac hypertrophy induced by angiotensin II. PLoS One 5: e14227

Molenaar P, Christ T, Hussain RI et al (2013) Phosphodiesterase PDE3, but not PDE4, reduces $ß_1$- and $ß_2$-adrenoceptor-mediated inotropic and lusitropic effects in failing ventricle from metoprolol-treated patients. Br J Pharmacol 169:528–538

Molina CE, Johnson DM, Mehel H et al (2014) Interventricular differences in ß-adrenergic responses in the canine heart: role of phosphodiesterases. J Am Heart Assoc 3:e000858. Doi:10.1161/JAHA.114.000858

Molina CE, Leroy J, Xie M et al (2012) Cyclic AMP phosphodiesterase type 4 protects against atrial arrhythmias. J Am Coll Cardiol 59:2182–2190

Mongillo M, McSorley T, Evellin S et al (2004) Fluorescence resonance energy transfer-based analysis of cAMP dynamics in live neonatal rat cardiac myocytes reveals distinct functions of compartmentalized phosphodiesterases. Circ Res 95:65–75

Mongillo M, Terrin A, Evellin S, Lissandron V, Zaccolo M (2005) Study of cyclic adenosine monophosphate microdomains in cells. Methods Mol Biol 307:1–13

Movsesian M (2015) New pharmacologic interventions to increase cardiac contractility: challenges and opportunities. Curr Opin Cardiol 30(3):285–291

Movsesian M, Wever-Pinzon O, Vandeput F (2011) PDE3 inhibition in dilated cardiomyopathy. Curr Opin Pharmacol 11(6):707–713

Nikolaev VO, Bunemann M, Hein L, Hannawacker A, Lohse MJ (2004) Novel single chain cAMP sensors for receptor-induced signal propagation. J Biol Chem 279:37215–37218

Nikolaev VO, Bunemann M, Schmitteckert E, Lohse MJ, Engelhardt S (2006a) Cyclic AMP imaging in adult cardiac myocytes reveals far-reaching $ß_1$-adrenergic but locally confined $ß_2$-adrenergic receptor-mediated signaling. Circ Res 99:1084–1091

Nikolaev VO, Gambaryan S, Engelhardt S, Walter U, Lohse MJ (2005) Real-time monitoring of live cell's PDE2 activity: hormone-stimulated cAMP hydrolysis is faster than hormone-stimulated cAMP synthesis. J Biol Chem 280:1716–1719

Nikolaev VO, Gambaryan S, Lohse MJ (2006b) Fluorescent sensors for rapid monitoring of intracellular cGMP. Nat Methods 3:23–25

Nikolaev VO, Moshkov A, Lyon AR et al (2010) $ß_2$-Adrenergic receptor redistribution in heart failure changes cAMP compartmentation. Science 327:1653–1657

Oikawa M, Wu M, Lim S et al (2013) Cyclic nucleotide phosphodiesterase 3A1 protects the heart against ischemia-reperfusion injury. J Mol Cell Cardiol 64:11–19

Osadchii OE (2007) Myocardial phosphodiesterases and regulation of cardiac contractility in health and cardiac disease. Cardiovasc Drugs Ther 21:171–194

Packer M, Carver JR, Rodeheffer RJ et al (1991) Effect of oral milrinone on mortality in severe chronic heart failure. The PROMISE Study Research Group. N Engl J Med 325:1468–1475

Perera RK, Sprenger JU, Steinbrecher JH et al (2015) Microdomain switch of cGMP-regulated phosphodiesterases leads to ANP-induced augmentation of beta-adrenoceptor-stimulated contractility in early cardiac hypertrophy. Circ Res 116:1304–1311

Pokreisz P, Vandenwijngaert S, Bito V et al (2009) Ventricular phosphodiesterase-5 expression is increased in patients with advanced heart failure and contributes to adverse ventricular remodeling after myocardial infarction in mice. Circulation 119:408–416

Rampersad SN, Ovens JD, Houston E et al (2010) Cyclic AMP phosphodiesterase 4D (PDE4D) tethers EPAC1 in a VE-cadherin (VECAD)-based signaling complex and controls cAMP-mediated vascular permeability. J Biol Chem 285:33614–33622

Redfield MM, Chen HH, Borlaug BA et al (2013) Effect of phosphodiesterase-5 inhibition on exercise capacity and clinical status in heart failure with preserved ejection fraction: a randomized clinical trial. JAMA 309:1268–1277

Rich TC, Fagan KA, Tse TE, Schaack J, Cooper DM, Karpen JW (2001) A uniform extracellular stimulus triggers distinct cAMP signals in different compartments of a simple cell. Proc Natl Acad Sci U S A 98:13049–13054

Richards M, Lomas O, Jalink K et al (2016) Intracellular tortuosity underlies slow cAMP diffusion in adult ventricular myocytes. Cardiovasc Res 110:395–407

Richter W, Day P, Agraval R et al (2008) Signaling from $ß_1$-and $ß_2$-adrenergic receptors is defined by differential interactions with PDE4. EMBO J 27:384–393

Richter W, Mika D, Blanchard E, Day P, Conti M (2013) $ß_1$-adrenergic receptor antagonists signal via PDE4 translocation. EMBO Rep 14:276–283

Rochais F, Abi-Gerges A, Horner K et al (2006) A specific pattern of phosphodiesterases controls the cAMP signals generated by different G_s-coupled receptors in adult rat ventricular myocytes. Circ Res 98:1081–1088

Rochais F, Vandecasteele G, Lefebvre F et al (2004) Negative feedback exerted by PKA and cAMP phosphodiesterase on subsarcolemmal cAMP signals in intact cardiac myocytes. An *in vivo* study using adenovirus-mediated expression of CNG channels. J Biol Chem 279:52095–52105

Ruiz-Hurtado G, Dominguez-Rodriguez A, Pereira L et al (2012) Sustained Epac activation induces calmodulin dependent positive inotropic effect in adult cardiomyocytes. J Mol Cell Cardiol 53:617–625

Rybalkin SD, Rybalkina IG, Shimizu-Albergine M, Tang XB, Beavo JA (2003) PDE5 is converted to an activated state upon cGMP binding to the GAF A domain. EMBO J 22:469–478

Sanada S, Kitakaze M, Papst PJ et al (2001) Cardioprotective effect afforded by transient exposure to phosphodiesterase III inhibitors -the role of protein kinase A and p38 mitogen-activated protein kinase. Circulation 104:705–710

Saucerman JJ, Greenwald EC, Polanowska-Grabowska R (2014) Perspectives on: cyclic nucleotide microdomains and signaling specificity: mechanisms of cyclic AMP compartmentation revealed by computational models. J Gen Physiol 143:39–48

Saucerman JJ, Zhang J, Martin JC et al (2006) Systems analysis of PKA-mediated phosphorylation gradients in live cardiac myocytes. Proc Natl Acad Sci U S A 103:12923–12928

Schindler RF, Brand T (2016) The Popeye domain containing protein family—a novel class of cAMP effectors with important functions in multiple tissues. Prog Biophys Mol Biol 120:28–36

Schultz JE, Dunkern T, Gawlitta-Gorka E, Sorg G (2011) The GAF-tandem domain of phosphodiesterase 5 as a potential drug target. Handb Exp Pharmacol:151–166

Sette C, Conti M (1996) Phosphorylation and activation of a cAMP-specific phosphodiesterase by the cAMP-dependent protein kinase -involvement of serine 54 in the enzyme activation. J Biol Chem 271:16526–16534

Shakur Y, Fong M, Hensley J et al (2002) Comparison of the effects of cilostazol and milrinone on cAMP-PDE activity, intracellular cAMP and calcium in the heart. Cardiovasc Drugs Ther 16:417–427

Sin YY, Edwards HV, Li X et al (2011) Disruption of the cyclic AMP phosphodiesterase-4 (PDE4)-HSP20 complex attenuates the beta-agonist induced hypertrophic response in cardiac myocytes. J Mol Cell Cardiol 50:872–883

Smith FD, Reichow SL, Esseltine JL et al (2013) Intrinsic disorder within an AKAP-protein kinase A complex guides local substrate phosphorylation. Elife 2:e01319

Sprenger JU, Perera RK, Steinbrecher JH et al (2015) *In vivo* model with targeted cAMP biosensor reveals changes in receptor-microdomain communication in cardiac disease. Nat Commun 6:6965

Steinberg SF, Brunton LL (2001) Compartmentation of G protein-coupled signaling pathways in cardiac myocytes. Annu Rev Pharmacol Toxicol 41:751–773
Sun B, Li H, Shakur Y et al (2007) Role of phosphodiesterase type 3A and 3B in regulating platelet and cardiac function using subtype-selective knockout mice. Cell Signal 19:1765–1771
Sutherland EW, Rall TW (1958) Fractionation and characterization of a cyclic adenine ribonucleotide formed by tissue particles. J Biol Chem 232:1077–1091
Takimoto E, Belardi D, Tocchetti CG et al (2007) Compartmentalization of cardiac ß-adrenergic inotropy modulation by phosphodiesterase type 5. Circulation 115:2159–2167
Takimoto E, Champion HC, Li M et al (2005) Chronic inhibition of cyclic GMP phosphodiesterase 5A prevents and reverses cardiac hypertrophy. Nat Med 11:214–222
Terrenoire C, Houslay MD, Baillie GS, Kass RS (2009) The cardiac I_{Ks} potassium channel macromolecular complex includes the phosphodiesterase PDE4D3. J Biol Chem 284:9140–9146
Terrin A, Di Benedetto G, Pertegato V et al (2006) PGE1 stimulation of HEK293 cells generates multiple contiguous domains with different [cAMP]: role of compartmentalized phosphodiesterases. J Cell Biol 175:441–451
Tosaka S, Makita T, Tosaka R et al (2007) Cardioprotection induced by olprinone, a phosphodiesterase III inhibitor, involves phosphatidylinositol-3-OH kinase-Akt and a mitochondrial permeability transition pore during early reperfusion. J Anesth 21:176–180
Tsai EJ, Kass DA (2009) Cyclic GMP signaling in cardiovascular pathophysiology and therapeutics. Pharmacol Ther 122:216–238
Vandeput F, Wolda SL, Krall J et al (2007) Cyclic nucleotide phosphodiesterase PDE1C in human cardiac myocytes. J Biol Chem 282:32749–32757
Verde I, Pahlke G, Salanova M et al (2001) Myomegalin is a novel protein of the Golgi/centrosome that interacts with a cyclic nucleotide phosphodiesterase. J Biol Chem 276:11189–11198
Vettel C, Lindner M, Dewenter M et al. (2016) Phosphodiesterase 2 protects against catecholamine-induced arrhythmias and preserves contractile function after myocardial infarction. Circ Res (in press)
Wagner M, Mehel H, Fischmeister R, El-Armouche A (2016) Phosphodiesterase 2: anti-adrenergic friend or hypertrophic foe in heart disease? Naunyn Schmiedeberg's Arch Pharmacol 389(11):1139–1141
Wechsler J, Choi YH, Krall J, Ahmad F, Manganiello VC, Movsesian MA (2002) Isoforms of cyclic nucleotide phosphodiesterase PDE3A in cardiac myocytes. J Biol Chem 277: 38072–38078
Wilson LS, Baillie GS, Pritchard LM et al (2011) A phosphodiesterase 3B-based signaling complex integrates exchange protein activated by cAMP 1 and phosphatidylinositol 3-kinase signals in human arterial endothelial cells. J Biol Chem 286:16285–16296
Wong W, Scott JD (2004) AKAP signalling complexes: focal points in space and time. Nat Rev Mol Cell Biol 5:959–970
Xiang Y, Naro F, Zoudilova M, Jin SL, Conti M, Kobilka B (2005) Phosphodiesterase 4D is required for $ß_2$ adrenoceptor subtype-specific signaling in cardiac myocytes. Proc Natl Acad Sci U S A 102:909–914
Yan C, Miller CL, Abe J (2007) Regulation of phosphodiesterase 3 and inducible cAMP early repressor in the heart. Circ Res 100:489–501
Yang L, Liu G, Zakharov SI, Bellinger AM, Mongillo M, Marx SO (2007) Protein kinase G phosphorylates Cav1.2 alpha1c and beta2 subunits. Circ Res 101:465–474
Zaccolo M, Pozzan T (2002) Discrete microdomains with high concentration of cAMP in stimulated rat neonatal cardiac myocytes. Science 295:1711–1715
Zoccarato A, Fields LH, Zaccolo M (2016) Response to Wagner et al.: phosphodiesterase-2-anti-adrenergic friend or hypertrophic foe in heart disease? Naunyn Schmiedeberg's Arch Pharmacol:1143–1145
Zoccarato A, Surdo NC, Aronsen JM et al (2015) Cardiac hypertrophy is inhibited by a local pool of cAMP regulated by phosphodiesterase 2. Circ Res 117:707–719

cAMP Compartmentalisation and Hypertrophy of the Heart: 'Good' Pools of cAMP and 'Bad' Pools of cAMP Coexist in the Same Cardiac Myocyte

7

Anna Zoccarato and Manuela Zaccolo

Contents

Abstract

Pathological growth of cardiomyocytes (hypertrophy) is a major determinant of heart failure, a leading medical cause of mortality worldwide. Initially, cardiac hypertrophy is a compensatory response aimed at increasing cardiac output. However, prolonged cardiac hypertrophy progresses to contractile dysfunction, cardiac decompensation, and finally to heart failure. Although chronic elevation of cardiac cAMP leads to pathologic sequelae, enhancement of particular aspects of cAMP/PKA signalling benefits the failing heart, suggesting that different components of this pathway may have different consequences on cardiac hypertrophy and failure. The finding that cAMP signalling is compartmentalised and

A. Zoccarato
Cardiovascular Division, King's College London BHF Centre of Research Excellence, London, UK

M. Zaccolo (✉)
Department of Physiology, Anatomy and Genetics, BHF Centre of Research Excellence, University of Oxford, Sherrington Building, Parks Road, Oxford OX1 3QX, UK
e-mail: manuela.zaccolo@dpag.ox.ac.uk

V. Nikolaev, M. Zaccolo (eds.), *Microdomains in the Cardiovascular System*, Cardiac and Vascular Biology 3, DOI 10.1007/978-3-319-54579-0_7

that distinct, spatially segregated pools of cAMP mediate different functional effects in the heart may provide a rationale for what appear to be contrasting effects of this pathway on cardiac physiology and pathophysiology. In this chapter, we review some of the evidence in support of compartmentalisation of cAMP/PKA signalling, and we summarise recent findings indicating that distinct pools of cAMP, under the control of different phosphodiesterases, have opposing effects on cardiac myocytes hypertrophic growth. The relevance of these findings for the potential development of innovative approaches to reverse the course of ventricular remodelling is also briefly discussed.

7.1 cAMP Signalling in the Heart

3′,5′-Cyclic adenosine monophosphate (cAMP) is an ubiquitous second messenger that regulates multiple physiological functions by translating signals conveyed by a large number of extracellular stimuli into specific intracellular responses. At any given time, the intracellular level of cAMP is determined by a finely tuned balance between its production and its degradation. cAMP is generated upon binding of an extracellular second messenger (neurotransmitter, hormone, chemokine, lipid mediator or drug) to a G protein-coupled receptor (GPCR) and the subsequent activation of adenylyl cyclases (AC), the enzymes that convert ATP into cAMP. The degradation of cAMP is mediated by phosphodiesterases (PDEs) which, by hydrolysing cAMP to 5′-adenosine monophosphate (AMP), terminate cAMP signals.

In the heart, cAMP is involved in a variety of key functions including the control of force of contraction (inotropy) and relaxation (lusitropy) (Bers 2008). Stimulation of β-adrenergic receptors (β-ARs) in ventricular myocytes activates ACs to generate cAMP, which in turn activates cAMP-dependent protein kinase (PKA). PKA then phosphorylates several proteins that are part of the excitation-contraction coupling (ECC) system. These include L-type Ca^{2+} channels (L-TCC) at the plasmalemma, ryanodine receptors (RyR) and phospholamban (PLB) at the sarcoplasmic reticulum (SR) and troponin I (TnI) and myosin-binding protein C (MyBP-C) on the myofilaments (Bers 2002). PKA-mediated phosphorylation of L-TCCs and RyRs modulates the open probability of these two channels leading to increased amount of Ca^{2+} available for contraction. PKA-dependent phosphorylation of PLB, a negative regulator of the sarcoplasmic reticulum Ca^{2+} ATPase (SERCA), increases SR Ca^{2+} reuptake, while phosphorylation of TnI promotes the dissociation of Ca^{2+} from myofilaments. Overall, the effect is to increase force of contraction and to accelerate cardiac relaxation (Bers 2008). In physiological conditions, catecholamines released from the sympathetic nervous system in response to stress or exercise rapidly increase the cardiac output by activating β-ARs and cAMP/PKA signalling. In this way, the heart can adjust its work to circumstances of increased demand. However, persistent stimulation by catecholamines is detrimental and results in cardiac hypertrophic growth, ventricular dysfunction and, in the long term, heart failure (HF) (Fowler et al. 1986; Lohse et al. 2003; Barry et al. 2008). One of the mechanisms responsible for catecholamine-induced cardiac hypertrophy involves activation of β-ARs, ACs activation, cAMP generation and intracellular Ca^{2+} increase via PKA-mediated mechanisms (Barry et al. 2008). The notion that prolonged β-AR signalling is detrimental to cardiac function is supported by clinical studies showing that β-ARs blockade improves survival in HF patients (Bristow 2000) and by several studies in transgenic mice, demonstrating that chronic activation of the cAMP cascade result in cardiomyopathy. For example, cardiac-specific overexpression of β1-AR in mice leads to progressive cardiac hypertrophy and HF (Engelhardt et al. 1999), and transgenic mice overexpressing the α-subunit of Gs develop a phenotype similar to the β1-receptor transgene characterised firstly by increased responsiveness to catecholamines but later by cardiac myocyte hypertrophy, fibrosis and ultimately HF (Iwase et al. 1996). Mice in which AC5 is knocked out show no changes in cardiac hypertrophy levels compared to WT littermates upon pressure overload induced by thoracic aortic constriction; however, cardiac function is preserved in $AC5^{-/-}$, and cardiomyocytes apoptosis is reduced as opposed to WT mice (Okumura et al. 2003).

In apparent contradiction with the above findings, several studies provide evidence that enhancement of specific components of the cAMP cascade is beneficial to the hypertrophic heart. For example, overexpression of β2-AR improves cardiac function and has anti-apoptotic effects (Milano et al. 1994; Zhu et al. 2001), intracoronary delivery of a recombinant adenovirus encoding for AC5 improves left ventricular function at baseline conditions (Lai et al. 2000), and overexpression of the Ca^{2+}-stimulated AC8 or of AC6 improves cardiac function and catecholamine responsiveness at baseline (Gao et al. 1999; Lipskaia et al. 2000) and in cardiomyopathy conditions (Roth et al. 1999, 2002). Contrasting results are present in the literature also on the role of PKA in the development of cardiac hypertrophy. Transgenic mice expressing the α-catalytic subunit (Cα) of PKA in the heart develop dilated cardiomyopathy with reduced cardiac contractility, arrhythmias and susceptibility to sudden death (Antos et al. 2001). These abnormalities are also accompanied by PKA-mediated hyperphosphorylation of RyR2 and PLB, increased RyR2 activity and altered Ca^{2+} fluxes (Antos et al. 2001). In line with these findings, mice lacking the catalytic β (Cβ) subunit of PKA are protected from angiotensin II-induced cardiac hypertrophy and dysfunction (Enns et al. 2010). In addition, a recent study suggests that PKA-mediated phosphorylation of Shp2, a component of the AKAP-Lbc complex, plays a role in the development of isoproterenol-induced cardiac myocytes hypertrophy (Burmeister et al. 2015). In contrast, other studies demonstrate that PKA activity has protective roles in the development of cardiac hypertrophy. For example, deficiencies in the PKA pathway have been linked to human cardiomyopathy due to reduced phosphorylation of downstream targets such as cardiac troponin I (Zakhary et al. 1999). More recently, Ha and colleagues showed that HDAC5 is a substrate for PKA, and PKA-mediated phosphorylation of HDAC5 prevents its export from the nucleus leading to repression of pro-hypertrophic gene transcription and inhibition of cardiomyocytes hypertrophy (Ha et al. 2010). PKA also counteracts calcium-/calmodulin-dependent protein kinase II (CaMKII)-mediated activation of the myocytes enhance factor 2 (MEF2), which is known to promote hypertrophy and pathological remodelling, by regulated proteolysis of HDAC4 (Backs et al. 2011). PKA phosphorylates HDAC4 leading to the generation of an N-terminal HDAC4 cleavage product (HDAC4-NT) which in turn selectively inhibits the activity of the pro-hypertrophic MEF2 (Backs et al. 2011). PKA also phosphorylates the nuclear factor of activated T-cells (NFAT) thus preventing its nuclear translocation and the consequent activation of pro-hypertrophic genes (Sheridan et al. 2002; Zoccarato et al. 2015). Overall, the role of cAMP/PKA signalling in the development of cardiac pathology appears somewhat discrepant and remains to be fully clarified. The conflicting data reported in the literature suggest that disruption of different components of the βAR/cAMP/PKA signalling cascade has different consequences on the pathophysiology of cardiac hypertrophy and failure. Compartmentalisation of cAMP signals, that is, the coexistence within the cells of multiple spatially segregated cAMP/PKA signalling pathways exerting different or even opposing effects, may be at the basis of these contrasting findings.

7.2 Compartmentalisation of the cAMP Signalling Pathway

Phosphorylation of protein components involved in ECC is just one of the many functions of cAMP/PKA signalling in the heart. Cardiac myocytes express other targets that are affected by cAMP including the small G protein exchange proteins activated by cAMP (Epac) (de Rooij et al. 1998) involved in cardiomyocytes growth (Morel et al. 2005; Metrich et al. 2008) and the cyclic nucleotide-gated channels (HCN) regulating heart automaticity. In addition, besides those directly involved in ECC, PKA can phosphorylate many other proteins, including several nuclear transcription factors involved in regulation of cardiac hypertrophy and metabolism (Muller et al. 2001; Sheridan et al. 2002; Markou et al. 2004). Moreover, in addition to β-ARs, many other GPCRs whose activation triggers cAMP generation in response to a variety of hormones (e.g. glucagon, glucagon-like peptide, prostaglandins, histamine and dopamine) are present at the plasmalemma (Hayes et al. 1979; Farah 1983; Vila Petroff et al. 2001; Di Benedetto et al. 2008). Activation of different cardiac GPCRs leads to distinct physiological outcomes even when comparable amounts of intracellular cAMP are generated (Hayes et al. 1979). How this is possible has been a long-standing open question in the pharmacology of the heart (Hayes et al. 1979). It is now clear that the hormonal specificity of cAMP signals relies on the confinement to specific locations of GPCRs, ACs, and cAMP effectors and regulators which are assembled in macromolecular complexes to allow selective activation of distinct subsets of PKA and phosphorylation of specific targets in response to specific stimuli (Di Benedetto et al. 2008). The distribution of different PDE isoforms within distinct intracellular compartments results in unique regulation of cAMP levels at those different sites and thus dictates which PKA targets are phosphorylated at any given time.

First observations of the hormonal specificity of cAMP activity were made several decades ago in hearts extracts and in isolated cardiac myocytes (Corbin et al. 1977; Hayes et al. 1979, 1980; Keely 1977; Buxton and Brunton 1983). In those studies, important differences were observed when comparing hearts perfused with the β-adrenergic receptors agonist isoproterenol or the prostanoid EP receptor agonist prostaglandin E1 (PGE1). Despite increasing cAMP levels to a similar extent, these two receptors elicited different effects; β-AR stimulation activated the particulate fraction of PKA, increased the activation state of phosphorylase kinase (Hayes et al. 1979), glycogen phosphorylase (Keely 1977) and troponin I (Brunton et al. 1979) inducing positive inotropic effects (Hayes et al. 1979). In contrast, the cAMP increase generated by PGE1 led to activation of soluble PKA and did not produce any change in the contractile activity or in the activities of PKA substrates related to glycogen metabolism (Brunton et al. 1979; Hayes et al. 1979, 1980). Importantly, the different effects observed in hearts perfused with different cAMP-raising stimuli were not due to selective stimulation of different populations of cells within the heart but relied on the ability of individual cells to discriminate between the two hormones, as similarly different functional effects were found in isolated cardiac myocytes (Hayes et al. 1982; Buxton and Brunton 1983). In addition, selective stimulation of different β-AR subtypes was shown to elicit different physiological effects in the heart and in isolated cardiac myocytes. β1-AR activation phosphorylates key proteins involved in ECC with positive inotropic and lusitropic effects (Xiao and Lakatta 1993). In contrast, β2-AR activation produces only small inotropic and no lusitropic effects (Xiao et al. 1994). In isolated cardiac myocytes, β1-AR activation induces pro-apoptotic signalling, as opposed to β2-AR that has anti-apoptotic effects (Communal et al. 1999; Zhu et al. 2001).

The identification of two PKA isoforms (PKA-type I and PKA-type II) that differ in their regulatory subunit (RI and RII) and subcellular distribution (Corbin et al. 1977) and the finding that activation of the particulate (PKA-type II) or the soluble (PKA-type I) fraction of PKA results in distinct outcomes (Hayes et al. 1980) led to the idea that spatial confinement of cAMP is required to selectively activate distinct subsets of PKA (Brunton et al. 1981). Based on these findings, Hayes and Brunton proposed the idea of compartmentalisation of cAMP signalling. This model postulates that, upon hormone-mediated activation of a GPCR, cAMP accumulation and PKA activation happens within spatially confined subcellular compartments resulting in the phosphorylation of only a subset of, rather than all, PKA substrates available in the cell (Hayes and Brunton 1982). The model implied the existence of physical compartments where receptors, ACs, PKA and its targets are clustered together in spatially confined region of the cells (Hayes and Brunton 1982).

At the time of these early studies, cAMP content in cells and tissues was measured using radioimmunoassay techniques based on anti-cAMP antibodies and radiolabelled cAMP as a tracer. These classical methods, although robust and sensitive, measure cAMP in total cell lysates or tissue homogenates thus affording limited temporal resolution and providing no spatial resolution. In more recent years, several ground-breaking techniques have been developed that allow monitoring of physiologically relevant cAMP changes in living cells with high resolution both in space and time. A now widely used method relies on fluorescence resonance energy transfer (FRET)-based biosensors (Adams et al. 1991; Nikolaev and Lohse 2006; Berrera et al. 2008; Gesellchen et al. 2011; Stangherlin et al. 2014). The first genetically encoded FRET probe for cAMP was based on PKA (Zaccolo et al. 2000; Zaccolo and Pozzan 2002) where fluorescent proteins that act as FRET donor and acceptor moieties are fused to the regulatory (RII) and catalytic (C) subunits of PKA, respectively. When PKA is in its holoenzyme conformation (R2C2), the two fluorescent proteins are close enough for energy transfer to occur. When cAMP binds to the regulatory subunits of PKA, the holoenzyme dissociates and the FRET signal decreases; thus a decrease in the FRET signal corresponds to an increase in intracellular cAMP (Zaccolo et al. 2000; Zaccolo and Pozzan 2002). Over the years, various new FRET-based sensors have been developed, many of which are based on nearly full-length Epac or Epac cAMP-binding domains (DiPilato et al. 2004; Nikolaev et al. 2004; Ponsioen et al. 2004) or on cyclic nucleotide-gated channels (CNGCs) (Rich et al. 2001; Nikolaev et al. 2006). Some of these probes have been targeted to specific subcellular compartments (Di Benedetto et al. 2008; Sin et al. 2011; Sprenger et al. 2015), have been used for adenovirus delivery into adult cardiac myocytes (Rochais et al. 2004; Lehnart et al. 2005; Warrier et al. 2005; Fields et al. 2015) and have been utilised to generate tissue-specific transgenic mice overexpressing the cAMP sensor (Nikolaev et al. 2006; Sprenger et al. 2015). Thanks to these important developments, direct evidence supporting the concept of compartmentalisation of cAMP signalling has been collected (Zaccolo and Pozzan 2002; Nikolaev et al. 2006; Di Benedetto et al. 2008; Stangherlin et al. 2011). It is now widely accepted that compartmentalisation of the cAMP signalling pathway is achieved via fine coordination of several mechanisms including segregation of specific GPCRs and ACs in confined and specialised domains at the plasma membrane and targeting of PKA subsets, of specific PDE isoforms and of PKA targets to distinct subcellular structures via binding to A-kinase anchoring protein (AKAPs) scaffolds. Within these domains, PDEs are critically involved in shaping the amplitude and duration of the cAMP signal. Importantly, disruption of any of these mechanisms may alter compartmentalisation and can affect cAMP signalling specificity and lead to cardiac diseases.

7.3 Compartmentalisation of GPCRs and ACs

The concept of compartmentalisation of cAMP signalling implies that cAMP, once generated at the plasma membrane, does not homogeneously equilibrate in the cell as, by doing so, it would activate all available PKA. Instead, its levels must be somehow differentially regulated in different regions of the cell, leading to activation of a selected subset of PKA enzymes and, in turn, to phosphorylation of a limited number of PKA targets. The existence of clusters of receptors, ACs and targets in spatially confined compartments of the plasma membrane is one mechanism to explain how specificity of response is achieved upon activation of a given receptor. Different GPCRs have been shown to localise in different domains of the plasma membrane. In cardiac cells, regions of the plasma membrane rich in cholesterol and sphingolipids called lipid rafts and caveolae, subset of lipid rafts enriched in the protein caveolin and forming small flask-shaped invaginations of the plasma membrane, represent the two domains where GPCRs are mostly concentrated (Patel et al. 2008). β1-ARs localise in both caveolae and non-caveolae membranes; β2-ARs are mainly restricted in caveolae/lipid-rafts membrane domains whereas the PGE2 receptor EP2 does not reside in caveolin-enriched regions (Rybin et al. 2000; Ostrom et al. 2001; Xiang et al. 2002). Recently, experiments performed using a combination of scanning ion conductance microscopy and FRET-based imaging of cAMP demonstrated that in adult cardiomyocytes, β2-ARs are localised in transverse tubules (T-tubules), whereas β1-ARs are more uniformly distributed throughout the plasma membrane (Nikolaev et al. 2010). The cAMP generated upon β1-AR selective activation diffuses throughout the cell as opposed to stimulation of β2-ARs which results in a cAMP signal that does not propagate far from the membrane but remains confined in proximity of the site where the stimulus is applied (Nikolaev et al. 2006). Interestingly, cardiomyocytes derived from a rat model of chronic HF presented redistribution of β2-AR from the T-tubules to the non-tubular plasma membrane, and stimulation of β2-ARs in detubulated areas of failing cardiomyocytes produced a diffuse cAMP signal that propagated throughout the cytosol. These findings not only confirm that different GPCRs localise to different domains of the plasma membrane contributing to the generation of localised cAMP pools but also indicate that disruption of such localisation associates with heart disease.

AC5 and AC6 are the two main adenylyl cyclase isoforms expressed in cardiac myocytes; they share 65% protein sequence identity, are Ca^{2+} sensitive and have been shown to be compartmentalised (Willoughby and Cooper 2007). Specifically, AC6 localises in caveolin-rich region of the membrane (Ostrom et al. 2001) and in plasma membrane region outside T-tubules (Timofeyev et al. 2013) and is mainly associated with β1-ARs (Ostrom et al. 2001; Timofeyev et al. 2013). In contrast, AC5 is mainly found in T-tubular region of the plasma membrane via direct interaction with caveolin-3 (Timofeyev et al. 2013). Interestingly, adenylyl cyclases also form complexes with AKAPs (Dessauer 2009). For example, mAKAP associates with AC activity in the heart and isolated cardiac myocytes (Kapiloff et al. 2009). mAKAP interacts with AC5 but not AC6, as demonstrated by complete lack of mAKAP-associated AC activity in AC5 knockout hearts (Kapiloff et al. 2009). Although many details regarding the molecular nature and topology of these interactions still remain to be defined, it appears that spatial co-localisation of ACs with GPCRs and other key signalling molecules significantly contributes to specificity and efficiency of the cell response to cAMP signals.

7.4 PKA Compartmentalisation via AKAPs

When Steinberg and Brunton illustrated the model of cAMP signalling compartmentalisation, they postulated that not all cAMP can access all PKA and not all PKA has access to all possible substrates (Steinberg and Brunton 2001). This implies that subsets of PKA and of PKA targets must somehow co-localise in spatially confined subcellular domains of the cell and be accessible to selected pools of cAMP. The discovery of AKAPs provided a molecular basis for the idea that activation of a restricted subset of PKA leads to a specific functional effect. AKAPs are a family of structurally diverse but functionally related proteins defined on the basis of their ability to bind to PKA. AKAPs present distinct protein-lipid or protein-protein targeting domains that allow tethering the AKAP-PKA complex to defined subcellular structures, membranes or organelles (Skroblin et al. 2010). Anchoring of PKA to AKAPs is accomplished in a PKA isoform-dependent manner by a 14–18 amino acid amphipathic α-helix on the AKAPs docking into a hydrophobic pocket formed by the dimerisation/docking (D/D) domains of the PKA regulatory subunit dimer (Wong and Scott 2004; Gold et al. 2006; Kinderman et al. 2006). Early studies described PKA-RII as the isoform mainly associated with the particulate fraction of cardiac myocyte lysates whereas PKA-RI was mostly found in the soluble fraction (Corbin and Keely 1977; Corbin et al. 1977). However, subsequent investigations showed that there are some AKAPs that can specifically anchor PKA-RI, for example, sphingosine kinase interacting protein (SKIP) which tether PKA-RI to mitochondria (Means et al. 2011) or small membrane AKAP (smAKAP) targeting RI to the plasma membrane (Burgers et al. 2012). AKAPs that bind both PKA isoforms have also been described. These include D-AKAP1 and D-AKAP2 (Huang et al. 1997), ezrin (Ruppelt et al. 2007) and OPA1 (Pidoux et al. 2011). The development of a variety of short peptides able to disrupt AKAP-PKA interactions has contributed to our understanding of the functional relevance of AKAPs in many different cellular processes. Generally, AKAP/PKA-disrupting peptides consist of a stretch of amino acids forming an amphipathic helix that mimics the AKAP region responsible for interacting with PKA. Binding of these disruptor peptides to the regulatory subunits of PKA impairs PKA-AKAP interaction (Carr et al. 1991; Carlson et al. 2006; Gold et al. 2006). The first peptide generated was derived from a human thyroid PKA anchoring protein named Ht31 (Carr et al. 1992). Being able to impair the interaction of both PKA-RI and PKA-RII isoforms with AKAPs, *in vitro* as well as in intact cells, Ht31 has been extensively used as a tool to study the physiological importance of PKA anchoring in specific subcellular compartments (Fink et al. 2001; Lygren et al. 2007; McConnell et al. 2009). Peptides able to specifically disrupt interactions between PKA-RI/AKAPs and PKA-RII/AKAPs have also been developed. These peptides, including SuperAKAP-IS (Gold et al. 2006) and the RI-anchoring disruptor (RIAD) (Carlson et al. 2006), selectively uncouple anchored signalling events that are carried out by the PKA type II or type I holoenzyme, respectively, thus allowing to dissect the individual contribution of anchored PKA-RII or PKA-RI in mediating different cellular responses (Di Benedetto et al. 2008; Stangherlin et al. 2011; Terrin et al. 2012). It should be noted that all these peptides have quite a broad impact on the PKA signalling pathway as once expressed or delivered into cells they will affect multiple AKAP/PKA interactions regardless of the type of AKAP, the targets anchored to it and the physiological role of these multi-protein complexes. However, they are useful tools to study and characterise AKAP interactions with different PKA isoforms. For example, in experiments where FRET sensors for cAMP fused to the D/D domain of the regulatory subunit of PKA-RI or of PKA-RII (RI_epac and RII_epac, respectively) were expressed in cardiac myocytes in the presence of short peptides able to disrupt AKAP-PKA interactions, it was demonstrated that selective activation of PKA-RI and PKA-RII by distinct local pools of cAMP relies on binding of PKA to AKAPs (Di Benedetto et al. 2008). Confocal analysis and fluorescent recovery after photobleaching (FRAP) experiments in cardiac myocytes expressing these sensors confuted the paradigm that PKA-RI is primarily cytosolic and confirmed that a substantial fraction of PKA-RI is anchored to subcellular structures (Di Benedetto et al. 2008). Differences in the ability of the two PKA isoforms to respond to different cAMP signals generated upon selective activation of specific GPCRs were also assessed using targeted FRET-based sensors and AKAP/PKA disruptor peptides. In these studies, activation of β-AR generated a cAMP signal that is preferentially detected in the PKA-RII compartment and is associated with specific phosphorylation of PLB, TnI, and β2-AR. In contrast, β-AR stimulation generated only a small cAMP increase in the PKA-RI compartment whereas PGE1, glucagon and glucagon-like peptide 1 stimulation induced a cAMP signal mainly detected in the PKA-RI compartment which was not associated with the phosphorylation state of PLB, TnI or β2-AR (Di Benedetto et al. 2008). Disruption of RI_epac or RII_epac interaction with RIAD and SuperAKAP-IS, respectively, resulted in the loss of compartmentalisation of the sensors and in the disruption of cAMP signal specificity, and the cAMP signal generated by both β-AR and PGE1 stimulation was similarly detected by both sensors (Di Benedetto et al. 2008). All together, these experiments indicate that specificity of response to a given stimulus is mediated by a spatially restricted pool of cAMP, by selective activation of a specific AKAP-anchored PKA subset and by the consequent specific phosphorylation of the PKA targets localised in proximity to the AKAP-PKA complex. This study also confirmed that anchoring of PKA to AKAPs is crucial for achieving cAMP signal specificity as when compartmentalisation of RI_epac and RII_epac sensors is disrupted by competing peptides, the ability of specific cAMP pools generated by distinct GPCRs to be specifically sensed by different PKA isoforms is lost (Di Benedetto et al. 2008).

Disruption of localisation of cAMP signalling components and changes in their expression levels are associated to human heart disease (Aye et al. 2012). Compared to healthy controls, heart samples from patients affected by dilated cardiomyopathy present a 50% reduction in PKA-RI and PKA-RII levels, an increase in PDE2A and in PKGIα levels and severe alterations of PKA-R association with different AKAPs. Specifically, an increased association of PKA-R with AKAP2, AKAP7 and SPHKAP but a significant decrease in PKA-R interaction with myofilaments proteins such as troponinT, troponin I, Titin, and cardiac myosin-binding protein C (cMyBPC3) were observed in diseased hearts (Aye et al. 2012). Overall, these data indicate that changes in localisation, expression levels and protein-protein interactions of various components of the cAMP signalling pathway occur in the setting of heart disease. These alterations will likely result in disruption of cAMP signalling compartmentalisation and of specificity of response. Although in some cases these changes may be adaptive, in other cases, they may contribute to the aetiology of the disease. In line with this idea is the observation that dramatic changes in the intracellular localisation of PKA-RI, PKA-RII, PDE4B and PDE4D occur in an in vitro model of cardiac hypertrophy (Fields et al. 2015). These changes in the compartmentalisation of key components of cAMP signalling pathways may contribute to the blunted cAMP response upon β-AR stimulation observed in these cells. In physiological conditions, β-AR stimulation of cardiac myocytes results in a significant increase in cAMP levels mainly in the compartments associated with PKA-RII (Di Benedetto et al. 2008; Fields et al. 2015); however on sustained exposure to catecholamines, the cAMP response to isoproterenol is reduced in this compartment, and the difference in cAMP levels in the PKA-RI and PKA-RII domains is abolished (Fields et al. 2015). Therefore, β-AR-mediated activation of PKA-RII and, as a consequence, regulation of ECC may be severely compromised. Further studies are required to elucidate which cAMP/PKA signalling components are altered during HF and whether they are compensatory or detrimental in the pathophysiology of the disease. Interestingly, when expression levels of ACs and PDE isoforms are quantified in explanted myocardium from adults with dilated cardiomyopathy (DCM), children with DCM and children with single ventricle congenital heart disease of right ventricular morphology (SRV), unique changes in expression profiles are observed (Nakano et al. 2016). For example, AC5 is exclusively upregulated in adult with DCM treated with PDE3 inhibitors; in paediatric DCM hearts, AC5 and AC6 are unchanged, AC2 is upregulated, and some PDE4 isoforms are downregulated, whereas paediatric SRV samples present higher levels of AC5, PDE1C, PDE3A1, PDE3B and PDE4D5 (Nakano et al. 2016). These findings suggest that alterations in the expression profile of specific cAMP signalling components can vary depending on the aetiology of the disease, the age and also the therapeutic treatment administered (Nakano et al. 2016).

7.5 PDEs and cAMP Compartmentalisation

To date, 21 genes encoding for about 60 different isoforms of PDEs have been identified in mammals (Bender and Beavo 2006; Lugnier 2006). PDEs are classified as belonging to one of 11 different families on the basis of their amino acid sequence, substrate selectivity, kinetics and their regulatory and pharmacological properties (Bender and Beavo 2006; Lugnier 2006). The extensive diversity of the PDE isoforms allows for individual variants to perform specific physiological roles, although there may be some redundancy among isoenzymes (Bender and Beavo 2006). Given that cAMP is a small and highly hydrophilic molecule that displays a diffusion coefficient of about 500 $\mu m^2/s$ (Nikolaev et al. 2004), ligand binding to any GPCRs is expected to generate a rapid flux of cAMP which should quickly fill the whole cell. However, this would lead to non-selective activation of all PKA subsets in spite of their localisation in distinct compartments, and specificity of response would be abolished. A number of mechanisms have been proposed to explain how equilibration of cAMP levels throughout the cell is prevented. One hypothesis suggests that a physical barrier formed by membranes or organelles such as the endoplasmic reticulum (ER) and plasma membrane invaginations is responsible for preventing cAMP diffusion into the bulk cytosol (Rich et al. 2000). Another hypothesis involves PKA-mediated buffering of cAMP. According to this hypothesis, cAMP diffusion may be limited by binding of cAMP to the regulatory subunits of PKA which is known to present low diffusivity (Saucerman et al. 2006) and to be expressed in excess of C subunits. Recently, Nikolaev and colleagues confirmed that localised subsets of PKA play indeed a role in preventing cAMP diffusion to other compartments (Nikolaev et al. 2010). In fact, displacement of PKA from its T-tubular localisation resulted in propagation of β2-AR induced cAMP signals to more distant parts of the cell (Nikolaev et al. 2010). However, as PKA-mediated phosphorylation of PDE4 can locally increase its activity, thus resulting in a negative feedback control of cAMP levels (Sette and Conti 1996), it is possible that displacement of PKA from its anchor sites may lead to a decreased PDE4 activity and therefore to increased cAMP diffusion from T-tubules. Accordingly, selective inhibition of PDE4 resulted in β2-AR-induced cAMP signals that propagated from locally activated T-tubules (Nikolaev et al. 2010). Therefore, although local buffering by PKA seems to play a role in preventing cAMP diffusion, local activity of PDEs is clearly involved in cAMP compartmentalisation.

Early evidence of a role of PDEs in maintaining cAMP signals compartmentalisation came from a study investigating local and distant effects of cAMP generated upon local β-AR receptors stimulation with isoproterenol (ISO) (Jurevicius and Fischmeister 1996). In the presence of the non-selective PDEs inhibitor 3-isobutyl-1-methylxanthine (IBMX), distant effects of ISO on L-type Ca^{2+} channel current were strongly potentiated as compared to ISO stimulation alone, thus suggesting that cAMP diffusion is limited by PDEs activity (Jurevicius and Fischmeister 1996). The role of PDEs in cAMP diffusion was further investigated using FRET-based sensors. These studies demonstrated that upon catecholamines stimulation, the level of cAMP almost exclusively increases in restricted microdomains within cardiac myocytes that correspond to sites where PKA-RII localises. Addition of the non-selective PDEs inhibitor IBMX dissipated the boundaries between the cAMP microdomains resulting in homogeneous distribution of cAMP in the cell (Zaccolo and Pozzan 2002). These studies provided the first direct evidence of the generation of spatially restricted microdomains of cAMP controlled and maintained by the activity of PDEs. Of note, stimulation of cardiac myocytes with saturating concentration of forskolin, a global ACs activator, generated an increase in cAMP that was also uniform throughout the cells. Inhibition of PDEs in the absence of any agonist also resulted in elevation of cAMP, suggesting that ACs are constantly active in basal conditions and the levels of cAMP are kept low thanks to PDEs activity (Zaccolo and Pozzan 2002).

The use of family-selective PDE inhibitors and of genetic knock-down or knock-out approaches has allowed to assess the contribution of individual PDE families in the compartmentalisation of cAMP signalling pathways in cardiac myocytes (Zaccolo and Pozzan 2002; Mongillo et al. 2004, 2006; Rochais et al. 2006; Di Benedetto et al. 2008; Stangherlin et al. 2011; Fields et al. 2015; Zoccarato et al. 2015). However, the pharmacological approach and, in several cases, the genetic ablation approach do not allow to discriminate between the contributions of individual PDE isoforms within each family. To overcome this limitation, dominant negative approaches have been developed. This strategy consists in the overexpression of a catalytically inactive mutant of the specific PDE isoform of interest which, by displacing the cognate endogenous PDE enzymes from their anchor sites, ablates the PDE activity at those specific sites resulting in a localised increase in cAMP (Baillie et al. 2003). This manoeuvre has been used in the past decade to specifically pinpoint the role of individual PDE isoforms in regulating locally confined cAMP signals (Baillie et al. 2003; Lynch et al. 2005; Stangherlin et al. 2011; Zoccarato et al. 2015).

The heart expresses multiple cAMP-degrading PDEs: PDE1, PDE2, PDE3, PDE4 and PDE8 (Lugnier 2006). Using a FRET-based approach, individual PDE families have been shown to have distinct physiological roles, and evidence supports the notion that the specific intracellular localisation of PDEs may be the mechanism by which distinct PDE isoforms can control cAMP gradients in specific functional compartments (Mongillo et al. 2004, 2006). For example, a striking difference was observed between the effect of PDE3 and PDE4 selective inhibition in the control of cAMP gradients generated upon β-AR stimulation in rat ventricular myocytes (Mongillo et al. 2004). Selective inhibition of about 10% of the total PDE4 activity resulted in a dramatic increase in cAMP, whereas total inhibition of PDE3 had only a marginal effect on cAMP levels generated upon β-AR stimulation. These findings demonstrate that PDE4, rather than PDE3, is functionally coupled with the pool of ACs activated in response to β-AR stimulation (Mongillo et al. 2004). These findings could not be explained by differences in the enzyme concentrations as PDE3 represent 30% of the total PDEs activity in these cells. Importantly, PDE3 and PDE4 enzymes were shown by immunostaining to localise to distinct compartments within the myocyte (Mongillo et al. 2004). Similarly, experiments performed in ventricular myocytes isolated from rat and mouse confirmed that PDE4 plays a key role in hydrolysing the cAMP pool generated in response to β-AR stimulation (Rochais et al. 2006) and in regulating PKA-mediated phosphorylation of key proteins involved in the excitation contraction coupling, including L-TCC, RyR2 and PLB (Verde et al. 2001; Lehnart et al. 2005; Kerfant et al. 2007).

PDE2 can hydrolyse both cAMP and GMP, and it is also referred to as cGMP-stimulated PDE given that PDE2 activity towards cAMP is enhanced upon cGMP binding to its regulatory GAF domains (Martinez et al. 2002). Despite contributing to only about 1% of the total PDEs activity in rat cardiac myocytes, PDE2 tightly controls the pool of cAMP generated in these cells upon β-AR stimulation (Mongillo et al. 2006). Using a FRET-based approach and a selective inhibitor, PDE2 was demonstrated to selectively shape the cAMP response to catecholamines via a mechanism involving activation of β3-ARs and activation of endothelial nitric oxide synthase (eNOS). The consequent NO-mediated stimulation of soluble guanylyl cyclases and synthesis of cGMP triggers a regulatory mechanism via activation of the cAMP hydrolysing activity of PDE2. Again PDE2 was shown to localise to distinct subcellular sites including the plasma membrane in correspondence of the cell-to-cell junctions and the sarcomeric Z-lines, thus indicating that PDE2 activity is highly compartmentalised to control specific pools of cAMP (Mongillo et al. 2006). Intriguingly, these data also suggest a mechanism of crosstalk between cGMP and cAMP signalling whereby cGMP modulates β-AR-generated cAMP by enhancing PDE2 activity. This mechanism was further explored in a subsequent study (Stangherlin et al. 2011). Using cGMP and cAMP FRET-based sensors specifically targeted to PKA-RI and PKA-RII-compartments, the authors demonstrated that in cardiac myocytes cGMP can strongly but differentially modulate cAMP pools in the PKA-RI or PKA-RII compartments via cGMP-mediated activation of PDE2- or cGMP-mediated inhibition of PDE3. Interestingly, this modulation differs between the two domains depending on the PDE isoforms associated with each compartment and on what type of guanylyl cyclase generates the cGMP (Stangherlin et al. 2011). Specifically, upon catecholamines stimulation, PDE2 activity is more associated with PKA-RII compartments as opposed to PDE3 activity which is preferentially coupled with PKA-RI compartments. Therefore, increase in cGMP levels upon activation of soluble guanylyl cyclases (sGC) results in cGMP-dependent inhibition of PDE3 in PKA-RI compartments and potentiation of PKA-RI activation. Conversely, cGMP-mediated activation of PDE2 in PKA-RII compartments results in a significantly attenuated activation of PKA-RII subsets (Stangherlin et al. 2011). In this way, the cGMP generated upon activation of soluble guanylyl cyclases (sGC) has distinct and opposite effects on cAMP pools in PKA-RI and PKA-RII domains. The study also shows that cGMP produced by atrial natriuretic peptide A (ANP)-activated particulate guanylyl cyclases (pGC) affects exclusively the cAMP pool that activates PKA-RII compartments while has no impact on cAMP pools in the PKA-RI compartments (Stangherlin et al. 2011). This demonstrates that modulation of cAMP pools generated on β-AR activation depends not only on the PDE isoforms associated with each compartments but also on the source of cGMP.

7.6 cAMP/PKA Signalling in Cardiac Hypertrophy: Opposing Effects of Spatially Distinct Pools of cAMP

Severe alterations in adrenergic signalling as well as changes in cAMP compartmentalisation have been described over the years as featuring in the pathogenesis of heart disease. Alterations in β-AR signal transduction are a well-established hallmark in the development of HF, characterised primarily by the downregulation of β1-ARs, desensitisation and reduced coupling of the receptors with the Gs-AC system as well as changes in β2-AR signalling and localisation (Bristow et al. 1982, 1986; Lefkowitz et al. 2000; Nikolaev et al. 2010). Downregulation of PKA-RI and PKA-RII (Zakhary et al. 1999; Aye et al. 2012) as well as alterations of PKA targets phosphorylation that are not associated with changes in PKA-C activity or overall reduction in intracellular cAMP concentration have also been described in diseased hearts (Zakhary et al. 1999). PDEs expression and activity levels were found to be changed both in *in vitro* and *in vivo* models of cardiac hypertrophy and failure (Abi-Gerges et al. 2009; Mokni et al. 2010; Aye et al. 2012), and disruptions of AKAP-PKA interactions and polymorphisms or mutations in AKAPs have also been associated with heart disease (Chen et al. 2007; Aye et al. 2012; Soni et al. 2014). Despite alterations in cAMP signalling in failing hearts are well established, whether these changes are beneficial, detrimental or simply secondary adaptive effects remains unclear (Movsesian 2004).

While substantial progress has been made in the pharmacological treatment of HF with the introduction of β-blockers therapy and its significant impact on morbidity and mortality in patients with left ventricular dysfunction and HF (Domanski et al. 2003), the administration of neurohormonal antagonists only delays, but does not halt or reverse, the progression of the disease. Positive inotropic treatments using PDE3 inhibitors such as vesnarinone, milrinone or enoximone have been used with the aim of improving myocardial contractility and the overall clinical outcome in patients with systolic HF. Although beneficial in the short term, such treatments are associated with severe adverse effects including increased mortality due to arrhythmias and sudden death (Uretsky et al. 1990; Packer et al. 1991; Cowley and Skene 1994; Cohn et al. 1998). The rationale for using PDE3 inhibitors is to improve myocardial contractility via restoration of cAMP content towards normal levels primarily in the SERCA compartment where both PDE3A and phospholamban localise (Movsesian et al. 1991). As mentioned above, PDE inhibitors, although selective for specific families, do not discriminate between different isoforms within a family. Given the compartmentalisation of cAMP signalling and of PDEs in heart cells, it is therefore possible that the non-selective inhibition of all PDE3 isoforms resulting from the use of these drugs may affect signalling at multiple compartments yielding both the beneficial short-term effects and the adverse long-term consequences observed in the clinic. Selective modulation of the activity of individual PDE isoforms at specific sites may, on the other hand, reduce side effects.

A detailed understanding of the localisation, regulation and function of individual PDE isoforms is required to explore whether novel therapeutics targeting these enzymes could more effectively treat cardiovascular disease via selective manipulation of specific pools of cAMP. The available data on the role of compartmentalised phosphodiesterases (PDEs) in the regulation of cAMP signals in cardiac hypertrophy and HF remains however limited and sometime contradictory (Abi-Gerges et al. 2009; Mokni et al. 2010; Wagner et al. 2016; Zoccarato et al. 2016). For example, a study showed that the expression and the activity of PDE3 and PDE4 are markedly reduced in cardiomyocytes isolated from a rat model of pressure overload (Abi-Gerges et al. 2009). In contrast, in a rat model of cardiac hypertrophy induced by angiotensin II infusion, PDE4 activity was overall increased despite no changes in PDE4A, PDE4B and PDE4C and decreased PDE4D mRNA levels (Mokni et al. 2010). The reason for these contradictory results is not clear, although the use of different models of hypertrophy as well as different approaches to measuring the contribution of PDEs to cAMP hydrolysis may be responsible for the discrepancies. Moreover, although indicating that changes in the activity and expression levels of different PDE isoforms occur in the settings of cardiomyocytes hypertrophy, most studies do not provide evidence supporting a beneficial or detrimental effect of these alterations in the pathophysiology of the disease nor on the specific cAMP compartment(s), cAMP signalling pathway and subset of targets that are preferentially affected by these changes.

A recent study has provided new evidence on the role of different PDEs in cardiomyocytes hypertrophic growth and has demonstrated the existence of distinct pools of cAMP under the control of specific PDEs which exert opposing effects on the development of catecholamine-induced cardiac hypertrophy (Zoccarato et al. 2015). In this study, the cAMP pools controlled by PDE3 and PDE4 were shown to mediate pro-hypertrophic responses as, when either of those two PDEs was selectively inhibited in otherwise unstimulated cells, hypertrophic growth was observed. In contrast, selective inhibition of PDE2 with the selective drug BAY 60–7550, despite elevating intracellular cAMP levels to the same extent as inhibition of PDE3, did not induce hypertrophy, but, strikingly, the drug was able to block the hypertrophy induced by β-AR stimulation (Zoccarato et al. 2015). In line with these findings, both *in vitro* and *in vivo* overexpression of PDE2, but not overexpression of PDE3 or PDE4, was shown to be sufficient to induce hypertrophy. These data challenged a previous report by Mehel et al. suggesting that adenoviral overexpression of PDE2 may attenuate the hypertrophy induced by *in vitro* exposure to catecholamines (Mehel et al. 2013). A more recent study from the same group, however, shows that transgenic mice overexpressing PDE2 have increased cardiac size compared to wild type and develop significantly greater hypertrophy after myocardial infarction (Vettel et al. 2016), supporting a pro-hypertrophic effect of PDE2 overexpression.

PDE2, PDE3 and PDE4 have been shown to localise at distinct subcellular sites both in neonatal and adult rat ventricular myocytes (Mongillo et al. 2004, 2006; Fields et al. 2015; Zoccarato et al. 2015). Therefore, it is reasonable to hypothesise that, by being localised in a specific intracellular compartment, PDE2 modulates hypertrophic responses by controlling a specific pool of cAMP and a specific subset of PKA and PKA targets which are spatially distinct from those controlled by PDE3 and PDE4. The importance of the specific subcellular localisation of PDEs in regulating cardiac hypertrophy was confirmed using a dominant negative approach. Using this strategy, displacement of endogenous PDE2 with a catalytically inactive mutant was shown to significantly reduce β-AR-induced hypertrophic growth. On the contrary, displacement of PDE3 and PDE4 from their intracellular anchor sites promoted hypertrophy. The cAMP pool controlled by PDE2 was found to mediate the anti-hypertrophic responses, at least in part, via PKA-mediated phosphorylation of the nuclear factor of activated T-cells (NFAT) (Zoccarato et al. 2015). NFAT is a transcription factor that when in the nucleus promotes activation of foetal cardiac genes and a pro-hypertrophic response (Molkentin et al. 1998). A key regulator of NFAT nuclear translocation is calcineurin, a calcium/calmodulin-regulated phosphatase which, once activated, dephosphorylates NFAT unmasking its nuclear localisation signal and promotes its translocation to the nucleus (Molkentin et al. 1998). PKA can phosphorylate NFAT at Ser245, Ser269 and Ser294 thus preventing its nuclear translocation (Sheridan et al. 2002). Immunoprecipitation experiments confirmed PKA-mediated phosphorylation of NFAT, reduced NFAT nuclear translocation and reduced transcription of NFAT-regulated genes, selectively on inhibition of PDE2 but not of PDE4 (Zoccarato et al. 2015).

The hypothesis that PDE2 activity plays a role in cardiac hypertrophy is supported by several other studies. PDE2 activity was shown to be significantly upregulated in pressure-overloaded ventricles of aortic-banded rats, although no significant changes in mRNA levels of PDE2 were detected (Yanaka et al. 2003). Moreover, PDE2 was recently shown to be sixfold upregulated in hearts from patients with dilated cardiomyopathy (Aye et al. 2012). PDE2 represents an interesting therapeutic target as, in addition to its role in controlling hypertrophic growth, its inhibition mediates positive inotropic effects (Mongillo et al. 2006; Mehel et al. 2013). The localisation of PDE2 at multiple sites within the cardiac myocyte (Mongillo et al. 2006) may underpin these multiple functions. Thus inhibition of a subset of PDE2 may regulate the amplitude of the Ca^{2+} transient in response to β-adrenergic stimulation (Mongillo et al. 2006; Mehel et al. 2013), and, at the same time, inhibition of a different subset of PDE2 may block the nuclear translocation of NFAT (Zoccarato et al. 2015).

As mentioned above, an increase in intracellular Ca^{2+} concentration can lead to CaN activation leading to dephosphorylation and nuclear translocation of NFAT. Inhibition of PDE2 and the consequent increase in cAMP levels is expected to raise intracellular Ca^{2+} concentration and therefore lead to CaN activation, in apparent contrast with the reported anti-hypertrophic effect of PDE2 inhibitors. It should be noted however that the source of Ca^{2+} that activates pathological hypertrophy is not clearly defined, and several studies suggest that the regulation of CaN/NFAT is not under the control of beat-to-beat calcium oscillations but rather relies on a more specialised micro-compartment of Ca^{2+}. It has been suggested, for example, that TRPC channels provide local Ca^{2+} increase in defined microdomains within cardiac myocytes and the inhibition of these channels significantly reduces the activity of the calcineurin-NFAT signalling both in transgenic mice and in cultured neonatal myocytes (Wu et al. 2010). In other studies, inhibition of L-TCCs within the caveolin-3 membrane domain was shown to block almost all of the Ca^{2+} influx-induced NFAT nuclear translocation without impacting on myocyte contractility (Makarewich et al. 2012). This finding is consistent with evidence suggesting the existence of a caveolin-3-enriched membrane domain in which L-TCCs are clustered together with AKAP150 and CaN (Nichols et al. 2010). CaN and NFAT could potentially be localised in specific subcellular compartments in close proximity to signalling complexes at the plasma membrane which also contain the channels that conduce the Ca^{2+} responsible for CaN activation. An earlier study demonstrates the co-localisation of CaN and α-actinin at the Z-line via binding to calsarcin-1, a member of the sarcomeric CaN-binding proteins (Frey et al. 2000). The Z-line, positioned at the interface between the cytoskeleton and the sarcomere, may play a role as mechanosensor coupling mechanical stresses induced by physical stretching with induction of hypertrophic gene expression (Sadoshima and Izumo 1997). The presence of CaN at the Z-line of cardiac cells could then link localised Ca^{2+} signals to pro-hypertrophic responses. A plausible hypothesis is therefore that PDE2 inhibition generates a pool of cAMP that enhances the Ca^{2+} transient upon adrenergic stimulation resulting in enhanced inotropy but does not affect the localised microdomain of Ca^{2+} that activates CaN. Concomitantly, PDE2 inhibition generates a distinct cAMP pool that leads to PKA-mediated phosphorylation of NFAT and reduced NFAT nuclear translocation (Zoccarato et al. 2015). Whether PDE2, CaN and NFAT are contained within the same subcellular domain and therefore whether PDE2 controls a pool of cAMP at a site where PKA-mediated phosphorylation of NFAT directly counteracts CaN-dependent dephosphorylation of NFAT remains an open question.

Given their essential role in regulating intracellular cyclic nucleotide concentrations and their multiple functional effects, PDEs have long been recognised as therapeutic targets to treat a variety of diseases, including HF (Maurice et al. 2014). For example, inhibition of PDE5 in rodents has been experimentally shown to attenuate and reverse cardiac hypertrophy, fibrosis and contractile dysfunction (Takimoto et al. 2005; Nagayama et al. 2009) and to promote cell survival (Salloum et al. 2003) via enhancement of cGMP/PKG signalling pathways. However, a randomised clinical trial testing the effects of PDE5 inhibition on exercise capacity and clinical status in HF gave disappointing results (Redfield et al. 2013). It has been suggested that a reasons for these negative results could be the lack of PDE5A upregulation in human HF and the fact that HF is associated with depressed NO signalling and generation of cGMP levels, which would make inhibition of a cGMP-degrading PDEs less effective (van Heerebeek et al. 2012). PDE2 may not suffer from the same drawbacks as PDE5. Unlike PDE5, the anti-hypertrophic effect of PDE2 inhibition relies on PKA activation and is PKG independent (Zoccarato et al. 2015). Therefore, low cGMP levels would not be a limitation. In addition, although cGMP can enhance PDE2 activity several folds, cGMP is not required for PDE2 enzymatic activity (Martins et al. 1982). Also, unlike PDE5, which has been demonstrated to degrade predominantly a cGMP pool generated by the soluble (NO-dependent) guanylyl cyclase, a number of studies (Castro et al. 2006; Stangherlin et al. 2011) have demonstrated that PDE2 is coupled to a pool of cGMP that is generated by the particulate (NO-independent) guanylyl cyclase; therefore, a deficit in NO signalling may not significantly impact on PDE2 activity.

Conclusions

Specificity of cAMP signalling is achieved via compartmentalisation. A large body of evidence indicates that spatial control of cAMP-PKA signals is critical for maintaining healthy hearts whereas alteration of β-AR expression and localisation, changes in AKAP-PKA interactions, alterations in the expression profile or subcellular localisation of different PDE isoforms have been associated with pathology. However, a systematic description of these changes and their individual contribution to the development of cardiac disease is still missing. The evidence that PDE2 controls a spatially and functionally distinct pool of cAMP with opposite effect in the pathophysiology of cardiac hypertrophy than the cAMP pools controlled by PDE3 and PDE4 opens new exciting possibilities for therapeutic intervention. To assess whether we can take advantage of compartmentalised cAMP signalling for therapeutic purposes, we are faced with the challenge of acquiring a detailed understanding of the molecular basis of the spatial organisation, of the signalling components and of the regulation of the relevant cAMP/PKA/PDE signalling microdomains. As we unravel the intricate particulars of such organisation, we may be able to devise ways we can target only relevant signalling cascades while leaving off-target signalling pathways unaffected.

Acknowledgements This study was supported by the British Heart Foundation (PG/10/75/28537 and RG/12/3/29423) and the BHF Centre of Research Excellence, Oxford (RE/08/004) to M. Z.

Compliance with Ethical Standards

Conflict of Interest Statement The authors declare that they have no conflict of interest.

References

Abi-Gerges A, Richter W, Lefebvre F, Mateo P, Varin A, Heymes C, Samuel JL, Lugnier C, Conti M, Fischmeister R, Vandecasteele G (2009) Decreased expression and activity of cAMP phosphodiesterases in cardiac hypertrophy and its impact on beta-adrenergic cAMP signals. Circ Res 105(8):784–792. doi:10.1161/CIRCRESAHA.109.197947

Adams SR, Harootunian AT, Buechler YJ, Taylor SS, Tsien RY (1991) Fluorescence ratio imaging of cyclic AMP in single cells. Nature 349(6311):694–697

Antos CL, Frey N, Marx SO, Reiken S, Gaburjakova M, Richardson JA, Marks AR, Olson EN (2001) Dilated cardiomyopathy and sudden death resulting from constitutive activation of protein kinase a. Circ Res 89(11):997–1004

Aye TT, Soni S, van Veen TA, van der Heyden MA, Cappadona S, Varro A, de Weger RA, de Jonge N, Vos MA, Heck AJ, Scholten A (2012) Reorganized PKA-AKAP associations in the failing human heart. J Mol Cell Cardiol 52(2):511–518. doi:10.1016/j.yjmcc.2011.06.003

Backs J, Worst BC, Lehmann LH, Patrick DM, Jebessa Z, Kreusser MM, Sun Q, Chen L, Heft C, Katus HA, Olson EN (2011) Selective repression of MEF2 activity by PKA-dependent proteolysis of HDAC4. J Cell Biol 195(3):403–415. doi:10.1083/jcb.201105063

Baillie GS, Sood A, McPhee I, Gall I, Perry SJ, Lefkowitz RJ, Houslay MD (2003) Beta-arrestin-mediated PDE4 cAMP phosphodiesterase recruitment regulates beta-adrenoceptor switching from Gs to Gi. Proc Natl Acad Sci U S A 100(3):940–945

Barry SP, Davidson SM, Townsend PA (2008) Molecular regulation of cardiac hypertrophy. Int J Biochem Cell Biol 40(10):2023–2039. doi:10.1016/j.biocel.2008.02.020

Bender AT, Beavo JA (2006) Cyclic nucleotide phosphodiesterases: molecular regulation to clinical use. Pharmacol Rev 58(3):488–520

Berrera M, Dodoni G, Monterisi S, Pertegato V, Zamparo I, Zaccolo M (2008) A toolkit for real-time detection of cAMP: insights into compartmentalized signaling. Handb Exp Pharmacol 186:285–298. doi:10.1007/978-3-540-72843-6_12

Bers DM (2002) Cardiac excitation-contraction coupling. Nature 415(6868):198–205. doi:10.1038/415198a

Bers DM (2008) Calcium cycling and signaling in cardiac myocytes. Annu Rev Physiol 70:23–49. doi:10.1146/annurev.physiol.70.113006.100455

Bristow MR (2000) Beta-adrenergic receptor blockade in chronic heart failure. Circulation 101(5):558–569

Bristow MR, Ginsburg R, Minobe W, Cubicciotti RS, Sageman WS, Lurie K, Billingham ME, Harrison DC, Stinson EB (1982) Decreased catecholamine sensitivity and beta-adrenergic-receptor density in failing human hearts. N Engl J Med 307(4):205–211. doi:10.1056/NEJM198207223070401

Bristow MR, Ginsburg R, Umans V, Fowler M, Minobe W, Rasmussen R, Zera P, Menlove R, Shah P, Jamieson S et al (1986) Beta 1- and beta 2-adrenergic-receptor subpopulations in nonfailing and failing human ventricular myocardium: coupling of both receptor subtypes to muscle contraction and selective beta 1-receptor down-regulation in heart failure. Circ Res 59(3):297–309

Brunton LL, Hayes JS, Mayer SE (1979) Hormonally specific phosphorylation of cardiac troponin I and activation of glycogen phosphorylase. Nature 280(5717):78–80

Brunton LL, Hayes JS, Mayer SE (1981) Functional compartmentation of cyclic AMP and protein kinase in heart. Adv Cyclic Nucleotide Res 14:391–397

Burgers PP, Ma Y, Margarucci L, Mackey M, van der Heyden MA, Ellisman M, Scholten A, Taylor SS, Heck AJ (2012) A small novel A-kinase anchoring protein (AKAP) that localizes specifically protein kinase A-regulatory subunit I (PKA-RI) to the plasma membrane. J Biol Chem 287(52):43789–43797. doi:10.1074/jbc.M112.395970

Burmeister BT, Wang L, Gold MG, Skidgel RA, O'Bryan JP, Carnegie GK (2015) Protein kinase a (PKA) phosphorylation of Shp2 protein inhibits its phosphatase activity and modulates ligand specificity. J Biol Chem 290(19):12058–12067. doi:10.1074/jbc.M115.642983

Buxton IL, Brunton LL (1983) Compartments of cyclic AMP and protein kinase in mammalian cardiomyocytes. J Biol Chem 258(17):10233–10239

Carlson CR, Lygren B, Berge T, Hoshi N, Wong W, Tasken K, Scott JD (2006) Delineation of type I protein kinase A-selective signaling events using an RI anchoring disruptor. J Biol Chem 281(30):21535–21545. doi:10.1074/jbc.M603223200

Carr DW, Hausken ZE, Fraser ID, Stofko-Hahn RE, Scott JD (1992) Association of the type II cAMP-dependent protein kinase with a human thyroid RII-anchoring protein. Cloning and characterization of the RII-binding domain. J Biol Chem 267(19):13376–13382

Carr DW, Stofko-Hahn RE, Fraser ID, Bishop SM, Acott TS, Brennan RG, Scott JD (1991) Interaction of the regulatory subunit (RII) of cAMP-dependent protein kinase with RII-anchoring proteins occurs through an amphipathic helix binding motif. J Biol Chem 266(22):14188–14192

Castro LR, Verde I, Cooper DM, Fischmeister R (2006) Cyclic guanosine monophosphate compartmentation in rat cardiac myocytes. Circulation 113(18):2221–2228. doi:10.1161/CIRCULATIONAHA.105.599241

Chen L, Marquardt ML, Tester DJ, Sampson KJ, Ackerman MJ, Kass RS (2007) Mutation of an A-kinase-anchoring protein causes long-QT syndrome. Proc Natl Acad Sci U S A 104(52):20990–20995. doi:10.1073/pnas.0710527105
Cohn JN, Goldstein SO, Greenberg BH, Lorell BH, Bourge RC, Jaski BE, Gottlieb SO, McGrew F 3rd, DeMets DL, White BG (1998) A dose-dependent increase in mortality with vesnarinone among patients with severe heart failure. Vesnarinone trial investigators. N Engl J Med 339(25):1810–1816. doi:10.1056/NEJM199812173392503
Communal C, Singh K, Sawyer DB, Colucci WS (1999) Opposing effects of beta(1)- and beta(2)-adrenergic receptors on cardiac myocyte apoptosis : role of a pertussis toxin-sensitive G protein. Circulation 100(22):2210–2212
Corbin JD, Keely SL (1977) Characterization and regulation of heart adenosine 3':5'-monophosphate-dependent protein kinase isozymes. J Biol Chem 252(3):910–918
Corbin JD, Sugden PH, Lincoln TM, Keely SL (1977) Compartmentalization of adenosine 3′:5′-monophosphate and adenosine 3′:5′-monophosphate-dependent protein kinase in heart tissue. J Biol Chem 252(11):3854–3861
Cowley AJ, Skene AM (1994) Treatment of severe heart failure: quantity or quality of life? A trial of enoximone. Enoximone Investigators Br Heart J 72(3):226–230
de Rooij J, Zwartkruis FJ, Verheijen MH, Cool RH, Nijman SM, Wittinghofer A, Bos JL (1998) Epac is a Rap1 guanine-nucleotide-exchange factor directly activated by cyclic AMP. Nature 396(6710):474–477
Dessauer CW (2009) Adenylyl cyclase—A-kinase anchoring protein complexes: the next dimension in cAMP signaling. Mol Pharmacol 76(5):935–941. doi:10.1124/mol.109.059345
Di Benedetto G, Zoccarato A, Lissandron V, Terrin A, Li X, Houslay MD, Baillie GS, Zaccolo M (2008) Protein kinase a type I and type II define distinct intracellular signaling compartments. Circ Res 103(8):836–844. doi:10.1161/CIRCRESAHA.108.174813
DiPilato LM, Cheng X, Zhang J (2004) Fluorescent indicators of cAMP and Epac activation reveal differential dynamics of cAMP signaling within discrete subcellular compartments. Proc Natl Acad Sci U S A 101(47):16513–16518
Domanski MJ, Krause-Steinrauf H, Massie BM, Deedwania P, Follmann D, Kovar D, Murray D, Oren R, Rosenberg Y, Young J, Zile M, Eichhorn E, Investigators B (2003) A comparative analysis of the results from 4 trials of beta-blocker therapy for heart failure: BEST, CIBIS-II, MERIT-HF, and COPERNICUS. J Card Fail 9(5):354–363
Engelhardt S, Hein L, Wiesmann F, Lohse MJ (1999) Progressive hypertrophy and heart failure in beta1-adrenergic receptor transgenic mice. Proc Natl Acad Sci U S A 96(12):7059–7064
Enns LC, Bible KL, Emond MJ, Ladiges WC (2010) Mice lacking the Cbeta subunit of PKA are resistant to angiotensin II-induced cardiac hypertrophy and dysfunction. BMC Res Notes 3:307. doi:10.1186/1756-0500-3-307
Farah AE (1983) Glucagon and the circulation. Pharmacol Rev 35(3):181–217
Fields LA, Koschinski A, Zaccolo M (2015) Sustained exposure to catecholamines affects cAMP/PKA compartmentalised signalling in adult rat ventricular myocytes. Cell Signal. doi:10.1016/j.cellsig.2015.10.003
Fink MA, Zakhary DR, Mackey JA, Desnoyer RW, Apperson-Hansen C, Damron DS, Bond M (2001) AKAP-mediated targeting of protein kinase a regulates contractility in cardiac myocytes. Circ Res 88(3):291–297
Fowler MB, Laser JA, Hopkins GL, Minobe W, Bristow MR (1986) Assessment of the beta-adrenergic receptor pathway in the intact failing human heart: progressive receptor down-regulation and subsensitivity to agonist response. Circulation 74(6):1290–1302
Frey N, Richardson JA, Olson EN (2000) Calsarcins, a novel family of sarcomeric calcineurin-binding proteins. Proc Natl Acad Sci U S A 97(26):14632–14637. doi:10.1073/pnas.260501097
Gao MH, Lai NC, Roth DM, Zhou J, Zhu J, Anzai T, Dalton N, Hammond HK (1999) Adenylyl cyclase increases responsiveness to catecholamine stimulation in transgenic mice. Circulation 99(12):1618–1622
Gesellchen F, Stangherlin A, Surdo N, Terrin A, Zoccarato A, Zaccolo M (2011) Measuring spatiotemporal dynamics of cyclic AMP signaling in real-time using FRET-based biosensors. Methods Mol Biol 746:297–316. doi:10.1007/978-1-61779-126-0_16

Gold MG, Lygren B, Dokurno P, Hoshi N, McConnachie G, Tasken K, Carlson CR, Scott JD, Barford D (2006) Molecular basis of AKAP specificity for PKA regulatory subunits. Mol Cell 24(3):383–395. doi:10.1016/j.molcel.2006.09.006

Ha CH, Kim JY, Zhao J, Wang W, Jhun BS, Wong C, Jin ZG (2010) PKA phosphorylates histone deacetylase 5 and prevents its nuclear export, leading to the inhibition of gene transcription and cardiomyocyte hypertrophy. Proc Natl Acad Sci U S A 107(35):15467–15472. doi:10.1073/pnas.1000462107

Hayes JS, Bowling N, King KL, Boder GB (1982) Evidence for selective regulation of the phosphorylation of myocyte proteins by isoproterenol and prostaglandin E1. Biochim Biophys Acta 714(1):136–142

Hayes JS, Brunton LL (1982) Functional compartments in cyclic nucleotide action. J Cyclic Nucleotide Res 8(1):1–16

Hayes JS, Brunton LL, Brown JH, Reese JB, Mayer SE (1979) Hormonally specific expression of cardiac protein kinase activity. Proc Natl Acad Sci U S A 76(4):1570–1574

Hayes JS, Brunton LL, Mayer SE (1980) Selective activation of particulate cAMP-dependent protein kinase by isoproterenol and prostaglandin E1. J Biol Chem 255(11):5113–5119

Huang LJ, Durick K, Weiner JA, Chun J, Taylor SS (1997) D-AKAP2, a novel protein kinase a anchoring protein with a putative RGS domain. Proc Natl Acad Sci U S A 94(21): 11184–11189

Iwase M, Bishop SP, Uechi M, Vatner DE, Shannon RP, Kudej RK, Wight DC, Wagner TE, Ishikawa Y, Homcy CJ, Vatner SF (1996) Adverse effects of chronic endogenous sympathetic drive induced by cardiac GS alpha overexpression. Circ Res 78(4):517–524

Jurevicius J, Fischmeister R (1996) cAMP compartmentation is responsible for a local activation of cardiac Ca2+ channels by beta-adrenergic agonists. Proc Natl Acad Sci U S A 93(1):295–299

Kapiloff MS, Piggott LA, Sadana R, Li J, Heredia LA, Henson E, Efendiev R, Dessauer CW (2009) An adenylyl cyclase-mAKAPbeta signaling complex regulates cAMP levels in cardiac myocytes. J Biol Chem 284(35):23540–23546. doi:10.1074/jbc.M109.030072

Keely SL (1977) Activation of cAMP-dependent protein kinase without a corresponding increase in phosphorylase activity. Res Commun Chem Pathol Pharmacol 18(2):283–290

Kerfant BG, Zhao D, Lorenzen-Schmidt I, Wilson LS, Cai S, Chen SR, Maurice DH, Backx PH (2007) PI3Kgamma is required for PDE4, not PDE3, activity in subcellular microdomains containing the sarcoplasmic reticular calcium ATPase in cardiomyocytes. Circ Res 101(4):400–408. doi:10.1161/CIRCRESAHA.107.156422

Kinderman FS, Kim C, von Daake S, Ma Y, Pham BQ, Spraggon G, Xuong NH, Jennings PA, Taylor SS (2006) A dynamic mechanism for AKAP binding to RII isoforms of cAMP-dependent protein kinase. Mol Cell 24(3):397–408. doi:10.1016/j.molcel.2006.09.015

Lai NC, Roth DM, Gao MH, Fine S, Head BP, Zhu J, McKirnan MD, Kwong C, Dalton N, Urasawa K, Roth DA, Hammond HK (2000) Intracoronary delivery of adenovirus encoding adenylyl cyclase VI increases left ventricular function and cAMP-generating capacity. Circulation 102(19):2396–2401

Lefkowitz RJ, Rockman HA, Koch WJ (2000) Catecholamines, cardiac beta-adrenergic receptors, and heart failure. Circulation 101(14):1634–1637

Lehnart SE, Wehrens XH, Reiken S, Warrier S, Belevych AE, Harvey RD, Richter W, Jin SL, Conti M, Marks AR (2005) Phosphodiesterase 4D deficiency in the ryanodine-receptor complex promotes heart failure and arrhythmias. Cell 123(1):25–35

Lipskaia L, Defer N, Esposito G, Hajar I, Garel MC, Rockman HA, Hanoune J (2000) Enhanced cardiac function in transgenic mice expressing a Ca(2+)-stimulated adenylyl cyclase. Circ Res 86(7):795–801

Lohse MJ, Engelhardt S, Eschenhagen T (2003) What is the role of beta-adrenergic signaling in heart failure? Circ Res 93(10):896–906. doi:10.1161/01.RES.0000102042.83024.CA

Lugnier C (2006) Cyclic nucleotide phosphodiesterase (PDE) superfamily: a new target for the development of specific therapeutic agents. Pharmacol Ther 109(3):366–398

Lygren B, Carlson CR, Santamaria K, Lissandron V, McSorley T, Litzenberg J, Lorenz D, Wiesner B, Rosenthal W, Zaccolo M, Tasken K, Klussmann E (2007) AKAP complex regulates Ca2+

re-uptake into heart sarcoplasmic reticulum. EMBO Rep 8(11):1061–1067. doi:10.1038/sj.embor.7401081
Lynch MJ, Baillie GS, Mohamed A, Li X, Maisonneuve C, Klussmann E, van Heeke G, Houslay MD (2005) RNA silencing identifies PDE4D5 as the functionally relevant cAMP phosphodiesterase interacting with beta arrestin to control the protein kinase a/AKAP79-mediated switching of the beta2-adrenergic receptor to activation of ERK in HEK293B2 cells. J Biol Chem 280(39):33178–33189
Makarewich CA, Correll RN, Gao H, Zhang H, Yang B, Berretta RM, Rizzo V, Molkentin JD, Houser SR (2012) A caveolae-targeted L-type Ca(2) + channel antagonist inhibits hypertrophic signaling without reducing cardiac contractility. Circ Res 110(5):669–674. doi:10.1161/CIRCRESAHA.111.264028
Markou T, Hadzopoulou-Cladaras M, Lazou A (2004) Phenylephrine induces activation of CREB in adult rat cardiac myocytes through MSK1 and PKA signaling pathways. J Mol Cell Cardiol 37(5):1001–1011. doi:10.1016/j.yjmcc.2004.08.002
Martinez SE, Wu AY, Glavas NA, Tang XB, Turley S, Hol WG, Beavo JA (2002) The two GAF domains in phosphodiesterase 2A have distinct roles in dimerization and in cGMP binding. Proc Natl Acad Sci U S A 99(20):13260–13265
Martins TJ, Mumby MC, Beavo JA (1982) Purification and characterization of a cyclic GMP-stimulated cyclic nucleotide phosphodiesterase from bovine tissues. J Biol Chem 257(4):1973–1979
Maurice DH, Ke H, Ahmad F, Wang Y, Chung J, Manganiello VC (2014) Advances in targeting cyclic nucleotide phosphodiesterases. Nat Rev Drug Discov 13(4):290–314. doi:10.1038/nrd4228
McConnell BK, Popovic Z, Mal N, Lee K, Bautista J, Forudi F, Schwartzman R, Jin JP, Penn M, Bond M (2009) Disruption of protein kinase A interaction with A-kinase-anchoring proteins in the heart in vivo: effects on cardiac contractility, protein kinase a phosphorylation, and troponin I proteolysis. J Biol Chem 284(3):1583–1592. doi:10.1074/jbc.M806321200
Means CK, Lygren B, Langeberg LK, Jain A, Dixon RE, Vega AL, Gold MG, Petrosyan S, Taylor SS, Murphy AN, Ha T, Santana LF, Tasken K, Scott JD (2011) An entirely specific type I A-kinase anchoring protein that can sequester two molecules of protein kinase a at mitochondria. Proc Natl Acad Sci U S A 108(48):E1227–E1235. doi:10.1073/pnas.1107182108
Mehel H, Emons J, Vettel C, Wittkopper K, Seppelt D, Dewenter M, Lutz S, Sossalla S, Maier LS, Lechene P, Leroy J, Lefebvre F, Varin A, Eschenhagen T, Nattel S, Dobrev D, Zimmermann WH, Nikolaev VO, Vandecasteele G, Fischmeister R, El-Armouche A (2013) Phosphodiesterase-2 is up-regulated in human failing hearts and blunts beta-adrenergic responses in cardiomyocytes. J Am Coll Cardiol 62(17):1596–1606. doi:10.1016/j.jacc.2013.05.057
Metrich M, Lucas A, Gastineau M, Samuel JL, Heymes C, Morel E, Lezoualc'h F (2008) Epac mediates beta-adrenergic receptor-induced cardiomyocyte hypertrophy. Circ Res 102(8):959–965. doi:10.1161/CIRCRESAHA.107.164947
Milano CA, Allen LF, Rockman HA, Dolber PC, McMinn TR, Chien KR, Johnson TD, Bond RA, Lefkowitz RJ (1994) Enhanced myocardial function in transgenic mice overexpressing the beta 2-adrenergic receptor. Science 264(5158):582–586
Mokni W, Keravis T, Etienne-Selloum N, Walter A, Kane MO, Schini-Kerth VB, Lugnier C (2010) Concerted regulation of cGMP and cAMP phosphodiesterases in early cardiac hypertrophy induced by angiotensin II. PLoS One 5(12):e14227. doi:10.1371/journal.pone.0014227
Molkentin JD, Lu JR, Antos CL, Markham B, Richardson J, Robbins J, Grant SR, Olson EN (1998) A calcineurin-dependent transcriptional pathway for cardiac hypertrophy. Cell 93(2):215–228
Mongillo M, McSorley T, Evellin S, Sood A, Lissandron V, Terrin A, Huston E, Hannawacker A, Lohse MJ, Pozzan T, Houslay MD, Zaccolo M (2004) Fluorescence resonance energy transfer-based analysis of cAMP dynamics in live neonatal rat cardiac myocytes reveals distinct functions of compartmentalized phosphodiesterases. Circ Res 95(1):67–75
Mongillo M, Tocchetti CG, Terrin A, Lissandron V, Cheung YF, Dostmann WR, Pozzan T, Kass DA, Paolocci N, Houslay MD, Zaccolo M (2006) Compartmentalized phosphodiesterase-2 activity blunts beta-adrenergic cardiac inotropy via an NO/cGMP-dependent pathway. Circ Res 98(2):226–234

Morel E, Marcantoni A, Gastineau M, Birkedal R, Rochais F, Garnier A, Lompre AM, Vandecasteele G, Lezoualc'h F (2005) cAMP-binding protein Epac induces cardiomyocyte hypertrophy. Circ Res 97(12):1296–1304. doi:10.1161/01.RES.0000194325.31359.86

Movsesian MA (2004) Altered cAMP-mediated signalling and its role in the pathogenesis of dilated cardiomyopathy. Cardiovasc Res 62(3):450–459. doi:10.1016/j.cardiores.2004.01.035

Movsesian MA, Smith CJ, Krall J, Bristow MR, Manganiello VC (1991) Sarcoplasmic reticulum-associated cyclic adenosine 5'-monophosphate phosphodiesterase activity in normal and failing human hearts. J Clin Invest 88(1):15–19. doi:10.1172/JCI115272

Muller FU, Boknik P, Knapp J, Linck B, Luss H, Neumann J, Schmitz W (2001) Activation and inactivation of cAMP-response element-mediated gene transcription in cardiac myocytes. Cardiovasc Res 52(1):95–102

Nagayama T, Hsu S, Zhang M, Koitabashi N, Bedja D, Gabrielson KL, Takimoto E, Kass DA (2009) Sildenafil stops progressive chamber, cellular, and molecular remodeling and improves calcium handling and function in hearts with pre-existing advanced hypertrophy caused by pressure overload. J Am Coll Cardiol 53(2):207–215. doi:10.1016/j.jacc.2008.08.069

Nakano SJ, Sucharov J, Van R, Cecil M, Nunley K, Wickers S, Karimpur-Fard A, Stauffer BL, Miyamoto SD, Sucharov CC (2016) Cardiac adenylyl cyclase and phosphodiesterase expression profiles vary by age, disease, and chronic phosphodiesterase inhibitor treatment. J Card Fail. doi:10.1016/j.cardfail.2016.07.429

Nichols CB, Rossow CF, Navedo MF, Westenbroek RE, Catterall WA, Santana LF, McKnight GS (2010) Sympathetic stimulation of adult cardiomyocytes requires association of AKAP5 with a subpopulation of L-type calcium channels. Circ Res 107(6):747–756. doi:10.1161/CIRCRESAHA.109.216127

Nikolaev VO, Bunemann M, Hein L, Hannawacker A, Lohse MJ (2004) Novel single chain cAMP sensors for receptor-induced signal propagation. J Biol Chem 279(36):37215–37218

Nikolaev VO, Bunemann M, Schmitteckert E, Lohse MJ, Engelhardt S (2006) Cyclic AMP imaging in adult cardiac myocytes reveals far-reaching beta1-adrenergic but locally confined beta2-adrenergic receptor-mediated signaling. Circ Res 99(10):1084–1091. doi:10.1161/01.RES.0000250046.69918.d5

Nikolaev VO, Lohse MJ (2006) Monitoring of cAMP synthesis and degradation in living cells. Physiology 21:86–92. doi:10.1152/physiol.00057.2005

Nikolaev VO, Moshkov A, Lyon AR, Miragoli M, Novak P, Paur H, Lohse MJ, Korchev YE, Harding SE, Gorelik J (2010) Beta2-adrenergic receptor redistribution in heart failure changes cAMP compartmentation. Science 327(5973):1653–1657. doi:10.1126/science.1185988

Okumura S, Takagi G, Kawabe J, Yang G, Lee MC, Hong C, Liu J, Vatner DE, Sadoshima J, Vatner SF, Ishikawa Y (2003) Disruption of type 5 adenylyl cyclase gene preserves cardiac function against pressure overload. Proc Natl Acad Sci U S A 100(17):9986–9990. doi:10.1073/pnas.1733772100

Ostrom RS, Gregorian C, Drenan RM, Xiang Y, Regan JW, Insel PA (2001) Receptor number and caveolar co-localization determine receptor coupling efficiency to adenylyl cyclase. J Biol Chem 276(45):42063–42069. doi:10.1074/jbc.M105348200

Packer M, Carver JR, Rodeheffer RJ, Ivanhoe RJ, DiBianco R, Zeldis SM, Hendrix GH, Bommer WJ, Elkayam U, Kukin ML et al (1991) Effect of oral milrinone on mortality in severe chronic heart failure. The PROMISE study research group. N Engl J Med 325(21):1468–1475. doi:10.1056/NEJM199111213252103

Patel HH, Murray F, Insel PA (2008) G-protein-coupled receptor-signaling components in membrane raft and caveolae microdomains. Handb Exp Pharmacol 186:167–184. doi:10.1007/978-3-540-72843-6_7

Pidoux G, Witczak O, Jarnaess E, Myrvold L, Urlaub H, Stokka AJ, Kuntziger T, Tasken K (2011) Optic atrophy 1 is an A-kinase anchoring protein on lipid droplets that mediates adrenergic control of lipolysis. EMBO J 30(21):4371–4386. doi:10.1038/emboj.2011.365

Ponsioen B, Zhao J, Riedl J, Zwartkruis F, van der Krogt G, Zaccolo M, Moolenaar WH, Bos JL, Jalink K (2004) Detecting cAMP-induced Epac activation by fluorescence resonance energy transfer: Epac as a novel cAMP indicator. EMBO Rep 5(12):1176–1180

Redfield MM, Chen HH, Borlaug BA, Semigran MJ, Lee KL, Lewis G, LeWinter MM, Rouleau JL, Bull DA, Mann DL, Deswal A, Stevenson LW, Givertz MM, Ofili EO, O'Connor CM, Felker GM, Goldsmith SR, Bart BA, McNulty SE, Ibarra JC, Lin G, Oh JK, Patel MR, Kim RJ, Tracy RP, Velazquez EJ, Anstrom KJ, Hernandez AF, Mascette AM, Braunwald E, Trial R (2013) Effect of phosphodiesterase-5 inhibition on exercise capacity and clinical status in heart failure with preserved ejection fraction: a randomized clinical trial. JAMA 309(12):1268–1277. doi:10.1001/jama.2013.2024

Rich TC, Fagan KA, Nakata H, Schaack J, Cooper DM, Karpen JW (2000) Cyclic nucleotide-gated channels colocalize with adenylyl cyclase in regions of restricted cAMP diffusion. J Gen Physiol 116(2):147–161

Rich TC, Tse TE, Rohan JG, Schaack J, Karpen JW (2001) In vivo assessment of local phosphodiesterase activity using tailored cyclic nucleotide-gated channels as cAMP sensors. J Gen Physiol 118(1):63–78

Rochais F, Abi-Gerges A, Horner K, Lefebvre F, Cooper DM, Conti M, Fischmeister R, Vandecasteele G (2006) A specific pattern of phosphodiesterases controls the cAMP signals generated by different Gs-coupled receptors in adult rat ventricular myocytes. Circ Res 98(8):1081–1088

Rochais F, Vandecasteele G, Lefebvre F, Lugnier C, Lum H, Mazet JL, Cooper DM, Fischmeister R (2004) Negative feedback exerted by cAMP-dependent protein kinase and cAMP phosphodiesterase on subsarcolemmal cAMP signals in intact cardiac myocytes: an in vivo study using adenovirus-mediated expression of CNG channels. J Biol Chem 279(50):52095–52105

Roth DM, Bayat H, Drumm JD, Gao MH, Swaney JS, Ander A, Hammond HK (2002) Adenylyl cyclase increases survival in cardiomyopathy. Circulation 105(16):1989–1994

Roth DM, Gao MH, Lai NC, Drumm J, Dalton N, Zhou JY, Zhu J, Entrikin D, Hammond HK (1999) Cardiac-directed adenylyl cyclase expression improves heart function in murine cardiomyopathy. Circulation 99(24):3099–3102

Ruppelt A, Mosenden R, Gronholm M, Aandahl EM, Tobin D, Carlson CR, Abrahamsen H, Herberg FW, Carpen O, Tasken K (2007) Inhibition of T cell activation by cyclic adenosine 5'-monophosphate requires lipid raft targeting of protein kinase a type I by the A-kinase anchoring protein ezrin. J Immunol 179(8):5159–5168

Rybin VO, Xu X, Lisanti MP, Steinberg SF (2000) Differential targeting of beta -adrenergic receptor subtypes and adenylyl cyclase to cardiomyocyte caveolae. A mechanism to functionally regulate the cAMP signaling pathway. J Biol Chem 275(52):41447–41457. doi:10.1074/jbc.M006951200

Sadoshima J, Izumo S (1997) The cellular and molecular response of cardiac myocytes to mechanical stress. Annu Rev Physiol 59:551–571. doi:10.1146/annurev.physiol.59.1.551

Salloum F, Yin C, Xi L, Kukreja RC (2003) Sildenafil induces delayed preconditioning through inducible nitric oxide synthase-dependent pathway in mouse heart. Circ Res 92(6):595–597. doi:10.1161/01.RES.0000066853.09821.98

Saucerman JJ, Zhang J, Martin JC, Peng LX, Stenbit AE, Tsien RY, McCulloch AD (2006) Systems analysis of PKA-mediated phosphorylation gradients in live cardiac myocytes. Proc Natl Acad Sci U S A 103(34):12923–12928. doi:10.1073/pnas.0600137103

Sette C, Conti M (1996) Phosphorylation and activation of a cAMP-specific phosphodiesterase by the cAMP-dependent protein kinase. Involvement of serine 54 in the enzyme activation. J Biol Chem 271(28):16526–16534

Sheridan CM, Heist EK, Beals CR, Crabtree GR, Gardner P (2002) Protein kinase a negatively modulates the nuclear accumulation of NF-ATc1 by priming for subsequent phosphorylation by glycogen synthase kinase-3. J Biol Chem 277(50):48664–48676

Sin YY, Edwards HV, Li X, Day JP, Christian F, Dunlop AJ, Adams DR, Zaccolo M, Houslay MD, Baillie GS (2011) Disruption of the cyclic AMP phosphodiesterase-4 (PDE4)-HSP20 complex attenuates the beta-agonist induced hypertrophic response in cardiac myocytes. J Mol Cell Cardiol 50(5):872–883. doi:10.1016/j.yjmcc.2011.02.006

Skroblin P, Grossmann S, Schafer G, Rosenthal W, Klussmann E (2010) Mechanisms of protein kinase a anchoring. Int Rev Cell Mol Biol 283:235–330. doi:10.1016/S1937-6448(10)83005-9

Soni S, Scholten A, Vos MA, van Veen TA (2014) Anchored protein kinase a signalling in cardiac cellular electrophysiology. J Cell Mol Med 18(11):2135–2146. doi:10.1111/jcmm.12365

Sprenger JU, Perera RK, Steinbrecher JH, Lehnart SE, Maier LS, Hasenfuss G, Nikolaev VO (2015) In vivo model with targeted cAMP biosensor reveals changes in receptor-microdomain communication in cardiac disease. Nat Commun 6:6965. doi:10.1038/ncomms7965

Stangherlin A, Gesellchen F, Zoccarato A, Terrin A, Fields LA, Berrera M, Surdo NC, Craig MA, Smith G, Hamilton G, Zaccolo M (2011) cGMP signals modulate cAMP levels in a compartment-specific manner to regulate catecholamine-dependent signaling in cardiac myocytes. Circ Res 108(8):929–939. doi:10.1161/CIRCRESAHA.110.230698

Stangherlin A, Koschinski A, Terrin A, Zoccarato A, Jiang H, Fields LA, Zaccolo M (2014) Analysis of compartmentalized cAMP: a method to compare signals from differently targeted FRET reporters. Methods Mol Biol 1071:59–71. doi:10.1007/978-1-62703-622-1_5

Steinberg SF, Brunton LL (2001) Compartmentation of G protein-coupled signaling pathways in cardiac myocytes. Annu Rev Pharmacol Toxicol 41:751–773. doi:10.1146/annurev.pharmtox.41.1.751

Takimoto E, Champion HC, Li M, Belardi D, Ren S, Rodriguez ER, Bedja D, Gabrielson KL, Wang Y, Kass DA (2005) Chronic inhibition of cyclic GMP phosphodiesterase 5A prevents and reverses cardiac hypertrophy. Nat Med 11(2):214–222

Terrin A, Monterisi S, Stangherlin A, Zoccarato A, Koschinski A, Surdo NC, Mongillo M, Sawa A, Jordanides NE, Mountford JC, Zaccolo M (2012) PKA and PDE4D3 anchoring to AKAP9 provides distinct regulation of cAMP signals at the centrosome. J Cell Biol 198(4):607–621. doi:10.1083/jcb.201201059

Timofeyev V, Myers RE, Kim HJ, Woltz RL, Sirish P, Heiserman JP, Li N, Singapuri A, Tang T, Yarov-Yarovoy V, Yamoah EN, Hammond HK, Chiamvimonvat N (2013) Adenylyl cyclase subtype-specific compartmentalization: differential regulation of L-type Ca2+ current in ventricular myocytes. Circ Res 112(12):1567–1576. doi:10.1161/CIRCRESAHA.112.300370

Uretsky BF, Jessup M, Konstam MA, Dec GW, Leier CV, Benotti J, Murali S, Herrmann HC, Sandberg JA (1990) Multicenter trial of oral enoximone in patients with moderate to moderately severe congestive heart failure. Lack of benefit compared with placebo. Enoximone Multicenter Trial Group Circulation 82(3):774–780

van Heerebeek L, Hamdani N, Falcao-Pires I, Leite-Moreira AF, Begieneman MP, Bronzwaer JG, van der Velden J, Stienen GJ, Laarman GJ, Somsen A, Verheugt FW, Niessen HW, Paulus WJ (2012) Low myocardial protein kinase G activity in heart failure with preserved ejection fraction. Circulation 126(7):830–839. doi:10.1161/CIRCULATIONAHA.111.076075

Verde I, Pahlke G, Salanova M, Zhang G, Wang S, Coletti D, Onuffer J, Jin SL, Conti M (2001) Myomegalin is a novel protein of the golgi/centrosome that interacts with a cyclic nucleotide phosphodiesterase. J Biol Chem 276(14):11189–11198

Vettel C, Lindner M, Dewenter M, Lorenz K, Schanbacher C, Riedel M, Lämmle S, Meinecke S, Mason FE, Sossalla S, Geerts A, Hoffmann M, Wunder F, Brunner FJ, Wieland T, Mehel H, Karam S, Lechêne P, Leroy J, Vandecasteele G, Wagner M, Fischmeister R, El-Armouche A (2016) Phosphodiesterase 2 Protects Against Catecholamine-Induced Arrhythmia and Preserves Contractile Function After Myocardial Infarction. Circ Res. CIRCRESAHA.116.310069. [Epub ahead of print]

Vila Petroff MG, Egan JM, Wang X, Sollott SJ (2001) Glucagon-like peptide-1 increases cAMP but fails to augment contraction in adult rat cardiac myocytes. Circ Res 89(5):445–452

Wagner M, Mehel H, Fischmeister R, El-Armouche A (2016) Phosphodiesterase 2: anti-adrenergic friend or hypertrophic foe in heart disease? Naunyn Schmiedeberg's Arch Pharmacol 389(11):1139–1141

Warrier S, Belevych AE, Ruse M, Eckert RL, Zaccolo M, Pozzan T, Harvey RD (2005) Beta-adrenergic- and muscarinic receptor-induced changes in cAMP activity in adult cardiac myocytes detected with FRET-based biosensor. Am J Physiol Cell Physiol 289(2):C455–C461

Willoughby D, Cooper DM (2007) Organization and Ca2+ regulation of adenylyl cyclases in cAMP microdomains. Physiol Rev 87(3):965–1010. doi:10.1152/physrev.00049.2006

Wong W, Scott JD (2004) AKAP signalling complexes: focal points in space and time. Nat Rev Mol Cell Biol 5(12):959–970

Wu X, Eder P, Chang B, Molkentin JD (2010) TRPC channels are necessary mediators of pathologic cardiac hypertrophy. Proc Natl Acad Sci U S A 107(15):7000–7005. doi:10.1073/pnas.1001825107

Xiang Y, Rybin VO, Steinberg SF, Kobilka B (2002) Caveolar localization dictates physiologic signaling of beta 2-adrenoceptors in neonatal cardiac myocytes. J Biol Chem 277(37):34280–34286. doi:10.1074/jbc.M201644200

Xiao RP, Hohl C, Altschuld R, Jones L, Livingston B, Ziman B, Tantini B, Lakatta EG (1994) Beta 2-adrenergic receptor-stimulated increase in cAMP in rat heart cells is not coupled to changes in Ca2+ dynamics, contractility, or phospholamban phosphorylation. J Biol Chem 269(29):19151–19156

Xiao RP, Lakatta EG (1993) Beta 1-adrenoceptor stimulation and beta 2-adrenoceptor stimulation differ in their effects on contraction, cytosolic Ca2+, and Ca2+ current in single rat ventricular cells. Circ Res 73(2):286–300

Yanaka N, Kurosawa Y, Minami K, Kawai E, Omori K (2003) cGMP-phosphodiesterase activity is up-regulated in response to pressure overload of rat ventricles. Biosci Biotechnol Biochem. 67(5):973–979

Zaccolo M, De Giorgi F, Cho CY, Feng L, Knapp T, Negulescu PA, Taylor SS, Tsien RY, Pozzan T (2000) A genetically encoded, fluorescent indicator for cyclic AMP in living cells. Nat Cell Biol 2(1):25–29

Zaccolo M, Pozzan T (2002) Discrete microdomains with high concentration of cAMP in stimulated rat neonatal cardiac myocytes. Science 295(5560):1711–1715

Zakhary DR, Moravec CS, Stewart RW, Bond M (1999) Protein kinase a (PKA)-dependent troponin-I phosphorylation and PKA regulatory subunits are decreased in human dilated cardiomyopathy. Circulation 99(4):505–510

Zhu WZ, Zheng M, Koch WJ, Lefkowitz RJ, Kobilka BK, Xiao RP (2001) Dual modulation of cell survival and cell death by beta(2)-adrenergic signaling in adult mouse cardiac myocytes. Proc Natl Acad Sci U S A 98(4):1607–1612. doi:10.1073/pnas.98.4.1607

Zoccarato A, Surdo NC, Aronsen JM, Fields LA, Mancuso L, Dodoni G, Stangherlin A, Livie C, Jiang H, Sin YY, Gesellchen F, Terrin A, Baillie GS, Nicklin SA, Graham D, Szabo-Fresnais N, Krall J, Vandeput F, Movsesian M, Furlan L, Corsetti V, Hamilton G, Lefkimmiatis K, Sjaastad I, Zaccolo M (2015) Cardiac hypertrophy is inhibited by a local pool of cAMP regulated by phosphodiesterase 2. Circ Res 117(8):707–719. doi:10.1161/CIRCRESAHA.114.305892

Zoccarato A, Fields LH, Zaccolo M (2016) Response to Wagner et al.: phosphodiesterase-2-anti-adrenergic friend or hypertrophic foe in heart disease? Naunyn Schmiedeberg's Arch Pharmacol 389(11):1143–1145

Subcellular Targeting of PDE4 in Cardiac Myocytes and Generation of Signaling Compartments

8

Marco Conti

Contents

Abstract

Of the 11 families of phosphodiesterases found in the human genome, three decades of pharmacological data have clearly implicated PDE3 in cardiac function. Conversely, much less was known about the PDE4 family of proteins expression and function in the heart. Indeed, PDE4-selective inhibitors were developed with the rationale that they would retain the beneficial properties of a nonselective inhibitor but would be devoid of cardiac effects. Yet in the last decade, a large body of work has been published on the expression and function of PDE4s in the hearts of humans and other mammals. To date, at least seven different PDE4 proteins encoded in the four PDE4 genes have been detected in cardiac cells. These PDE4s are targeted to different subcellular compartments of cardiac myocytes through interaction with several classes of scaffold proteins. This PDE4 sequestration in macromolecular complexes limits their access to

M. Conti
Department of Obstetrics, University of California, San Francisco, San Francisco, CA, USA
e-mail: Marco.Conti@ucsf.edu

V. Nikolaev, M. Zaccolo (eds.), *Microdomains in the Cardiovascular System*, Cardiac and Vascular Biology 3, DOI 10.1007/978-3-319-54579-0_8

cAMP in confined subcellular domains including the dyad, the sarcoplasmic reticulum, and the sarcomere. This molecular organization defines the distinctive role of the different PDE4 proteins in the different aspects of excitation/contraction coupling. In this chapter, the molecular properties and function of PDE4s in the different compartments of the cardiac myocytes will be reviewed.

8.1 Introduction

Decades of drug development and patient treatment constitutes persuasive evidence that cyclic nucleotide phosphodiesterases (PDEs) play a central role in the control of cardiac contractility. As an extension of this concept, their pharmacological manipulation is thought to have considerable potential for rescuing the failing heart. Understanding the exact role of PDEs in cardiac function, however, has become increasingly challenging as an unexpected complexity of the PDE family has emerged during the last three decades. It is now established that 11 families of PDEs and 23 genes are expressed in the human body. In addition, emerging concepts on cAMP signaling have made the precise function of these enzymes even more challenging to dissect. Two decades ago, the idea was proposed that PDE3 enzymes have an important role in the control of cardiac muscle contractility, and compounds targeting these PDEs have shown promise for their inotropic and lusitropic activity (Maurice et al. 2014). Eventually PDE3 inhibitors lost favor because of the poor outcome of chronic treatment with these compounds (Packer et al. 1991). PDE5 also has a surprising role in the failing heart (Kumar et al. 2009). Conversely, less is known about the PDE4 role in cardiac function, and the data have been somewhat controversial. While a role in the human heart is debated (Eschenhagen 2013), a clear role of PDE4 in cardiac homeostasis in model organisms has been amply documented. Here I will review the expression and localization of PDE4s and their involvement in the control of cardiac homeostasis. In the context of the concept that cyclic nucleotide signaling is highly compartmentalized, I will also review available information on the subcellular compartments and the macromolecular complexes involving different PDE4 proteins, as the cardiac model has been at the forefront as the experimental model to define compartmentalized signaling.

8.2 A Brief Description of the PDE4 Family of Enzymes

The PDE4 family is composed of four genes termed PDE4A, PDE4B, PDE4C, and PDE4D (Conti and Beavo 2007; Swinnen et al. 1989; Colicelli et al. 1989). Each gene is characterized by a cluster of exons that code for the catalytic domain. An additional cluster codes for the dimerization domain of this enzyme but not all transcripts include these exons (Monaco et al. 1994). Upstream from this second cluster are additional first exons, each under the control of a specific promoter. This exon arrangement is conserved in all four genes. As a consequence of this gene structure, a considerable number of PDE4 proteins are generated by alternate promoter usage. To date, there are 11 variants described for PDE4A, five for PDE4B, seven for PDE4C, and nine to 11 variants described for PDE4D depending on the species. As a consequence of inclusion of the dimerization domain in some but not all transcripts, long forms of PDE4 are dimers, whereas short forms are monomers (Richter and Conti 2002, 2004; Cedervall et al. 2015). Upstream from the dimerization domain, amino termini of different lengths serve to stabilize interactions with other proteins or as an anchor to the membrane bilayer. PDE4s undergo numerous phosphorylations (Mika and Conti 2016) which can ultimately increase the catalytic activity (Sette and Conti 1996; Sette et al. 1994), decrease it, or serve as priming phosphorylation for other kinases (Mika et al. 2015), as well as to modulate protein/protein interaction (Dodge et al. 2001).

8.3 PDE4 Expressed in Cardiac Myocytes: Comparison of Different Species

Being widely expressed in most tissues and cells, PDE4s are also detected in the heart extract as well as isolated cardiac atrial or ventricular myocytes. The first demonstration of PDE4 cardiac expression comes from chromatographic separation of PDEs from extracts of rabbit, guinea pig, and dog total heart or ventricles (Lugnier et al. 1993; Muller et al. 1992; Reeves et al. 1987). Even though the complexity of PDE4 isoforms was not appreciated at that time, these studies provided biochemical evidence that a cAMP hydrolyzing activity, sensitive to rolipram and insensitive to cGMP, is present in the heart and that could be clearly separated from the other PDEs on ion exchange chromatography. In these early studies, PDE4s appeared to be mainly cytosolic enzymes in dog myocardium, with cytosolic PDE4 activity being several times higher as compared to particulate PDE4 (Weishaar et al. 1987a, b). More recent functional data provide a different view on PDE4 subcellular localization (see below).

With the cloning of different PDE4 cDNAs, additional and more powerful tools have become available to monitor PDE4 expression in cardiac cells. Presently, PDE4 expression can be assessed by measuring mRNA levels, by Western blot analysis with subtype- or variant-specific antibodies or, in a more quantitative fashion, by IP with selective antibodies and measurement of cAMP hydrolytic activity (Richter et al. 2005). Using a combination of these assays, the emerging general consensus is that PDE4A, PDE4B, and PDE4D are expressed in the heart of most species, although with significant species differences that will be discussed below. Less clear is the expression of PDE4C, which is the least understood PDE4 isoenzyme and detected mostly as mRNA, for instance, in the rat (Johnson et al. 2012). Using IP and PDE assay, Richter et al. (2011) noted that the total PDE activity in the human heart is several fold higher than in the rat or mouse because of very high expression of PDE3, PDE2, and PDE1. As a consequence, human PDE4 activity is less than 10% of the total activity, whereas in rodents it reaches 30–50% of the PDE activity. Others have confirmed this large difference in overall cAMP-PDE expression between the human and rat (Johnson et al. 2012). Of note and in absolute terms (i.e., pmoles/min/mg protein in the extract), the PDE4 activity recovered from the human heart is comparable to that present in rodents, with PDE4D being even more abundant than in rodents (see below).

PDE4A, PDE4B, and PDE4D expressions can be detected both as mRNA and by Western blot, and additional information is available for the expression of different splicing variants from the three genes. An often observed and puzzling finding is that although short form mRNAs can be amplified, the presence of the corresponding protein is generally difficult to detect by Western blot analysis.

PDE4A is expressed as a protein of 115 kDa, which may correspond to the PDE4A4, PDE4A10 or PDE4A11 splicing variants with PDE4A10 being induced by cAMP elevation (McCahill et al. 2008a). At least in the mouse, the identity of this immunoreactive band is confirmed by its absence in the PDE4A KO mouse. However, with the exception of a report documenting a PDE4A variant in complex with PI3K (Ghigo et al. 2012), few studies have focused on the function of PDE4A in the heart of any species. No overt cardiac phenotype could be observed in mice deficient in PDE4A (Leroy et al. 2011).

As for PDE4B isozyme, in rat neonatal cardiac myocytes, Mongillo et al., have reported an immunoreactive PDE4B protein of molecular weight of around 75 kDa with electrophoretic mobility similar to that of the recombinant PDE4B2 short form (Mongillo et al. 2004). Earlier studies also detected the short form of PDE4B in rat cardiac myocytes (Kostic et al. 1997). Conversely, Richter et al. reported in rat, mouse, and human a protein of higher molecular weight of about 93 kDa which corresponds to the mobility of PDE4B3 a long form (Richter et al. 2011). The expression of a long PDE4B3 form in the mouse heart is also consistent with the immunoprecipitation with PDE4B3-specific antibodies, with the fact that this form is activated by PKA (short forms do not), and the absence of this immunoreactive band in PDE4B-KO mice (Mika et al. 2014). The origin of these discrepancies by these different reports is unclear but could be due to species differences or different physiological states of the cardiac muscle as PDE4B3 may be induced during prostacyclin preconditioning of the heart (Kostic et al. 1997).

Two PDE4D electrophoretic species have been consistently detected by Western blot of heart and cardiac myocyte extracts, with molecular weights of around 105 and 93–95 kDa. With the availability of variant-specific antibodies against the N terminus of these proteins, it has been determined that the 105 kDa species corresponds to PDE4D5, whereas the 93–95 kDa band is a mixture of three splicing variants PDE4D3, PDE4D8, and PDE4D9 (Richter et al. 2011). Surprisingly, and even allowing for different affinity of the antibody used, PDE4D3 may not be the major form as it is almost undetectable in the mouse and present at low levels in rat and human hearts. Thus, a given splicing variant cannot be identified only on the basis of electrophoretic mobility on Western blot. mRNA analysis of heart extracts has also detected the presence of PDE4D short forms such as PDE4D1/D2 (Richter et al. 2005; Kostic et al. 1997). However and as mentioned above, the corresponding proteins have seldom been detected in heart extracts.

8.4 Subcellular Localization of PDE4s

Biochemical evidence that PDE4s may be associated with different organelles was available early on as some rolipram-sensitive activity is detected in particulate fractions of the heart after homogenization. However, there is general agreement that much of the PDE4 anchoring is lost upon cell or tissue homogenization. For instance, early studies detected rolipram-sensitive activity in the particulate fraction of heart homogenate (Lugnier et al. 1993; Muller et al. 1992). Nuclear membrane PDE4B and PDE4D were detected in extracts of the ovine heart. Immunolocalization of different PDE4s has been reported by several laboratories.

Verde et al. reported immunolocalization of PDE4 in correspondence to the Z lines (Verde et al. 2001). According to Mongillo et al., rat neonatal cardiac myocytes show PDE4B2 localized at the M lines and intercalated with the striated pattern of α-actinin, whereas PDE4D immunolocalization indicated a striated pattern superimposable with that of α-actinin, suggestive of localization to the Z line. In mouse adult cardiac myocytes an additional report, instead, observed the PDE4B immunoreactivity overlapping with the Z band and therefore in the proximity of T tubules (Leroy et al. 2011).

Given the considerable body of data suggesting PDE4 presence in complex with other signaling molecules and scaffold proteins (see below), one would expect PDE4 localization in specialized sarcolemma domains. For instance, caveolae are enriched in components of signal transduction machinery, including G proteins, GPCRs, and effector molecules (Fu and Xiang 2015). Immunofluorescence microscopy demonstrated colocalization of βAR with caveolin, indicating a nonrandom distribution of βAR in the plasma membrane. Surprisingly, there is little information available as to whether PDE4s are associated with caveolae in the heart in spite of the fact that a PDE4 association has been reported in T cells (Bjorgo et al. 2010).

8.5 Molecular Basis of PDE4 Targeting to Different Subcellular Structures in Cardiac Myocytes

Although biochemical evidence for PDE4s localization to the particulate fraction in the heart is sometimes inconclusive and inconsistent, the existence of such a subcellular targeting of PDE4s to organelles and different membrane compartments is clearly supported by numerous reports of macromolecular complexes involving PDE4s (summarized in Fig. 8.1). I will briefly review some of these complexes. This topic has been the subject of extensive reviews (McCahill et al. 2008b; Klussmann 2016; Calejo and Tasken 2015).

β-adrenergic receptors. Immunoprecipitation of β_1-AR from mouse neonatal cardiomyocytes recovers PDE4 activity in the IP pellet (Richter et al. 2008). The use of cardiomyocytes deficient in the different PDE4s indicates that PDE4D is the only PDE4 interacting with this receptor because PDE activity continued to be present in the IP pellet of the PDE4A- and PDE4B-deficient cells. Similarly, immunoprecipitation of β_1-AR from the human heart also detects PDE4D activity, confirming that the endogenous receptor is in complex with this PDE (Richter et al. 2011). The interaction is direct because similar complexes could be demonstrated with purified proteins, and it continues to be present in β-arrestin-deficient MEFs. PDE4D8 binds better than PDE4D9 and PDE4D3/PDE4D5; however, the site of PDE4 interaction with the receptor is not completely understood even though deletion mutagenesis suggests that the amino terminus of PDE4D is required for this interaction. Interestingly, it has been shown that ligation of the β_1-AR leads to dissociation of

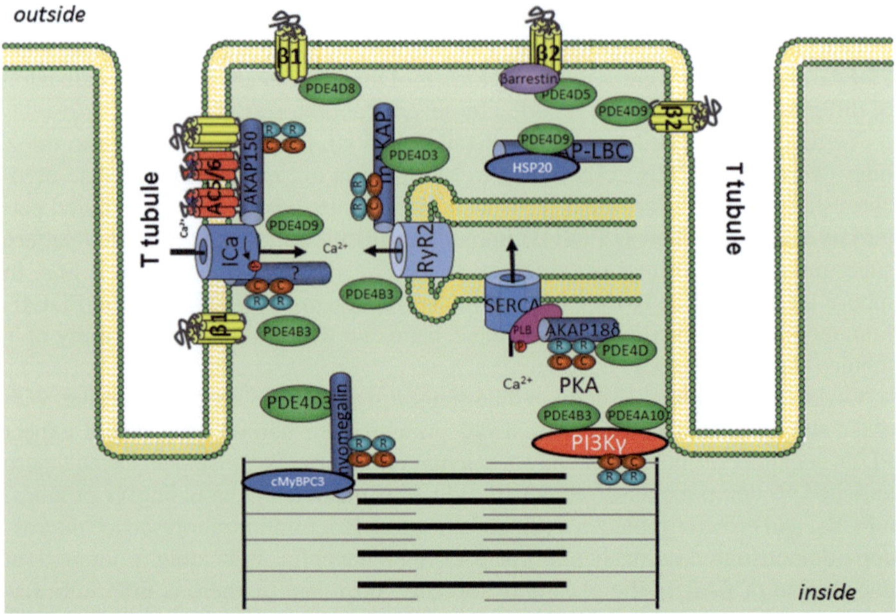

Fig. 8.1 Scheme summarizing the macromolecular complexes described in cardiac myocytes. The dyad is delimited by two T tubules. PDE4s are reported as *green* objects. Scaffold are represented as rods, channels as hollow cylinders, receptors as a bundle of seven transmembrane domains, and PKA as a tetramer of 2C and 2R subunits. One sarcomere unit is reported as lines of different thickness. The scheme is not drawn to scale

the PDE4D-β_1-AR complex (Richter et al. 2008). De Arcangelis et al. have confirmed the interaction of β_1-AR with PDE4D8 but reported that PDE4s bind also β_2-AR (De Arcangelis et al. 2009). Thus, an interaction of PDE4D with β_2-AR may be both direct as well as indirect through β-arrestin recruitment (see below). These authors proposed that simultaneous stimulation of both β_1- and β_2-ARs can lead to switching PDE4D8 from β_1-AR to β_2-AR via an arrestin-dependent mechanism, which cooperates to produce a localized increase in cAMP in the proximity of the activated β1-ARs.

Although immunofluorescence data on membrane localization are inconclusive, this interaction of PDE4D with the β-adrenergic receptors strongly suggests that in cardiac myocytes, a pool of PDE4D is tethered to the membrane, a finding consistent with FRET and cyclic nucleotide-gated channel measurements of cAMP at the plasma membrane level (Leroy et al. 2008). The functional consequence of this complex has been inferred by the effect of PDE4D ablation on phosphorylation of membrane-bound proteins such as $Ca_v1.2$ (Leroy et al. 2011). An additional attempt to dissect the function of the PDE4D/β_1-AR complex was made by monitoring cAMP levels using a FRET probe directly tethered to the β_1-AR. These measurements have uncovered an unexpected effect of antagonists which also disrupt the PDE4D/β_1-AR interaction (Richter et al. 2013).

β-arrestins. The PDE4 localization to the membrane domain may be highly dynamic. Studies in transacted cells have led to the conclusion that β-arrestins coordinate the termination of the adrenergic signal by recruiting PDEs to activate β2-adrenergic receptors in the plasma membrane (Perry et al. 2002). This translocation limits the activation of membrane-associated cAMP-activated protein kinase (PKA) by simultaneously decreasing the rate of cAMP production through receptor desensitization and increasing the rate of its degradation at the membrane via this PDE4D translocation (Perry et al. 2002). Following the initial report that PDE4D interacts with β-arrestin and is recruited to the membrane upon β2-AR activity, several additional reports have dissected the function of this complex. In cardiac myocytes, the β2-AR also switches from Gs to Gi coupling. The stimulation of primary cardiac myocytes with isoprenaline induces recruitment of PDE4D3 and PDE4D5 to membranes and activates ERK1/2 (Baillie et al. 2003). Adenovirus-mediated expression of the dominant negative mutant PDE4D5-D556A potentiates ERK1/2 activation. Although several PDE4D variants co-immunoprecipitate (co-IP) with arrestins, reconstitution experiments in HEK293 cells, as well as siRNA knockdown, indicate that the PDE4D5 splicing variant is preferentially associated with β-arrestin 2 (Lynch et al. 2005).

AKAPs. The often reported observation of PKA/PDE4 co-immunoprecipitation (Raymond et al. 2009; Dodge et al. 2001) is reconciled by the finding that PDE4 interacts with several A-kinase anchoring proteins (AKAPs) (Diviani et al. 2011). Thus far, complexes of PDE4D with mAKAP (Dodge et al. 2001), AKAP9/AKAP450/Yotiao (Tasken et al. 2001; Terrenoire et al. 2009), AKAP18α (McSorley et al. 2006; Smith et al. 2013), and AKAP-Lbc have been documented. PDE4D also binds myomegalin (Verde et al. 2001) which has been proposed to function as an AKAP (Uys et al. 2011). The presence of PDE4s in AKAP complexes implies that these enzymes control cAMP access to associate PKA and termination of the cAMP signal locally. Thus, cAMP may activate PKA, which in turn phosphorylates and activates PDE4D in the complex (Sette and Conti 1996; Dodge et al. 2001); this in turn decreases cAMP, promoting the PKA return to basal state. The presence of this local feedback in the spatially confined complex should increase the efficacy of the feedback in space and time. However, another cAMP-binding protein, EPAC, may also be included in some of these complexes. Indeed, it has been reported that a supramolecular mAKAP/PDE4D3/EPAC/ERK5 complex modulates cardiomyocyte hypertrophy (Dodge-Kafka et al. 2005). In addition to the PDE4D/mAKAP complex, PDE4D3 binds to Epac1 and ERK5 and controls the activation of the associated ERK. RNA interference of mAKAP or disruptor peptides competing for the mAKAP/perinuclear membrane binding site blocks the cytokine-induced cardiomyocyte growth.

Wang et al. have identified a peptide termed UCR1 which activates PDE4D3 bound to AKAP-Lbc in mouse heart extracts (Wang et al. 2015). This activation leads to a decrease in local PKA activity. The expression of UCRC1 in cardiac myocytes prevents hypertrophy providing a novel potential strategy to slow maladaptive changes in the failing heart.

Myomegalin is a large scaffold protein initially identified as a PDE4D-interacting protein by yeast two-hybrid screening (Verde et al. 2001). Numerous splicing variants of myomegalin have been identified as well as related gene called myomegalin-like (MMGL) (Soejima et al. 2001). Subsequently, it has been shown that several myomegalin isoforms interact with cardiac myosin-binding protein-C (cMyBP-C). An interaction of myomegalin with the PKA regulatory subunit was also detected by yeast two hybrid and fluorescent 3D colocalization (Uys et al. 2011). This complex positions the PDE4D/PKA local feedback in the vicinity of important regulators of sarcomere shortening such as cMyBP-C and cardiac troponin I (cTNI).

Other PDE4 interactors. The small heat shock protein HSP20 plays an important role in cell function and is known to be cardioprotective under conditions of stress (Fan and Kranias 2011; Fan et al. 2006). The mechanism underlying its protective abilities depends on its phosphorylation of Ser16 by PKA. It has been recently shown that the phosphorylation state of HSP20 is dependent on the presence of PDE4D in a complex with this heat shock protein (Sin et al. 2011). HSP20 binds directly to PDE4 within a region of the conserved catalytic domain. The disruption of the HSP20-PDE4 complex using this peptide is sufficient to induce phosphorylation of HSP20 by PKA and to protect against the hypertrophic response measured in neonatal cardiac myocytes following chronic β-adrenergic stimulation. Edwards et al. (2012a, b) showed the involvement of AKAP-Lbc, which anchors PKA and HSP20, and is responsible for directing PKA phosphorylation of HSP20.

PI3Kγ. Emerging evidence suggests that phosphoinositide 3-kinase (PI3K) may modulate cardiac inotropy. PI3K modulates myocardial contractility via a cAMP-dependent mechanism that regulates the catalytic activity of PDE4 (Gregg et al. 2010) (Ghigo et al. 2012; Kerfant et al. 2007). Mice lacking PI3Kγ (PI3Kγ$^{-/-}$) have propensity to undergo episodes of premature ventricular contractions after adrenergic stimulation which can be rescued by a selective β_2-AR antagonists (Ghigo et al. 2012). These mice also develop sustained ventricular tachycardia after transverse aortic constriction, which eventually causes their demise, a phenotype of arrhythmias reminiscent of PDE4 ablation. Indeed in using FRET sensors for cAMP, these authors could demonstrate abnormal cAMP accumulation after β_2-AR activation in PI3Kγ$^{-/-}$ cardiomyocytes which was dependent on the loss of the PI3K scaffold but not on the catalytic activity of PI3Kγ. Downstream from β-adrenergic receptors, PI3Kγ was found to participate in large macromolecular complex linking PKA to PDE3A, PDE4A, and PDE4B but not PDE4D. These PI3Kγ-regulated PDEs lower cAMP and limit PKA-mediated phosphorylation of L-type calcium channel $Ca_v1.2$ and phospholamban (PLB). Indeed, $Ca_v1.2$ and phospholamban were hyperphosphorylated in PI3Kγ$^{-/-}$ cardiomyocytes. Furthermore, an increase Ca^{2+} spark and amplitude after adrenergic stimulation as well as spontaneous Ca^{2+} release could be detected in PI3Kγ null mice, confirming a function for this PDE4/PI3Kγ complex. Of note, a report using PDE3 and PDE4 inhibition of PI3K$^{-/-}$ cardiac myocytes came to a different conclusion that only PDE4 is responsible for the phenotype caused by the loss of the scaffold (Kerfant et al. 2007).

8.6 PDE4 Control of Distinct Pools of cAMP in Cardiac Myocytes and Different Ca^{2+} Handling Components

The increase in cardiac myocyte contraction strength, termed inotropic effects induced by an acute stimulation of β-ARs (sympathetic stimulus), is largely mediated by PKA-dependent phosphorylation of several key Ca^{2+}-handling proteins. PKA phosphorylates both the $Ca_v1.2$ and the ryanodine receptor (RYR) channels (Bers 2008), leading to the amplification of Ca-induced Ca^{2+} release from the sarcoplasmic reticulum (SR). Conversely, phosphorylation of phospholamban by PKA enhances Ca^{2+} uptake by the SR Ca^{2+} ATPase (i.e., SERCA2a) thereby accelerating Ca^{2+} return to the basal level and enhancing the relaxation of the muscle (termed lusitropy) (Brittsan and Kranias 2000). Relaxation is also promoted by faster dissociation of Ca^{2+} from myofilaments of the sarcomere as a consequence of troponin I phosphorylation. Given this multidimensional involvement of cAMP signaling in fine-tuning the excitation contraction coupling, it is reasonable to assume that PDE4s play a critical role in fine-tuning this homeostasis. Along the same line, a widely accepted explanation for the presence of the large number of slightly different PDE4 isoenzymes in cardiomyocytes is that PDE4 variants, by virtue of their selective interactions with different Ca^{2+} handling proteins, contribute to cAMP homeostasis and excitation contraction coupling in small, confined domains. The PDE4 role in creating these cAMP domains are at present debated as PDE4 may function as sink of cAMP or as barriers to cAMP diffusion (Conti et al. 2014; Stangherlin and Zaccolo 2012). At present, the size of these domains is also unclear, but as resolution increases, these microdomains have decreased in size to the point that they may coincide with small macromolecular complexes rather than a large subcellular domains. A summary of the different excitation contraction components differentially affected by PDE4B and PDE4D is reported in Table 8.1.

Using PKA activity (Corbin et al. 1977; Brunton et al. 1979) or channel (Jurevicius and Fischmeister 1996) as readout, early reports had indicated that cAMP in cardiac myocytes does not equilibrate instantaneously throughout the cell; rather, it accumulates at different levels in slowly equilibrating compartments. A more direct proof of this behavior of the cyclic nucleotide came when a PKA-based FRET reporter was used to probe the cAMP concentration throughout a neonatal cardiac myocyte (Zaccolo and Pozzan 2002). This novel approach uncovered peaks and troughs of cAMP concentration with a period similar to the Z bands. Treatments with nonselective PDE inhibitors abolished these patterns, suggesting that PDEs are at least in part responsible for stabilizing these different cAMP concentrations, perhaps by functioning as a barrier to diffusion. Along the same line, it has been reported that β_1-AR and β_2-AR cAMP signals differ in their properties (Nikolaev et al. 2006), with the β_1-AR signal propagating at a distance from the membrane, whereas β_2-AR signals remain confined in close proximity to the membrane. Since some of the IBMX effect could be reproduced with a PDE4-selective inhibitor, further studies have focused on defining which PDE4 is involved. Using neonatal rat cardiac myocytes, PDE4, rather than

Table 8.1 Subcellular targeting and functional consequences of macromolecular complexes involving PDE4B and PDE4D[a]

Subcellular domain	Macromolecular complex	PDE4B		PDE4D	
		Neonatal	Adult	Neonatal	Adult
Membrane		PKA-S		PKA-S	
	β1-AR	FRET (Mika et al. 2014; Mongillo et al. 2004) PDE-p (Mika et al. 2014)	NA	Co-IP (Richter et al. 2008) FRET	Co-IP (Richter et al. 2008)
	β2-AR/ β-Arrestin	No		Co-IP (Perry et al. 2002; De Arcangelis et al. 2009) FRET	Co-IP
	Cav1.2	PDE-p (Mika et al. 2014) PKA-S (Mika et al. 2014)	CO-IP (Leroy et al. 2011), ICal (Leroy et al. 2011)	NA	CO-IP (Leroy et al. 2011)
	IKs channel/Yotiao	No	No		CO-IP (Terrenoire et al. 2009)
	PI3K-γ	PKA-S (Ghigo et al. 2012)		No	No
Dyad space					
	RYR/mAKAP	PKA-S (Mika et al. 2014)	FRET[b]	NA	Co-IP (Lehnart et al. 2005) FRET
Sarcoplasmic reticulum	PLB/SERCA/ AKAP18δ	No	No	PKA-S	PKA-S (Beca et al. 2011)
Sarcomere/myofilament					
	Myomegalin/cMyBPC	No	No	NA	Co-IP (Verde et al. 2001) IF
Nuclear membrane	mAKAP			CO-iP (Dodge et al. 2001)	

Co-IP co-immunoprecipitation of endogenous complexes, *ICal* PDE4 function detected and changes in I_{cal} currents, *SR* PDE4 function detected as Sarcoplasmic Ca^{2+} load, *PDE-p* PDE4 phosphorylation and activation, *FRET* effect of PDE4 monitored as local cAMP accumulation measured by FRET, *PKA-S* PDE4KO/KD and PDE4inhibitor effect on phosphorylation of local PKA substrates, *IF* immunofluorescence, *No* no effect detected, *NA* data not available, *Hum* human data available in support of rodent findings

[a]Note: little info is available on macromolecular complexes involving PDE4A, and its function and is not included in the table

[b]Note: unpublished data provided by Dr. V.O. Nikolaev

PDE3, appears to be responsible for modulating the amplitude and duration of the cAMP response to β-agonists (Mongillo et al. 2004). A dominant negative approach, where catalytically inactive PDE4s are overexpressed to displace the endogenous ones, indicated that PDE4B2 rather than PDE4D5 alters cAMP signaling in cardiac myocytes and increases PKA activation by 50% (Mongillo et al. 2004). The limited effect of this dominant negative approach may be in part due to the fact that the two forms targeted are present at low levels in rat cardiac myocytes. Lehnart et al. (2005) identified a complex of PDE4D3 and the ryanodine receptor in mouse and human ventricular myocytes. The disruption of this complex in the PDE4D KO mice leads to aberrant levels of cAMP in the Z band and was associated with a late-onset cardiac myopathy. Parallel studies in a failing human heart also showed that the PDE4D-RYR complex is lost, suggesting that the PDE4D targeting to the RYR also has a local effect on signaling in humans. Further studies on adult rat cardiac myocytes using an EPAC FRET probe showed that PDE4 inhibition, rather than PDE3 inhibition, had a major effect on the time course of β-adrenergic signaling (Leroy et al. 2008).

Subsequent studies have provided better resolving power in dissecting the role of PDE4A, PDE4B, and PDE4D subtypes and related variants in excitation contraction coupling. Although present in cardiac myocytes, no effect on Ca^{2+} handling and contractility could be observed in cardiac myocytes deficient in PDE4A (Leroy et al. 2011). This is a puzzling finding that needs further investigation, as PDE4A activity is comparable to that of PDE4B or PDE4D (see above).

Given the array of interaction with scaffold and other proteins described above, the overall picture emerging from numerous observations is that PDE4D has a broad impact on cyclic AMP levels and does so in different subcellular compartments (Fig. 8.1 and Table 8.1). This broad effect is consistent with the propensity for arrhythmias and late-onset cardiac failure in the PDE4D-null mice. PDE4D ablation affects cAMP responses measured with both a cytosolic and a membrane probe. This is due to the fact that PDE4D variants reside at the membrane, likely within or proximal to the so-called signalosome organized around both β_1- and β_2-ARs. In addition, after adrenergic stimulation, PDE4D5 is recruited to the membrane in a complex with β-arrestin. Although it co-immunoprecipitates with the $Ca_v1.2$, a surprising finding is that PDE4D ablation does not affect I_{cal} either under basal or after β-adrenergic stimulation (Leroy et al. 2011; Beca et al. 2011). It is possible that compensation follows chronically altered Ca^{2+} handling in these PDE4D compartments. It has also been speculated that PDE4D, although likely part of the signalosome, functions at a certain distance from the $Ca_v1.2$ (Leroy et al. 2011). Together with the above observations, the finding that PDE4D ablation affects PKA-mediated phosphorylation of membrane-localized substrates including the $Ca_v1.2$ strongly supports the conclusion that PDE4D controls cAMP access to PKA at the sarcolemma.

PDE4D3 is likely functioning in the SR membrane facing the T tubules, as indicated by the presence of its complexes with RYR and mAKAP (Lehnart et al. 2005), the altered phosphorylation of RYR at putative PKA sites after PDE4D ablation, and as suggested by the Ca^{2+} leak from the SR in PDE4D KO myocytes. It should be noted though that some reports locate most of the RYR-mAKAP complex to the nuclear membrane rather than the SR (Kapiloff et al. 2001). Thus, additional scaffolds may bring PDE4D3 in close proximity of the RYR present in the SR.

An additional subdomain where PDE4D is functioning is the complex in the vicinity of the SERCA2/PLB (phospholamban). This has been demonstrated functionally with data showing increased Ca^{2+} reuptake and increased phosphorylation of PLB in the PDE4D-deficient myocytes (Beca et al. 2011). These latter findings are consistent with the reports of PDE4D being present in macromolecular complexes assembling PKA, AKAP18δ, PLB, and SERCA2 to these SR (Lygren et al. 2007; Kerfant et al. 2007). Complexes identified also suggest that PDE4D may function in concert with PDE3 isoenzymes targeted to the same complex (Beca et al. 2013; Ahmad et al. 2015). Using a transgenic mouse model expressing an EPAC sensor targeted to SERCA/PLB, it has been shown that under basal of β_1-AR activation, cAMP in this compartment accumulates at higher levels than in the bulk cytosol and that this difference is abolished by the inhibition of PDE activity (Sprenger et al. 2015). Together with PDE3, PDE4 exerts an important constraint on cAMP levels in this SERCA2/PLB compartment. Given the fact that PDE4D but not PDE4B is present in complex with PLB, it is most likely that PDE4D is the major PDE4-controlling cAMP around the PLB/SERCA complex. It has been proposed that the increased cAMP concentration in this compartment may be due to PDE4, and likely PDE4D, providing a barrier to diffusion. It should be also noted that the assumption that all cAMP is generated at the plasma membrane may be no longer entirely valid, as increasing evidence suggests receptor signaling from internal organelles (Irannejad et al. 2013; Irannejad and von Zastrow 2014). How one can distinguish between PDE4D functioning locally or broadly to prevent cAMP diffusion is still unclear. Strategies are now available to deduce PDE activity in a subcellular domain by measuring the rate of the decrease of cAMP in this specific compartment (Conti et al. 2014). It will then be important to do these measurements to determine how local PDE4 concentrations contribute to different cAMP concentrations.

One observation to accommodate in the overall picture of PDE4D control of cAMP signaling is that PGE2 pretreatment completely desensitizes the β-adrenergic responses even in the SERCA/PLB compartment (Liu et al. 2012). Pretreatment with PGE2 prevents PKA activation induced by a β-adrenergic agonist at the SR and blocks isoproterenol-induced PKA phosphorylation of PLB and contractile responses in myocytes. It has been proposed that this is due to PGE2 signals leading to the activation of PDE4D which prevents cAMP diffusion from the membrane to the SERCA/PLB compartment (Liu et al. 2012).

Another subcellular compartment where PDE4D functions is around the nuclear membrane in complex with mAKAP and several signaling proteins as described above (Dodge et al. 2001). The function of this PDE4 in this compartment seems to be critical for the control of cardiac hypertrophy.

In contrast to the broad function of PDE4D in different locales within the cell, PDE4B functions in a very restricted subcompartment of ventricular cardiac myocytes (Fig. 8.1). In neonatal cardiac myocytes, PDE4B ablation affects a pool of cAMP measured by FRET at the membrane but not in the cytosol (Mika et al. 2014). Of note, PDE4B is not recruited to the membrane and is not associated with β_1/β_2 AR (Richter et al. 2008). However, PDE4B is activated by ligation of β_1-AR but not

β_2 AR, suggesting that this PDE is located in the vicinity of the β_1 receptor but in a restricted subdomain (Mika et al. 2014). PDE4B-KO mice show altered β-adrenergic-dependent I_{cal} (Leroy et al. 2011), suggesting that this PDE functions in the proximity of the channel. Indeed, PDE4B co-immunoprecipitates with $Ca_v1.2$, by monitoring phosphorylation of different PKA substrates present in different domains within a cardiac ventricular myocytes, Mika et al. has shown that PDE4B ablation affects phosphorylation of $Ca_v1.2$ and RYR, but not PLB or filament-associated proteins (Mika et al. 2014). Importantly, the PDE4B-KO phenotype could be reproduced by inhibiting PDE4B with selective inhibitors (Mika et al. 2014). This acute PDE4B inhibition paradigm rules out the possibility that the phenotypes observed are not directly dependent the PDE4B function or due to pitfalls in the KO strategy as suggested by others (Houslay et al. 2007). Thus, these data strongly suggest that PDE4B function is confined to the dyadic space. A function of PDE4B in the vicinity of membrane structures is consistent with observations made in other cells (Blackman et al. 2011; Terrin et al. 2006).

A question still to be addressed is why the PDE4B function remains so restricted to this domain. Even though interaction with the Disc1 scaffold has been reported in the CNS (Millar et al. 2005) and with few additional observations based on co-IP, little is known about the mechanisms of PDE4B anchoring in the heart. Further studies are necessary to define how PDE4B is tethered to the membrane domain. Also puzzling is the fact that the detected increase in cAMP in the vicinity of the membrane resulting from PDE4B ablation remains so localized, and no changes in cAMP after PDE4B ablation can be detected close to the SERCA2/PLB SR region. One may speculate that PDE4D is responsible for preventing diffusion of this increased signal. This is a testable hypothesis, for instance, by monitoring the effect of PDE4B-selective inhibitors on a PDE4D-null background.

In spite of some unclear details, the body of work summarized above clearly shows that PDE4D and PDE4B play a distinct role in excitation contraction coupling. Both PDE4D- and PDE4B-KO mice show propensity to arrhythmias and disrupted Ca^{2+} handling. The fact that in one case, this is likely due to increased peak I_{cal}, as in the PDE4B KO, and in the PDE4D case due to altered Ca^{2+} SR loading or Ca^{2+} leak due to RYR hyperphosphorylation strongly confirms the view that the two PDEs function in different macromolecular complexes and in exciding small subcellular domains.

The final question is how can these observations made in rodents as model organisms be extrapolated to the human heart. Although PDE4 inhibition by itself may not have major inotropic, chronotropic, or lusitropic effects, a function of PDE4 in β-adrenergic signaling in humans is suggested by observations made on contractility after PDE3 inhibition. Under these conditions, PDE4s do have an effect on human cardiomyocyte contraction and channel permeability (Molenaar et al. 2013). In addition, effects of PDE4 inhibition on atrial cardiac myocyte contraction have been reported (Molina et al. 2012), suggesting a protective effect of PDE4 against atrial fibrillation. Together with the fact that cAMP functions in restricted compartments of signaling, these findings open the possibility that PDE4s are bound to have specific functions also in the human heart.

Compliance with Ethical Standards

Conflict of Interest Statement The author declares no conflict of interest.

References

Ahmad F, Shen W, Vandeput F, Szabo-Fresnais N, Krall J, Degerman E, Goetz F, Klussmann E, Movsesian M, Manganiello V (2015) Regulation of sarcoplasmic reticulum Ca2+ ATPase 2 (SERCA2) activity by phosphodiesterase 3A (PDE3A) in human myocardium: phosphorylation-dependent interaction of PDE3A1 with SERCA2. J Biol Chem 290(11):6763–6776. doi:10.1074/jbc.M115.638585

Baillie GS, Sood A, McPhee I, Gall I, Perry SJ, Lefkowitz RJ, Houslay MD (2003) Beta-Arrestin-mediated PDE4 cAMP phosphodiesterase recruitment regulates beta-adrenoceptor switching from Gs to Gi. Proc Natl Acad Sci U S A 100(3):940–945. doi:10.1073/pnas.262787199

Beca S, Helli PB, Simpson JA, Zhao D, Farman GP, Jones PP, Tian X, Wilson LS, Ahmad F, Chen SR, Movsesian MA, Manganiello V, Maurice DH, Conti M, Backx PH (2011) Phosphodiesterase 4D regulates baseline sarcoplasmic reticulum Ca2+ release and cardiac contractility, independently of L-type Ca2+ current. Circ Res 109(9):1024–1030. doi:CIRCRESAHA.111.250464 [pii]

Beca S, Ahmad F, Shen W, Liu J, Makary S, Polidovitch N, Sun J, Hockman S, Chung YW, Movsesian M, Murphy E, Manganiello V, Backx PH (2013) Phosphodiesterase type 3A regulates basal myocardial contractility through interacting with sarcoplasmic reticulum calcium ATPase type 2a signaling complexes in mouse heart. Circ Res 112(2):289–297. doi:10.1161/CIRCRESAHA.111.300003

Bers DM (2008) Calcium cycling and signaling in cardiac myocytes. Annu Rev Physiol 70:23–49. doi:10.1146/annurev.physiol.70.113006.100455

Bjorgo E, Solheim SA, Abrahamsen H, Baillie GS, Brown KM, Berge T, Okkenhaug K, Houslay MD, Tasken K (2010) Cross talk between phosphatidylinositol 3-kinase and cyclic AMP (cAMP)-protein kinase a signaling pathways at the level of a protein kinase B/beta-arrestin/cAMP phosphodiesterase 4 complex. Mol Cell Biol 30(7):1660–1672. doi:10.1128/MCB.00696-09

Blackman BE, Horner K, Heidmann J, Wang D, Richter W, Rich TC, Conti M (2011) PDE4D and PDE4B function in distinct subcellular compartments in mouse embryonic fibroblasts. J Biol Chem 286(14):12590–12601. doi:M110.203604 [pii]

Brittsan AG, Kranias EG (2000) Phospholamban and cardiac contractile function. J Mol Cell Cardiol 32(12):2131–2139. doi:10.1006/jmcc.2000.1270

Brunton LL, Hayes JS, Mayer SE (1979) Hormonally specific phosphorylation of cardiac troponin I and activation of glycogen phosphorylase. Nature 280(5717):78–80

Calejo AI, Tasken K (2015) Targeting protein-protein interactions in complexes organized by a kinase anchoring proteins. Front Pharmacol 6:192. doi:10.3389/fphar.2015.00192

Cedervall P, Aulabaugh A, Geoghegan KF, McLellan TJ, Pandit J (2015) Engineered stabilization and structural analysis of the autoinhibited conformation of PDE4. Proc Natl Acad Sci U S A 112(12):E1414–E1422. doi:10.1073/pnas.1419906112

Colicelli J, Birchmeier C, Michaeli T, O'Neill K, Riggs M, Wigler M (1989) Isolation and characterization of a mammalian gene encoding a high-affinity cAMP phosphodiesterase. Proc Natl Acad Sci U S A 86:3599–3603

Conti M, Beavo J (2007) Biochemistry and physiology of cyclic nucleotide phosphodiesterases: essential components in cyclic nucleotide signaling. Annu Rev Biochem 76:481–511. doi:10.1146/annurev.biochem.76.060305.150444

Conti M, Mika D, Richter W (2014) Cyclic AMP compartments and signaling specificity: role of cyclic nucleotide phosphodiesterases. J Gen Physiol 143(1):29–38. doi:10.1085/jgp.201311083

Corbin JD, Sugden PH, Lincoln TM, Keely SL (1977) Compartmentalization of adenosine 3':5'-monophosphate and adenosine 3':5'-monophosphate-dependent protein kinase in heart tissue. J Biol Chem 252(11):3854–3861

De Arcangelis V, Liu R, Soto D, Xiang Y (2009) Differential association of phosphodiesterase 4D isoforms with beta2-adrenoceptor in cardiac myocytes. J Biol Chem 284(49):33824–33832. doi:10.1074/jbc.M109.020388

Diviani D, Dodge-Kafka KL, Li J, Kapiloff MS (2011) A-kinase anchoring proteins: scaffolding proteins in the heart. Am J Phys Heart Circ Phys 301(5):H1742–H1753. doi:10.1152/ajpheart.00569.2011

Dodge KL, Khouangsathiene S, Kapiloff MS, Mouton R, Hill EV, Houslay MD, Langeberg LK, Scott JD (2001) mAKAP assembles a protein kinase a/PDE4 phosphodiesterase cAMP signaling module. EMBO J 20(8):1921–1930

Dodge-Kafka KL, Soughayer J, Pare GC, Carlisle Michel JJ, Langeberg LK, Kapiloff MS, Scott JD (2005) The protein kinase a anchoring protein mAKAP coordinates two integrated cAMP effector pathways. Nature 437(7058):574–578. doi:10.1038/nature03966

Edwards HV, Scott JD, Baillie GS (2012a) The A-kinase-anchoring protein AKAP-Lbc facilitates cardioprotective PKA phosphorylation of Hsp20 on Ser(16). Biochem J 446(3):437–443. doi:10.1042/BJ20120570

Edwards HV, Scott JD, Baillie GS (2012b) PKA phosphorylation of the small heat-shock protein Hsp20 enhances its cardioprotective effects. Biochem Soc Trans 40(1):210–214. doi:10.1042/BST20110673

Eschenhagen T (2013) PDE4 in the human heart - major player or little helper? Br J Pharmacol 169(3):524–527. doi:10.1111/bph.12168

Fan GC, Kranias EG (2011) Small heat shock protein 20 (HspB6) in cardiac hypertrophy and failure. J Mol Cell Cardiol 51(4):574–577. doi:10.1016/j.yjmcc.2010.09.013

Fan GC, Yuan Q, Song G, Wang Y, Chen G, Qian J, Zhou X, Lee YJ, Ashraf M, Kranias EG (2006) Small heat-shock protein Hsp20 attenuates beta-agonist-mediated cardiac remodeling through apoptosis signal-regulating kinase 1. Circ Res 99(11):1233–1242. doi:10.1161/01.RES.0000251074.19348.af

Fu Q, Xiang YK (2015) Trafficking of beta-adrenergic receptors: implications in intracellular receptor signaling. Prog Mol Biol Transl Sci 132:151–188. doi:10.1016/bs.pmbts.2015.03.008

Ghigo A, Perino A, Mehel H, Zahradnikova A Jr, Morello F, Leroy J, Nikolaev VO, Damilano F, Cimino J, De Luca E, Richter W, Westenbroek R, Catterall WA, Zhang J, Yan C, Conti M, Gomez AM, Vandecasteele G, Hirsch E, Fischmeister R (2012) PI3Kgamma protects against catecholamine-induced ventricular arrhythmia through PKA-mediated regulation of distinct phosphodiesterases. Circulation. doi:CIRCULATIONAHA.112.114074 [pii]

Gregg CJ, Steppan J, Gonzalez DR, Champion HC, Phan AC, Nyhan D, Shoukas AA, Hare JM, Barouch LA, Berkowitz DE (2010) beta2-adrenergic receptor-coupled phosphoinositide 3-kinase constrains cAMP-dependent increases in cardiac inotropy through phosphodiesterase 4 activation. Anesth Analg 111(4):870–877. doi:10.1213/ANE.0b013e3181ee8312

Houslay MD, Baillie GS, Maurice DH (2007) cAMP-specific phosphodiesterase-4 enzymes in the cardiovascular system: a molecular toolbox for generating compartmentalized cAMP signaling. Circ Res 100(7):950–966. doi:10.1161/01.RES.0000261934.56938.38

Irannejad R, von Zastrow M (2014) GPCR signaling along the endocytic pathway. Curr Opin Cell Biol 27:109–116. doi:10.1016/j.ceb.2013.10.003

Irannejad R, Tomshine JC, Tomshine JR, Chevalier M, Mahoney JP, Steyaert J, Rasmussen SG, Sunahara RK, El-Samad H, Huang B, von Zastrow M (2013) Conformational biosensors reveal GPCR signalling from endosomes. Nature 495(7442):534–538. doi:10.1038/nature12000

Johnson WB, Katugampola S, Able S, Napier C, Harding SE (2012) Profiling of cAMP and cGMP phosphodiesterases in isolated ventricular cardiomyocytes from human hearts: comparison with rat and Guinea pig. Life Sci 90(9–10):328–336. doi:10.1016/j.lfs.2011.11.016

Jurevicius J, Fischmeister R (1996) cAMP compartmentation is responsible for a local activation of cardiac Ca2+ channels by beta -adrenergic agonists pnas.93.1.295. PNAS 93(1):295–299

Kapiloff MS, Jackson N, Airhart N (2001) mAKAP and the ryanodine receptor are part of a multicomponent signaling complex on the cardiomyocyte nuclear envelope. J Cell Sci 114(Pt 17):3167–3176

Kerfant BG, Zhao D, Lorenzen-Schmidt I, Wilson LS, Cai S, Chen SR, Maurice DH, Backx PH (2007) PI3Kgamma is required for PDE4, not PDE3, activity in subcellular microdomains containing the sarcoplasmic reticular calcium ATPase in cardiomyocytes. Circ Res 101(4): 400–408. doi:10.1161/CIRCRESAHA.107.156422

Klussmann E (2016) Protein-protein interactions of PDE4 family members - functions, interactions and therapeutic value. Cell Signal 28(7):713–718. doi:10.1016/j.cellsig.2015.10.005

Kostic MM, Erdogan S, Rena G, Borchert G, Hoch B, Bartel S, Scotland G, Huston E, Houslay MD, Krause EG (1997) Altered expression of PDE1 and PDE4 cyclic nucleotide phosphodiesterase isoforms in 7-oxo-prostacyclin-preconditioned rat heart. J Mol Cell Cardiol 29(11): 3135–3146. doi:10.1006/jmcc.1997.0544

Kumar P, Francis GS, Tang WH (2009) Phosphodiesterase 5 inhibition in heart failure: mechanisms and clinical implications. Nat Rev Cardiol 6(5):349–355. doi:10.1038/nrcardio.2009.32

Lehnart SE, Wehrens XH, Reiken S, Warrier S, Belevych AE, Harvey RD, Richter W, Jin SL, Conti M, Marks AR (2005) Phosphodiesterase 4D deficiency in the ryanodine-receptor complex promotes heart failure and arrhythmias. Cell 123(1):25–35

Leroy J, Abi-Gerges A, Nikolaev VO, Richter W, Lechene P, Mazet JL, Conti M, Fischmeister R, Vandecasteele G (2008) Spatiotemporal dynamics of beta-adrenergic cAMP signals and L-type Ca2+ channel regulation in adult rat ventricular myocytes: role of phosphodiesterases. Circ Res 102(9):1091–1100. doi:CIRCRESAHA.107.167817 [pii]

Leroy J, Richter W, Mika D, Castro LR, Abi-Gerges A, Xie M, Scheitrum C, Lefebvre F, Schittl J, Mateo P, Westenbroek R, Catterall WA, Charpentier F, Conti M, Fischmeister R, Vandecasteele G (2011) Phosphodiesterase 4B in the cardiac L-type Ca(2)(+) channel complex regulates Ca(2)(+) current and protects against ventricular arrhythmias in mice. J Clin Invest 121(7): 2651–2661

Liu S, Li Y, Kim S, Fu Q, Parikh D, Sridhar B, Shi Q, Zhang X, Guan Y, Chen X, Xiang YK (2012) Phosphodiesterases coordinate cAMP propagation induced by two stimulatory G protein-coupled receptors in hearts. Proc Natl Acad Sci U S A 109(17):6578–6583

Lugnier C, Muller B, Le Bec A, Beaudry C, Rousseau E (1993) Characterization of indolidan- and rolipram-sensitive cyclic nucleotide phosphodiesterases in canine and human cardiac microsomal fractions. J Pharmacol Exp Ther 265(3):1142–1151

Lygren B, Carlson CR, Santamaria K, Lissandron V, McSorley T, Litzenberg J, Lorenz D, Wiesner B, Rosenthal W, Zaccolo M, Tasken K, Klussmann E (2007) AKAP complex regulates Ca2+ re-uptake into heart sarcoplasmic reticulum. EMBO Rep 8(11):1061–1067. doi:10.1038/sj.embor.7401081

Lynch MJ, Baillie GS, Mohamed A, Li X, Maisonneuve C, Klussmann E, van Heeke G, Houslay MD (2005) RNA silencing identifies PDE4D5 as the functionally relevant cAMP phosphodiesterase interacting with beta arrestin to control the protein kinase a/AKAP79-mediated switching of the beta2-adrenergic receptor to activation of ERK in HEK293B2 cells. J Biol Chem 280(39):33178–33189. doi:10.1074/jbc.M414316200

Maurice DH, Ke H, Ahmad F, Wang Y, Chung J, Manganiello VC (2014) Advances in targeting cyclic nucleotide phosphodiesterases. Nat Rev Drug Discov 13(4):290–314. doi:10.1038/nrd4228

McCahill A, Campbell L, McSorley T, Sood A, Lynch MJ, Li X, Yan C, Baillie GS, Houslay MD (2008a) In cardiac myocytes, cAMP elevation triggers the down-regulation of transcripts and promoter activity for cyclic AMP phosphodiesterase-4A10 (PDE4A10). Cell Signal 20(11):2071–2083. doi:10.1016/j.cellsig.2008.07.017

McCahill AC, Huston E, Li X, Houslay MD (2008b) PDE4 associates with different scaffolding proteins: modulating interactions as treatment for certain diseases. Handb Exp Pharmacol 186:125–166. doi:10.1007/978-3-540-72843-6_6

McSorley T, Stefan E, Henn V, Wiesner B, Baillie GS, Houslay MD, Rosenthal W, Klussmann E (2006) Spatial organisation of AKAP18 and PDE4 isoforms in renal collecting duct principal cells. Eur J Cell Biol 85(7):673–678. doi:10.1016/j.ejcb.2006.01.005

Mika D, Conti M (2016) PDE4D phosphorylation: a coincidence detector integrating multiple signaling pathways. Cell Signal 28(7):719–724. doi:10.1016/j.cellsig.2015.11.001

Mika D, Richter W, Westenbroek RE, Catterall WA, Conti M (2014) PDE4B mediates local feedback regulation of beta(1)-adrenergic cAMP signaling in a sarcolemmal compartment of cardiac myocytes. J Cell Sci 127(Pt 5):1033–1042. doi:10.1242/jcs.140251
Mika D, Richter W, Conti M (2015) A CaMKII/PDE4D negative feedback regulates cAMP signaling. Proc Natl Acad Sci U S A 112(7):2023–2028. doi:10.1073/pnas.1419992112
Millar JK, Pickard BS, Mackie S, James R, Christie S, Buchanan SR, Malloy MP, Chubb JE, Huston E, Baillie GS, Thomson PA, Hill EV, Brandon NJ, Rain JC, Camargo LM, Whiting PJ, Houslay MD, Blackwood DH, Muir WJ, Porteous DJ (2005) DISC1 and PDE4B are interacting genetic factors in schizophrenia that regulate cAMP signaling. Science 310 (5751): 1187–1191. doi:310/5751/1187 [pii]
Molenaar P, Christ T, Hussain RI, Engel A, Berk E, Gillette KT, Chen L, Galindo-Tovar A, Krobert KA, Ravens U, Levy FO, Kaumann AJ (2013) PDE3, but not PDE4, reduces beta(1) - and beta(2)-adrenoceptor-mediated inotropic and lusitropic effects in failing ventricle from metoprolol-treated patients. Br J Pharmacol 169(3):528–538. doi:10.1111/bph.12167
Molina CE, Leroy J, Richter W, Xie M, Scheitrum C, Lee IO, Maack C, Rucker-Martin C, Donzeau-Gouge P, Verde I, Llach A, Hove-Madsen L, Conti M, Vandecasteele G, Fischmeister R (2012) Cyclic adenosine monophosphate phosphodiesterase type 4 protects against atrial arrhythmias. J Am Coll Cardiol 59(24):2182–2190
Monaco L, Vicini E, Conti M (1994) Structure of two rat genes coding for closely related rolipram-sensitive cAMP-phosphodiesterases. J Biol Chem 269:347–357
Mongillo M, McSorley T, Evellin S, Sood A, Lissandron V, Terrin A, Huston E, Hannawacker A, Lohse MJ, Pozzan T, Houslay MD, Zaccolo M (2004) Fluorescence resonance energy transfer-based analysis of cAMP dynamics in live neonatal rat cardiac myocytes reveals distinct functions of compartmentalized phosphodiesterases. Circul Res 95(1):67–75
Muller B, Stoclet JC, Lugnier C (1992) Cytosolic and membrane-bound cyclic nucleotide phosphodiesterases from Guinea pig cardiac ventricles. Eur J Pharmacol 225(3):263–272
Nikolaev VO, Bunemann M, Schmitteckert E, Lohse MJ, Engelhardt S (2006) Cyclic AMP imaging in adult cardiac myocytes reveals far-reaching {beta}1-adrenergic but locally confined {beta}2-adrenergic receptor-mediated signaling. Circul Res 99(10):1084–1091
Packer M, Carver JR, Rodeheffer RJ, Ivanhoe RJ, DiBianco R, Zeldis SM, Hendrix GH, Bommer WJ, Elkayam U, Kukin ML et al (1991) Effect of oral milrinone on mortality in severe chronic heart failure. The PROMISE study research group. N Engl J Med 325(21):1468–1475. doi:10.1056/NEJM199111213252103
Perry SJ, Baillie GS, Kohout TA, McPhee I, Magiera MM, Ang KL, Miller WE, McLean AJ, Conti M, Houslay MD, Lefkowitz RJ (2002) Targeting of cyclic AMP degradation to beta 2-adrenergic receptors by beta-arrestins. Science 298(5594):834–836
Raymond DR, Carter RL, Ward CA, Maurice DH (2009) Distinct phosphodiesterase-4D variants integrate into protein kinase A-based signaling complexes in cardiac and vascular myocytes. Am J Phys Heart Circ Phys 296(2):H263–H271. doi:10.1152/ajpheart.00425.2008
Reeves ML, Leigh BK, England PJ (1987) The identification of a new cyclic nucleotide phosphodiesterase activity in human and Guinea-pig cardiac ventricle. Implications for the mechanism of action of selective phosphodiesterase inhibitors. Biochem J 241(2):535–541
Richter W, Conti M (2002) Dimerization of the type 4 cAMP-specific phosphodiesterases is mediated by the upstream conserved regions (UCRs). J Biol Chem 277(43):40212–40221
Richter W, Conti M (2004) The oligomerization state determines regulatory properties and inhibitor sensitivity of type 4 cAMP-specific phosphodiesterases. J Biol Chem 279(29):30338–30348
Richter W, Jin SL, Conti M (2005) Splice variants of the cyclic nucleotide phosphodiesterase PDE4D are differentially expressed and regulated in rat tissue. Biochem J 388(Pt 3):803–811
Richter W, Day P, Agrawal R, Bruss MD, Granier S, Wang YL, Rasmussen SG, Horner K, Wang P, Lei T, Patterson AJ, Kobilka B, Conti M (2008) Signaling from beta1- and beta2-adrenergic receptors is defined by differential interactions with PDE4. EMBO J 27(2):384–393
Richter W, Xie M, Scheitrum C, Krall J, Movsesian MA, Conti M (2011) Conserved expression and functions of PDE4 in rodent and human heart. Basic Res Cardiol 106(2):249–262. doi:10.1007/s00395-010-0138-8

Richter W, Mika D, Blanchard E, Day P, Conti M (2013) beta1-adrenergic receptor antagonists signal via PDE4 translocation. EMBO Rep 14(3):276–283. doi:10.1038/embor.2013.4
Sette C, Conti M (1996) Phosphorylation and activation of a cAMP-specific phosphodiesterase by the cAMP-dependent protein kinase. Involvement of serine 54 in the enzyme activation. J Biol Chem 271:16526–16534
Sette C, Iona S, Conti M (1994) The short-term activation of a rolipram-sensitive, cAMP-specific phosphodiesterase by thyroid-stimulating hormone in thyroid FRTL- 5 cells is mediated by a cAMP-dependent phosphorylation. J BiolChem 269:9245–9252
Sin YY, Edwards HV, Li X, Day JP, Christian F, Dunlop AJ, Adams DR, Zaccolo M, Houslay MD, Baillie GS (2011) Disruption of the cyclic AMP phosphodiesterase-4 (PDE4)-HSP20 complex attenuates the beta-agonist induced hypertrophic response in cardiac myocytes. J Mol Cell Cardiol 50(5):872–883. doi:10.1016/j.yjmcc.2011.02.006
Smith FD, Reichow SL, Esseltine JL, Shi D, Langeberg LK, Scott JD, Gonen T (2013) Intrinsic disorder within an AKAP-protein kinase a complex guides local substrate phosphorylation. eLife 2:e01319. doi:10.7554/eLife.01319
Soejima H, Kawamoto S, Akai J, Miyoshi O, Arai Y, Morohka T, Matsuo S, Niikawa N, Kimura A, Okubo K, Mukai T (2001) Isolation of novel heart-specific genes using the BodyMap database. Genomics 74(1):115–120. doi:10.1006/geno.2001.6527
Sprenger JU, Perera RK, Steinbrecher JH, Lehnart SE, Maier LS, Hasenfuss G, Nikolaev VO (2015) In vivo model with targeted cAMP biosensor reveals changes in receptor-microdomain communication in cardiac disease. Nat Commun 6:6965. doi:10.1038/ncomms7965
Stangherlin A, Zaccolo M (2012) Phosphodiesterases and subcellular compartmentalized cAMP signaling in the cardiovascular system. Am J Phys Heart Circ Phys 302(2):H379–H390. doi:ajpheart.00766.2011 [pii]
Swinnen JV, Joseph DR, Conti M (1989) Molecular cloning of rat homologues of the *Drosophila melanogaster* dunce cAMP phosphodiesterase: evidence for a family of genes. Proc Natl Acad Sci U S A 86:5325–5329
Tasken KA, Collas P, Kemmner WA, Witczak O, Conti M, Tasken K (2001) Phosphodiesterase 4D and protein kinase a type II constitute a signaling unit in the centrosomal area. J Biol Chem 276(25):21999–22002
Terrenoire C, Houslay MD, Baillie GS, Kass RS (2009) The cardiac IKs potassium channel macromolecular complex includes the phosphodiesterase PDE4D3. J Biol Chem 284(14):9140–9146
Terrin A, Di Benedetto G, Pertegato V, Cheung Y-F, Baillie G, Lynch MJ, Elvassore N, Prinz A, Herberg FW, Houslay MD, Zaccolo M (2006) PGE1 stimulation of HEK293 cells generates multiple contiguous domains with different [cAMP]: role of compartmentalized phosphodiesterases. J Cell Biol 175(3):441–451
Uys GM, Ramburan A, Loos B, Kinnear CJ, Korkie LJ, Mouton J, Riedemann J, Moolman-Smook JC (2011) Myomegalin is a novel A-kinase anchoring protein involved in the phosphorylation of cardiac myosin binding protein C. BMC Cell Biol 12:18. doi:1471-2121-12-18 [pii]
Verde I, Pahlke G, Salanova M, Zhang G, Wang S, Coletti D, Onuffer J, Jin SL, Conti M (2001) Myomegalin is a novel protein of the golgi/centrosome that interacts with a cyclic nucleotide phosphodiesterase. J Biol Chem 276(14):11189–11198
Wang L, Burmeister BT, Johnson KR, Baillie GS, Karginov AV, Skidgel RA, O'Bryan JP, Carnegie GK (2015) UCR1C is a novel activator of phosphodiesterase 4 (PDE4) long isoforms and attenuates cardiomyocyte hypertrophy. Cell Signal 27(5):908–922. doi:10.1016/j.cellsig.2015.02.003
Weishaar RE, Kobylarz-Singer DC, Kaplan HR (1987a) Subclasses of cyclic AMP phosphodiesterase in cardiac muscle. J Mol Cell Cardiol 19(10):1025–1036
Weishaar RE, Kobylarz-Singer DC, Steffen RP, Kaplan HR (1987b) Subclasses of cyclic AMP-specific phosphodiesterase in left ventricular muscle and their involvement in regulating myocardial contractility. Circ Res 61(4):539–547
Zaccolo M, Pozzan T (2002) Discrete microdomains with high concentration of cAMP in stimulated rat neonatal cardiac myocytes. Science 295(5560):1711–1715

Cardiac cAMP Microdomains and Their Modulation Using Disruptor Peptides

9

Lauren Wills, Bracy A. Fertig, and George S. Baillie

Contents

Abstract

PDEs are an important superfamily of enzymes that regulate dynamic gradients of cyclic nucleotides. The plethora of isoforms that can be generated from the 11 families are selectively expressed and targeted in cellular microdomains, with each isoform's function underpinned by its location, affinity/specificity of its cyclic nucleotide substrate, and ability to be modified by posttranslational modifications. The best way to define the function of a particular "pool" of PDE within a cell is to displace the active protein from its site of action. As many PDE isoforms have multiple functions within the same or closely related cell types, pharmacological inhibition, silencing with siRNA, or dominant negative approaches can only reveal functional changes that are a product of the attenuation of the activity of an isoform in all its possible locations. To allow a more detailed appreciation of the role of cyclic nucleotide signaling in cellular

L. Wills • B.A. Fertig • G.S. Baillie (✉)
Institute of Cardiovascular and Medical Sciences, College of Veterinary Medical and Life Sciences, Glasgow University, Glasgow G128QQ, UK
e-mail: George.Baillie@glasgow.ac.uk; l.wills.1@research.gla.ac.uk

V. Nikolaev, M. Zaccolo (eds.), *Microdomains in the Cardiovascular System*, Cardiac and Vascular Biology 3, DOI 10.1007/978-3-319-54579-0_9

microdomains, peptides directed at sequences that serve to anchor individual enzymes to discrete cellular locations have been devised to displace endogenous, active PDE "pools." As this approach is not directed at the active site of the PDE, global PDE activity within the cell remains unaltered, while confined microdomains experience increases in cyclic nucleotide concentration that may alter PKA-directed modifications of local substrates or promote inappropriate EPAC activation. This chapter reviews recent developments in the disruption of cAMP signalosomes using specific disruptor peptides.

9.1 Introduction

Phosphodiesterases have many vital functions in a range of biological situations, as they are the only family of enzymes that degrade cyclic nucleotides (cAMP and cGMP) (Houslay et al. 2007). The ability of PDEs to shape cAMP gradients that originate from membrane associated adenylate cyclase following GPCR activation allows the specific downstream relay of cellular signaling processes that ensure specificity of receptor response. In short, compartmentalization of the enzymes that synthesize, are activated by, and break down cAMP allows coordinated and specific responses to the generation of a ubiquitous second messenger (McCormick and Baillie 2014). Effectors for cAMP are few. To date, we know of only four proteins which are activated by physical interaction with cAMP. These are protein kinase A (PKA) (Taylor et al. 2013), exchange factor directly activated by cAMP (EPAC) (Bos 2006), cyclic nucleotide-gated ion channels (CNGC), and popeye proteins (Schindler and Brand 2016). Each activates different signaling cascades and are localized by a number of interacting proteins which allow them to sense cAMP concentrations that breach their threshold of activation. The key to the fine control of cAMP response is the positioning of PDEs in the vicinity of cAMP effectors. As cAMP drives functional consequences in many tissues, cell types, and cellular microdomains, evolution has dictated that 11 families (containing 21 different genes) of structurally similar but functionally distinct PDEs are sufficient to tailor the cAMP responses in every possible situation (Conti and Beavo 2007). The function, structure, and pharmacology of the PDEs have recently been described in a comprehensive review and will not be described here (Maurice et al. 2014). It is highly pertinent to our review that almost all PDE isoforms have been shown to exist in a functional signalosome that contains a cAMP effector and other signaling proteins that act in concert to coordinate localized signaling. PDE proteins are often expressed in cardiac tissue at levels which are hard to detect as their function is dictated by their location. It is this facet that makes disruption of PDE targeting the perfect method to unpick the function of individual isoforms (see Fig. 9.1) (Lee et al. 2013b). As much of the early work on the localization of PDEs was done on the PDE4 family, most of this review will concentrate on PDE4 disruptors, though we suspect that this technology will be used more widely in the future to confirm isoform functions of other PDE types. We have also covered literature on the disruption of other cAMP signalosomes that contain the cAMP effectors PKA and EPAC.

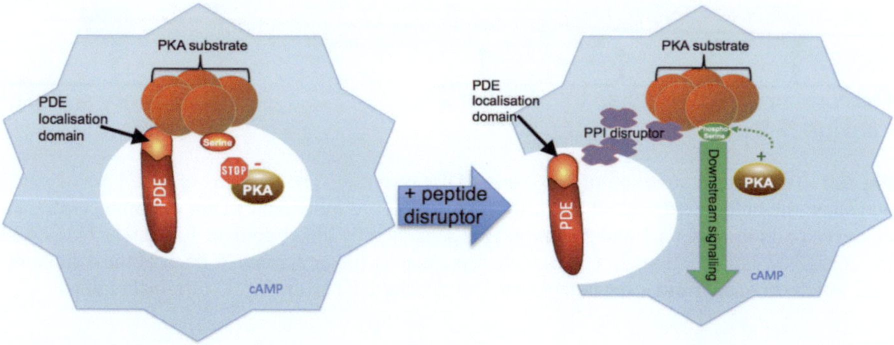

Fig. 9.1 Disruption of PDE localization results in activation of local PKA, which phosphorylates substrates within cAMP microdomain. Phosphodiesterases (*red*) associate with protein complexes that contain cAMP effectors by virtue of a localization domain (*orange*). Local PKA (*brown*) remains inactive, as cyclic AMP (*blue*) concentrations are low because of high levels of associated PDE activity. Disruption of the anchored PDE pool using cell-permeable peptides (*purple*) that mimic one surface of the PDE-binding domain allows local cAMP to rise, activating PKA to facilitate phosphorylation of the PKA substrate. Downstream signaling that induces physiological change is initiated allowing function to be attributed to PDE within that microdomain

9.2 PDE4 Localization

The notion that individual isoforms of PDE4 could be sequestered in microdomains was hatched from the study of PDE4A1 (Scotland and Houslay 1995). Molecular biology was used to prove that the unique N-terminus of this PDE was required for membrane localization, as mutants lacking this region were completely cytosolic. A more in depth study later showed that a short amino acid motif in the N-terminus of the PDE4 conferred calcium-dependent insertion of the enzyme to membrane compartments that contained phosphatidic acid (Baillie et al. 2002). As sequence differences between enzymes in any subfamily of PDE4 (consists of four genes, A-D) often only exist in the N-terminal region, it was easy to conclude that this region acts as a targeting device to enable anchoring of PDE4s to signaling complexes where cAMP is the activating influence (Houslay and Adams 2003). Indeed, the fact that the helical PDE4A1 membrane insertion sequence was only 11 amino acids in length made this concept more believable as many of the unique PDE4 N-terminal regions are short (Baillie et al. 2002). We now know that certain PDE4 N-terminal regions are important for targeting RACK1, arrestin (Bolger et al. 2006), and phosphatidic acid (Baillie et al. 2002); however, it is becoming clearer that almost all of the modular regions (see Figs. 9.1 and 9.2) have been implicated in at least one protein-protein interaction (PPI) that acts to tether a PDE4 to a signaling complex. For example, the neurotrophin receptor, p75 NTR, interacts with PDE4A4 in linker region 1, the catalytic domain, and the subfamily-specific C-terminus (Sachs et al. 2007), whereas the immunophilin XAP2 binds the N-terminal of PDE4A4 and a region in UCR2 (Bolger et al. 2003a), and the WD repeat protein LIS1 binds the UCR1 region of PDE4D (Murdoch et al. 2011).

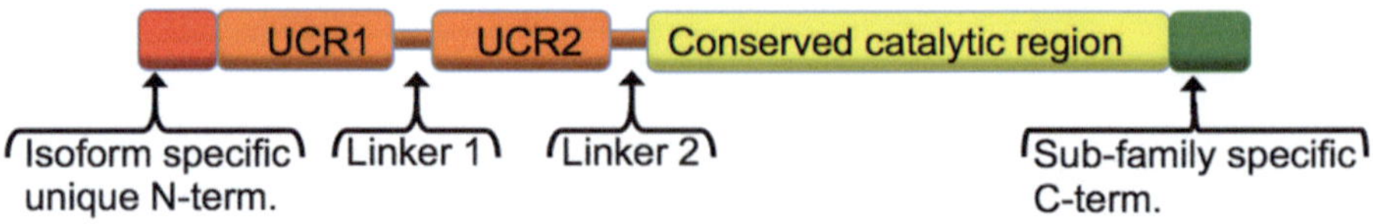

Fig. 9.2 The domain structure of long-form PDE4 enzymes. The domains of PDE4 are depicted and show the unique N-terminal (*red*), which is distinct to individual PDE4 isoforms. Upstream conserved regions (UCR) 1 and 2 (*orange*) are separated by linker domain 1, whereas USC2 and the conserved catalytic domain (*yellow*) are separated by linker domain 2. Each of the 4 subfamilies, which originate from a separate gene (A, B, C, and D), has distinct C-terminal regions

The physical nature of a PDE4 PPI surface dictates how amenable the connection is to disruption using peptides that mimic one of the PPI docking sites. If it is a short, contiguous sequence, then disruption is possible; if, on the other hand, it is a complex three-dimensional interaction involving multiple sites of interaction involving a variety of faces, then it is unlikely that one peptide could unhook such an interaction. One method which can be used to identify short, linear binding motifs that enable PPIs is peptide array (Li & Wu 2009). This technique conjugates short peptides of known sequence to cellulose so that the peptide can be "displayed" in spots to be exposed to a purified protein partner. Using a "far-western" technique, interaction points between immobilized peptides and proteins can be deduced and the information utilized to design high-affinity, cell-permeable disruptor peptides (Bialek et al. 2003). The validity of such peptide disruptors of signaling complexes involving PDEs has been demonstrated using immunoprecipitation experiments with endogenous or purified proteins, surface plasmon resonance, confocal microscopy, ELISA, and cAMP FRET (Brown et al. 2012; Sin et al. 2011). This review summarizes instances where such peptides have been used to manipulate cAMP microdomains in the cardiovascular area and discusses future directions for this technique.

9.3 The Arrestin-PDE4D5 Complex

The fist PDE4 isoform to which a biological function was attributed was PDE4D5, following the discovery that it was associated with the multifunctional scaffold protein β-arrestin (Perry et al. 2002). β-Arrestins are versatile adaptors that are vital for desensitization, sequestration, and termination of G-protein signaling. Therefore, the association with PDE4 was intriguing as it suggested a mechanism in which a mobile pool of cAMP-hydrolyzing activity could be delivered to the β-adrenergic receptor (β-AR), and at the same time receptor signaling at the membrane was being dampened by the physical interaction with β-arrestin. In effect, a "double whammy" that promoted signal termination by concomitantly limiting receptor-G-protein communication and further cAMP diffusion into the cell (Baillie and Houslay 2005). The latter action significantly reduces PKA phosphorylation of the receptor and attenuates "switching" of the β-AR from G_s to G_i signaling (Baillie et al. 2003).

A number of reports utilizing multiple methods, such as siRNA (Lynch et al. 2005), dominant negative mutants (Baillie et al. 2003), or pharmacological inhibition (Smith et al. 2007) of PDE4 to disrupt this pool of PDE4D5 has resulted in an elevation in receptor PKA phosphorylation that logically promotes the "switch" from G_s to G_i signaling. Additionally, mutant forms of β-arrestin, which could not bind to PDE4D5, were still capable of translocating to the receptor. However, these did not aid degradation of cAMP at the membrane and also resulted in elevated PKA phosphorylation of the β_2-adrenergic receptor (β_2-AR) (Baillie et al. 2007). The PDE4D5-β-arrestin complex is also influenced by the posttranslational modification, ubiquitination (Li et al. 2009a), which occurs at the unique N-terminal of the PDE and is promoted by a pool of ubiquitinating enzyme MDM2, which is tethered to β-arrestin. PDE4D5 interacts readily with both β-arrestin and the scaffold protein, RACK1, but in the ubiquitinated form, it preferentially interacts with β-arrestin; therefore, the posttranslational change promotes the PDE4D5-β-arrestin complex in cardiac myocytes (Li et al. 2009b).

Definition of the β-arrestin-binding site on PDE4D5 using yeast two-hybrid and site-directed mutagenesis pinpointed the unique N-terminal region of the enzyme (Bolger et al. 2003a, b) as the β-arrestin-docking site. Peptide array confirmed the interaction domain on the PDE (Bolger et al. 2003b), which was subsequently modeled onto the NMR structure of the PDE4D5 N-terminal region (Smith et al. 2007). This information was used to devise cell penetrating peptides that could selectively disrupt the β-arrestin-PDE4D5 interaction to amplify the cAMP response of the β_2-AR, leading to increased receptor phosphorylation by PKA and a more robust ERK MAP kinase signal as a result of elevated receptor "switching" (Smith et al. 2007). The possible therapeutic uses for such a peptide include counteracting the hyper-desensitization of β-ARs in the airways of asthma patients which causes upregulation of PDE4D5 in response to the chronic elevation of cAMP caused by salbutamol (Nino et al. 2009; Willis and Baillie 2014). Such a response results in a higher dose of the drug being required to achieve adequate cAMP concentrations to promote airway relaxation.

9.4 The HSP20-PDE4D Complex

Small heat shock proteins (HSP) are a highly conserved family of molecular chaperones, which are expressed in response to stressful environmental conditions (Taylor and Benjamin 2005). One member of the family, HSP20, is activated following phosphorylation at serine 16, which is part of a protein kinase A/protein kinase G (PKA/PKG) consensus motif (Beall et al. 1999). Phosphorylation of HSP20 by PKA induces its cardio-protective properties, and much has been written about the chaperone's protective capabilities in the areas of cardiac ischemia, apoptosis, and hypertrophy (reviewed in Edwards et al. 2012b; Fan et al. 2005; Martin et al. 2014a). HSP20 exists in complex with the PKA anchoring protein, AKAP-Lbc (Edwards et al. 2012a). The association between AKAP-Lbc and HSP20 positively regulates the chaperone's cardio-protective character by bringing PKA into close proximity to HSP20. In contrast, AKAP-Lbc is also known to bind PDE4 (Wang et al. 2015). This interaction suggests that AKAP-Lbc also acts to coordinate the spatial and temporal activation of HSP20 by positioning the complex in areas where the PDE4 activity can be swamped following conditions of elevated cAMP, allowing the concentration of the second messenger to breach the threshold of activation of the tethered PKA pool, promoting HSP20 phosphorylation.

The direct association between PDE4 and HSP20 has also been shown (Sin et al. 2011). HSP20 binds to all PDE4 isoforms via a binding site in the enzyme's conserved catalytic domain. This serves to dramatically reduce cAMP concentrations in the vicinity of HSP20 following β-adrenergic stimulation, which in turn puts a brake on HSP20 phosphorylation at serine 16. A cell-permeable peptide based on the HSP20-binding site on PDE4 is sufficient to promote disruption of PDE4 from HSP20 and to induce HSP20 phosphorylation in cardiac myocytes where the basal concentrations of cAMP are high (Sin et al. 2011). Such an action triggers the anti-hypertrophic action of HSP20 which prevents increases in cultured neonatal cardiac myocyte size and attenuates expression of the fetal gene cassette following chronic treatment with isoprenaline. When tested in a mouse model of pressure overload mediated hypertrophy (Martin et al. 2014b), the HSP20-PDE4 disruptor attenuated action potential prolongation and provided significant protection against adverse cardiac remodeling. The benefits of peptide injection included improved contractility, reduced heart size, and diminished interstitial fibrosis. All of these gains are linked with increased levels of phospho-HSP20.

9.5 Disruptors of AKAP-PKA in the Heart

As PKA phosphorylates many of the intracellular signaling intermediates that orchestrate excitation-contraction coupling, the role of AKAPs in ensuring a coordinated response to β-adrenergic signaling is vital (reviewed in Mauban et al. 2009; McConnachie et al. 2006). Many of the early experiments that highlighted the importance of PKA localization in the cardiac setting used a 24-amino acid PKA R subunit sequence to outcompete the AKAP-PKA interaction (Fink et al. 2001). The disruptor peptide, Ht31, corresponded to the RII-binding domain of AKAP-Lbc. The Ht31 peptide exhibited little selectivity for RII vs RI subunits. Treatment of cardiac myocytes with Ht31 affected PKA phosphorylation of PKA substrates such as troponin I, myosin-binding protein C, and phospholamban under basal conditions, which manifested as changes in calcium dynamics that affected cell shortening and relaxation rate (Fink et al. 2001).

Using bioinformatic analysis of PKA-binding regions on a variety of AKAPs, the Scott group discovered an RII-selective PKA localization disruptor peptide (AKAP-IS) that was more effective than the previous attempts (Alto et al. 2003). AKAP-IS has proved a valuable tool in the determination of cAMP-dependent regulation of HERG channel activity (Li et al. 2008), β_2-adrenergic signaling, and airway relaxation (Horvat et al. 2012). An RI selective disruptor was discovered by the same group in 2006 (Carlson et al. 2006). This peptide inhibited the negative PKA control of T-cell signaling and could act to inhibit steroid synthesis in mitochondria of progesterone producing cells (Carlson et al. 2006). Other functions attributed to RI PKA anchoring using this peptide include regulation of PGE2 control of COX2 expression in macrophages (Lee et al. 2013a) and lipid raft-specific signaling in platelets which diminishes aggregation (Raslan et al. 2015).

More recently, cell-permeable peptides that encompass the PKA-binding region of AKAP10 have been used to disrupt both AKAP-RI and AKAP-RII interactions in isolated cardiac myocytes and perfused whole hearts. Relocalization of PKA in this manner served to depress phosphorylation of PKA substrates during isoprenaline stimulation, which resulted in negative effects on contractility and developed pressure (Patel et al. 2010). Finally, a cell-permeable peptide based on the RII-binding region of AKAP7 delta has been used to compete with RII-binding interactions in cells (Hundsrucker et al. 2006). This peptide could displace PKA from aquaporin-bearing vesicles and affect the PKA regulation of the L-type calcium channel in cardiac myocytes following isoprenaline treatment.

9.6 EPAC Disruptors

In addition to disruptors of the cAMP effector PKA, peptides that relocate EPAC have also been discovered. These peptidic agents have been used to unpick functions of EPAC in the cardiovascular system. Paradoxically, one of these spatial regulators of EPAC is the selective disruptor of the β-arrestin-4D5 interaction as described above (Smith et al. 2007). EPAC can be activated by cAMP produced as a result of β-adrenergic signaling (Rangarajan et al. 2003), but the signaling axis via EPAC to H-Ras via PLC is only induced by β_1-adrenergic receptors (β_1-AR) and not β_2-AR (Berthouze-Duquesnes et al. 2012). Therefore, investigation into the localized EPAC complexes of both receptor types has been undertaken (Berthouze-Duquesnes et al. 2012). Following agonist treatment, EPAC is recruited by arrestin proteins to the β_1-AR to initiate signaling to Rap2b through PLC to H-Ras. This acts as a trigger for the phosphorylation of HDAC4 by CaM kinase 2, which in turn induces a pro-hypertrophic gene expression event (Berthouze-Duquesnes et al. 2012). EPAC is prevented from undertaking similar signaling at the β_2-AR as the EPAC-binding site on the β_2-AR pool of arrestin is occluded by PDE4D5. Disturbance of the arrestin-PDE4D5 complex using a cell-permeable peptide allows arrestin to load EPAC instead of PDE4D5 and recruit this cargo to the β_2-AR to mimic the hypertrophic signaling observed at the β_1-AR (Berthouze-Duquesnes et al. 2012). In this manner, replacement of a cAMP-hydrolyzing enzyme (PDE4D5) with a cAMP effector (EPAC) is sufficient to reverse the opposing effects of β-AR subtypes.

Another EPAC complex that has been amenable to disruption by peptides is the association between EPAC and PDE4D in vascular endothelial cells (VECs) (Rampersad et al. 2010). As PDE4D regulates both the activity of EPAC (by greatly increasing the amount of cAMP needed in the vicinity to breech the enzyme's threshold of activation) and its location (by acting as an EPAC anchor), disruption of the complex impacts the integrity of EPAC signaling at VE-cadherin adhesions and interferes with the coordinated control of VEC permeability (Rampersad et al. 2010). Finally, the EPAC interaction with PDE3B in human arterial endothelial cells (HAECs) is vital for cAMP-dependent regulation of HAEC adhesion (Wilson et al. 2011). Disruption of this signalosome, which also includes the p84 regulatory subunit of PI3 kinase gamma, attenuates HAEC spreading and tubule construction and hinders angiogenesis.

9.7 Conclusions and Future Directions

Many of the examples cited within this review highlight the importance of discretely localized cAMP signalosomes in the shaping of receptor function. In addition, these interactions illustrate the utility of disruptor peptides in the development of our understanding of cAMP signaling circuitry. As peptides are notoriously difficult to develop into therapeutic agents (Ahrens et al. 2012), the obvious next step would be to develop small molecules that could have positive outcomes in cardiovascular diseases that are underpinned by aberrant cAMP signaling via their manipulation of discrete cAMP domains rather than global cAMP concentrations. Although this concept is in its infancy, a few examples demonstrate that this approach may prove effective. The first example concerns the phosphorylation of the aforementioned chaperone HSP20. It is known that PKA phosphorylation of HSP20 results in relaxation of both airway and vascular smooth muscle (Komalavilas et al. 2008; Woodrum et al. 2003) by a mechanism that involves actin cytoskeletal changes resulting from sequestration of 14-3-3 proteins (by phospho-HSP20) away from the actin-depolymerizing protein cofilin (Dreiza et al. 2005). In an attempt to find small molecules that could selectively inhibit the ability of human smooth muscle cells to generate contractile force via the same mechanism, An and colleagues (An et al. 2011) devised a fluorescence polarization (FP) assay using a fluorescent peptide containing the phosphorylated PKA site of HSP20 (serine 16) and purified 14-3-3 proteins. The FP assay was used to screen a small molecule library to find compounds that could cause relaxation of isolated human airway smooth muscle (ASM) cells in vitro and attenuate the development of force in intact tissue ex vivo (An et al. 2011). Reassuringly, from 268 compounds that were found to inhibit the HSP20 peptide/14-33 interaction, a handful of candidate molecules belonging to two structurally related scaffolds were shown to have the desired physiological effects on ASM cells; however, the report did not investigate whether the effects on muscle relaxation were due to disturbance of the HSP20/14-3-3 protein-protein interaction or some other unexpected mechanism.

The second example involves discovery of a small molecule that can actively inhibit PKA-AKAP interactions (Christian et al. 2011). Using an ELISA-based assay that evaluated the interaction between AKAP18δ and RII, Christian and coworkers discovered FMP-API-1, a compound that could bind in allosterically to PKA regulatory subunits of PKA. Binding of FMP-API-1 to R subunits was shown to have a dual effect. It not only disrupted PKA-AKAP interactions but also activated PKA concomitantly. Both actions interfered with localized cAMP signaling to promote PKA phosphorylation of proteins such as phospholamban and cardiac troponin I that positively influence calcium-dependent contractility of cardiac myocytes (Christian et al. 2011). As a result, FMP-API-1 could be considered a "prototype" therapeutic agent for diseases such as chronic heart failure, though the lack of specificity in the disruption of specific AKAP-RII complexes could possibly manifest in side effects.

Clearly, the target-selective nature of PPI disruptors makes them an attractive strategy not only to investigate microdomain architecture of cAMP signaling complexes but also as a conduit for the development of new therapeutics. It is expected that these agents, in concert with novel technology to image localized cAMP signaling processes (Gorelik et al. 2013; Nikolaev et al. 2010), will provide a better understanding of how cAMP microdomains underpin receptor specificity of response and help abrogate detrimental changes in cAMP signaling during disease.

Compliance with Ethical Standards

Conflict of Interest Statement The authors declare that they have no conflict of interest.

References

Ahrens VM, Bellmann-Sickert K, Beck-Sickinger AG (2012) Peptides and peptide conjugates: therapeutics on the upward path. Future Med Chem 4(12):1567–1586

Alto NM, Soderling SH, Hoshi N, Langeberg LK, Fayos R, Jennings PA, Scott JD (2003) Bioinformatic design of A-kinase anchoring protein-in silico: a potent and selective peptide antagonist of type II protein kinase A anchoring. Proc Natl Acad Sci U S A 100(8):4445–4450

An SS, Askovich PS, Zarembinski TI, Ahn K, Peltier JM, Von Rechenberg M, Sahasrabudhe S, Fredberg JJ (2011) A novel small molecule target in human airway smooth muscle for potential treatment of obstructive lung diseases: a staged high-throughput biophysical screening. Respir Res 12:8

Baillie GS, Houslay MD (2005) Arrestin times for compartmentalised cAMP signalling and phosphodiesterase-4 enzymes. Curr Opin Cell Biol 17(2):129–134

Baillie GS, Huston E, Scotland G, Hodgkin M, Gall I, Peden AH, MacKenzie C, Houslay ES, Currie R, Pettitt TR, Walmsley AR, Wakelam MJ, Warwicker J, Houslay MD (2002) TAPAS-1, a novel microdomain within the unique N-terminal region of the PDE4A1 cAMP-specific phosphodiesterase that allows rapid, Ca^{2+}-triggered membrane association with selectivity for interaction with phosphatidic acid. J Biol Chem 277(31):28298–28309

Baillie GS, Sood A, McPhee I, Gall I, Perry SJ, Lefkowitz RJ, Houslay MD (2003) beta-Arrestin-mediated PDE4 cAMP phosphodiesterase recruitment regulates beta-adrenoceptor switching from Gs to Gi. Proc Natl Acad Sci U S A 100(3):940–945

Baillie GS, Adams DR, Bhari N, Houslay TM, Vadrevu S, Meng D, Li X, Dunlop A, Milligan G, Bolger GB, Klussmann E, Houslay MD (2007) Mapping binding sites for the PDE4D5 cAMP-specific phosphodiesterase to the N- and C-domains of beta-arrestin using spot-immobilized peptide arrays. Biochem J 404(1):71–80

Beall A, Bagwell D, Woodrum D, Stoming TA, Kato K, Suzuki A, Rasmussen H, Brophy CM (1999) The small heat shock-related protein, HSP20, is phosphorylated on serine 16 during cyclic nucleotide-dependent relaxation. J Biol Chem 274(16):11344–11351

Berthouze-Duquesnes M, Lucas A, Sauliere A, Sin YY, Laurent AC, Gales C, Baillie GS, Lezoualc'h F (2012) Specific interactions between Epac1, beta-arrestin2 and PDE4D5 regulate beta-adrenergic receptor subtype differential effects on cardiac hypertrophic signaling. Cell Signal 25(4):970–980

Bialek K, Swistowski A, Frank R (2003) Epitope-targeted proteome analysis: towards a large-scale automated protein-protein-interaction mapping utilizing synthetic peptide arrays. Anal Bioanal Chem 376(7):1006–1013

Bolger GB, McCahill A, Huston E, Cheung YF, McSorley T, Baillie GS, Houslay MD (2003a) The unique amino-terminal region of the PDE4D5 cAMP phosphodiesterase isoform confers preferential interaction with beta-arrestins. J Biol Chem 278(49):49230–49238

Bolger GB, Peden AH, Steele MR, MacKenzie C, McEwan DG, Wallage DA, Huston E, Baillie GS, Houslay MD (2003b) Attenuation of the activity of the cAMP-specific phosphodiesterase PDE4A5 by interaction with the immunophilin XAP2. J Biol Chem 278(35): 33351–33363

Bolger GB, Baillie Gs, Li X, Lynch MJ, Herzyk P, Mohamed A, Mitchell LH, McCahill A, Hundsrucker C, Klussman E, Adams DR, Christian F, Szaszak M, Friedl S, Drewianka S, Lorenz D, Goncalves A, Furkert J, Vargas C, Schmieder P, Gotz F, Zuhlke K, Moutty M, Gottert H, Joshi M, Reif B, Haase H, Morano I, Grossmann S, Klukovits A, Verli J, Gaspar R, Noack C, Bergmann M, Kass R, Hampel K, Kashin D, Genieser HG, Herberg FW, Willoughby D, Cooper DM, Baillie GS, Houslay MD, von Kries JP, Zimmermann B, Rosenthal W, Klussmann E (2011) Small molecule AKAP-protein kinase A (PKA) interaction disruptors that activate PKA interfere with compartmentalized cAMP signaling in cardiac myocytes. J Biol Chem 286(11):9079–9096

Bos JL (2006) Epac proteins: multi-purpose cAMP targets. Trends Biochem Sci 31(12): 680–686

Brown KM, Lee LC, Findlay JE, Day JP, Baillie GS (2012) Cyclic AMP-specific phosphodiesterase, PDE8A1, is activated by protein kinase A-mediated phosphorylation. FEBS Lett 586(11):1631–1637

Carlson CR, Lygren B, Berge T, Hoshi N, Wong W, Tasken K, Scott JD (2006) Delineation of type I protein kinase A-selective signaling events using an RI anchoring disruptor. J Biol Chem 281(30):21535–21545

Christian F. et al. (2011) Small molecule AKAP-protein kinase A (PKA) interaction disruptors that activate PKA interfere with compartmentalized cAMP signaling in cardiac myocytes. J Biol Chem 286(11):9079–9096

Conti M, Beavo J (2007) Biochemistry and physiology of cyclic nucleotide phosphodiesterases: essential components in cyclic nucleotide signaling. Annu Rev Biochem 76:481–511

Dreiza CM, Brophy CM, Komalavilas P, Furnish EJ, Joshi L, Pallero MA, Murphy-Ullrich JE, von Rechenberg M, Ho YS, Richardson B, Xu N, Zhen Y, Peltier JM, Panitch A (2005) Transducible heat shock protein 20 (HSP20) phosphopeptide alters cytoskeletal dynamics. FASEB J 19(2):261–263

Edwards HV, Scott JD, Baillie GS (2012a) The A-kinase-anchoring protein AKAP-Lbc facilitates cardioprotective PKA phosphorylation of Hsp20 on Ser(16). Biochem J 446(3): 437–443

Edwards HV, Scott JD, Baillie GS (2012b) PKA phosphorylation of the small heat-shock protein Hsp20 enhances its cardioprotective effects. Biochem Soc Trans 40(1):210–214

Fan GC, Chu G, Kranias EG (2005) Hsp20 and its cardioprotection. Trends Cardiovasc Med 15(4):138–141

Fink MA, Zakhary DR, Mackey JA, Desnoyer RW, Apperson Hansen C, Damron DS, Bond M (2001) AKAP-mediated targeting of protein kinase A regulates contractility in cardiac myocytes. Circ Res 88(3):291–297

Gorelik J, Wright PT, Lyon AR, Harding SE (2013) Spatial control of the betaAR system in heart failure: the transverse tubule and beyond. Cardiovasc Res 98(2):216–224

Horvat SJ, Deshpande DA, Yan H, Panettieri RA, Codina J, DuBose TD, Xin W, Rich TC, Penn RB (2012) A-kinase anchoring proteins regulate compartmentalized cAMP signaling in airway smooth muscle. FASEB J 26(9):3670–3679

Houslay MD, Adams DR (2003) PDE4 cAMP phosphodiesterases: modular enzymes that orchestrate signalling cross-talk, desensitization and compartmentalization. Biochem J 370(Pt 1):1–18

Houslay MD, Baillie GS, Maurice DH (2007) cAMP-Specific phosphodiesterase-4 enzymes in the cardiovascular system: a molecular toolbox for generating compartmentalized cAMP signaling. Circ Res 100(7):950–966

Hundsrucker C, Krause G, Beyermann M, Prinz A, Zimmermann B, Diekmann O, Lorenz D, Stephan E, Nedvetsky P, Dathe M, Christian F, McSorley T, Krause E, McConnachie G, Herberg FW, Scott JD, Rosenthal W, Klussmann E (2006) High-affinity AKAP7delta-protein kinase A interaction yields novel protein kinase A-anchoring disruptor peptides. Biochem J 396(2):297–306

Kaupp UB, Seifert R (2002) Cyclic nucleotide-gated ion channels. Physiol Rev 82(3):769–824

Komalavilas P, Penn RB, Flynn CR, Thresher J, Lopes LB, Furnish EJ, Guo M, Pallero MA, Murphy-Ulrich JE, Brophy CM (2008) The small heat shock-related protein, HSP20, is a cAMP-dependent protein kinase substrate that is involved in airway smooth muscle relaxation. Am J Physiol Lung Cell Mol Physiol 294(1):L69–L78

Lee JH, Jung NH, Lee BH, Kim SH, Jun JH (2013a) Suppression of Heme Oxygenase-1 by prostaglandin E2-protein kinase A-A-kinase anchoring protein signaling is central for augmented cyclooxygenase-2 expression in lipopolysaccharide-stimulated RAW 264.7 macrophages. Allergy Asthma Immunol Res 5(5):329–336

Lee LC, Maurice DH, Baillie GS (2013b) Targeting protein-protein interactions within the cyclic AMP signaling system as a therapeutic strategy for cardiovascular disease. Future Med Chem 5(4):451–464

Li SS, Wu C (2009) Using peptide array to identify binding motifs and interaction networks for modular domains. Methods Mol Biol 570:67–76

Li Y, Sroubek J, Krishnan Y, McDonald TV (2008) A-kinase anchoring protein targeting of protein kinase A and regulation of HERG channels. J Membr Biol 223(2):107–116

Li X, Baillie GS, Houslay MD (2009) Mdm2 directs the ubiquitination of beta-arrestin-sequestered cAMP phosphodiesterase-4D5. J Biol Chem 284(24):16170–16182

Lynch MJ, Baillie GS, Mohamed A, Maisonneuve C, Klussmann E, van Heeke G, Houslay MD (2005) RNA silencing identifies PDE4D5 as the functionally relevant cAMP phosphodiesterase interacting with beta arrestin to control the protein kinase A/AKAP79-mediated switching of the beta2-adrenergic receptor to activation of ERK in HEK293B2 cells. J Biol Chem 280(39):33178–33189

Martin TP, Currie S, Baillie GS (2014a) The cardioprotective role of small heat-shock protein 20. Biochem Soc Trans 42(2):270–273

Martin TP, Hortingon-Vinagre MP, Findlay JE, Elliott C, Currie S, Baillie GS (2014b) Targeted disruption of the heat shock protein 20-phosphodiesterase 4D (PDE4D) interaction protects against pathological cardiac remodelling in a mouse model of hypertrophy. FEBS Open Bio 4:923–927

Mauban JR, O'Donnell M, Warrier S, Manni S, Bond M (2009) AKAP-scaffolding proteins and regulation of cardiac physiology. Physiology (Bethesda) 24:78–87

Maurice DH, Ke H, Ahmad F, Wang Y, Chung J, Manganiello VC (2014) Advances in targeting cyclic nucleotide phosphodiesterases. Nat Rev Drug Discov 13(4):290–314

McConnachie G, Langeberg LK, Scott JD (2006) AKAP signaling complexes: getting to the heart of the matter. Trends Mol Med 12(7):317–323

McCormick K, Baillie GS (2014) Compartmentalisation of second messenger signalling pathways. Curr Opin Genet Dev 27:20–25

Murdoch H, Vadrevu S, Prinz A, Dunlop AJ, Klussmann E, Bolger GB, Norman JC, Houslay MD (2011) Interaction between LIS1 and PDE4, and its role in cytoplasmic dynein function. J Cell Sci 124(Pt 13):2253–2266

Nikolaev VO, Moshkov A, Lyon AR, Miragoli M, Novak P, Paur H, Lohse MJ, Korchev YE, Harding SE, Gorelik J (2010) Beta2-adrenergic receptor redistribution in heart failure changes cAMP compartmentation. Science 327(5973):1653–1657

Nino G, Hu A, Grunstein JS, Grunstein MM (2009) Mechanism regulating proasthmatic effects of prolonged homologous beta2-adrenergic receptor desensitization in airway smooth muscle. Am J Physiol Lung Cell Mol Physiol 297(4):L746–L757

Patel HH, Hamuro LL, Chun BJ, Kawaraguchi Y, Quick A, Rebolledo B, Pennypacker J, Thurston J, Rodriquez-Pinto N, Self C, Olson G, Insel PA, Giles WR, Taylor SS, Roth DM (2010) Disruption of protein kinase A localization using a trans-activator of transcription (TAT)-conjugated A-kinase-anchoring peptide reduces cardiac function. J Biol Chem 285(36): 27632–27640

Perry SJ, Baillie GS, Kohout TA, McPhee I, Magiera MM, Ang KL, Miller WE, McLean AJ, Conti M, Houslay MD, Lefkowitz RJ (2002) Targeting of cyclic AMP degradation to beta 2-adrenergic receptors by beta-arrestins. Science 298(5594):834–836

Rampersad SN, Ovens JD, Huston E, Umana MB, Wilson LS, Netherton SJ, Lynch MJ, Baillie GS, Houslay MD, Maurice DH (2010) Cyclic AMP phosphodiesterase 4D (PDE4D) Tethers EPAC1 in a vascular endothelial cadherin (VE-Cad)-based signaling complex and controls cAMP-mediated vascular permeability. J Biol Chem 285(44):33614–33622

Rangarajan S, Enserink JM, Kuiperij HB, de Rooij J, Price LS, Schwede F, Bos JL (2003) Cyclic AMP induces integrin-mediated cell adhesion through Epac and Rap1 upon stimulation of the beta 2-adrenergic receptor. J Cell Biol 160(4):487–493

Raslan Z, Magwenzi S, Aburima A, Tasken K, Naseem KM (2015) Targeting of type I protein kinase A to lipid rafts is required for platelet inhibition by the 3',5'-cyclic adenosine monophosphate-signaling pathway. J Thromb Haemost 13(9):1721–1734

Sachs BD, Baillie GS, McCall JR, Passino MA, Schachtrup C, Wallace DA, Dunlop AJ, MacKenzie KF, Klussmann E, Lynch MJ, Sikorski SL, Nuriel T, Tsigelny I, Zhang J, Houslay MD, Chao MV, Akassoglou K (2007) p75 neurotrophin receptor regulates tissue fibrosis through inhibition of plasminogen activation via a PDE4/cAMP/PKA pathway. J Cell Biol 177(6): 1119–1132

Schindler RF, Brand T (2016) The Popeye domain containing protein family—a novel class of cAMP effectors with important functions in multiple tissues. Prog Biophys Mol Biol 120(1–3): 28–36

Scotland G, Houslay MD (1995) Chimeric constructs show that the unique N-terminal domain of the cyclic AMP phosphodiesterase RD1 (RNPDE4A1A; rPDE-IVA1) can confer membrane association upon the normally cytosolic protein chloramphenicol acetyltransferase. Biochem J 308(Pt 2):673–681

Sin YY, Edwards HV, Li X, Day JP, Christian F, Dunlop AJ, Adams DR, Zaccolo M, Houslay MD, Baillie GS (2011) Disruption of the cyclic AMP phosphodiesterase-4 (PDE4)-HSP20 complex attenuates the beta-agonist induced hypertrophic response in cardiac myocytes. J Mol Cell Cardiol 50(5):872–883

Smith KJ, Baillie GS, Hyde EI, Li X, Houslay TM, McCahill A, Dunlop AJ, Bolger GB, Klussman E, Adams DR, Houslay MD (2007) 1H NMR structural and functional characterisation of a cAMP-specific phosphodiesterase-4D5 (PDE4D5) N-terminal region peptide that disrupts PDE4D5 interaction with the signalling scaffold proteins, beta-arrestin and RACK1. Cell Signal 19(12):2612–2624

Taylor RP, Benjamin IJ (2005) Small heat shock proteins: a new classification scheme in mammals. J Mol Cell Cardiol 38(3):433–444

Taylor SS, Zhang P, Steichen JM, Keshwani MM, Kornev AP (2013) PKA: lessons learned after twenty years. Biochim Biophys Acta 1834(7):1271–1278

Wang L, Burmeister BT, Johnson KR, Baillie GS, Karginov AV, Skidgel RA, O'Bryan JP, Carnegie GK (2015) UCR1C is a novel activator of phosphodiesterase 4 (PDE4) long isoforms and attenuates cardiomyocyte hypertrophy. Cell Signal 27(5):908–922

Willis MJ, Baillie GS (2014) Arrestin-dependent localization of phosphodiesterases. Handb Exp Pharmacol 219:293–307

Wilson LS, Baillie GS, Pritchard LM, Umana B, Terrin A, Zaccolo M, Houslay MD, Maurice DH (2011) A phosphodiesterase 3B-based signaling complex integrates exchange protein activated by cAMP 1 and phosphatidylinositol 3-kinase signals in human arterial endothelial cells. J Biol Chem 286(18):16285–16296

Woodrum D, Pipkin W, Tessier D, Komalavilas P, Brophy CM (2003) Phosphorylation of the heat shock-related protein, HSP20, mediates cyclic nucleotide-dependent relaxation. J Vasc Surg 37(4):874–881

Computational Modeling of Cyclic Nucleotide Signaling Mechanisms in Cardiac Myocytes

10

Claire Y. Zhao

Contents

C.Y. Zhao
Department of Biomedical Engineering, Institute for Computational Medicine,
The Johns Hopkins University School of Medicine and Whiting School of Engineering,
3400 N Charles Street, Baltimore, MD 21218, USA

Acute Care Solutions, Philips Research North America,
2 Canal Park, 3rd floor, Cambridge, MA 02141, USA
e-mail: clairezhao1@gmail.com

V. Nikolaev, M. Zaccolo (eds.), *Microdomains in the Cardiovascular System*,
Cardiac and Vascular Biology 3, DOI 10.1007/978-3-319-54579-0_10

Abstract

The balanced signaling between the two cyclic nucleotides (cNs), cAMP and cGMP, in the cN signaling system plays a critical role in regulating cardiac contractility. Many therapeutic agents have been developed to selectively inhibit or stimulate proteins in the cN signaling system in the attempt to manage and treat heart diseases. Nonetheless, it has been challenging to obtain a comprehensive, system-level understanding of the signal transduction mechanisms, in part because of the participation of multiple phosphodiesterases (PDEs) in the common task of cN degradation, the complex interactions between the signaling proteins, and the large number of cN regulated targets in the tightly coupled excitation-contraction (EC) coupling process. Multi-scale, biophysically detailed, and experimentally validated computational models are well suited to dissect the underlying mechanisms in these nonlinear and intertwined reaction networks. By precisely defining and quantifying biochemical reactions involved, data-driven and integrative modeling bridge causal gaps across spatiotemporal scale, from the characteristics of individual molecular components to the collective responses of the entire signaling network. Through predictive modeling and in-depth analysis, these computational models are powerful in providing insights into cellular mechanisms, formulating novel hypothesis, and proposing possible future experiments. This review focuses on the development of mechanistic models, the close interplay between modeling and experimentation, and the identification of opportunities for future modeling research in the cardiac myocyte cN signaling system.

10.1 Introduction

The cyclic nucleotides (cNs), cyclic adenosine-3′, 5′-monophosphate (cAMP), and cyclic guanosine-3′, 5′-monophosphate (cGMP) are second messengers that regulate the response of cardiac myocyte to external and internal stimuli (Antos et al. 2009; Cohen 2002; Graves and Krebs 1999). The signal transduction mechanisms that regulate the cellular dynamics of cAMP and cGMP are together referred to as the cN signaling system (Beavo and Brunton 2002). They are among the earliest identified signal transduction systems (Maurice et al. 2014), with key discoveries recognized by the award of several Nobel Prizes (Nobelprize.org 2014). Today, new therapeutic agents have been developed on the basis of their ability to potently and selectively target the constituent cN signaling components (Conti and Beavo 2007; Lugnier 2006). On the other hand, the precise mechanisms by which the cN signaling system modulates tissue-specific intracellular signaling remain to be established (Lugnier 2006; Fischmeister et al. 2006; Hofmann et al. 2009; Conti and Beavo 2007; Francis et al. 2009). As such, research on cN signal transduction mechanisms continue to expand after over half a century of scientific investigation (Beavo and Brunton 2002).

As information quickly accumulates on the various aspects of cN signaling, it has become increasingly important to obtain a comprehensive and integrated understanding of this intricately interconnected signaling system. As a result, the need for data-driven mechanistic and integrative modeling is growing. Through in-depth interpretation of experimental observations and generation of testable biological hypotheses, computational modeling of the cN signaling system aims to reveal the interactions between signal components within the cN system, to uncover synchronization of the cN system with other biological processes, and to bridge gaps in casual links within and between spatiotemporal scales. This review focuses on the close interplay between experiments and mechanistic models of the cN signaling system and identifies opportunities for future research in this area.

10.1.1 Overview of the cN Signaling System

The principle signaling pathways that form the backbone of the cN signaling system are the β-adrenergic signaling pathway for cAMP regulation, among other G-protein-coupled receptor (GPCR) pathways that also participate in cAMP regulation, and the nitric oxide/cGMP/protein kinase G (NO/cGMP/PKG) and natriuretic peptide/cGMP/PKG (NP/cGMP/PKG) signaling pathways for cGMP regulation (Fig. 10.1). All three pathways show parallel in their overall structure, which warrants overall similarity in their corresponding model equations. Initially, a first messenger molecule, usually a hormone or neurotransmitter, triggers signal transduction (pink background) by binding to a trans-membrane receptor (blue background). Then, a cyclase and multiple families of phosphodiesterases (PDEs) are activated, respectively, for second messenger production and degradation (green background), shaping the dynamics of the cN signals (yellow background). Finally, a kinase is activated by the cNs (gray background), which in turn regulates downstream effectors via phosphorylation of these targets. The information encoded in the cN second messengers can be amplified many folds through the responses of downstream targets, such that cellular responses are able to change dramatically upon minute alternations to the cN signals (Antos et al. 2009; Cohen 2002; Graves and Krebs 1999).

10.1.2 cN Signaling Pathways and Cardiac Function

Among the various signaling mechanisms capable of exerting positive inotropic effects in humans, the β-adrenergic signaling pathway is the most powerful through which heart rate and contractility are physiologically regulated and maintained (Dzimiri 1999). Following its discovery in 1957, the development of accessible assays that provide meaningful measurement of cAMP in 1971 brought an explosion of activity and interest in characterizing the β-adrenergic pathway (Beavo and Brunton 2002). Based on this research, the second messenger signaling concept (Sect. 10.1.1

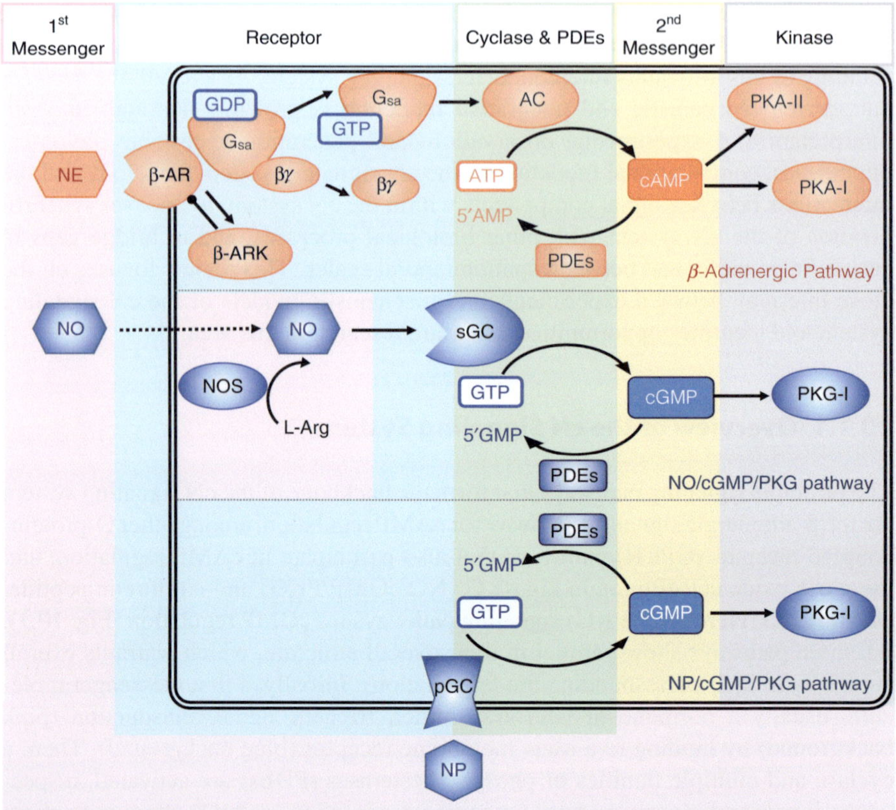

Fig. 10.1 cN signaling system in cardiac myocytes. In cardiac myocytes, the synthesis and degradation of cAMP are regulated by GPCR pathways, primarily the β-adrenergic signaling pathway, and that of cGMP are largely regulated by the NO/cGMP/PKG and NP/cGMP/PKG signaling pathways. The three signaling pathways show similarities in their overall structure with a first messenger (*pink background*) to activate a receptor (*blue background*), a cyclase and various PDEs (*green background*) for the production and degradation of the second messengers respectively (*yellow background*), and activation of a kinase (*gray background*) to phosphorylate target downstream. The representation in this schematic illustrates the key components comprising the signaling cascades, but does not account for signaling cross-talk or microdomains/compartments.

and Fig. 10.1), where extracellular first messengers bind to cell-surface receptors and initiates the production of intracellular second messengers, gradually emerged as a signaling paradigm commonly demployed by the cell (Beavo and Brunton 2002). As shown in Fig. 10.1 (top row), synthesis of cAMP (dark red oval) is primarily governed by the β-adrenergic pathway (red-shaded background) in response to elevated catecholamines (e.g., norepinephrine and epinephrine) (Antos et al. 2009; Cohen 2002; Graves and Krebs 1999). These ligands bind to and activate β-adrenergic receptors (β-ARs), which via a G-protein (Gs)-mediated process activate adenylyl cyclase (AC), the enzyme which catalyzes cAMP synthesis (Saucerman and McCulloch 2006; Bers 2002). Subsequent to degradation by PDEs, the net cAMP signal controls PKA activation (Saucerman et al. 2003), and hence, PKA-mediated phosphorylation of downstream targets that regulate contraction and relaxation of cardiac myocytes (Omori and Kotera 2007; Bers 2002). More specifically, PKA-I and PKA-II are the predominant PKA isoforms in cardiac myocytes (Zhao et al. 2015, 2016b).

Conversely, increased cGMP concentration ([cGMP]) is associated with attenuation of cardiac contractility (Tsai and Kass 2009; Boerrigter et al. 2009; Antos et al. 2009; Massion et al. 2005; Hammond and Balligand 2012). As shown in Fig. 10.1 (middle row), synthesis of cGMP is catalyzed by the intracellular soluble guanylate cyclase (sGC) in response to increased NO concentration ([NO]) (Hammond and Balligand 2012; Tsai and Kass 2009) (Fig. 10.1, middle schematic). In the 1960s, cGMP and the enzymes for its synthesis and degradation were discovered (Kots et al. 2009). On the other hand, it was not until the 1980s that the hormone which stimulated the synthesis of cGMP was discovered, first identified as endothelial-derived relaxant factor (EDRF), and later realized to be NO (Ignarro et al. 1987). Research on the NO/cGMP/PKG pathway in the heart exploded in the 1990s, following the discovery that cardiac myocytes constitutively expressed NO synthase (NOS), an enzyme that catalyzes NO synthesis (Balligand et al. 1993). In addition to NO-derived cGMP, cGMP synthesis in cardiac myocytes can be triggered by NP, both atrial NP (ANP) and brain NP (BNP), through its activation of membrane-bound particular guanylate cyclase (pGC), or more specifically GC-A (Roy et al. 2008) (Fig. 10.1, bottom row). In fact, soon following the discoveries of ANP and BNP in 1981 and 1988 respectively, it became apparent that they exert cardiovascular responses (Kuhn 2004; Nishikimi et al. 2006). Both NO- and NP-derived cGMP activate PKG, more specifically isoform PKG-I for cardiac myocytes (Zhao et al. 2015, 2016b), which then regulate downstream targets via protein phosphorylation (Lee and Kass 2012).

10.1.3 Overview of Mechanistic Models of cN Signaling Pathways

A widely adapted mechanistic model of the β-adrenergic signaling pathway in the cardiac myocyte was developed by Saucerman et al. (2003). Later models have subsequently included multiple compartments of cAMP, in the attempt to study spatial localization of cAMP and its effect on myocyte contraction (Iancu et al. 2007; Heijman et al. 2011; Bondarenko 2014). Other modeling studies dissected the mechanisms contributing to cAMP localization, often based on experiments with simpler cellular systems, such as HEK cells (Xin et al. 2008; Oliveira et al. 2010). Nonetheless, the majority of studies focused on studying β-adrenergic responses of cardiac myocytes, connecting existing models of the β-adrenergic signaling pathway with those of electrophysiology. These studies have elucidated the effects of β-adrenergic stimulation on myofilament contraction (Negroni et al. 2015) and regulation of ion channels (Terrenoire et al. 2009), calcium (Ca^{2+}) cycling (Yang and Saucerman 2012; Song et al. 2001), and ion homeostasis (Kuzumoto et al. 2008). More recent models explored disease mechanisms in relation to the β-adrenergic pathway, such as the initiation of alternans (Hammer et al. 2015) and early afterdepolarizations (Xie et al. 2013), the consequences of signaling disturbances due to gene mutations (Saucerman et al. 2004; Terrenoire et al. 2005), and the development of cardiac hypertrophy (Yang et al. 2014; Ryall et al. 2012).

Comparatively, fewer modeling works have investigated the NO/cGMP/PKG and NP/cGMP/PKG pathways, especially with respect to cardiac myocytes. Models of NO-derived cGMP regulation have primarily been constructed for non-cardiac cells, such as vascular smooth muscles (Kapela et al. 2008; Yang et al. 2005; Held and Dostmann 2012; Cawley et al. 2007), neuronal cells (Bellamy et al. 2000; Philippides et al. 2000), platelets (Roy and Garthwaite 2006; Mo et al. 2004), and HEK cells (Batchelor et al. 2010). Additional modeling efforts have concentrated on understanding NO, a freely diffusible, free radical gas (Hall and Garthwaite 2009), with regard to its diffusion (Schmidt et al. 1997; Ramamurthi and Lewis 1997; Kar and Kavdia 2011), bio-transport (Chen and Popel 2007; Tsoukias et al. 2004; Buerk 2001; Tsoukias 2008), and synthesis via NOS (Heinzen and Pollack 2003; Chen and Popel 2006). Despite of their non-cardiac origins, these models provide valuable molecular insights to the biochemistry of the cGMP signaling pathways in cardiac myocytes.

Due to the data-intensive nature of their modeling approach, mechanistically detailed, biochemically based models that investigate the effects of simultaneous activations of multiple signaling pathways are relatively rare (Saucerman and McCulloch 2004). Soltis et al. (Soltis and Saucerman 2010) investigated the synergy between β-adrenergic and Ca^{2+}-/calmodulin-dependent protein kinase II (CaMKII) pathways via modeling phosphorylation of their common EC coupling substrates. Using logic-based differential equations, where activation or inhibition reactions are represented by normalized Hill functions and cross-talks are computed with logical AND and OR gates, Ryall et al. modeled 14 established pathways regulating cardiac myocyte growth, including all of the three pathways of the cN signaling system (Ryall et al. 2012). Otherwise, despite increasing realization that signaling is highly integrated (Saucerman and McCulloch 2004), the pathways of the cN signaling system (Fig. 10.1) have primarily been modeled in isolation of each other. Overall, investigation of the cross communication and synergistic effects of simultaneous activation of cN pathways still await incorporation of further mechanistically details.

10.2 Modeling Multiple PDE Interactions in Cardiac Myocytes

Cyclic nucleotide phosphodiesterase isoenzymes (PDEs) degrade cAMP and cGMP. They are ubiquitous in mammalian cells (Beavo 1995; Conti et al. 2014; Francis et al. 2011). Although early research has primarily focused on cN synthesis, recent studies have revealed that PDEs are critical to the regulation of numerous physiological processes, such as cell signal transduction, proliferation, and differentiation, apoptosis, and metabolism (Beavo 1995; Conti et al. 2014; Francis et al. 2011). In the cardiovascular system, distinct PDE isoenzymes regulate contractility and relaxation, cell growth/survival, and cardiac structural remodeling (Miller and Yan 2010; Zaccolo and Movsesian 2007; Omori and Kotera 2007). The molecular diversity of PDEs was recognized shortly after its discovery in 1958, as gel filtration and ion exchange chromatography of tissue extracts revealed multiple peaks of PDE activities (Maurice et al. 2014). The present-day nomenclature of PDEs is based on the PDEs' biochemical properties, regulation mechanisms, and sensitivity to pharmacological agents, as well as the genes they are products of (Lugnier 2006). Currently, PDEs are classified to Classes I, II, and III, with mammalian PDEs belonging to Class I, which is further organized into 11 structurally related but functionally distinct PDE families, PDEs 1–11 (Francis et al. 2011).

10.2.1 Diverse PDE Families in Cardiac Myocytes

The cytoplasm of cardiac myocytes primarily contains PDEs 1–5, each with its own unique biochemical characteristics and regulatory mechanisms (Kass et al. 2007b; Zhang et al. 2008; Omori and Kotera 2007; Beavo 1995; Francis et al. 2009; Lugnier 2006; Fischmeister et al. 2006; Zaccolo and Movsesian 2007). Recently, PDE9 has also been identified to regulate NP-derived cGMP in cardiac myocytes (Lee et al. 2015), independent of NO-derived cGMP. As such, multiple distinctively regulated PDEs participated in the common task of cN degradation. This gives rise to complex interactions between the PDEs themselves and between the PDEs and cNs. Consequently, it has been challenging to obtain a quantitative understanding of the role of each PDE in modulating intracellular signaling (Lugnier 2006; Conti and Beavo 2007; Fischmeister et al. 2006; Beavo and Brunton 2002; Maurice et al. 2014; Beavo 1995).

The significance of PDEs regulating contraction in an isoform-specific manner is highlighted by the prominent effects resulting from alteration of a specific PDE isoform (Ding et al. 2005a; Abi-Gerges et al. 2009; Mehel et al. 2013; Lehnart et al. 2005; Marín-García 2010). For instance, ablation of specific PDE activities through pharmacological inhibition or gene depletion is observed to promote cardiac apoptosis (Ding et al. 2005b), accelerate development of HF (Lehnart et al. 2005), and increase likelihood of cardiac arrhythmias (Lehnart et al. 2005; Molina et al. 2012). On the other hand, drugs that restore specific PDE activities (Knight and Yan 2012), such as PDE3 activity in ischemic and dilated cardiomyopathies (Yan et al. 2007a) and PDE1 and PDE4 activities in cardiac ischemia (Kostic et al. 1997), have cardioprotective effects. In addition, PDE5 inhibition is shown to be beneficial in various cardiac pathologies, such as heart failure (HF), cardiac hypertrophy, and ventricular arrhythmias (Takimoto 2012; Guazzi 2008; Kass et al. 2007a; Zhang et al. 2008).

10.2.2 Mechanistic Models of PDEs

As a first step in understanding multiple PDE actions, Zhao et al. (2015, 2016b) modeled the cN-mediated molecular mechanisms of PDEs 1–5 in detail (Fig. 10.2a–e). Additionally, a PDE9 model is proposed in this review (Fig. 10.2f). The columns of Fig. 10.2 categorize the PDEs by their selectivity for cNs: PDEs 1, 2, and 3 are dual specific, that is, capable of hydrolyzing both cNs (first column), PDE4 is specific to cAMP (second column), and PDEs 5 and 9 are selective for cGMP (third column). Each of the PDE isoenzymes is modeled as dimers of two identical subunits; for

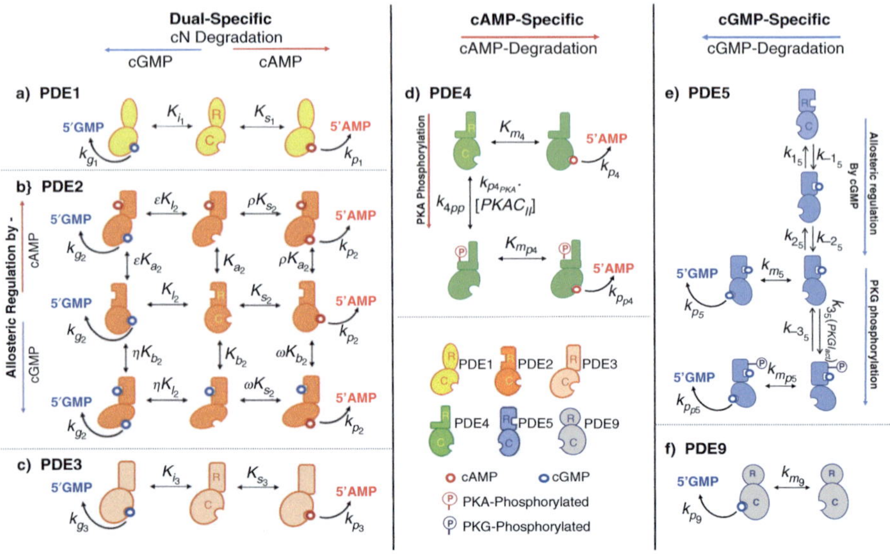

Fig. 10.2 PDEs in cardiac myocytes. PDE monomer subunits are shown with catalytic domains (ovals, denoted by "C") and regulatory domains (varied shapes, denoted by "R") for PDEs 1–5 and 9 in (a–f), respectively. Active sites within catalytic domains that bind cAMP and/or cGMP are represented by semicircular sockets. (**a–c**) Competitive binding of cAMP and cGMP to PDEs 1–3, respectively. (**b**) The GAF-B regulatory domain of PDE2 is represented by open rectangular socket. Its occupancy by either cNs allosterically activates PDE2 by increasing its catalytic domain's binding affinity to both cNs. (**d**) cAMP hydrolysis by PDE4 is regulated by PKA-mediated phosphorylation. (**e**) PDE5 is allosterically activated by cGMP binding to its GAF-A regulatory domain (*open rectangular sockets*). PKG-phosphorylated PDE5 has increased catalytic rate and increased cGMP affinity compared to the non-phosphorylated active species. Model adapted from that of Batchelor et al. (2010). (**f**) PDE9 is specific to degradation of NP-regulated cGMP. Figure adapted from Zhao et al. (2015, 2016b).

simplicity, each panel of Fig. 10.2 shows only one of the two subunits. All mammalian PDE subunits are made up of a catalytic and a regulatory domain (Conti and Beavo 2007; Lugnier 2006; Francis et al. 2011), denoted by the letters "C" and "R," respectively. The catalytic domain (oval) contains a conserved active site (semicircular socket) that can bind either cAMP or cGMP (Lugnier 2006; Francis et al. 2011; Conti and Beavo 2007). The regulatory domains differ markedly among PDEs (Francis et al. 2011) and consequently are denoted by symbols of different shapes. In addition, for each PDE, cN catalysis reactions are represented by horizontal transitions, and additional regulatory reactions, if any, are shown as vertical transitions. Reversible and irreversible reactions are denoted by double-headed and single-headed arrows, respectively. Reactions consisting of binding of cNs to PDEs, PDE conformational changes, and cN degradation via breakage of the 3′-cyclic phosphate bond are assumed to reach equilibrium rapidly with respect to the time scale of other signaling reactions. The parameters indicated by capital "K" denote binding affinities, whereas those with lowercase "k" denote rate constants. The final degraded products, 5′-AMP and 5′-GMP, are inactive in cN signaling pathways (Francis et al. 2011).

For dual-specific PDEs 1–3 (Fig. 10.2a–c respectively), both cNs compete for occupancy of cN-binding domains. The cNs competitively bind to the catalytic domains of PDEs 1 and 3 (Fig. 10.2a and c, respectively) and to both the regulatory (rectangular socket) and catalytic domains (semicircular socket) of PDE2 (Fig. 10.2b). Binding of either cNs to the regulatory domain of PDE2 induces allosteric activation of the enzyme (vertical transitions) by increasing the binding affinity of cNs to the catalytic domain (horizontal transitions) (Zoraghi et al. 2004; Francis et al. 2011; Martinez et al. 2002). The above binding schemes replicate and explain experimentally observed PDE cN hydrolysis rates. For PDEs 1 and 3, cAMP and cGMP hydrolysis are progressively suppressed by increasing levels of the other cN (Yan et al. 1996; He et al. 1998). This is because the other cN replaces the said cN in the catalytic domains of the PDEs. On the other hand, PDE2 hydrolysis rates for both cNs are bimodal against increasing concentrations of the other cN, resembling a dome shape (Prigent et al. 1988; Russell et al. 1973). More specifically, for PDE2 cAMP hydrolysis, although cGMP stimulates cAMP degradation rate upon binding to PDE2 regulatory domain, high cGMP can suppress cAMP hydrolysis by preventing cAMP binding at the PDE2 catalytic site. Consequently, PDE2 cAMP hydrolysis rate increases with increasing cGMP until ~3 μM cGMP, above which the rate decreases until it is suppressed below that measured without cGMP (Zhao et al. 2015). A similar bimodal dome shape is observed for PDE2 cGMP hydrolysis rate against cAMP, but with a much less pronounced cAMP-mediated rate stimulation due to differing cGMP affinity to PDE2 domains compared to cAMP (Zhao et al. 2015).

The rest of the PDEs shows much higher binding affinity to one of the cNs (Fig. 10.2d–f). As shown in Fig. 10.2d, PDE4 only appreciably degrades cAMP, but not cGMP. PKA-mediated phosphorylation of PDE4 induces a conformational change (vertical transitions), allowing the PKA-phosphorylated form to hydrolyze cAMP at a faster rate than the non-phosphorylated form (horizontal transitions). On the other hand, the catalytic domains of PDEs 5 and 9 are much more selective for cGMP. As shown in Fig. 10.2e, PDE5 cGMP catalytic rate can be increased by allosteric binding of cGMP to its regulatory domain (Batchelor et al. 2010) and PDE5 phosphorylation by PKG (MacKenzie et al. 2002; Leroy et al. 2008; Corbin et al. 2000; Rybalkin et al. 2002; Castro et al. 2010; Sette and Conti 1996). PDE9, specifically PDE9A, has recently been shown to regulate NP-derived cGMP in cardiac myocytes (Lee et al. 2015). It has even higher cGMP selectivity than PDE5 but lacks the cGMP/PKG stimulatory regulatory domains found in PDE5 (Soderling et al. 1998) (Fig. 10.2f). It is interesting to note that, unlike the other PDEs, PDE9A is not inhibited by xanthine derivatives, such as 3-isobutyl-1-methylxanthine (IBMX), but a number of selective inhibitors have been developed (Lugnier 2006; Bender and Beavo 2006).

10.2.3 Cross-Talk between cN Signaling Pathways

As shown in Fig. 10.2, the cN cross-talk signaling network is composed of the β-adrenergic pathway (red background), the NO/cGMP/PKG signaling pathway (blue background), and the cross-talk between them (yellow background) (Zhao et al. 2016b). NP, such as ANP, elicits distinct responses in cGMP dynamics than that produced by NO (Castro et al. 2006) and does not affect β-adrenergic-stimulated contractility (Takimoto et al. 2007; Perera et al. 2015). As a result, the NP/cGMP/PKG pathway and PDE9, which is specific to NP-derived cGMP (Lee et al. 2015), are currently not included in the cross-talk network (Fig. 10.2). Individual stimulation of the β-adrenergic and NO/cGMP/PKG pathways exert opposing effects, with the former enhancing cardiac inotropy and lusitropy (Katz 2011; Bers 2002) and the latter attenuating contractility (Tsai and Kass 2009; Boerrigter et al. 2009; Antos et al. 2009; Massion et al. 2005; Hammond and Balligand 2012) and antagonizing β-adrenergic tone (Zaccolo and Movsesian 2007; Stangherlin et al. 2011; Champion et al. 2004; Balligand 1999). As a result, the cN cross-talk signaling network (Fig. 10.1) can act to maintain the delicate balance between the cAMP and cGMP signals required for normal cardiac contraction (Stangherlin et al. 2011; Takimoto et al. 2005a; Zaccolo and Movsesian 2007; Senzaki et al. 2001; Mongillo et al. 2006; Stangherlin and Zaccolo 2012; Weiss et al. 1999; Moalem et al. 2006; Abi-Gerges et al. 2009; Mehel et al. 2013).

The cross-talk between the β-adrenergic and NO/cGMP/PKG pathways (Fig. 10.3, yellow background) can be interpreted as a network phenomenon arising from the molecular selectivity of PDEs to cAMP and cGMP (Fig. 10.1). As shown in Figure 10.3b, cAMP can stimulate its own degradation through activation of PDEs 2 and 4 (green arrows) in the form of negative feedback (Zhao et al. 2015).

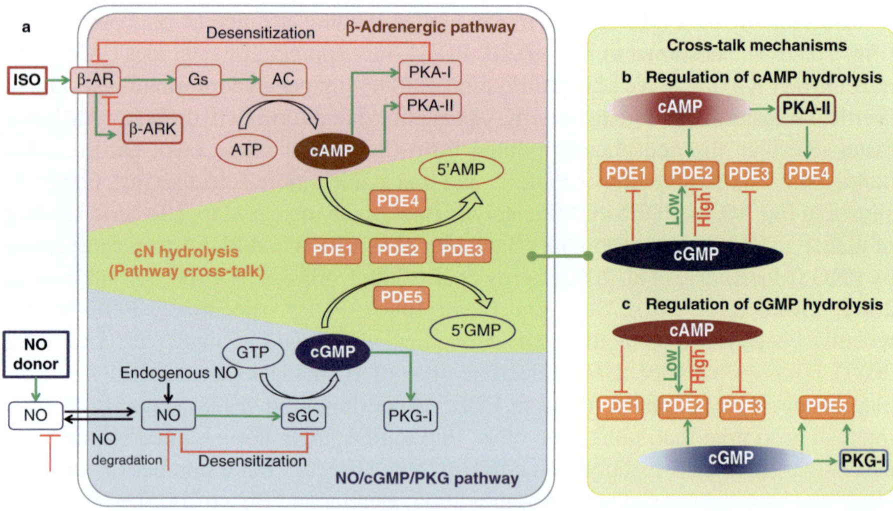

Fig. 10.3 cN cross-talk signaling network in cardiac myocytes. **a** The cN cross-talk signaling network model is composed of the β-adrenergic pathway (*red background*), the NO/cGMP/PKG signaling pathway (*blue background*), and cross-talk between them (*yellow background*) mediated by PDEs 1–5. In the regulation of cAMP- (b) and cGMP- (c) hydrolysis, cNs exert positive (*green arrows*) or negative (*red arrows*) regulation of PDE activities. To avoid crowding the figure, the hydrolysis reactions of cNs are omitted in (b) and (c), which would have been drawn as red arrows originating from each PDE to cAMP in (b) and cGMP in (c). Hydrolysis of cAMP and cGMP are, respectively, represented by ovals of *faded red* in (b) and *faded blue* in (c). Figure adapted from Zhao et al. (2016b)

On the other hand, the presence of cGMP can potentially increase cAMP concentration ([cAMP]) by inhibiting cAMP hydrolysis rates of PDEs 1 and 3 (red arrows) (Zhao et al. 2015). In addition, depending on its concentration ([cGMP]), cGMP can either inhibit or potentiate [cAMP] by regulating PDE2 cAMP hydrolysis activity (red and green arrows) (Zhao et al. 2015). Similarly, as shown in Figure 10.3c, negative feedback on cGMP is accomplished by cAMP- and cGMP-dependent activation of PDE2 and cGMP-dependent activation of PDE5 (Castro et al. 2010; Kass et al. 2007b; Zhang et al. 2008; Francis et al. 2009). On the other hand, the presence of cAMP can potentially increase cGMP by inhibiting cGMP-degrading activities of PDEs 1 and 3, while either inhibiting or potentiating cGMP by regulating PDE2 cGMP hydrolysis activity depending on cAMP (Kass et al. 2007b; Francis et al. 2009). Consequently, the complex interactions comprising cN cross-talk and the participation of multiple PDEs in the common task of cN degradation make understanding the nature of these regulatory mechanisms challenging (Lugnier 2006; Conti and Beavo 2007; Fischmeister et al. 2006; Beavo and Brunton 2002).

The underlying mechanisms of and interactions between PDEs in this nonlinear, tightly coupled reaction system can be revealed by analysis of computational models of the cN cross-talk network that mechanistically replicates experimentally observed activation-response relationships and temporal dynamics (Zhao et al. 2015, 2016b). Modeling reveals that a reduction in the activity of one PDE is compensated by the remaining PDEs, a behavior referred to as coupling (Zhao et al. 2015, 2016b). This model result indicates that the interpretation of experiments investigating the roles of multiple PDEs by measuring cN in response to application of selective blockers can be confounded by network interactions between the different PDEs. It is also discovered that PDE2 exhibits strong coupling with PDE4 in cAMP hydrolysis and with PDE5 for cGMP hydrolysis (Zhao et al. 2015, 2016b). Such coupling between PDE isoenzymes may be an important mechanism in stabilizing cAMP dynamics in the heart (Abi-Gerges et al. 2009; Marín-García 2010; Yan et al. 2007b), including disease settings where alterations in isoenzyme-specific PDE expression and/or activity have been implicated (Menniti et al. 2006; Maurice et al. 2014; Rahnama'i et al. 2013; Yan et al. 2007b; Marín-García 2010; Ding et al. 2005a; Mehel et al. 2013; Lehnart et al. 2005).

10.3 Modeling Regulation of Cardiac Electrophysiology by the cN Signaling System

The availability of increasingly mechanistic models of cN signal transduction system now opens the door for in-depth investigation of neurohormonal regulation of cardiac electrophysiology. Cardiac electrophysiology is a discipline with a rich history of close interplay between experiments and computational modeling since the 1960s (Winslow et al. 2011). Comparatively, signaling networks remain poorly characterized, in part because they are tightly interwoven with a variety of physiological processes in the cell (Saucerman and McCulloch 2004). The cN signaling system regulates the activity of many proteins involved in shaping cardiac contractility via PKA- and PKG-mediated phosphorylation of these proteins. Functional integration of models of the cN signaling system into a whole-cell myocyte model will help clarify and dissect the way by which cN signaling regulates Ca^{2+} cycling and AP morphology.

10.3.1 Overview of Models of Cardiac EC Coupling

EC coupling (Fig. 10.4) is the process by which electrical excitation leads to action potential (AP) formation, intracellular calcium (Ca^{2+}) cycling, and mechanical contraction of the myofilaments (Fig. 10.4). Briefly, sodium (Na^+) channels activate upon membrane depolarization, leading to the upstroke of AP. The subsequent increase in membrane depolarization allows L-type Ca^{2+} channels (LTCCs) to open, letting Ca^{2+} into the cell (I_{CaL}). As this trigger Ca^{2+} binds to and prompts the opening of ryanodine receptors (RyRs), Ca^{2+} from the sarcoplasmic reticulum (SR) Ca^{2+} store is released into the cytoplasm—a process commonly referred to as Ca^{2+}-induced Ca^{2+} release (CICR) (Bers 2001). The increased Ca^{2+} concentration ($[Ca^{2+}]$) in the cytosol allows it to bind to troponin I (TnI), inducing a conformational change in TnI and initiating contraction through shortening of the myofilaments (Bers 2002). The plateau of the AP is shaped by activation of potassium (K^+) channels, where the resulting outward K^+ current (I_K) opposes I_{CaL} (Tomaselli and Marbán 1999). As LTCC gradually

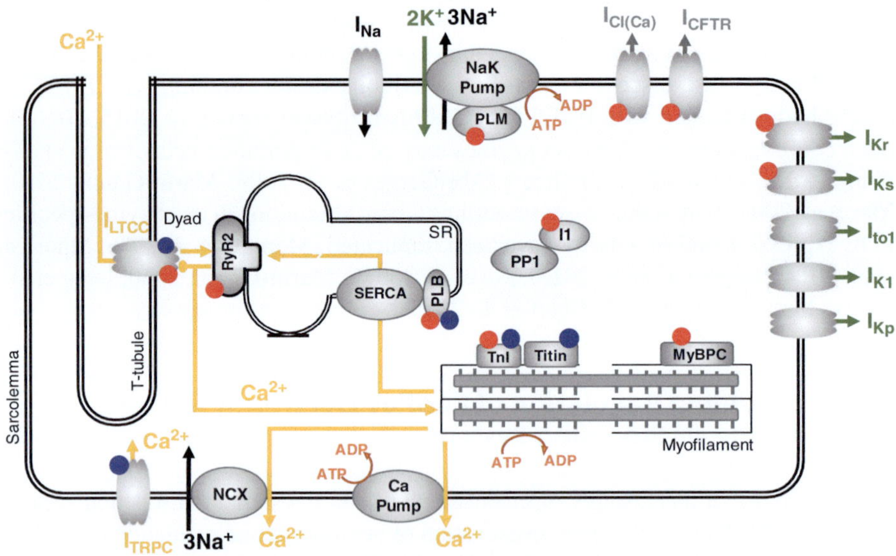

Fig. 10.4 Regulation of EC coupling by the cN signaling system in cardiac myocytes. Via PKA- and PKG-mediated phosphorylation of several proteins involved (denoted by *red* and *blue dots*, respectively), the cN signaling system regulates the EC coupling process, from Ca^{2+} cycling (*orange lines*) to contraction of the myofilaments. Phosphorylation targets shown (*dots*) are summarized from prior literature (Bers 2002; Takimoto 2012; Saucerman and McCulloch 2006; Tamargo et al. 2010; Feldman et al. 2005).

inactivates, I_K dominates over I_{CaL}, driving AP back to rest (Tomaselli and Marbán 1999). Cytosolic Ca^{2+} declines as Ca^{2+} is extruded from the cell primarily via Na^+/Ca^{2+} exchanger (NCX) or is pumped back into the SR through the SR Ca^{2+} (SERCA) pump (Tomaselli and Marbán 1999). As Ca^{2+} declines, Ca^{2+} dissociates from the myofilaments, and the myocyte begins to relax (Bers 2002).

Many computational models of EC coupling process have been developed in varying biophysical detail, with many in-depth reviews on this topic (Roberts et al. 2012; Winslow et al. 2011; Fink et al. 2011; Trayanova and Rice 2011; Williams et al. 2010; Smith et al. 2007; Richard 2001; Noble and Rudy 2001). Briefly, these models are typically composed of a system of ordinary differential equations (ODEs) and/or stochastic processes that depict the opening and closing of channels, transporters, pumps, and exchangers and the resulting ionic fluxes. Starting from deterministic common-pool models where CICR and Ca^{2+} cycling are modeled in a single compartment, later models incorporated increasingly detailed temporal and spatial profiles of Ca^{2+} dynamics by incorporating multiple Ca^{2+} compartments or even finite element meshes to spatially divide the cell, resulting in improved model behavior when compared to emerging experiments, and consequently provided enhanced mechanistic insights, such as those regarding SR release and interval-force relations. Additionally, new functional components are being developed and integrated into existing whole-cell models in pace with new experimental findings, such as models of EC coupling proteins with increased biophysical details, mitochondrial energetics, and myofilament contractions (Winslow et al. 2016). To meet the computational demands of increasingly detailed and integrated models, a spectrum of techniques to improve algorithmic and computational efficiency has also been developed, such as model reduction methods, numerical algorithms, and parallel computing approaches.

10.3.2 PKA-Mediated Phosphorylation of EC Coupling Proteins

Among the three primary pathways in the cN signaling system (Fig. 10.1), phosphorylation targets of the β-adrenergic pathway, their phosphorylation mechanisms, and the effects of phosphorylation contributing to sympathetic responses are the most extensively studied (Fig. 10.4, red dots). Phosphorylation of phospholamban (PLB) and TnI, respectively, speeds up SR Ca^{2+} re-uptake and dissociation of Ca^{2+} from the myofilaments, leading to the lusitropic effects of PKA (Li et al. 2000). Additionally, PKA-mediated phosphorylation of LTCC underlies *ICaL* potentiation in response to β-adrenergic stimulation; however, research is still ongoing to understand the functional link between LTCC gating changes and phosphorylation of the putative PKA sites (Yue et al. 1990; Harvey and Hell 2013). Under β-adrenergic stimulation, the greatly enhanced Ca^{2+} transient amplitude due to I_{CaL} potentiation more than offsets the reduction in myofilament Ca^{2+} sensitivity caused by TnI phosphorylation, which by itself alone would have reduced force (Li et al. 2000). At the same time, the faster SR Ca^{2+} re-uptake contributes to increased SR Ca^{2+} content. Together, the aforementioned mechanisms ultimately lead to the inotropic effect of PKA.

PKA is reported to phosphorylate RyRs, but results are mixed with decrease, increase, and lack of change in RyR open probability (Reiken et al. 2003; Bers 2006). Results are similarly varied concerning whether RyR phosphorylation alters the intrinsic responsiveness of SR Ca^{2+} release that triggers Ca^{2+}entry from LTCC. This is in part because of the challenges in isolation of intrinsic RyR effects from the highly interwoven CICR process (Bers 2002; Bers 2006). Besides RyRs, PKA phosphorylation of the corresponding channels has reported to modulate the kinetics of K^+ currents, I_{Ks} (Chen and Kass 2011; Kurokawa et al. 2009) and I_{Kr} (Harmati et al. 2011), as well as chloride (Cl^-) currents, $I_{Cl(Ca)}$ (Guo et al. 2008) and I_{CFTR} (Hwang et al. 1993). PKA phosphorylation of phospholemman (PLM) is reported to increase NaK pump Na^+ sensitivity and activation (Despa et al. 2008; Despa et al. 2005) and act as a feed-forward regulator of Ca^{2+} (Yang and Saucerman 2012). Furthermore, protein phosphatase inhibitor 1 (I1) phosphorylation by PKA enhances its inhibition of PP1 and may thereby augment β-adrenergic signaling (Gupta et al. 1996). Finally, PKA phosphorylates myosin-binding protein C (MyBPC) in addition to TnI at the myofilaments, but the precise contractile effects of this phosphorylation are still much debated (Marston and de Tombe 2008; Solaro and Kobayashi 2011).

10.3.3 PKG-Mediated Phosphorylation of EC Coupling Proteins

Many physiological effects modulated by cGMP, such as attenuating contractility, accelerating relaxation, and improving myocyte stiffness, are likely to be mediated by direct PKG phosphorylation of the respective proteins (Takimoto 2012) (Fig. 10.4, blue dots). On the other hand, much less is known of PKG-mediated phosphorylation compared to that of PKA. Studies have demonstrated PKG phosphorylation sites on LTCC (Jiang et al. 2000; Yang et al. 2007). Activation of the NO/cGMP/PKG pathway reduces LTCC open probability (Tohse and Sperelakis 1991; Fiedler et al. 2002; Tohse et al. 1995; Schröder et al. 2003) through prolonged channel closed time with no effects on channel open time or single-channel conductance (Tohse and Sperelakis 1991; Fiedler et al. 2002). Additionally, PKG phosphorylation of myofilament on TnI and titin sites has been shown to induce negative inotropic effects and accelerate relaxation (Shah et al. 1994; Krüger et al. 2009; Lee et al. 2010). Furthermore, PLB can be phosphorylated by PKG in vitro, which may serve to reverse the inhibition of PLB upon SERCA and thus increase SR Ca^{2+} uptake and produce positive inotropic and lusitropic effects in mammalian hearts (Huggins et al. 1989; Pierkes et al. 2002). Finally, PKG phosphorylation of transient receptor potential canonical 6 (TRPC6), a receptor-operated Ca^{2+} channel, suppresses its conductance and results in inhibition of calcineurin (CaN)-NFAT signaling (Koitabashi et al. 2010).

10.3.4 Effects of Dual Phosphorylation by PKA and PKG

As shown in Fig. 10.4 (red and blue dots), several proteins shaping EC coupling can be phosphorylated by both PKA and PKG, including LTCC, TnI, and PLB. While some phosphorylation sites are shared by both PKA and PKG, other targets possess distinct PKA and PKG phosphorylation sites. For instance, serine 16 on PLB is a common PKA and PKG phosphorylation site, and LTCC has both common and distinct sites (Colyer 1998; Wegener et al. 1989). Consequently, some of these EC coupling proteins may be simultaneously phosphorylated by both kinases, and each of them may theoretically undergo four distinct phosphorylation states: non-, PKA-, PKG-, and PKA- and PKG phosphorylated. In most cases, the behavior of the target protein under each phosphorylation state, especially under dual phosphorylation, awaits further investigation. Comparatively, more efforts has been focused on LTCC. The cN cross-talk signaling network is shown to exert both stimulatory and inhibitory regulatory actions on LTCCs via PKA- and PKG-mediated phosphorylation (Zhao et al. 2016a). Suppression of pre-stimulated I_{CaL} via NO/cGMP/PKG pathway activation is consistently observed by various groups (Méry et al. 1991; Sumii and Sperelakis 1995; Levi et al. 1994; Shirayama and Pappano 1996; Wahler and Dollinger 1995; Mubagwa et al. 1993; Imai et al. 2001; Levi et al. 1989; Abi-Gerges et al. 2001; Ziolo et al. 2003). On the other hand, under basal, non-stimulated conditions, NO/cGMP/PKG pathway activation has yielded complex results on I_{CaL}, including increased (Kumar et al. 1997), decreased (Grushin et al. 2008; Ziolo et al. 2001), or unchanged currents (Levi et al. 1994; Abi-Gerges et al. 2001). At the single-channel level, it is known that PKA increases LTCC currents by promoting more channels to gate in a high-activity gating mode (Yue et al. 1990). On the other hand, further studies are required to elucidate the impact of PKA and PKG phosphorylation on LTCC current kinetics, distribution of gating modes, gating characteristics of each gating mode, and Ca^{2+}-dependent inactivation (CDI) (Catterall 2000; Josephson et al. 2010).

10.4 Modeling Spatially Resolved cN Signaling in Cardiac Myocytes

The compartmentation hypothesis is proposed to explain the context-dependent specificity of cN signaling, which often manifests as distinct physiological responses via manipulation of the same signaling molecule. In the late 1970s, data emerged demonstrating distinct cardiac functions caused by cAMP activated by different hormone receptors, indicating that cAMP selectively activates its downstream targets (Hayes et al. 1979; Brunton et al. 1979). On the other hand, our understanding of the subcellular formation, localization, and dynamics of cN signals has lagged far behind that of Ca^{2+} (Rich et al. 2014). Traditionally, activation of the cN signaling system is measured by monitoring downstream physiological events or by membrane-based biochemical assays (Beavo and Brunton 2002). The former method is unable to resolve signaling events from downstream regulatory events, and the latter alters the native structure of the cell. As a result, these techniques do not permit detailed kinetic or spatial assessment of signaling events, especially in the micron-second spatial-temporal scale. With advancements in techniques for real-time and spatiotemporally resolved recording of the cN signaling system (Lohse et al. 2008; Herget et al. 2008; Sprenger and Nikolaev 2013; Rich et al. 2014), it is now possible to quantify the dynamics and spatial distribution of cN signaling in cardiac myocytes. Leveraging upon these experimental insights, spatially resolved computational modeling can help in understanding the diversification of cN signals in subcellular microdomains and communication between the proposed cellular compartments (Saucerman et al. 2014).

10.4.1 Compartmentation of cN Signaling

Figure 10.5 summarizes the proposed localization of cN signaling proteins with cardiac myocyte membrane ultrastructures (yellow), EC coupling proteins (green), and functional signaling axes (red) identified in literature. The vertical columns of Fig. 10.5 represent possible microdomains and/or signaling hubs. As shown, the sarcolemma compartment contains AC and pGC due to their membrane association (Laflamme and Becker 1999; Kuhn 2003), as well as PDE2 (Simmons and Hartzell

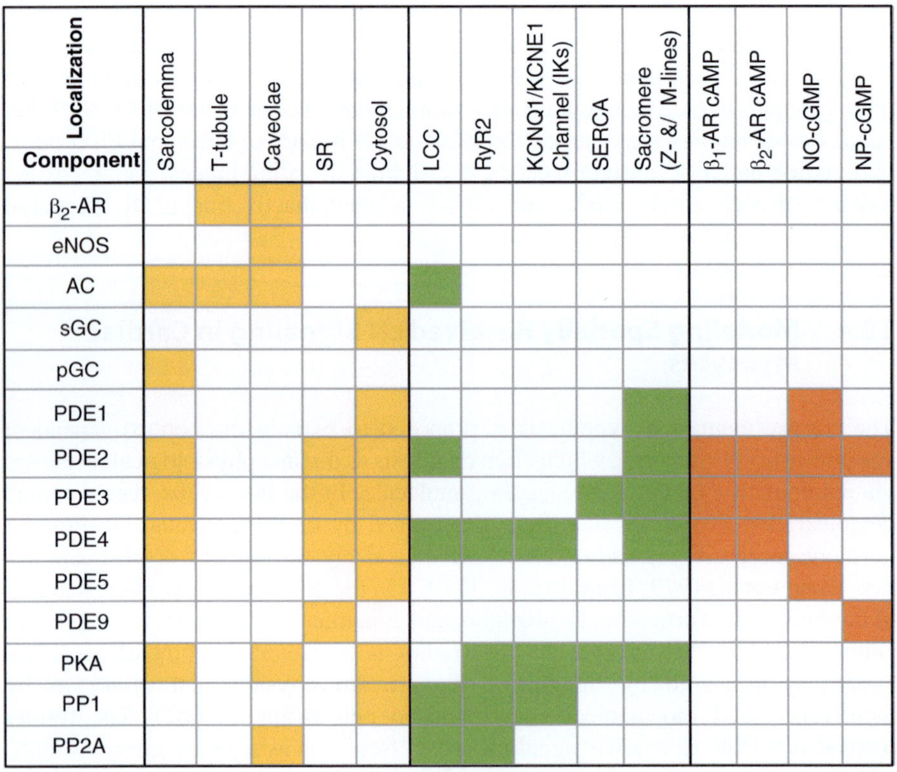

Fig. 10.5 Spatial organization of cN signaling components in cardiac myocytes. Key cN signaling components (*rows*) localized within potential microdomains/compartments (*columns*) in cardiac myocytes are identified by color-filled cells, with different colors identifying localization with myocyte ultrastructures (*yellow*), macromolecular signaling complexes (*green*), and functional signaling pathways (*red*).

1988), PDE3 (Weishaar et al. 1987), PDE4 (Baillie et al. 2002; Mika et al. 2014), and PKA, primarily PKA-II (Di Benedetto et al. 2008). Enrichments of β_2-AR (Nikolaev et al. 2006; Nikolaev et al. 2010) and AC (Laflamme and Becker 1999) are also found in T-tubular membranes. Recently, caveolae have emerged as a structural hub for localized signaling (Calaghan et al. 2008). Through their association with caveolin, a protein marker for caveolae, β_2-AR, AC, PKA, and PP2A (Balijepalli et al. 2006; Rybin et al. 2000), as well as eNOS (Calaghan et al. 2008), are shown to reside in this microdomain. PDEs 2 (Mongillo et al. 2006; Fischmeister et al. 2006), 3 (Ahmad et al. 2015), 4 (Lehnart et al. 2005), and 9 (Lee et al. 2015) are reported to associate with the SR membrane. Besides the above, most signaling proteins have a fraction that is diffused throughout the cytosol (Fig. 10.5) (Omori and Kotera 2007; Fischmeister et al. 2006).

Evidence also emerged for the existence of macromolecular signaling complexes, or signalosomes, in structuring localized cN signaling by organizing signaling components into spatial and/or functional aggregates (Fig. 10.5, green). For instance, subpopulations of LTCC is believed to be differentially distributed and regulated within the cardiac myocyte (Balijepalli et al. 2006; Christ et al. 2009; Shaw and Colecraft 2013), with tethered PP2A and PP2B (Xu et al. 2010) and distinct patterns of PDE regulation (Leroy et al. 2008; Wang et al. 2009; Benitah et al. 2010; Warrier et al. 2007). In addition, PDE4D is found to be an integral component of the RyR2 complex at the SR membrane, which contributes to localized control of Ca^{2+} release and SR Ca^{2+} store depletion (Lehnart et al. 2005; Beca et al. 2011). It is also found in association with the KCNQ1/KCNE1 K^+ channel, responsible for localized I_{Ks} regulation (Terrenoire et al. 2009). PDE3A1 has also recently been shown to participate in the SERCA/PLB/AKAP signalosome, where phosphorylation by PKA promotes its targeting to the signalosome and where it may modulate cAMP in a highly localized manner (Ahmad et al. 2015). In addition, via localization in lipid rafts, compartmentalized PDE2 activity is shown to attenuate cAMP signals via a NO/cGMP-dependent pathway (Mongillo et al. 2006). Finally, the sarcomere is also shown to be co-localized with a variety of PDEs, with PDEs 1–4 all showing a Z- and/or M-line enrichment (Mika et al. 2012; Hammond and Balligand 2012; Fischmeister et al. 2006).

The functional segregation of the cN signaling system has gradually been shaped in literature (Fig. 10.5, red). Specific targeting of cAMP is first to be recognized (Hayes et al. 1979; Brunton et al. 1979). In addition to cAMP microdomains created by signalosomes (Fig. 10.5, green), cAMP synthesized by β_1-AR stimulation creates a gradient that propagated throughout the cell, whereas that by β_2-AR stimulation does not markedly diffuse (Nikolaev et al. 2006). More recently, NP-derived cGMP is shown to trigger different responses than that of NO-derived cGMP (Kuhn 2015; Hammond and Balligand 2012) and is proposed to reside in distinct subcellular compartments (Castro et al. 2006; Lee et al. 2015; Piggott et al. 2006). The NP-derived cGMP pool is also shown to be primarily regulated by PDEs 2, 5, and 9 (Castro et al. 2006; Kass et al. 2007a; Lee et al. 2015), whereas PDEs 1, 2, 3, and 5 all contribute to degrading the NO-derived pool (Castro et al. 2006).

10.4.2 PDE Localization as a Mechanism Underlying cN Compartmentation

Localized cN degradation by PDEs is by far the most recognized mechanism of cN compartmentation, among others such as localized synthesis, physical barriers, cAMP buffering, cell shape, and cAMP export (Saucerman et al. 2014; Mika et al. 2012; Ziolo et al. 2008; Fischmeister et al. 2006; Vandecasteele et al. 2006). As shown in the rows of Fig. 10.5, PDEs are localized with many anatomical and functional structures of the cardiac myocytes. Compartmentation of PDE4 isoforms is mediated by their unique N-terminal domains (Houslay and Adams 2003) that can bind anchor/scaffold proteins, myomegalin (Verde et al. 2001), mAKAP (Dodge et al. 2001), and β-arrestins (Perry et al. 2002), which enables their targeting to specific cellular membranes. On the other hand, mechanisms of localization of other PDE isoforms await further investigation (Fischmeister et al. 2006; Omori and Kotera 2007). In addition, precise measurements of the concentrations of the PDEs in each proposed compartment are lacking, together with that of other residing signaling proteins.

10.5 Modeling cN Signaling System in Heart Failure

Disease progression of the hypertrophied or remodeled heart inevitably proceeds to heart failure (HF) (Tsai and Kass 2009). At the cellular level, HF is characterized by impaired inotropic signaling, Ca^{2+} mismanagement, and altered myofilament function (Mudd and Kass 2008; Marín-García 2010). Prior research has concentrated on the remodeling of ion currents in HF (Tomaselli and Marbán 1999; Aiba and Tomaselli 2010; Janse 2004; Nattel et al. 2007), but recently remodeling of the cN signaling system begins to receive more attention as a possible contributing factor (Maurice et al. 2014; Kass 2012; Marín-García 2010; Mudd and Kass 2008; Bender and Beavo 2006; Beavo and Brunton 2002; Wollert and Drexler 2002). Prominent remodeling of the components comprising the cN signaling network has been observed, including changes in expression levels and spatial reorganization of signaling proteins, that causes altered and/or untargeted cN signals and aberrant phosphorylation of target proteins (Marín-García 2010; Saucerman and McCulloch 2006; Wollert and Drexler 2002; Perera and Nikolaev 2013; Lohse et al. 2003). These perturbations disturb the delicate balance between the cNs and may ultimately lead to cardiac dysfunction (Marín-García 2010; Saucerman and McCulloch 2006; Wollert and Drexler 2002; Perera and Nikolaev 2013; Lohse et al. 2003). The construction of computational models for disease conditions can be viewed as sensitivity analyses of baseline models of normal physiology, with specific parameters perturbed as informed by experimental finding. These disease models will help understand and compare the effects of observed molecular changes, dissect the underlying mechanisms of physiological changes, and establish causal links between pathophysiological changes observed at varying spatiotemporal scales.

10.5.1 Changes in cN Synthesis in the Failing Heart

The changes affecting cN synthesis are summarized in Table 10.1 below. Studies show down-regulation of β_1-ARs in myocardial membranes (Bristow et al. 1982) and a decrease in functional coupling of the remaining β_1-ARs and β_2-ARs to the G_{as}-AC system in HF (Bristow et al. 1989). This uncoupling is likely to be exacerbated by enhanced expression of β-adrenergic receptor kinase (β-ARK) (Ungerer et al. 1993). Together with decreased expression of AC (Damy et al. 2004) and increased tonic inhibition by G_i proteins (Neumann et al. 1988; Böhm et al. 1990), these changes lead to attenuation of the systolic tension response to β-agonist (Lamba and Abraham 2000). In addition, the synthesis and secretion of ANP and BNP by ventricular myocytes are increased with the development of HF, leading to elevated plasma levels (Yasue et al. 1994), which are useful biomarkers indicative

Table 10.1 Remodeling of cN synthesis in the failing heart

Component	Changes	HF model	Species	Ref.
β_1-AR	Density ↓ 50%	DCM	Human	Bristow et al. (1982)
	Number ↓ 60%	DCM	Human	Bristow et al. (1986)
	Number and mRNA ↓ 50%	DCM and ICM	Human	Ungerer et al. (1993)
β_2-AR	Uncoupling from AC	DCM	Human	Bristow et al. (1989)
β-ARK	mRNA ↑ 3 fold; activity ↑	DCM and ICM	Human	Ungerer et al. (1993)
	Abundance ↑	--	Human	Koch et al. (1995)
G_i	Protein ↑ 40%	DCM	Human	Neumann et al. (1988)
	Protein ↑ 40%	DCM	Human	Böhm et al. (1990)
AC	Activity ↓ 60%	PI	Canine	Marzo et al. (1991)
	mRNA ↓ 40%; activity ↓ 40%	PI	Canine	Ishikawa et al. (1994)
	AC6 mRNA and activities ↓ 35%	MI	Rat	Espinasse et al. (1999)
	Expression and activity ↓	DCM	Human	Damy et al. (2004)
ANP	Synthesis and secretion by ventricular myocytes ↑	DCM	Human	Yasue et al. (1994)
BNP	Synthesis and secretion by ventricular myocytes ↑	DCM	Human	Yasue et al. (1994)
GC-A	Protein ↓	CHF	Human	Tsutamoto et al. (1992)
	↓ cGMP production	CHF	Human	Kuhn et al. (2002)
eNOS	Protein ↓	CHF	Human	Loyer et al. (2008)
	Uncouples	TAC	Mouse	Takimoto et al. (2005b)
sGC	↓ NO-stimulated activity	TAC	Mouse	Tsai et al. (2012)

CHF congestive HF, *DCM* dilated cardiomyopathy, *ICM* ischemic cardiomyopathy, *MI* myocardial infarction, *PI* pacing induced, *TAC* transverse aortic constriction, -- various etiologies of cardiac pathology (McKie and Burnett 2005). The reduction of GC-A (Tsutamoto et al. 1992) and/or its decreased responsiveness to ANP (Kuhn et al. 2002) have been associated with the blunted vasodilatory and diuretic response to ANP in HF (Riegger et al. 1988). Regarding the NO/cGMP/PKG pathway, it has been reported that reduced expression and uncoupling of eNOS in hypertrophic heart decrease NO bioavailability and promote the production of reactive oxygen species (ROS) (Takimoto et al. 2005b; Loyer et al. 2008). Furthermore, decreased cGMP production from sGC upon NO stimulation is observed in mice with transverse aortic constriction (TAC)-induced cardiac hypertrophy and dysfunction (Tsai et al. 2012).

10.5.2 Changes in Expressions and Activities of PDEs in the Failing Heart

Mathematical modeling is well suited for revealing the alterations in PDE interactions caused by the isoform-specific changes in PDEs in HF, summarized in Table 10.2. Miller et al. (2009) reported that PDE1A expression was significantly upregulated in hypertrophy and contributes to the reduction of cGMP-PKG signaling. Furthermore, PDE2 upregulation in the failing heart is observed to attenuate β-adrenergic signaling (Mehel et al. 2013), decreased PDE3 activity promotes cardiac myocyte apoptosis (Ding et al. 2005a), and PDE4 downregulation is associated with arrhythmias in cardiac hypertrophy and HF (Lehnart et al. 2005). Most studies report an increase in PDE5 expression and activity in the failing heart (Nagendran et al. 2007; Lu et al. 2010; Pokreisz et al. 2009), which may arise from increased

Table 10.2 Remodeling of PDEs in the failing heart

Component	Changes	HF model	Species	Ref.
PDE1	mRNA and protein ↑	Hypertrophy	Rat	Miller et al. (2009)
PDE2	Protein ↑ 3 fold	DCM	Human	Aye et al. (2012)
	Expression and activity ↑ 2 fold	Advanced HF	Human	Mehel et al. (2013)
PDE3	Expression and activity ↓	Apoptosis	Human	Ding et al. (2005a)
	Activity ↓	PI	Canine	Sato et al. (1999)
	mRNA and protein ↓	PI	Canine	Smith et al. (1998)
PDE4	Expression ↓ 40% in RyR2 complex	--	Human	Lehnart et al. (2005), Abi-Gerges et al. (2009)
PDE5	mRNA and protein ↑	Hypertrophy	Human	Nagendran et al. (2007)
	Expression and activity ↑	DCM and ICM	Human	Pokreisz et al. (2009)
	Protein ↑ 4.5 fold	CHF	Human	Lu et al. (2010)
PDE9	Protein ↑ 1 fold	DCM and HFpEF	Human	Lee et al. (2015)

CHF congestive HF, *DCM* dilated cardiomyopathy, *HFpEF* HF with preserved ejection fraction, *ICM* ischemic cardiomyopathy, *PI* pacing induced, *TAC* transverse aortic constriction, -- various etiologies myocardial oxidative stress (Lu et al. 2010) and contributes to adverse ventricular remodeling (Pokreisz et al. 2009). Finally, PDE9A activity is upregulated in hypertrophy and HF, and its inhibition is reported to reverse pre-established heart disease independent of NOS (Lee et al. 2015).

10.5.3 Changes in Spatial Localization of cN Signaling in the Failing Heart

Recent studies revealed spatially resolved remodeling of the cN signaling system in the failing heart. For instance, via nanoscale live-cell scanning ion conductance and fluorescence resonance energy transfer (FRET) microscopy techniques, Nikolaev et al. discovered that β_2-AR-mediated cAMP synthesis, which is normally limited to T-tubules (Nikolaev et al. 2006), redistributes to outer sarcolemma in HF (Nikolaev et al. 2010), while functional β_1-AR remain distributed throughout the cell membrane. Regarding the cGMP pathways in HF, PDE5 is observed to be retargeted to hydrolyze NP-derived instead of NO-derived cGMP (Zhang et al. 2012), together with a loss of Z-band localization (Senzaki et al. 2001). In addition, it has been reported that sGC β_1 subunits shift out of caveolin-enriched microdomains in HF that lead to reduced NO-stimulated activity in sGC (Tsai et al. 2012). Furthermore, in cardiomyocytes from hypertrophic hearts, Perera et al. revealed that β-AR-stimulated contractility is amplified by NP/cGMP signaling due to spatial redistribution of PDEs 2 and 3 (Perera et al. 2015).

10.5.4 Changes in Phosphorylation Status of EC Coupling Proteins in the Failing Heart

Prolongation of AP duration (APD) is the electrophysiological hallmark of myocytes isolated from failing hearts regardless of disease etiology (Aiba and Tomaselli 2010). Remodeling of ionic currents, such as downregulation of K^+ currents and alternation in Na^+ and Ca^{2+} currents, is a key change in HF and is reviewed in detail by prior works (Aiba and Tomaselli 2010; Janse 2004; Nattel et al. 2007; Tomaselli and Marbán 1999). Models of diseased cellular electrophysiology have been developed primarily based on altering the ionic fluxes of EC coupling proteins, often in isolation of signaling networks (Winslow et al. 1998; Winslow et al. 2001). Changes in the phosphorylation status of the corresponding EC coupling proteins in HF may contribute to electrical remodeling, although further studies are needed to ascertain these results. For instance, in human HF, single-channel studies suggest a reduction in LTCC number with an increase in open probability that may arise from altered channel phosphorylation (Chen et al. 2002; Schröder et al. 1998). Hyperphosphorylation of RyR by PKA may increase diastolic Ca^{2+} leak and generate spontaneous Ca^{2+} waves underlying triggered arrhythmias in HF (Marx et al. 2000). Additionally, evidence suggests that the reduction in I-1 protein and its phosphorylation by PKA in the failing heart lead to increased PP1 activity, which in turn may result in reduced phosphorylation of downstream proteins, such as PLB (El-Armouche et al. 2004). It is also interesting to note that the kinases themselves are remodeled in HF, which may be an underlying cause in the observed alterations in target protein phosphorylation. For instance, protein expressions of the regulatory subunits of PKA, types PKA-RI and PKA-RII, are reported to decrease by ~ 50% (Aye et al. 2012; Zakhary et al. 1999). In addition, interaction between PKA and AKAP SPHKAP is observed to increase by six fold (Aye et al. 2012). Finally, PKGIα protein expression is observed to increase three fold (Aye et al. 2012), but PKGI activity has been reported to decrease in some HF patients (van Heerebeek et al. 2012).

10.6 Multi-Type Data Integration and Fusion via Modeling

Biophysically based and experimentally validated computational modeling is a powerful tool for dissecting the integrated behavior of complex biological systems (Winslow et al. 2012). When coupled with experiments, this approach is particularly insightful for the distillation of critical relationships in complex, intertwined, and often times nonintuitive (patho-)physiological phenomena (Winslow et al. 2012). Models allow for the alteration of internal concentrations and rates not necessarily well controlled or even accessible in experiments, therefore providing a tightly controlled experimental setup for predictions unattainable by current technology. By integrating wide ranges of experimental data obtained via a variety of experimental techniques, mechanistic models provide a common and coherent framework for the data sets incorporated, which would often be otherwise interpreted in isolation. These model-integrated data, in turn, strengthen the predictive power of the models. Model simulation, thus, can be viewed as a form of data mining that seeks to extrapolate new knowledge from existing data sets.

Because of the numerous and fast-accumulating experimental findings in various aspects of the cN signaling system (Sects. 10.2–10.5), the mechanisms by which cN signaling influence cardiac function have been difficult to conceptualize in a unified manner. Modeling via an integrated systems approach is well suited for comprehensive and in-depth understanding of the regulatory pathways. As an example, the cN cross-talk signaling network model by Zhao et al. (2015, 2016b) fused a multitude of experimental data, effectively leveraging upon the extensive knowledge in published literature (Fig. 10.6), such that biomedical reactions can be represented in mechanistic detail. As shown in Fig. 10.6 (lower level), gel filtration and ion exchange chromatography of tissue extracts and radioimmunoassay with specific PDE inhibitors helped confirm the presence and activity of the specific PDE isoenzyme in cardiac myocytes (Weishaar et al. 1987; Bode et al. 1991; Movsesian et al. 1991; Bethke et al. 1992; Méry et al. 1995). The states of each PDE models (Zhao et al. 2015) were extracted from structural data from X-ray crystallography and amino acid sequencing and electron density maps (Pandit et al. 2009; Wu et al. 2004; Martinez et al. 2002). Quantification of binding affinities is derived from biochemical assays of purified PDEs, with the behavior of cN competition characterized by assays having both cNs in the reaction system (Yan et al. 1996; Prigent

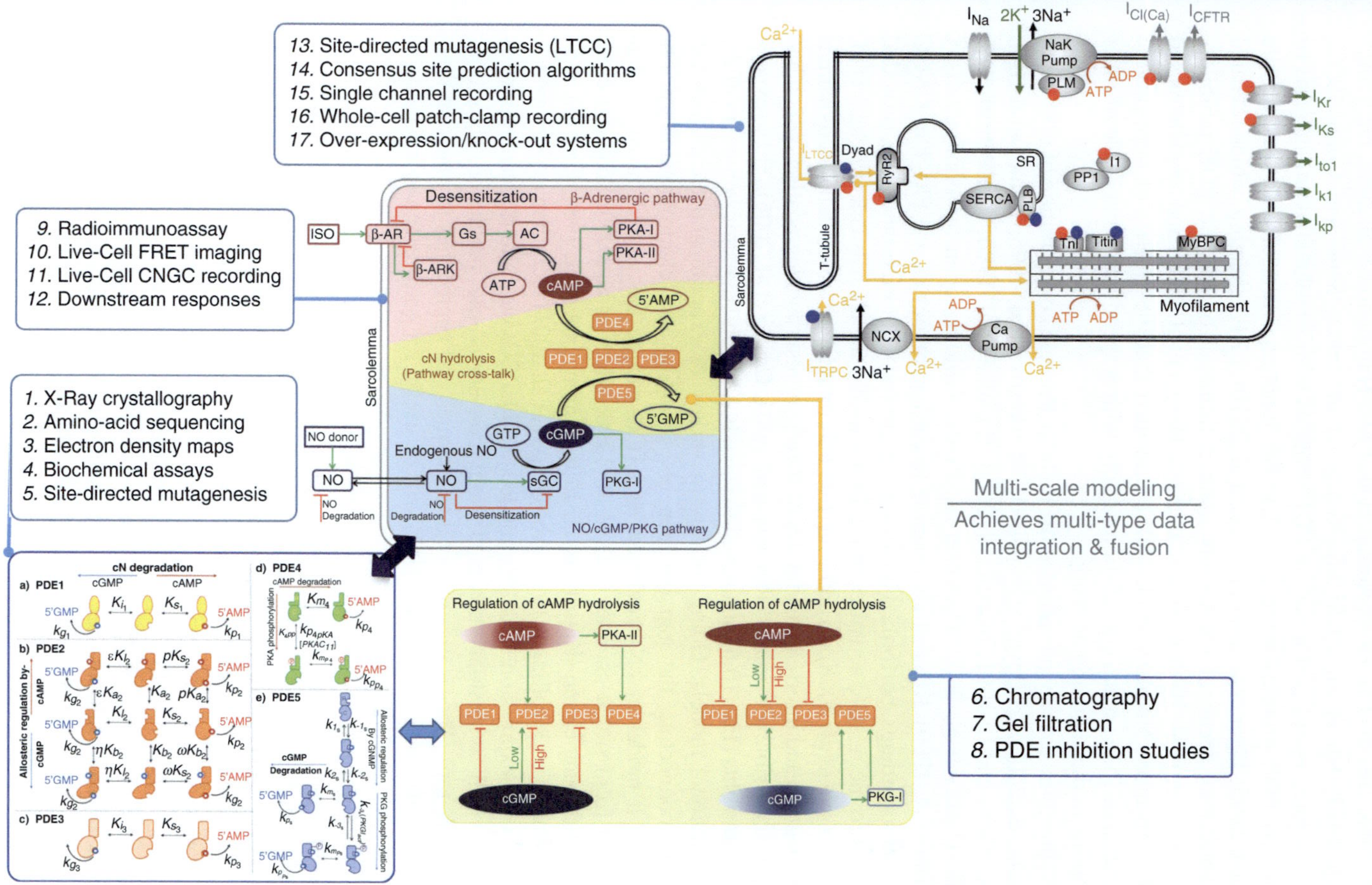

Fig. 10.6 Multi-scale modeling of cN signaling network. Experimental techniques informing models of the cN cross-talk signaling network (Zhao et al. 2015, 2016a, b) are shown at their respective spatiotemporal scales

et al. 1988; Beavo et al. 1971; Komas et al. 1991; Hambleton et al. 2005; Degerman et al. 1997; He et al. 1998; MacKenzie et al. 2002; Sette and Conti 1996). Furthermore, assays performed on PDE proteins with site-directed mutagenesis or truncated PDE proteins serve to isolate the functional role of each binding site (Beavo 1995; Jäger et al. 2010; Francis et al. 2010). FRET imaging, which revealed instantaneous decay of cN signals upon withdrawal of specific PDE inhibitors (Fischmeister et al. 2006; Castro et al. 2006; Herget et al. 2008) and upon increased PDE activation (Nikolaev et al. 2005), informed that the kinetics of intra-protein reactions of PDEs reach equilibrium rapidly with respect to the time scale of other reactions in the signaling network.

Radioimmunoassay pin-pointed the cellular-average concentration of cNs under varying extents of pathway stimulation (Kuznetsov et al. 1995; Kameyama et al. 1985; Hohl and Li 1991; Iancu et al. 2008; Katsube et al. 1996; Vila-Petroff et al. 1999; Castro et al. 2006; Castro et al. 2010). Building upon radioimmunoassay (Verde et al. 1999), live-cell cN imaging in cardiac myocytes, via FRET or CNGC channel recordings, performed under various protocols of PDE inhibition, provided insights to the relative activities of the PDEs within the cN network (Rochais et al. 2006; Castro et al. 2006). In addition, live-cell imaging revealed detailed dynamics of cN signals and inferred the kinetics of their synthesis and degradation, which had been inaccessible previously (Rochais et al. 2004; Castro et al. 2006). These experimental insights (Fig. 10.6) quantified the steady-state and dynamic behaviors of the computational models. Going forward, a variety of experimental techniques can help in connecting the gap between cN signaling and electrophysiology. For instance, a model of cN regulation of LTCC can be formulated based on data obtained from single channel (Yue et al. 1990; Klein et al. 2000; Schröder et al. 2003) and patch clamp (Katsube et al. 1996; Kameyama et al. 1985; Grushin et al. 2008; Wahler and Dollinger 1995; Abi-Gerges et al. 2001) recording experiments in control, over-expression, knockout models, and LTCCs with site-directed mutagenesis, as well as insights obtained from consensus site prediction algorithms (Fig. 10.6, top). Theses experimental techniques and approaches to integrate them in a computational model (Fig. 10.6) will be useful for guiding future modeling as well as experimental research.

As demonstrated in Fig. 10.6, integrative modeling unified many physiological quantities across spatiotemporal scale, providing a comprehensive framework for experimental findings to be viewed in perspective of each other. Thus, model analysis can provide in-depth interpretations of experiments which are often confounded by complex compensatory network interactions (Zhao et al. 2015, 2016b, a). For the cN cross-talk signaling network, the PDE models revealed the molecular mechanisms by which the PDEs decode information carried in the cGMP signal in β-adrenergic pathway regulation (Zhao et al. 2015). A precise quantitative definition of cN cross-talk and the ways by which cGMP and cAMP signals influence each other were obtained by analyzing the interaction of individual signaling elements in the context of the network architecture (Zhao et al. 2015, 2016b). PDE interactions arising from dynamic cN cross-talk within the cN cross-talk signaling network were delineated (Zhao et al. 2015, 2016b). The way by which LTCC simultaneously interact with both kinases was hypothesized (Zhao et al. 2016a). As such, integrated computational modeling is well suited for identifying biological mechanisms and predicting downstream consequences.

Conclusion

Therapeutic potential of the cN signaling network is highlighted in literature (Cohen 2002; Amanfu and Saucerman 2011; Boerrigter et al. 2009; Bender and Beavo 2006; Francis et al. 2009; Lugnier 2006); however, the extent to which pathological changes intervene with the normal operations of the entire signaling system, the cardiac cell, and ultimately the whole heart remains to be carefully studied. Leveraging more than 50 years of experimental insights (Sect. 10.1), mechanistic models of the cN signaling systems has expanded over the last decade (Sect. 10.1.3). Analyses of these models have provided quantitative and system-level insights and mechanistic predictions within each and across multiple biophysical hierarchies (Sects. 10.1 and 10.2). Data-driven, multi-scale modeling proves to be a powerful tool that links the characteristics of individual proteins, to the signaling network, and to the interaction between signaling and electrophysiology (Sects. 10.1, 10.2, and 10.6). Further functional integration of these models with electrophysiological models and into higher-dimensional tissue models will continue to advance the understanding of the formation and propagation of Ca^{2+} signals and AP in the heart (Sect. 10.3). In addition, recent breakthroughs in spatiotemporally resolved live-cell imaging of cNs in cardiac myocytes have begun to provide new insights in cN compartmentation and will offer valuable guidance for the development of future computational models (Sect. 10.4). Spatially resolved modeling is well suited for studying how local and global signaling is orchestrated to achieve coherent cellular communication. As the understanding of normal physiology gradually matures, new frontiers emerge for modeling diseased cardiac myocyte, which will help establish the causal link between altered signaling components, remodeled signal transduction process, and pathophysiological responses (Sect. 10.5). On the other hand, significant challenges remain. Divergent effects of neurohormonal stimulation observed in experiments remain a hurdle to the construction and validation of computation models. Open questions persist in the fundamental biophysics of the cN signaling network, such as precise quantification of the contribution of each signaling protein in each proposed cellular microdomains/compartments and mechanisms and functional consequences of protein phosphorylation. Nonetheless, computational modeling serves to provide in-depth analysis of experimental observations, form new hypothesis, and identify experimental conditions for future studies of the cN signaling system.

Acknowledgments This work was supported by Natural Sciences and Engineering Research Council (NSERC) of Canada scholarships, CGS M-377616-2009 and PGSD3-405041-2011, awarded to C.Y.Z.

Compliance with Ethical Standards

Conflict of Interest Statement The authors declare that they have no conflict of interest.

References

Abi-Gerges A, Richter W, Lefebvre F, Mateo P, Varin A, Heymes C, Samuel J-L, Lugnier C, Conti M, Fischmeister R (2009) Decreased expression and activity of cAMP phosphodiesterases in cardiac hypertrophy and its impact on β-adrenergic cAMP signals. Circ Res 105(8): 784–792

Abi-Gerges N, Fischmeister R, Méry PF (2001) G protein-mediated inhibitory effect of a nitric oxide donor on the L-type Ca^{2+} current in rat ventricular myocytes. J Physiol 531(1):117–130

Ahmad F, Shen W, Vandeput F, Szabo-Fresnais N, Krall J, Degerman E, Goetz F, Klussmann E, Movsesian M, Manganiello V (2015) Regulation of SERCA2 activity by PDE3A in human myocardium: Phosphorylation-dependent interaction of PDE3A1 with SERCA2. Journal of Biological Chemistry:jbc. M115. 638585

Aiba T, Tomaselli GF (2010) Electrical remodeling in the failing heart. Curr Opin Cardiol 25(1):29

Amanfu RK, Saucerman JJ (2011) Cardiac models in drug discovery and development: a review. 39(5):379–395. doi:10.1615/CritRevBiomedEng.v39.i5.30

Antos LLK, Schmidt HH, Hoffmann F, Stasch J-P (2009) cGMP: generators, effectors and therapeutic implications, vol 191. Springer

Aye T-T, Soni S, van Veen TAB, van der Heyden MAG, Cappadona S, Varro A, de Weger RA, de Jonge N, Vos MA, Heck AJR, Scholten A (2012) Reorganized PKA-AKAP associations in the failing human heart. J Mol Cell Cardiol 52(2):511–518. doi:10.1016/j.yjmcc.2011.06.003

Baillie GS, Huston E, Scotland G, Hodgkin M, Gall I, Peden AH, MacKenzie C, Houslay ES, Currie R, Pettitt TR (2002) TAPAS-1, a novel microdomain within the unique N-terminal region of the PDE4A1 cAMP-specific phosphodiesterase that allows rapid, Ca^{2+}-triggered membrane association with selectivity for interaction with phosphatidic acid. J Biol Chem 277(31):28298–28309

Balijepalli RC, Foell JD, Hall DD, Hell JW, Kamp TJ (2006) Localization of cardiac L-type Ca^{2+} channels to a caveolar macromolecular signaling complex is required for β2-adrenergic regulation. Proc Natl Acad Sci 103(19):7500–7505

Balligand J-L (1999) Regulation of cardiac β-adrenergic response by nitric oxide. Cardiovasc Res 43(3):607–620. doi:10.1016/s0008-6363(99)00163-7

Balligand J-L, Kelly RA, Marsden PA, Smith TW, Michel T (1993) Control of cardiac muscle cell function by an endogenous nitric oxide signaling system. Proc Natl Acad Sci 90(1):347–351

Batchelor AM, Bartus K, Reynell C, Constantinou S, Halvey EJ, Held KF, Dostmann WR, Vernon J, Garthwaite J (2010) Exquisite sensitivity to subsecond, picomolar nitric oxide transients conferred on cells by guanylyl cyclase-coupled receptors. Proc Natl Acad Sci 107(51):22060–22065. doi:10.1073/pnas.1013147107

Beavo J, Hardman JG, Sutherland EW (1971) Stimulation of adenosine 3', 5'-monophosphate hydrolysis by guanosine 3', 5'-monophosphate. J Biol Chem 246(12):3841–3846

Beavo JA (1995) Cyclic nucleotide phosphodiesterases: functional implications of multiple isoforms. Physiol Rev 75(4):725–748

Beavo JA, Brunton LL (2002) Cyclic nucleotide research -- still expanding after half a century. Nat Rev Mol Cell Biol 3(9):710–718

Beca S, Helli PB, Simpson JA, Zhao D, Farman GP, Jones PP, Tian X, Wilson LS, Ahmad F, Chen SRW, Movsesian MA, Manganiello V, Maurice DH, Conti M, Backx PH (2011) Phosphodiesterase 4D regulates baseline sarcoplasmic reticulum Ca^{2+} release and cardiac contractility, independently of L-type Ca^{2+} current. Circ Res 109(9):1024–1030. doi:10.1161/circresaha.111.250464

Bellamy TC, Wood J, Goodwin DA, Garthwaite J (2000) Rapid desensitization of the nitric oxide receptor, soluble guanylyl cyclase, underlies diversity of cellular cGMP responses. Proc Natl Acad Sci 97(6):2928–2933

Bender AT, Beavo JA (2006) Cyclic nucleotide phosphodiesterases: molecular regulation to clinical use. Pharmacol Rev 58(3):488–520

Benitah J-P, Alvarez JL, Gómez AM (2010) L-type Ca 2+ current in ventricular cardiomyocytes. J Mol Cell Cardiol 48(1):26–36

Bers DM (2001) Excitation contraction coupling and cardiac contractile force, 2nd edn. Kluwer Academic Publishers, Dordrecht

Bers DM (2002) Cardiac excitation–contraction coupling. Nature 415(6868):198–205

Bers DM (2006) Cardiac ryanodine receptor phosphorylation: target sites and functional consequences. Biochem J 396(1). doi:10.1042/bj20060377

Bethke T, Meyer W, Schmitz W, Scholz H, Stein B, Thomas K, Wenzlaff H (1992) Phosphodiesterase inhibition in ventricular cardiomyocytes from Guinea-pig hearts. Br J Pharmacol 107(1):127–133

Bode DC, Kanter JR, Brunton LL (1991) Cellular distribution of phosphodiesterase isoforms in rat cardiac tissue. Circ Res 68(4):1070–1079

Boerrigter G, Lapp H, Burnett JC (2009) Modulation of cGMP in heart failure: a new therapeutic paradigm. In: cGMP: generators, effectors and therapeutic implications. Springer. pp 485–506

Böhm M, Gierschik P, Jakobs K-H, Pieske B, Schnabel P, Ungerer M, Erdmann E (1990) Increase of Gi alpha in human hearts with dilated but not ischemic cardiomyopathy. Circulation 82(4):1249–1265

Bondarenko VE (2014) A compartmentalized mathematical model of the β 1-adrenergic signaling system in mouse ventricular myocytes. PLoS One 9(2):e89113

Bristow M, Hershberger R, Port JD, Minobe W, Rasmussen R (1989) Beta 1-and beta 2-adrenergic receptor-mediated adenylate cyclase stimulation in nonfailing and failing human ventricular myocardium. Mol Pharmacol 35(3):295–303

Bristow MR, Ginsburg R, Minobe W, Cubicciotti RS, Sageman WS, Lurie K, Billingham ME, Harrison DC, Stinson EB (1982) Decreased catecholamine sensitivity and β-adrenergic-receptor density in failing human hearts. N Engl J Med 307(4):205–211

Bristow MR, Ginsburg R, Umans V, Fowler M, Minobe W, Rasmussen R, Zera P, Menlove R, Shah P, Jamieson S (1986) Beta 1-and beta 2-adrenergic-receptor subpopulations in nonfailing and failing human ventricular myocardium: coupling of both receptor subtypes to muscle contraction and selective beta 1-receptor down-regulation in heart failure. Circ Res 59(3):297–309

Brunton LL, Hayes JS, Mayer SE (1979) Hormonally specific phosphorylation of cardiac troponin I and activation of glycogen phosphorylase.

Buerk DG (2001) Can we model nitric oxide biotransport? A survey of mathematical models for a simple diatomic molecule with surprisingly complex biological activities. Annu Rev Biomed Eng 3(1):109–143

Calaghan S, Kozera L, White E (2008) Compartmentalisation of cAMP-dependent signalling by caveolae in the adult cardiac myocyte. J Mol Cell Cardiol 45(1):88–92. doi:10.1016/j.yjmcc.2008.04.004

Castro LRV, Schittl J, Fischmeister R (2010) Feedback control through cGMP-dependent protein kinase contributes to differential regulation and compartmentation of cGMP in rat cardiac myocytes. Circ Res 107(10):1232–1240. doi:10.1161/circresaha.110.226712

Castro LRV, Verde I, Cooper DMF, Fischmeister R (2006) Cyclic guanosine monophosphate compartmentation in rat cardiac myocytes. Circulation 113(18):2221–2228. doi:10.1161/circulationaha.105.599241

Catterall WA (2000) Structure and regulation of voltage-gated Ca^{2+} channels. Annu Rev Cell Dev Biol 16(1):521–555

Cawley SM, Sawyer CL, Brunelle KF, van der Vliet A, Dostmann WR (2007) Nitric oxide-evoked transient kinetics of cyclic GMP in vascular smooth muscle cells. Cell Signal 19(5):1023–1033. doi:10.1016/j.cellsig.2006.11.012

Champion HC, Skaf MW, Hare JM (2004) Role of nitric oxide in the pathophysiology of heart failure. In: The role of nitric oxide in heart failure. Springer, pp 81–92

Chen K, Popel AS (2006) Theoretical analysis of biochemical pathways of nitric oxide release from vascular endothelial cells. Free Radic Biol Med 41(4):668–680

Chen K, Popel AS (2007) Vascular and perivascular nitric oxide release and transport: biochemical pathways of neuronal nitric oxide synthase (NOS1) and endothelial nitric oxide synthase (NOS3). Free Radic Biol Med 42(6):811–822

Chen L, Kass RS (2011) A-kinase anchoring protein 9 and I(Ks) channel regulation. J Cardiovasc Pharmacol 58(5):459–413. doi:10.1097/FJC.0b013e318232c80c

Chen X, Piacentino V, Furukawa S, Goldman B, Margulies KB, Houser SR (2002) L-type Ca^{2+} channel density and regulation are altered in failing human ventricular myocytes and recover after support with mechanical assist devices. Circ Res 91(6):517–524. doi:10.1161/01.res.0000033988.13062.7c

Christ T, Galindo-Tovar A, Thoms M, Ravens U, Kaumann AJ (2009) Inotropy and L-type Ca^{2+} current, activated by β1- and β2-adrenoceptors, are differently controlled by phosphodiesterases 3 and 4 in rat heart. Br J Pharmacol 156(1):62–83. doi:10.1111/j.1476-5381.2008.00015.x

Cohen P (2002) Protein kinases—the major drug targets of the twenty-first century? Nat Rev Drug Discov 1(4):309–315

Colyer J (1998) Phosphorylation states of Phospholamban. Ann N Y Acad Sci 853(1):79–91

Conti M, Beavo J (2007) Biochemistry and physiology of cyclic nucleotide phosphodiesterases: essential components in cyclic nucleotide signaling. Annu Rev Biochem 76:481–511

Conti M, Mika D, Richter W (2014) Cyclic AMP compartments and signaling specificity: role of cyclic nucleotide phosphodiesterases. J Gen Physiol 143(1):29–38. doi:10.1085/jgp.201311083

Corbin JD, Turko IV, Beasley A, Francis SH (2000) Phosphorylation of phosphodiesterase-5 by cyclic nucleotide-dependent protein kinase alters its catalytic and allosteric cGMP-binding activities. Eur J Biochem 267(9):2760–2767

Damy T, Ratajczak P, Shah AM, Camors E, Marty I, Hasenfuss G, Marotte F, Samuel J-L, Heymes C (2004) Increased neuronal nitric oxide synthase-derived NO production in the failing human heart. Lancet 363(9418):1365–1367

Degerman E, Belfrage P, Manganiello VC (1997) Structure, localization, and regulation of cGMP-inhibited phosphodiesterase (PDE3). J Biol Chem 272(11):6823–6826. doi:10.1074/jbc.272.11.6823

Despa S, Bossuyt J, Han F, Ginsburg KS, Jia L-G, Kutchai H, Tucker AL, Bers DM (2005) Phospholemman-phosphorylation mediates the β-adrenergic effects on Na/K pump function in cardiac myocytes. Circ Res 97(3):252–259. doi:10.1161/01.RES.0000176532.97731.e5

Despa S, Tucker AL, Bers DM (2008) Phospholemman-mediated activation of Na/K-ATPase limits [Na]i and inotropic state during β-adrenergic stimulation in mouse ventricular myocytes. Circulation 117(14):1849–1855. doi:10.1161/circulationaha.107.754051

Di Benedetto G, Zoccarato A, Lissandron V, Terrin A, Li X, Houslay MD, Baillie GS, Zaccolo M (2008) Protein kinase a type I and type II define distinct intracellular signaling compartments. Circ Res 103(8):836–844. doi:10.1161/circresaha.108.174813

Ding B, J-i A, Wei H, Huang Q, Walsh RA, Molina CA, Zhao A, Sadoshima J, Blaxall BC, Berk BC, Yan C (2005a) Functional role of phosphodiesterase 3 in cardiomyocyte apoptosis: implication in heart failure. Circulation 111(19):2469–2476. doi:10.1161/01.cir.0000165128.39715.87

Ding B, J-i A, Wei H, Xu H, Che W, Aizawa T, Liu W, Molina CA, Sadoshima J, Blaxall BC, Berk BC, Yan C (2005b) A positive feedback loop of phosphodiesterase 3 (PDE3) and inducible cAMP early repressor (ICER) leads to cardiomyocyte apoptosis. Proc Natl Acad Sci U S A 102(41):14771–14776. doi:10.1073/pnas.0506489102

Dodge KL, Khouangsathiene S, Kapiloff MS, Mouton R, Hill EV, Houslay MD, Langeberg LK, Scott JD (2001) mAKAP assembles a protein kinase a/PDE4 phosphodiesterase cAMP signaling module. EMBO J 20(8):1921–1930

Dzimiri N (1999) Regulation of β-adrenoceptor signaling in cardiac function and disease. Pharmacol Rev 51(3):465–502

El-Armouche A, Pamminger T, Ditz D, Zolk O, Eschenhagen T (2004) Decreased protein and phosphorylation level of the protein phosphatase inhibitor-1 in failing human hearts. Cardiovasc Res 61(1):87–93. doi:10.1016/j.cardiores.2003.11.005

Espinasse I, Iourgenko V, Richer C, Heimburger M, Defer N, Bourin M-C, Samson F, Pussard E, Giudicelli J-F, Michel J-B, Hanoune J, Mercadier J-J (1999) Decreased type VI adenylyl cyclase mRNA concentration and Mg2+−dependent adenylyl cyclase activities and unchanged type V adenylyl cyclase mRNA concentration and Mn2+−dependent adenylyl cyclase activities in the left ventricle of rats with myocardial inf…. Cardiovascular Research 42 (1):87–98. doi:10.1016/s0008-6363(98)00283-1

Feldman DS, Carnes CA, Abraham WT, Bristow MR (2005) Mechanisms of disease: [beta]-adrenergic receptors[mdash]alterations in signal transduction and pharmacogenomics in heart failure. Nat Clin Pract Cardiovasc Med 2(9):475–483

Fiedler B, Lohmann SM, Smolenski A, Linnemüller S, Pieske B, Schröder F, Molkentin JD, Drexler H, Wollert KC (2002) Inhibition of calcineurin-NFAT hypertrophy signaling by cGMP-dependent protein kinase type I in cardiac myocytes. Proc Natl Acad Sci 99(17): 11363–11368

Fink M, Niederer SA, Cherry EM, Fenton FH, Koivumäki JT, Seemann G, Thul R, Zhang H, Sachse FB, Beard D, Crampin EJ, Smith NP (2011) Cardiac cell modelling: observations from the heart of the cardiac physiome project. Prog Biophys Mol Biol 104(1–3):2–21. doi:10.1016/j.pbiomolbio.2010.03.002

Fischmeister R, Castro LR, Abi-Gerges A, Rochais F, Jurevičius J, Leroy J, Vandecasteele G (2006) Compartmentation of cyclic nucleotide signaling in the heart the role of cyclic nucleotide phosphodiesterases. Circ Res 99(8):816–828

Francis SH, Blount MA, Corbin JD (2011) Mammalian cyclic nucleotide phosphodiesterases: molecular mechanisms and physiological functions. Physiol Rev 91(2):651–690

Francis SH, Busch JL, Corbin JD (2010) cGMP-dependent protein kinases and cGMP phosphodiesterases in nitric oxide and cGMP action. Pharmacol Rev 62(3):525–563. doi:10.1124/pr.110.002907

Francis SH, Corbin JD, Bischoff E (2009) Cyclic GMP-hydrolyzing phosphodiesterases. In: cGMP: generators, effectors and therapeutic implications. Springer, pp 367–408

Graves JD, Krebs EG (1999) Protein phosphorylation and signal transduction. Pharmacol Ther 82(2–3):111–121. doi:10.1016/S0163-7258(98)00056-4

Grushin K, Nenov M, Dynnik V, Semushina S, Pakhomova I, Murashev A, Kokoz YM (2008) Role of the NO-cGMP cascade in regulation of L-type Ca^{2+} currents in isolated cardiomyocytes. Biochem (Moscow) Suppl Ser A Membr Cell Biol 2(3):243–252

Guazzi M (2008) Clinical use of phosphodiesterase-5 inhibitors in chronic heart failure. Circ Heart Fail 1(4):272–280. doi:10.1161/circheartfailure.108.802116

Guo D, Young L, Patel C, Jiao Z, Wu Y, Liu T, Kowey PR, Yan G-X (2008) Calcium-activated chloride current contributes to action potential alternations in left ventricular hypertrophy rabbit. Am J Physiol Heart Circ Physiol 295(1):H97–H104. doi:10.1152/ajpheart.01032.2007

Gupta RC, Neumann J, Watanabe AM, Lesch M, Sabbah HN (1996) Evidence for presence and hormonal regulation of protein phosphatase inhibitor-1 in ventricular cardiomyocyte. Am J Physiol Heart Circ Physiol 270(4):H1159–H1164

Hall CN, Garthwaite J (2009) What is the real physiological NO concentration in vivo? Nitric Oxide 21(2):92–103. doi:10.1016/j.niox.2009.07.002

Hambleton R, Krall J, Tikishvili E, Honeggar M, Ahmad F, Manganiello VC, Movsesian MA (2005) Isoforms of cyclic nucleotide phosphodiesterase PDE3 and their contribution to cAMP hydrolytic activity in subcellular fractions of human myocardium. J Biol Chem 280(47):39168–39174. doi:10.1074/jbc.M506760200

Hammer KP, Ljubojevic S, Ripplinger CM, Pieske BM, Bers DM (2015) Cardiac myocyte alternans in intact heart: influence of cell–cell coupling and β-adrenergic stimulation. J Mol Cell Cardiol 84:1–9

Hammond J, Balligand J-L (2012) Nitric oxide synthase and cyclic GMP signaling in cardiac myocytes: from contractility to remodeling. J Mol Cell Cardiol 52(2):330–340. doi:10.1016/j.yjmcc.2011.07.029

Harmati G, Banyasz T, Barandi L, Szentandrássy N, Horvath B, Szabo G, Szentmiklósi J, Szenasi G, Nanasi P, Magyar J (2011) Effects of β-adrenoceptor stimulation on delayed rectifier K+ currents in canine ventricular cardiomyocytes. Br J Pharmacol 162(4):890–896

Harvey RD, Hell JW (2013) Ca< sub> V 1.2 signaling complexes in the heart. J Mol Cell Cardiol 58:143–152

Hayes JS, Brunton LL, Brown JH, Reese JB, Mayer SE (1979) Hormonally specific expression of cardiac protein kinase activity. Proc Natl Acad Sci 76(4):1570–1574

He R, Komas N, Ekholm D, Murata T, Taira M, Hockman S, Degerman E, Manganiello VC (1998) Expression and characterization of deletion recombinants of two cGMP-inhibited cyclic nucleotide phosphodiesterases (PDE-3). Cell Biochem Biophys 29(1–2):89–111

Heijman J, Volders PG, Westra RL, Rudy Y (2011) Local control of β-adrenergic stimulation: effects on ventricular myocyte electrophysiology and Ca^{2+}-transient. J Mol Cell Cardiol 50(5):863–871
Heinzen EL, Pollack GM (2003) Pharmacokinetics and pharmacodynamics of L-arginine in rats: a model of stimulated neuronal nitric oxide synthesis. Brain Res 989(1):67–75
Held KF, Dostmann WR (2012) Sub-nanomolar sensitivity of nitric oxide mediated regulation of cGMP and vasomotor reactivity in vascular smooth muscle. Front Pharmacol:3. doi:10.3389/fphar.2012.00130
Herget S, Lohse MJ, Nikolaev VO (2008) Real-time monitoring of phosphodiesterase inhibition in intact cells. Cell Signal 20(8):1423–1431. doi:10.1016/j.cellsig.2008.03.011
Hofmann F, Bernhard D, Lukowski R, Weinmeister P (2009) cGMP regulated protein kinases (cGK). In: Schmidt HHW, Hofmann F, Stasch J-P (eds) cGMP: generators, effectors and therapeutic implications, Handbook of experimental pharmacology, vol 191. Springer, Berlin, pp 137–162. doi:10.1007/978-3-540-68964-5_8
Hohl CM, Li Q (1991) Compartmentation of cAMP in adult canine ventricular myocytes. Relation to single-cell free Ca^{2+} transients. Circ Res 69(5):1369–1379
Houslay MD, Adams DR (2003) PDE4 cAMP phosphodiesterases: modular enzymes that orchestrate signalling cross-talk, desensitization and compartmentalization. Biochem J 370(1):1–18. doi:10.1042/bj20021698
Huggins JP, Cook EA, Piggott JR, Mattinsley T, England P (1989) Phospholamban is a good substrate for cyclic GMP-dependent protein kinase in vitro, but not in intact cardiac or smooth muscle. Biochem J 260(3):829–835
Hwang TC, Horie M, Gadsby DC (1993) Functionally distinct phospho-forms underlie incremental activation of protein kinase-regulated Cl- conductance in mammalian heart. J Gen Physiol 101(5):629–650. doi:10.1085/jgp.101.5.629
Iancu RV, Jones SW, Harvey RD (2007) Compartmentation of cAMP signaling in cardiac myocytes: a computational study. Biophys J 92(9):3317–3331
Iancu RV, Ramamurthy G, Warrier S, Nikolaev VO, Lohse MJ, Jones SW, Harvey RD (2008) Cytoplasmic cAMP concentrations in intact cardiac myocytes. Am J Phys Cell Phys 295(2):C414
Ignarro LJ, Buga GM, Wood KS, Byrns RE, Chaudhuri G (1987) Endothelium-derived relaxing factor produced and released from artery and vein is nitric oxide. Proc Natl Acad Sci 84(24):9265–9269
Imai Y, Jiang B, Pappano AJ (2001) Mechanism for muscarinic inhibition of ICa (L) is determined by the path for elevating cyclic AMP in cardiac myocytes. Cardiovasc Res 51(2):331–343
Ishikawa Y, Sorota S, Kiuchi K, Shannon RP, Komamura K, Katsushika S, Vatner DE, Vatner SF, Homcy CJ (1994) Downregulation of adenylylcyclase types V and VI mRNA levels in pacing-induced heart failure in dogs. J Clin Investig 93(5):2224–2229
Jäger R, Schwede F, Genieser H-G, Koesling D, Russwurm M (2010) Activation of PDE2 and PDE5 by specific GAF ligands: delayed activation of PDE5. Br J Pharmacol 161(7):1645–1660. doi:10.1111/j.1476-5381.2010.00977.x
Janse MJ (2004) Electrophysiological changes in heart failure and their relationship to arrhythmogenesis. Cardiovasc Res 61(2):208–217. doi:10.1016/j.cardiores.2003.11.018
Jiang L, Gawler D, Hodson N, Milligan C, Pearson H, Porter V, Wray D (2000) Regulation of cloned cardiac L-type calcium channels by cGMP-dependent protein kinase. J Biol Chem 275(9):6135–6143
Josephson IR, Guia A, Lakatta EG, Lederer WJ, Stern MD (2010) Ca^{2+}-dependent components of inactivation of unitary cardiac L-type Ca^{2+} channels. J Physiol 588(1):213–223
Kameyama M, Hofmann F, Trautwein W (1985) On the mechanism of β-adrenergic regulation of the Ca channel in the Guinea-pig heart. Pflugers Arch 405(3):285–293
Kapela A, Bezerianos A, Tsoukias NM (2008) A mathematical model of Ca 2+ dynamics in rat mesenteric smooth muscle cell: agonist and NO stimulation. J Theor Biol 253(2):238–260
Kar S, Kavdia M (2011) Modeling of biopterin-dependent pathways of eNOS for nitric oxide and superoxide production. Free Radic Biol Med 51(7):1411–1427
Kass DA (2012) Heart failure: a PKGarious balancing act. Circulation:CIRCULATIONAHA 112:124909

Kass DA, Champion HC, Beavo JA (2007a) Phosphodiesterase type 5 expanding roles in cardiovascular regulation. Circ Res 101(11):1084–1095
Kass DA, Takimoto E, Nagayama T, Champion HC (2007b) Phosphodiesterase regulation of nitric oxide signaling. Cardiovasc Res 75(2):303–314
Katsube Y, Yokoshiki H, Nguyen L, Sperelakis N (1996) Differences in isoproterenol stimulation of Ca^{2+} current of rat ventricular myocytes in neonatal compared to adult. Eur J Pharmacol 317(2):391–400
Katz A (2011) Physiology of the heart. Wolters Kluwer Health,
Klein G, Drexler H, Schröder F (2000) Protein kinase G reverses all isoproterenol induced changes of cardiac single L-type calcium channel gating. Cardiovasc Res 48(3):367–374
Knight W, Yan C (2012) Cardiac cyclic nucleotide phosphodiesterases: function, regulation, and therapeutic prospects. Hormone and metabolic research= Hormon-und Stoffwechselforschung= Hormones et metabolisme 44 (10):766
Koch WJ, Rockman HA, Samama P, Hamilton R, Bond RA, Milano CA, Lefkowitz RJ (1995) Cardiac function in mice overexpressing the beta-adrenergic receptor kinase or a beta ARK inhibitor. Science 268(5215):1350–1353
Koitabashi N, Aiba T, Hesketh GG, Rowell J, Zhang M, Takimoto E, Tomaselli GF, Kass DA (2010) Cyclic GMP/PKG-dependent inhibition of TRPC6 channel activity and expression negatively regulates cardiomyocyte NFAT activation: novel mechanism of cardiac stress modulation by PDE5 inhibition. J Mol Cell Cardiol 48(4):713–724
Komas N, Le Bec A, Stoclet J-C, Lugnier C (1991) Cardiac cGMP-stimulated cyclic nucleotide phosphodiesterases: effects of cGMP analogues and drugs. Eur J Pharmacol Mol Pharmacol 206(1):5–13
Kostic M, Erdogan S, Rena G, Borchert G, Hoch B, Bartel S, Scotland G, Huston E, Houslay M, Krause E-G (1997) Altered expression of PDE1 and PDE4 cyclic nucleotide phosphodiesterase isoforms in 7-oxo-prostacyclin-preconditioned rat heart. J Mol Cell Cardiol 29(11):3135–3146
Kots AY, Martin E, Sharina IG, Murad F (2009) A short history of cGMP, guanylyl cyclases, and cGMP-dependent protein kinases. In: cGMP: generators, effectors and therapeutic implications. Springer, pp 1–14
Krüger M, Kötter S, Grützner A, Lang P, Andresen C, Redfield MM, Butt E, dos Remedios CG, Linke WA (2009) Protein kinase G modulates human myocardial passive stiffness by phosphorylation of the Titin Springs. Circ Res 104(1):87–94. doi:10.1161/circresaha.108.184408
Kuhn M (2003) Structure, regulation, and function of mammalian membrane guanylyl cyclase receptors, with a focus on guanylyl cyclase-a. Circ Res 93(8):700–709
Kuhn M (2004) Molecular physiology of natriuretic peptide signalling. Basic Res Cardiol 99(2):76–82. doi:10.1007/s00395-004-0460-0
Kuhn M (2015) Cardiology: a big-hearted molecule. Nature 519(7544):416–417. doi:10.1038/nature14373
Kuhn M, Holtwick R, Baba H, Perriard J, Schmitz W, Ehler E (2002) Progressive cardiac hypertrophy and dysfunction in atrial natriuretic peptide receptor (GC-A) deficient mice. Heart 87(4):368–374
Kumar R, Namiki T, Joyner RW (1997) Effects of cGMP on L-type calcium current of adult and newborn rabbit ventricular cells. Cardiovasc Res 33(3):573–582
Kurokawa J, Bankston JR, Kaihawa A, Chen L, Furukawa T, Kass RS (2009) KCNE variants reveal a critical role of the beta subunit carboxyl terminus in PKA-dependent regulation of the IKs potassium channel. Channels 3(1):16–24. doi:10.4161/chan.3.1.7387
Kuznetsov V, Pak E, Robinson RB, Steinberg SF (1995) β2-adrenergic receptor actions in neonatal and adult rat ventricular myocytes. Circ Res 76(1):40–52. doi:10.1161/01.res.76.1.40
Kuzumoto M, Takeuchi A, Nakai H, Oka C, Noma A, Matsuoka S (2008) Simulation analysis of intracellular Na+ and Cl− homeostasis during β1-adrenergic stimulation of cardiac myocyte. Prog Biophys Mol Biol 96(1–3):171–186. doi:10.1016/j.pbiomolbio.2007.07.005
Laflamme MA, Becker PL (1999) Gs and adenylyl cyclase in transverse tubules of heart: implications for cAMP-dependent signaling. Am J Physiol Heart Circ Physiol 277(5):H1841–H1848
Lamba MS, Abraham MWT (2000) Alterations in adrenergic receptor signaling in heart failure. Heart Fail Rev 5(1):7–16
Lee DI, Kass DA (2012) Phosphodiesterases and cyclic GMP regulation in heart muscle. Physiology 27(4):248–258. doi:10.1152/physiol.00011.2012

Lee DI, Vahebi S, Tocchetti CG, Barouch LA, Solaro RJ, Takimoto E, Kass DA (2010) PDE5A suppression of acute β-adrenergic activation requires modulation of myocyte beta-3 signaling coupled to PKG-mediated troponin I phosphorylation. Basic Res Cardiol 105(3):337–347. doi:10.1007/s00395-010-0084-5

Lee DI, Zhu G, Sasaki T, Cho G-S, Hamdani N, Holewinski R, Jo S-H, Danner T, Zhang M, Rainer PP, Bedja D, Kirk JA, Ranek MJ, Dostmann WR, Kwon C, Margulies KB, Van Eyk JE, Paulus WJ, Takimoto E, Kass DA (2015) Phosphodiesterase 9A controls nitric-oxide-independent cGMP and hypertrophic heart disease. Nature 519 (7544):472–476. doi:10.1038/nature14332. http://www.nature.com/nature/journal/v519/n7544/abs/nature14332.html - supplementary-information

Lehnart SE, Wehrens XH, Reiken S, Warrier S, Belevych AE, Harvey RD, Richter W, Jin S-LC, Conti M, Marks AR (2005) Phosphodiesterase 4D deficiency in the ryanodine-receptor complex promotes heart failure and arrhythmias. Cell 123(1):25–35

Leroy J, Abi-Gerges A, Nikolaev VO, Richter W, Lechêne P, Mazet J-L, Conti M, Fischmeister R, Vandecasteele G (2008) Spatiotemporal dynamics of β-adrenergic cAMP signals and L-type Ca^{2+} channel regulation in adult rat ventricular myocytes role of phosphodiesterases. Circ Res 102(9):1091–1100

Levi RC, Alloatti G, Fischmeister R (1989) Cyclic GMP regulates the Ca-channel current in Guinea pig ventricular myocytes. Pflügers Archiv-Eur J Physiol 413(6):685–687

Levi RC, Alloatti G, Penna C, Gallo MP (1994) Guanylate-cyclase-mediated inhibition of cardiac I Ca by carbachol and sodium nitroprusside. Pflugers Arch 426(5):419–426

Li L, Desantiago J, Chu G, Kranias EG, Bers DM (2000) Phosphorylation of phospholamban and troponin I in β-adrenergic-induced acceleration of cardiac relaxation. Am J Physiol Heart Circ Physiol 278(3):H769–H779

Lohse MJ, Engelhardt S, Eschenhagen T (2003) What is the role of β-adrenergic signaling in heart failure? Circ Res 93(10):896–906

Lohse MJ, Nikolaev VO, Hein P, Hoffmann C, Vilardaga J-P, Bünemann M (2008) Optical techniques to analyze real-time activation and signaling of G-protein-coupled receptors. Trends Pharmacol Sci 29(3):159–165. doi:10.1016/j.tips.2007.12.002

Loyer X, Heymes C, Samuel JL (2008) Constitutive nitric oxide synthases in the heart from hypertrophy to failure. Clin Exp Pharmacol Physiol 35(4):483–488

Lu Z, Xu X, Hu X, Lee S, Traverse JH, Zhu G, Fassett J, Tao Y, Zhang P, dos Remedios C (2010) Oxidative stress regulates left ventricular PDE5 expression in the failing heart. Circulation 121(13):1474–1483

Lugnier C (2006) Cyclic nucleotide phosphodiesterase (PDE) superfamily: a new target for the development of specific therapeutic agents. Pharmacol Ther 109(3):366–398

MacKenzie SJ, Baillie GS, McPhee I, MacKenzie C, Seamons R, McSorley T, Millen J, Beard MB, Heeke G, Houslay MD (2002) Long PDE4 cAMP specific phosphodiesterases are activated by protein kinase A-mediated phosphorylation of a single serine residue in upstream conserved region 1 (UCR1). Br J Pharmacol 136(3):421–433

Marín-García J (2010) Cyclic nucleotides signaling (second messengers) and control of myocardial function: effects of heart failure. In: heart failure. contemporary cardiology. Humana Press, pp 161–169. doi:10.1007/978-1-60761-147-9_8

Marston SB, de Tombe PP (2008) Point/counterpoint troponin phosphorylation and myofilament Ca(2+)-sensitivity in heart failure: increased or decreased? J Mol Cell Cardiol 45(5):603–607. doi:10.1016/j.yjmcc.2008.07.004

Martinez SE, Wu AY, Glavas NA, Tang X-B, Turley S, Hol WGJ, Beavo JA (2002) The two GAF domains in phosphodiesterase 2A have distinct roles in dimerization and in cGMP binding. Proc Natl Acad Sci 99(20):13260–13265. doi:10.1073/pnas.192374899

Marx SO, Reiken S, Hisamatsu Y, Jayaraman T, Burkhoff D, Rosemblit N, Marks AR (2000) PKA phosphorylation dissociates FKBP12.6 from the calcium Release Channel (ryanodine receptor): defective regulation in failing hearts. Cell 101(4):365–376. doi:10.1016/S0092-8674(00)80847-8

Marzo KP, Frey MJ, Wilson JR, Liang BT, Manning DR, Lanoce V, Molinoff PB (1991) Beta-adrenergic receptor-G protein-adenylate cyclase complex in experimental canine congestive heart failure produced by rapid ventricular pacing. Circ Res 69(6):1546–1556

Massion PB, Pelat M, Belge C, Balligand J-L (2005) Regulation of the mammalian heart function by nitric oxide. Comp Biochem Physiol A Mol Integr Physiol 142(2):144–150

Maurice DH, Ke H, Ahmad F, Wang Y, Chung J, Manganiello VC (2014) Advances in targeting cyclic nucleotide phosphodiesterases. Nat Rev Drug Discov 13(4):290–314

McKie PM, Burnett JC Jr (2005) B-type natriuretic peptide as a biomarker beyond heart failure: speculations and opportunities. Mayo Clin Proc 80(8):1029–1036. doi:10.4065/80.8.1029

Mehel H, Emons J, Vettel C, Wittköpper K, Seppelt D, Dewenter M, Lutz S, Sossalla S, Maier LS, Lechêne P (2013) Phosphodiesterase-2 is up-regulated in human failing hearts and blunts β-adrenergic responses in cardiomyocytes. J Am Coll Cardiol 62(17):1596–1606

Menniti FS, Faraci WS, Schmidt CJ (2006) Phosphodiesterases in the CNS: targets for drug development. Nat Rev Drug Discov 5(8):660–670

Méry PF, Lohmann SM, Walter U, Fischmeister R (1991) Ca^{2+} current is regulated by cyclic GMP-dependent protein kinase in mammalian cardiac myocytes. Proc Natl Acad Sci 88(4):1197–1201. doi:10.1073/pnas.88.4.1197

Méry PF, Pavoine C, Pecker F, Fischmeister R (1995) Erythro-9-(2-hydroxy-3-nonyl)adenine inhibits cyclic GMP-stimulated phosphodiesterase in isolated cardiac myocytes. Mol Pharmacol 48(1):121–130

Mika D, Leroy J, Vandecasteele G, Fischmeister R (2012) PDEs create local domains of cAMP signaling. J Mol Cell Cardiol 52(2):323–329

Mika D, Richter W, Westenbroek RE, Catterall WA, Conti M (2014) PDE4B mediates local feedback regulation of β1-adrenergic cAMP signaling in a sarcolemmal compartment of cardiac myocytes. J Cell Sci 127(5):1033–1042. doi:10.1242/jcs.140251

Miller CL, Oikawa M, Cai Y, Wojtovich AP, Nagel DJ, Xu X, Xu H, Florio V, Rybalkin SD, Beavo JA, Chen Y-F, Li J-D, Blaxall BC, J-i A, Yan C (2009) Role of Ca^{2+}/calmodulin-stimulated cyclic nucleotide phosphodiesterase 1 in mediating cardiomyocyte hypertrophy. Circ Res 105(10):956–964. doi:10.1161/circresaha.109.198515

Miller CL, Yan C (2010) Targeting cyclic nucleotide phosphodiesterase in the heart: therapeutic implications. J Cardiovasc Transl Res 3(5):507–515

Mo E, Amin H, Bianco IH, Garthwaite J (2004) Kinetics of a cellular nitric oxide/cGMP/phosphodiesterase-5 pathway. J Biol Chem 279(25):26149–26158

Moalem J, Weiss HR, Davidov T, Rodriguez R, Molino B, Lazar MJ, Scholz PM (2006) Heart failure reduces both the effects and interaction between cyclic GMP and cyclic AMP. J Surg Res 134(2):300–306. doi:10.1016/j.jss.2006.01.015

Molina CE, Leroy J, Richter W, Xie M, Scheitrum C, Lee I-O, Maack C, Rucker-Martin C, Donzeau-Gouge P, Verde I, Llach A, Hove-Madsen L, Conti M, Vandecasteele G, Fischmeister R (2012) Cyclic adenosine monophosphate phosphodiesterase type 4 protects against atrial arrhythmias. J Am Coll Cardiol 59(24):2182–2190. doi:10.1016/j.jacc.2012.01.060

Mongillo M, Tocchetti CG, Terrin A, Lissandron V, Cheung Y-F, Dostmann WR, Pozzan T, Kass DA, Paolocci N, Houslay MD (2006) Compartmentalized phosphodiesterase-2 activity blunts β-adrenergic cardiac inotropy via an NO/cGMP-dependent pathway. Circ Res 98(2):226–234

Movsesian MA, Smith CJ, Krall J, Bristow MR, Manganiello VC (1991) Sarcoplasmic reticulum-associated cyclic adenosine 5'-monophosphate phosphodiesterase activity in normal and failing human hearts. J Clin Invest 88(1):15–19. doi:10.1172/jci115272

Mubagwa K, Shirayama T, Moreau M, Pappano A (1993) Effects of PDE inhibitors and carbachol on the L-type Ca current in Guinea pig ventricular myocytes. Am J Phys Heart Circ Phys 265(4):H1353–H1363

Mudd JO, Kass DA (2008) Tackling heart failure in the twenty-first century. Nature 451(7181):919–928

Nagendran J, Archer SL, Soliman D, Gurtu V, Moudgil R, Haromy A, Aubin CS, Webster L, Rebeyka IM, Ross DB (2007) Phosphodiesterase type 5 is highly expressed in the hypertrophied human right ventricle, and acute inhibition of phosphodiesterase type 5 improves contractility. Circulation 116(3):238–248

Nattel S, Maguy A, Le Bouter S, Yeh Y-H (2007) Arrhythmogenic Ion-Channel remodeling in the heart: heart failure, myocardial infarction, and atrial fibrillation. Physiol Rev 87(2):425–456. doi:10.1152/physrev.00014.2006

Negroni JA, Morotti S, Lascano EC, Gomes AV, Grandi E, Puglisi JL, Bers DM (2015) β-adrenergic effects on cardiac myofilaments and contraction in an integrated rabbit ventricular myocyte model. J Mol Cell Cardiol 81:162–175

Neumann J, Scholz H, Döring V, Schmitz W, Meyerinck L, Kalmár P (1988) INCREASE IN MYOCARDIAL G< sub> i–PROTEINS IN HEART FAILURE. Lancet 332(8617): 936–937

Nikolaev VO, Bünemann M, Schmitteckert E, Lohse MJ, Engelhardt S (2006) Cyclic AMP imaging in adult cardiac myocytes reveals far-reaching β1-adrenergic but locally confined β2-adrenergic receptor–mediated signaling. Circ Res 99(10):1084–1091. doi:10.1161/01.RES.0000250046.69918.d5

Nikolaev VO, Gambaryan S, Engelhardt S, Walter U, Lohse MJ (2005) Real-time monitoring of the PDE2 activity of live cells: HORMONE-STIMULATED cAMP HYDROLYSIS IS FASTER THAN HORMONE-STIMULATED cAMP SYNTHESIS. J Biol Chem 280(3):1716–1719. doi:10.1074/jbc.C400505200

Nikolaev VO, Moshkov A, Lyon AR, Miragoli M, Novak P, Paur H, Lohse MJ, Korchev YE, Harding SE, Gorelik J (2010) β2-adrenergic receptor redistribution in heart failure changes cAMP compartmentation. Science 327(5973):1653–1657

Nishikimi T, Maeda N, Matsuoka H (2006) The role of natriuretic peptides in cardioprotection. Cardiovasc Res 69(2):318–328. doi:10.1016/j.cardiores.2005.10.001

Nobelprize.org. (2014. Web. 1 Apr 2016.) Nobel Media AB. http://www.nobelprize.org/faq/questions_in_categoryphp?id=22

Noble D, Rudy Y (2001) Models of cardiac ventricular action potentials: iterative interaction between experiment and simulation. Philos Trans Royal Soc London A Mathematical Phys Eng Sci 359(1783):1127–1142. doi:10.1098/rsta.2001.0820

Oliveira RF, Terrin A, Di Benedetto G, Cannon RC, Koh W, Kim M, Zaccolo M, Blackwell KT (2010) The role of type 4 phosphodiesterases in generating microdomains of cAMP: large scale stochastic simulations. PLoS One 5(7):e11725

Omori K, Kotera J (2007) Overview of PDEs and their regulation. Circ Res 100(3):309–327

Pandit J, Forman MD, Fennell KF, Dillman KS, Menniti FS (2009) Mechanism for the allosteric regulation of phosphodiesterase 2A deduced from the X-ray structure of a near full-length construct. Proc Natl Acad Sci 106(43):18225–18230. doi:10.1073/pnas.0907635106

Perera RK, Nikolaev VO (2013) Compartmentation of cAMP signalling in cardiomyocytes in health and disease. Acta Physiologica

Perera RK, Sprenger JU, Steinbrecher JH, Hübscher D, Lehnart SE, Abesser M, Schuh K, El-Armouche A, Nikolaev VO (2015) Microdomain switch of cGMP-regulated phosphodiesterases leads to ANP-induced augmentation of β-adrenoceptor-stimulated contractility in early cardiac hypertrophy. Circ Res 116(8):1304–1311. doi:10.1161/circresaha.116.306082

Perry SJ, Baillie GS, Kohout TA, McPhee I, Magiera MM, Ang KL, Miller WE, McLean AJ, Conti M, Houslay MD (2002) Targeting of cyclic AMP degradation to β2-adrenergic receptors by β-arrestins. Science 298(5594):834–836

Philippides A, Husbands P, O'Shea M (2000) Four-dimensional neuronal signaling by nitric oxide: a computational analysis. The. J Neurosci 20(3):1199–1207

Pierkes M, Gambaryan S, Bokník P, Lohmann SM, Schmitz W, Potthast R, Holtwick R, Kuhn M (2002) Increased effects of C-type natriuretic peptide on cardiac ventricular contractility and relaxation in guanylyl cyclase A-deficient mice. Cardiovasc Res 53(4):852–861

Piggott LA, Hassell KA, Berkova Z, Morris AP, Silberbach M, Rich TC (2006) Natriuretic peptides and nitric oxide stimulate cGMP synthesis in different cellular compartments. J Gen Physiol 128(1):3–14. doi:10.1085/jgp.200509403

Pokreisz P, Vandenwijngaert S, Bito V, Van den Bergh A, Lenaerts I, Busch C, Marsboom G, Gheysens O, Vermeersch P, Biesmans L (2009) Ventricular phosphodiesterase-5 expression is increased in patients with advanced heart failure and contributes to adverse ventricular remodeling after myocardial infarction in mice. Circulation 119(3):408–416

Prigent A, Fougier S, Nemoz G, Anker G, Pacheco H, Lugnier C, Lebec A, Stoclet J (1988) Comparison of cyclic nucleotide phosphodiesterase isoforms from rat heart and bovine aorta: separation and inhibition by selective reference phosphodiesterase inhibitors. Biochem Pharmacol 37(19):3671–3681

Rahnama'i MS, Ückert S, Hohnen R, van Koeveringe GA (2013) The role of phosphodiesterases in bladder pathophysiology. Nat Rev Urol 10(7):414–424

Ramamurthi A, Lewis RS (1997) Measurement and modeling of nitric oxide release rates for nitric oxide donors. Chem Res Toxicol 10(4):408–413. doi:10.1021/tx960183w

Reiken S, Gaburjakova M, Guatimosim S, Gomez AM, D'Armiento J, Burkhoff D, Wang J, Vassort G, Lederer WJ, Marks AR (2003) Protein kinase a phosphorylation of the cardiac calcium Release Channel (ryanodine receptor) in normal and failing hearts: ROLE OF PHOSPHATASES AND RESPONSE TO ISOPROTERENOL. J Biol Chem 278(1):444–453. doi:10.1074/jbc.M207028200

Rich TC, Webb KJ, Leavesley SJ (2014) Can we decipher the information content contained within cyclic nucleotide signals? J Gen Physiol 143(1):17–27. doi:10.1085/jgp.201311095

Richard HC (2001) Computational models of normal and abnormal action potential propagation in cardiac tissue: linking experimental and clinical cardiology. Physiol Meas 22(3):R15

Riegger G, Elsner D, Kromer E, Daffner C, Forssmann WG, Muders F, Pascher E, Kochsiek K (1988) Atrial natriuretic peptide in congestive heart failure in the dog: plasma levels, cyclic guanosine monophosphate, ultrastructure of atrial myoendocrine cells, and hemodynamic, hormonal, and renal effects. Circulation 77(2):398–406

Roberts BN, Yang P-C, Behrens SB, Moreno JD, Clancy CE (2012) Computational approaches to understand cardiac electrophysiology and arrhythmias. Am J Physiol Heart Circ Physiol 303(7):H766–H783. doi:10.1152/ajpheart.01081.2011

Rochais F, Abi-Gerges A, Horner K, Lefebvre F, Cooper DM, Conti M, Fischmeister R, Vandecasteele G (2006) A specific pattern of phosphodiesterases controls the cAMP signals generated by different Gs-coupled receptors in adult rat ventricular myocytes. Circ Res 98(8):1081–1088

Rochais F, Vandecasteele G, Lefebvre F, Lugnier C, Lum H, Mazet J-L, Cooper DMF, Fischmeister R (2004) Negative feedback exerted by cAMP-dependent protein kinase and cAMP phosphodiesterase on Subsarcolemmal cAMP signals in intact cardiac myocytes: AN IN VIVO STUDY USING ADENOVIRUS-MEDIATED EXPRESSION OF CNG CHANNELS. J Biol Chem 279(50):52095–52105. doi:10.1074/jbc.M405697200

Roy B, Garthwaite J (2006) Nitric oxide activation of guanylyl cyclase in cells revisited. Proc Natl Acad Sci 103(32):12185–12190

Roy B, Halvey EJ, Garthwaite J (2008) An enzyme-linked receptor mechanism for nitric oxide-activated guanylyl cyclase. J Biol Chem 283(27):18841–18851

Russell TR, Terasaki WL, Appleman MM (1973) Separate phosphodiesterases for the hydrolysis of cyclic adenosine 3',5'-monophosphate and cyclic guanosine 3',5'-monophosphate in rat liver. J Biol Chem 248(4):1334–1340

Ryall KA, Holland DO, Delaney KA, Kraeutler MJ, Parker AJ, Saucerman JJ (2012) Network reconstruction and systems analysis of cardiac myocyte hypertrophy signaling. J Biol Chem 287(50):42259–42268. doi:10.1074/jbc.M112.382937

Rybalkin SD, Rybalkina IG, Feil R, Hofmann F, Beavo JA (2002) Regulation of cGMP-specific phosphodiesterase (PDE5) phosphorylation in smooth muscle cells. J Biol Chem 277(5):3310–3317

Rybin VO, Xu X, Lisanti MP, Steinberg SF (2000) Differential targeting of β-adrenergic receptor subtypes and adenylyl cyclase to cardiomyocyte caveolae a mechanism to functionally regulate the cAMP signaling pathway. J Biol Chem 275(52):41447–41457

Sato N, Asai K, Okumura S, Takagi G, Shannon RP, Fujita-Yamaguchi Y, Ishikawa Y, Vatner SF, Vatner DE (1999) Mechanisms of desensitization to a PDE inhibitor (milrinone) in conscious dogs with heart failure. Am J Physiol Heart Circ Physiol 276(5):H1699–H1705

Saucerman JJ, Brunton LL, Michailova AP, McCulloch AD (2003) Modeling β-adrenergic control of cardiac myocyte contractility in Silico. J Biol Chem 278(48):47997–48003. doi:10.1074/jbc.M308362200

Saucerman JJ, Greenwald EC, Polanowska-Grabowska R (2014) Mechanisms of cyclic AMP compartmentation revealed by computational models. J Gen Physiol 143(1):39–48. doi:10.1085/jgp.201311044

Saucerman JJ, Healy SN, Belik ME, Puglisi JL, McCulloch AD (2004) Proarrhythmic consequences of a KCNQ1 AKAP-binding domain mutation: computational models of whole cells and heterogeneous tissue. Circ Res 95(12):1216–1224. doi:10.1161/01.RES.0000150055.06226.4e

Saucerman JJ, McCulloch AD (2004) Mechanistic systems models of cell signaling networks: a case study of myocyte adrenergic regulation. Prog Biophys Mol Biol 85(2):261–278

Saucerman JJ, McCulloch AD (2006) Cardiac β-adrenergic signaling. Ann N Y Acad Sci 1080(1):348–361. doi:10.1196/annals.1380.026

Schmidt K, Desch W, Klatt P, Kukovetz WR, Mayer B (1997) Release of nitric oxide from donors with known half-life: a mathematical model for calculating nitric oxide concentrations in aerobic solutions. Naunyn Schmiedeberg's Arch Pharmacol 355(4):457–462. doi:10.1007/pl00004969

Schröder F, Handrock R, Beuckelmann DJ, Hirt S, Hullin R, Priebe L, Schwinger RHG, Weil J, Herzig S (1998) Increased availability and open probability of single L-type calcium channels from failing compared with nonfailing human ventricle. Circulation 98(10):969–976. doi:10.1161/01.cir.98.10.969

Schröder F, Klein G, Fiedler B, Bastein M, Schnasse N, Hillmer A, Ames S, Gambaryan S, Drexler H, Walter U (2003) Single L-type Ca^{2+} channel regulation by cGMP-dependent protein kinase type I in adult cardiomyocytes from PKG I transgenic mice. Cardiovasc Res 60(2):268–277

Senzaki H, Smith CJ, Juang GJ, Isoda T, Mayer SP, Ohler A, Paolocci N, Tomaselli GF, Hare JM, Kass DA (2001) Cardiac phosphodiesterase 5 (cGMP-specific) modulates β-adrenergic signaling in vivo and is down-regulated in heart failure. FASEB J 15(10):1718–1726. doi:10.1096/fj.00-0538com

Sette C, Conti M (1996) Phosphorylation and activation of a cAMP-specific phosphodiesterase by the cAMP-dependent protein kinase INVOLVEMENT OF SERINE 54 IN THE ENZYME ACTIVATION. J Biol Chem 271(28):16526–16534

Shah AM, Spurgeon HA, Sollott SJ, Talo A, Lakatta EG (1994) 8-bromo-cGMP reduces the myofilament response to Ca^{2+} in intact cardiac myocytes. Circ Res 74(5):970–978

Shaw RM, Colecraft HM (2013) L-type calcium channel targeting and local signalling in cardiac myocytes. Cardiovasc Res 98(2):177–186. doi:10.1093/cvr/cvt021

Shirayama T, Pappano AJ (1996) Biphasic effects of intrapipette cyclic guanosine monophosphate on L-type calcium current and contraction of Guinea pig ventricular myocytes. J Pharmacol Exp Ther 279(3):1274–1281

Simmons MA, Hartzell HC (1988) Role of phosphodiesterase in regulation of calcium current in isolated cardiac myocytes. Mol Pharmacol 33(6):664–671

Smith CJ, He J, Ricketts SG, Ding J-Z, Moggio RA, Hintze TH (1998) Downregulation of right ventricular phosphodiesterase PDE-3A mRNA and protein before the development of canine heart failure. Cell Biochem Biophys 29(1):67–88. doi:10.1007/bf02737829

Smith NP, Crampin EJ, Niederer SA, Bassingthwaighte JB, Beard DA (2007) Computational biology of cardiac myocytes: proposed standards for the physiome. J Exp Biol 210(9):1576–1583. doi:10.1242/jeb.000133

Soderling SH, Bayuga SJ, Beavo JA (1998) Identification and characterization of a novel family of cyclic nucleotide phosphodiesterases. J Biol Chem 273(25):15553–15558

Solaro RJ, Kobayashi T (2011) Protein phosphorylation and signal transduction in cardiac thin filaments. J Biol Chem 286(12):9935–9940. doi:10.1074/jbc.R110.197731

Soltis AR, Saucerman JJ (2010) Synergy between CaMKII substrates and β-adrenergic signaling in regulation of cardiac myocyte Ca 2+ handling. Biophys J 99(7):2038–2047

Song L-S, Wang S-Q, Xiao R-P, Spurgeon H, Lakatta EG, Cheng H (2001) β-adrenergic stimulation synchronizes intracellular Ca^{2+} release during excitation-contraction coupling in cardiac myocytes. Circ Res 88(8):794–801. doi:10.1161/hh0801.090461

Sprenger JU, Nikolaev VO (2013) Biophysical techniques for detection of cAMP and cGMP in living cells. Int J Mol Sci 14(4):8025–8046. doi:10.3390/ijms14048025

Stangherlin A, Gesellchen F, Zoccarato A, Terrin A, Fields LA, Berrera M, Surdo NC, Craig MA, Smith G, Hamilton G (2011) cGMP signals modulate cAMP levels in a compartment-specific manner to regulate catecholamine-dependent signaling in cardiac MyocytesNovelty and significance. Circ Res 108(8):929–939

Stangherlin A, Zaccolo M (2012) cGMP-cAMP interplay in cardiac myocytes: a local affair with far-reaching consequences for heart function. Biochem Soc Trans 40(1):11–14. doi:10.1042/BST20110655

Sumii K, Sperelakis N (1995) cGMP-dependent protein kinase regulation of the L-type Ca^{2+} current in rat ventricular myocytes. Circ Res 77(4):803–812. doi:10.1161/01.res.77.4.803

Takimoto E (2012) Cyclic GMP-dependent signaling in cardiac myocytes. Circ J 76(8):1819–1825

Takimoto E, Belardi D, Tocchetti CG, Vahebi S, Cormaci G, Ketner EA, Moens AL, Champion HC, Kass DA (2007) Compartmentalization of cardiac β-adrenergic inotropy modulation by phosphodiesterase type 5. Circulation 115(16):2159–2167

Takimoto E, Champion HC, Belardi D, Moslehi J, Mongillo M, Mergia E, Montrose DC, Isoda T, Aufiero K, Zaccolo M, Dostmann WR, Smith CJ, Kass DA (2005a) cGMP catabolism by phosphodiesterase 5A regulates cardiac adrenergic stimulation by NOS3-dependent mechanism. Circ Res 96(1):100–109. doi:10.1161/01.res.0000152262.22968.72

Takimoto E, Champion HC, Li M, Ren S, Rodriguez ER, Tavazzi B, Lazzarino G, Paolocci N, Gabrielson KL, Wang Y (2005b) Oxidant stress from nitric oxide synthase–3 uncoupling stimulates cardiac pathologic remodeling from chronic pressure load. J Clin Invest 115(5):1221–1231

Tamargo J, Caballero R, Gómez R, Delpón E (2010) Cardiac electrophysiological effects of nitric oxide. Cardiovasc Res 87(4):593–600. doi:10.1093/cvr/cvq214

Terrenoire C, Clancy CE, Cormier JW, Sampson KJ, Kass RS (2005) Autonomic control of cardiac action potentials role of Potassium Channel kinetics in response to sympathetic stimulation. Circ Res 96(5):e25–e34

Terrenoire C, Houslay MD, Baillie GS, Kass RS (2009) The cardiac IKs Potassium Channel macromolecular complex includes the phosphodiesterase PDE4D3. J Biol Chem 284(14):9140–9146. doi:10.1074/jbc.M805366200

Tohse N, Nakaya H, Kanno YT (1995) Cyclic GMP-mediated inhibition of L-type Ca^{2+} channel activity by human natriuretic peptide in rabbit heart cells. Br J Pharmacol 114(5):1076–1082

Tohse N, Sperelakis N (1991) cGMP inhibits the activity of single calcium channels in embryonic chick heart cells. Circ Res 69(2):325–331. doi:10.1161/01.res.69.2.325

Tomaselli GF, Marbán E (1999) Electrophysiological remodeling in hypertrophy and heart failure. Cardiovasc Res 42(2):270–283. doi:10.1016/s0008-6363(99)00017-6

Trayanova NA, Rice JJ (2011) Cardiac electromechanical models: from cell to organ. Front Physiol 2. doi:10.3389/fphys.2011.00043

Tsai EJ, Kass DA (2009) Cyclic GMP signaling in cardiovascular pathophysiology and therapeutics. Pharmacol Ther 122(3):216–238. doi:10.1016/j.pharmthera.2009.02.009

Tsai EJ, Liu Y, Koitabashi N, Bedja D, Danner T, Jasmin J-F, Lisanti MP, Friebe A, Takimoto E, Kass DA (2012) Pressure-overload–induced subcellular Relocalization/oxidation of soluble guanylyl cyclase in the heart modulates enzyme stimulation. Circ Res 110(2):295–303

Tsoukias NM (2008) Nitric oxide bioavailability in the microcirculation: insights from mathematical models. Microcirculation 15(8):813–834

Tsoukias NM, Kavdia M, Popel AS (2004) A theoretical model of nitric oxide transport in arterioles: frequency- vs. amplitude-dependent control of cGMP formation. Am J Physiol Heart Circ Physiol 286(3):H1043–H1056. doi:10.1152/ajpheart.00525.2003

Tsutamoto T, Kanamori T, Wada A, Kinoshita M (1992) Uncoupling of atrial natriuretic peptide extraction and cyclic guanosine monophosphate production in the pulmonary circulation in patients with severe heart failure. J Am Coll Cardiol 20(3):541–546

Ungerer M, Böhm M, Elce J, Erdmann E, Lohse M (1993) Altered expression of beta-adrenergic receptor kinase and beta 1-adrenergic receptors in the failing human heart. Circulation 87(2):454–463

van Heerebeek L, Hamdani N, Falcão-Pires I, Leite-Moreira AF, Begieneman MP, Bronzwaer JG, van der Velden J, Stienen GJ, Laarman GJ, Somsen A (2012) Low myocardial protein kinase G activity in heart failure with preserved ejection fraction. Circulation 126(7):830–839

Vandecasteele G, Rochais F, Abi-Gerges A, Fischmeister R (2006) Functional localization of cAMP signalling in cardiac myocytes. Biochem Soc Trans 34(4):484–488

Verde I, Pahlke G, Salanova M, Zhang G, Wang S, Coletti D, Onuffer J, Jin S-LC, Conti M (2001) Myomegalin is a novel protein of the Golgi/centrosome that interacts with a cyclic nucleotide phosphodiesterase. J Biol Chem 276(14):11189–11198. doi:10.1074/jbc.M006546200

Verde I, Vandecasteele G, Lezoualc'h F, Fischmeister R (1999) Characterization of the cyclic nucleotide phosphodiesterase subtypes involved in the regulation of the L-type Ca^{2+} current in rat ventricular myocytes. Br J Pharmacol 127(1):65–74

Vila-Petroff MG, Younes A, Egan J, Lakatta EG, Sollott SJ (1999) Activation of distinct cAMP-dependent and cGMP-dependent pathways by nitric oxide in cardiac myocytes. Circ Res 84(9):1020–1031

Wahler GM, Dollinger SJ (1995) Nitric oxide donor SIN-1 inhibits mammalian cardiac calcium current through cGMP-dependent protein kinase. Am J Phys Cell Phys 268(1):C45–C54

Wang H, Kohr MJ, Traynham CJ, Ziolo MT (2009) Phosphodiesterase 5 restricts NOS3/soluble guanylate cyclase signaling to L-type Ca^{2+} current in cardiac myocytes. J Mol Cell Cardiol 47(2):304–314. doi:10.1016/j.yjmcc.2009.03.021

Warrier S, Ramamurthy G, Eckert RL, Nikolaev VO, Lohse MJ, Harvey RD (2007) cAMP microdomains and L-type Ca^{2+} channel regulation in Guinea-pig ventricular myocytes. J Physiol 580(3):765–776

Wegener A, Simmerman H, Lindemann J, Jones LR (1989) Phospholamban phosphorylation in intact ventricles. Phosphorylation of serine 16 and threonine 17 in response to beta-adrenergic stimulation. J Biol Chem 264(19):11468–11474

Weishaar RE, Kobylarz-Singer DC, Steffen RP, Kaplan HR (1987) Subclasses of cyclic AMP-specific phosphodiesterase in left ventricular muscle and their involvement in regulating myocardial contractility. Circ Res 61(4):539–547

Weiss HR, Gong GX, Straznicka M, Yan L, Tse J, Scholz PM (1999) Cyclic GMP and cyclic AMP induced changes in control and hypertrophic cardiac myocyte function interact through cyclic GMP affected cyclic-AMP phosphodiesterases. Can J Physiol Pharmacol 77(6):422–431

Williams GSB, Smith GD, Sobie EA, Jafri MS (2010) Models of cardiac excitation–contraction coupling in ventricular myocytes. Math Biosci 226(1):1–15. doi:10.1016/j.mbs.2010.03.005

Winslow RL, Cortassa S, O'Rourke B, Hashambhoy YL, Rice JJ, Greenstein JL (2011) Integrative modeling of the cardiac ventricular myocyte. Wiley Interdiscip Rev Syst Biol Med 3(4): 392–413

Winslow RL, Greenstein JL, Tomaselli GF, O'Rourke B (2001) Computational models of the failing myocyte: relating altered gene expression to cellular function. Philos Trans R Soc London, Ser A 359(1783):1187–1200. doi:10.1098/rsta.2001.0825

Winslow RL, Rice J, Jafri S (1998) Modeling the cellular basis of altered excitation–contraction coupling in heart failure. Prog Biophys Mol Biol 69(2–3):497–514. doi:10.1016/S0079-6107(98)00022-4

Winslow RL, Trayanova N, Geman D, Miller MI (2012) Computational medicine: translating models to clinical care. Science Translational Medicine 4 (158):158rv111-158rv111

Winslow RL, Walker MA, Greenstein JL (2016) Modeling calcium regulation of contraction, energetics, signaling, and transcription in the cardiac myocyte. Wiley Interdiscip Rev Syst Biol Med 8(1):37–67

Wollert KC, Drexler H (2002) Regulation of cardiac remodeling by nitric oxide: focus on cardiac myocyte hypertrophy and apoptosis. Heart Fail Rev 7(4):317–325

Wu AY, Tang X-B, Martinez SE, Ikeda K, Beavo JA (2004) Molecular determinants for cyclic nucleotide binding to the regulatory domains of phosphodiesterase 2A. J Biol Chem 279(36):37928–37938. doi:10.1074/jbc.M404287200

Xie Y, Grandi E, Puglisi JL, Sato D, Bers DM (2013) β-adrenergic stimulation activates early afterdepolarizations transiently via kinetic mismatch of PKA targets. J Mol Cell Cardiol 58:153–161

Xin W, Tran TM, Richter W, Clark RB, Rich TC (2008) Roles of GRK and PDE4 activities in the regulation of β2 adrenergic signaling. J Gen Physiol 131(4):349–364. doi:10.1085/jgp.200709881

Xu H, Ginsburg KS, Hall DD, Zimmermann M, Stein IS, Zhang M, Tandan S, Hill JA, Horne MC, Bers D (2010) Targeting of protein phosphatases PP2A and PP2B to the C-terminus of the L-type calcium channel Cav1. 2. Biochemistry 49(48):10298–10307

Yan C, Ding B, Shishido T, Woo C-H, Itoh S, Jeon K-I, Liu W, Xu H, McClain C, Molina CA (2007a) Activation of extracellular signal-regulated kinase 5 reduces cardiac apoptosis and dysfunction via inhibition of a phosphodiesterase 3A/inducible cAMP early repressor feedback loop. Circ Res 100(4):510–519

Yan C, Miller CL, J-i A (2007b) Regulation of phosphodiesterase 3 and inducible cAMP early repressor in the heart. Circ Res 100(4):489–501

Yan C, Zhao AZ, Bentley JK, Beavo JA (1996) The calmodulin-dependent phosphodiesterase Gene PDE1C encodes several functionally different splice variants in a tissue-specific manner. J Biol Chem 271(41):25699–25706

Yang J, Clark JW, Bryan RM, Robertson CS (2005) Mathematical modeling of the nitric oxide/cGMP pathway in the vascular smooth muscle cell. Am J Phys Heart Circ Phys 289(2): H886–H897

Yang JH, Polanowska-Grabowska RK, Smith JS, Shields Iv CW, Saucerman JJ (2014) PKA catalytic subunit compartmentation regulates contractile and hypertrophic responses to β-adrenergic signaling. J Mol Cell Cardiol 66:83–93. doi:10.1016/j.yjmcc.2013.11.001

Yang JH, Saucerman JJ (2012) Phospholemman is a negative feed-forward regulator of Ca^{2+} in β-adrenergic signaling, accelerating β-adrenergic inotropy. J Mol Cell Cardiol 52(5):1048–1055

Yang L, Liu G, Zakharov SI, Bellinger AM, Mongillo M, Marx SO (2007) Protein kinase G phosphorylates Cav1. 2 α1c and β2 subunits. Circ Res 101(5):465–474

Yasue H, Yoshimura M, Sumida H, Kikuta K, Kugiyama K, Jougasaki M, Ogawa H, Okumura K, Mukoyama M, Nakao K (1994) Localization and mechanism of secretion of B-type natriuretic peptide in comparison with those of A-type natriuretic peptide in normal subjects and patients with heart failure. Circulation 90(1):195–203. doi:10.1161/01.cir.90.1.195

Yue DT, Herzig S, Marban E (1990) Beta-adrenergic stimulation of calcium channels occurs by potentiation of high-activity gating modes. Proc Natl Acad Sci 87(2):753–757

Zaccolo M, Movsesian MA (2007) cAMP and cGMP signaling cross-talk role of phosphodiesterases and implications for cardiac pathophysiology. Circ Res 100(11):1569–1578

Zakhary DR, Moravec CS, Stewart RW, Bond M (1999) Protein kinase a (PKA)-dependent troponin-I phosphorylation and PKA regulatory subunits are decreased in human dilated cardiomyopathy. Circulation 99(4):505–510

Zhang M, Koitabashi N, Nagayama T, Rambaran R, Feng N, Takimoto E, Koenke T, O'Rourke B, Champion HC, Crow MT, Kass DA (2008) Expression, activity, and pro-hypertrophic effects of PDE5A in cardiac myocytes. Cell Signal 20(12):2231–2236. doi:10.1016/j.cellsig.2008.08.012

Zhang M, Takimoto E, Lee D-i, Santos CX, Nakamura T, Hsu S, Jiang A, Nagayama T, Bedja D, Yuan Y (2012) Pathological cardiac hypertrophy alters intracellular targeting of phosphodiesterase type 5 from nitric oxide synthase-3 to natriuretic peptide signaling. Circulation 126(8):942–951

Zhao CY, Greenstein JL, Winslow RL (2015) Interaction between phosphodiesterases in the regulation of the cardiac β-adrenergic pathway. J Mol Cell Cardiol 88:29–38. doi:10.1016/j.yjmcc.2015.09.011

Zhao CY, Greenstein JL, Winslow RL (2016a) Regulation of the Cardiac L-Type Calcium Channel by the Cyclic Nucleotide Cross-Talk Signaling Network. Biophysical Journal 110 (3):359a. doi:10.1016/j.bpj.2015.11.1937

Zhao CY, Greenstein JL, Winslow RL (2016b) Roles of phosphodiesterases in the regulation of the cardiac cyclic nucleotide cross-talk signaling network. J Mol Cell Cardiol 91:215–227. doi:10.1016/j.yjmcc.2016.01.004

Ziolo MT, Harshbarger CH, Roycroft KE, Smith JM, Romano FD, Sondgeroth KL, Wahler GM (2001) Myocytes isolated from rejecting transplanted rat hearts exhibit a nitric oxide-mediated reduction in the calcium current. J Mol Cell Cardiol 33(9):1691–1699

Ziolo MT, Kohr MJ, Wang H (2008) Nitric oxide signaling and the regulation of myocardial function. J Mol Cell Cardiol 45(5):625–632

Ziolo MT, Lewandowski SJ, Smith JM, Romano FD, Wahler GM (2003) Inhibition of cyclic GMP hydrolysis with zaprinast reduces basal and cyclic AMP-elevated L-type calcium current in Guinea-pig ventricular myocytes. Br J Pharmacol 138(5):986–994

Zoraghi R, Corbin JD, Francis SH (2004) Properties and functions of GAF domains in cyclic nucleotide phosphodiesterases and other proteins. Mol Pharmacol 65(2):267–278

Signalling Microdomains: The Beta-3 Adrenergic Receptor/NOS Signalosome

11

J. Hammond and J.-L. Balligand

Contents

Abstract

Plasma membrane "signalosomes" are highly compartmentalised, as required for signalling specificity to diverse downstream effectors from particular surface receptors. This is achieved by spatial confinement of receptors and effectors in specific membrane locales, such as caveolae, but also by receptor translocation to particular plasma membrane locations upon ligand stimuli, cellular stress or pathophysiologic conditions, thereby changing downstream coupling. Beta-adrenoreceptors in car-

J. Hammond • J.-L. Balligand (✉)
Institut de Recherche Experimentale et Clinique (IREC), Pole of Pharmacology and Therapeutics (FATH), Université catholique de Louvain, B1.53.09, 52 avenue Mounier 1200 Brussels, Belgium
e-mail: jl.balligand@uclouvain.be

V. Nikolaev, M. Zaccolo (eds.), *Microdomains in the Cardiovascular System*, Cardiac and Vascular Biology 3, DOI 10.1007/978-3-319-54579-0_11

diac myocytes are no exception. In this review, we will illustrate the case of cardiac beta-3 adrenergic receptors (β_3ARs) coupled to nitric oxide synthases and guanylyl cyclase/cyclic GMP and its effects on cardiac remodelling.

11.1 Introduction

Classically in adipose tissue and smooth muscle, the β_3AR is coupled through $G_{\alpha s}$ proteins to activate cAMP-dependent signalling pathways (Emorine 1989; Fujimura et al. 1999; Strosberg 1997). The adipose β_3AR has also been demonstrated to activate extracellular regulated kinase 1/2 (ERK1/2) which has variably been reported to depend on coupling to $G_{\alpha s}$ (Lindquist et al. 2000) or $G_{\alpha i}$ (Soeder et al. 1999); the latter involves Src-dependent (but β-arrestin-independent) epidermal growth factor receptor (EGFR) activation of ERK1/2 signalling (Robidoux et al. 2006). As discussed below, in cardiac myocytes, to date most evidence points towards the cardiac β_3AR coupling to $G_{\alpha i}$; however, data from $\beta_1\beta_2$AR double KO mice suggests that the cardiac β_3AR can dually couple to both $G_{\alpha s}$ and $G_{\alpha i}$ (Devic and Xiang 2001), similar to the cardiac β_2AR. The functional significance of this is unknown but raises the intriguing possibility that the β_3AR may couple to cAMP signalling in a tightly compartmentalised manner as has been described for the cardiac β_2AR (Wright et al. 2014).

11.2 Identification of a $G_{\alpha i}$-NOS Signalling Pathway Downstream of Cardiac β_3AR Activation

In the seminal paper describing a cardiac myocyte β_3AR in the ventricular endomyocardial biopsies, the negative inotropic effect of β_3AR stimulation was inhibited by pertussis toxin (PTX) pretreatment, indicative of the β_3AR coupling to $G_{\alpha i}$ activation (Gauthier et al. 1996). This has subsequently been reaffirmed in isolated cardiac myocytes from many species including isolated adult rat heart preparations (Angelone et al. 2008). Furthermore, in adult rat ventricular myocytes (ARVMs), PTX pretreatment prevents the inhibitory effects of β_3AR simulation on $I_{Ca\text{-}L}$ (Zhang et al. 2005). Likewise, in nuclear extracts, β_3AR-mediated activation of transcription has been shown to be dependent on $G_{\alpha i}$-dependent signalling (Vaniotis et al. 2013).

Prior to the identification of the cardiac β_3AR, nitric oxide (NO), produced endogenously in cardiac myocytes by NO synthase (NOS) enzymes, had been demonstrated to cause inhibition of cardiac contractility in both animal models and humans (Balligand et al. 1993a; Hare et al. 1995); hence, further research focused on identifying whether a NO signalling pathway lays downstream of the β_3AR and if it was responsible for the negative inotropic effect associated with β_3AR stimulation. It was first shown that pharmacological NOS inhibition attenuated the negative inotropic effect of β_3AR stimulation in human ventricular endomyocardial biopsies (Gauthier et al. 1998). Furthermore, in this setting, β_3AR stimulation increased NO production, as shown in Fig. 11.1. Subsequent studies have confirmed that an NO-dependent signalling pathway contributes towards the acute functional effects of cardiac β_3AR stimulation in many species. Additional supporting evidence comes

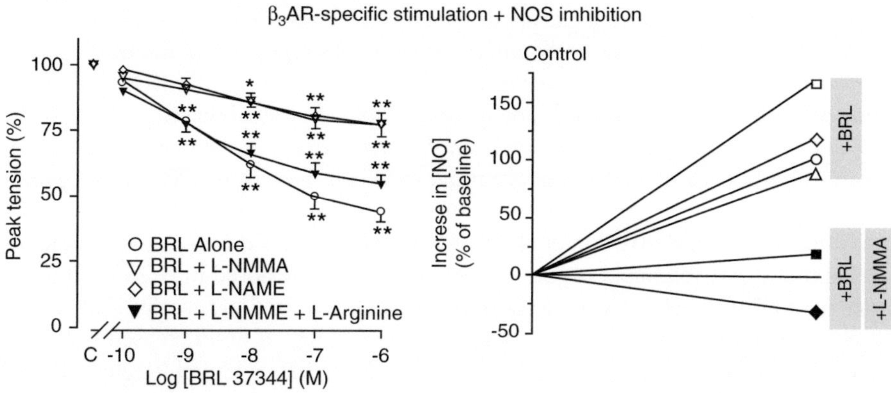

Fig. 11.1 Involvement of NO in the negative inotropic effect of β_3AR stimulation in human heart. *Left*: Dose-response curves for the negative inotropic effect of the β_3AR agonist BRL 37344 (BRL) in the presence of the NOS inhibitors, L-NMMA and L-NAME, in human endomyocardial biopsies. β_3AR-specific stimulation produces a dose-dependent negative inotropic effect that is reversed by NOS inhibition. *Right*: Increase in NO production in human endomyocardial biopsies induced by a β_3AR agonist measured by an electrochemical sensor. Reproduced from Gauthier et al. (1998)

from β_3AR KO mice; in WT mice, NOS inhibition causes an increase in β-AR-stimulated inotropy, but this effect is absent in β_3AR KO mice (Varghese et al. 2000). A NOS-dependent signalling pathway has also been demonstrated to lie downstream of cardiac nuclear β_3ARs (Vaniotis et al. 2013).

It is interesting to note that in most studies involving PTX and NOS inhibition, the β_3AR-mediated functional effects were prevented by PTX treatment, yet only attenuated by NOS inhibition (e.g. Zhang et al. 2005). Assuming equal inhibition of $G_{\alpha i}$ and NOS, this may indicate the cardiac β_3AR activates additional NOS-independent signalling pathways downstream of $G_{\alpha i}$.

At the time of the publication of the first paper demonstrating that the negative inotropic effect of β_3AR stimulation was NOS dependent, only NOS3 (endothelial NOS) had been identified in healthy cardiac myocytes (Balligand et al. 1995), with NOS2 (inducible NOS) being expressed only following inflammatory cytokine exposure (Balligand et al. 1993b) (Balligand et al. 1994). Hence, NOS3 was presumed to be the NOS isoform activated by β_3AR stimulation. This notion was subsequently reinforced by the observation that β_3AR KO mice and NOS3 KO mice demonstrate a similar magnitude of enhanced β-AR-stimulated inotropy (Varghese et al. 2000). Nevertheless, shortly thereafter, NOS1 (neuronal NOS) was also identified to be constitutively expressed in cardiac myocytes (Xu et al. 1999) and to play a role in the regulation of cardiac contractility (Ashley et al. 2002). Yet early studies ruled out a role for NOS1 in β_3AR-mediated negative inotropy, as it was demonstrated that the negative inotropic response observed following β_3AR-specific stimulation in isolated WT cardiac myocytes was preserved in NOS1 KO cardiac myocytes, whilst it was absent in NOS3 KO cardiac myocytes (Barouch et al. 2002), thus further reinforcing the hypothesis that the cardiac β_3AR couples to NOS3 activation. Despite these early observations, subsequent publications have linked the cardiac β_3AR to the activation of either NOS3 (Calvert et al. 2011) or NOS1 (Watts et al. 2013) or both isoforms (Belge et al. 2014).

In an attempt to resolve the issue of which NOS isoform lies downstream of the cardiac myocyte β_3AR, a recently published study has repeated contractility experiments in isolated cardiac myocytes from WT, NOS3 KO and NOS1 KO mice (Idigo et al. 2012); a lack of β_3AR-mediated negative inotropy was observed in both NOS3 KO and NOS1 KO myocytes as well as in WT myocytes pretreated with the NOS1-specific inhibitor S-methyl-L-thiocitrulline (SMTC). Given that NOS1 had previously been shown to play an antioxidant role in cardiac myocytes (through the inhibition of xanthine oxidoreductase (XOR)) (Khan et al. 2004), it was demonstrated that β_3AR-mediated negative inotropy could be restored in NOS1 KO myocytes by inhibiting XOR (with oxypurinol). Further Investation revealed that lack of NOS1 essentially results in a functional lack of NOS3, as the increased $O_2^{\bullet-}$ produced by XOR in NOS1 KO myocytes results in the S-glutathionylation[1] and uncoupling of NOS3, thereby turning NOS3 into a $O_2^{\bullet-}$-generating enzyme.

11.3 Signalling Elements in the NO/cGMP Pathway

11.3.1 Nitric Oxide Synthases

Since the identification of NO as the endothelium-derived relaxing factor nearly 30 years ago (Palmer et al. 1987), a wealth of information has accumulated on the role of NO as a signalling moiety (Pacher et al. 2007). NO is synthesised by the NOS family of enzymes—NOS1, NOS2 and NOS3, all three of which are expressed in cardiac myocytes (Balligand et al. 1994, 1995; Xu et al. 1999). Inflammatory cytokine exposure induces NOS2 expression in the cytoplasm, whereas NOS3 (located at the sarcolemmal membrane in caveolae (Feron et al. 1996)) and NOS1 (located at the sarcoplasmic reticulum (Xu et al. 1999)) in association with the ryanodine receptor (RyR) (Barouch et al. 2002; Bendall et al. 2004) and XOR (Khan et al. 2004) and also found in sarcolemmal caveolae (in association with the plasma membrane Ca^{2+} ATPase (PMCA) (Mohamed et al. 2011; Oceandy et al. 2007)) are constitutively expressed and play a role in the regulation of excitation-contraction (EC) coupling (reviewed in Sears et al. 2004; Seddon et al. 2007; Belge et al. 2005; Balligand et al. 2009a; Massion et al. 2003) as well as cardiac remodelling (discussed in Sect. 11.4.1).

NOS enzymes are homodimeric, with each monomer consisting of a multidomain C-terminal reductase domain (which binds NADPH, FAD and FMN) and an N-terminal oxygenase domain (containing binding sites for haem, tetrahydrobiopterin (BH_4), L-arginine and O_2), separated by a calmodulin (CaM)-binding domain; the dimer structure is stabilised by Zn^{2+} ions positioned between the interface of the two oxygenase domains (Alderton et al. 2001). NO synthesis is stimulated by Ca^{2+}/CaM binding, which allows the flow of NADPH-donated electrons through FAD and FMN in the reductase domain of one NOS monomer, to the oxidase domain of the second NOS monomer, where they interact with haem iron and BH_4 to catalyse the conversion of L-arginine and oxygen to citrulline and NO, as illustrated in Fig. 11.2.

[1] S-glutathionylation = post-translational modification of cysteine residue by addition of glutathione

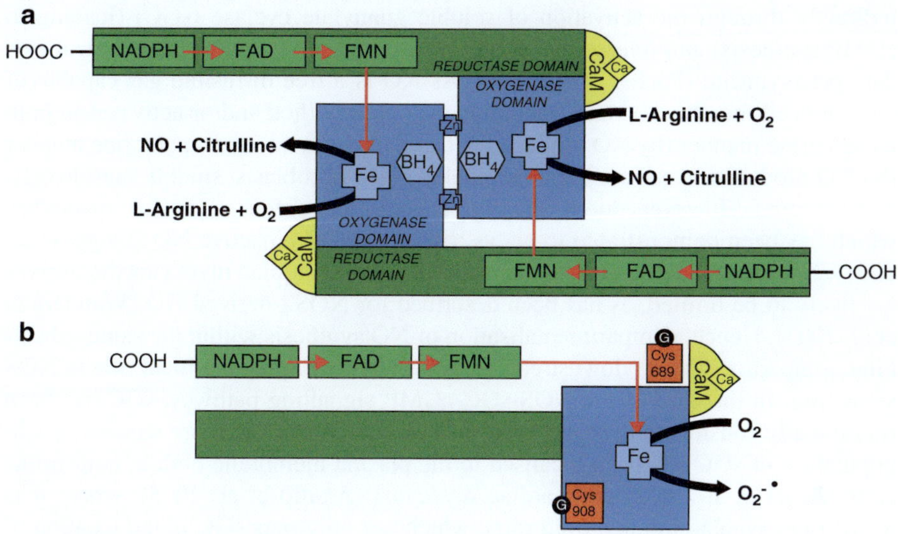

Fig. 11.2 Schematic structure and electron flow in coupled and uncoupled NOS. (**a**) Coupled NOS: In coupled (dimeric) NOS, NADPH-donated electrons (*red arrows*) flow through FAD and FMN in the reductase domain of one NOS monomer, to the oxidase domain of the second NOS monomer, where they interact with haem iron and BH_4 to catalyse the conversion of L-arginine and oxygen to citrulline and NO. CaM binding is required for electron flow (NOS1 and NOS3 require Ca^{2+}-bound CaM). (**b**) Uncoupled NOS: When L-arginine or BH_4 becomes limiting or when NOS is exposed to oxidative stress (resulting in BH_4 oxidation and/or destabilisation of the zinc-thiolate complex and/or glutathionylation of key cysteine residues), electrons flow from the reductase domain to oxygenase domain in the same monomer and catalyse the formation of superoxide. Figure adapted from Alderton et al. (2001) and Munzel et al. (2005)

In the absence of sufficient L-arginine (Xia et al. 1996) or BH_4 (Vásquez-Vivar et al. 1998) or when peroxynitrite production causes BH_4 oxidation (Landmesser et al. 2003) or disruption of the dimer-stabilising zinc-thiolate complex (Zou et al. 2002), NOS may become "uncoupled" and produce superoxide ($O_2^{\bullet-}$) as opposed to NO. There is also evidence that S-glutathionylation of NOS3 results in its uncoupling (Idigo et al. 2012; Zweier et al. 2011), and, in certain settings, this can result in superoxide production by the reductase domain (which may be insensitive to L-NAME inhibition) (Chen et al. 2010; Zweier et al. 2011). In the cardiovascular field, greatest attention has been focused on NOS3 uncoupling, largely in the context of its role in endothelial dysfunction (Munzel et al. 2005; Förstermann and Sessa 2012), yet relevant to cardiac myocytes, NOS1 is also prone to uncoupling (Pou et al. 1992, 1999) and, at least in vitro, has the greatest propensity to catalyse the uncoupled reaction (Alderton et al. 2001).

NO exerts its effects both directly through the S-nitrosylation (the addition of a nitrosyl group to a free thiol on a cysteine residue of target proteins (Lima et al. 2010; Sun and Murphy 2010), including the RyR, the α_1 subunit of the L-type Ca^{2+} channel (LTCC), sarcoplasmic/endoplasmic reticulum Ca^{2+} ATPase (SERCA) and mitochondrial proteins such as F1 ATPase $\alpha 1$ subunit and mitochondrial complex I (Gonzalez et al. 2007; Sun et al. 2007), amongst many (Kohr et al. 2011)) and indirectly through the activation of soluble guanylate cyclase (sGC) (leading to cGMP synthesis) and interactions with $O_2^{\bullet -}$ to form the (largely) pathological oxidant peroxynitrite (Pacher et al. 2007). As NO is a free diffusible gas capable of crossing cell membranes, target proteins may be modified and/or activated in both an autocrine manner (by NO produced in cardiac myocytes) and a paracrine manner (by NO produced in surrounding endothelial cells, fibroblasts, smooth muscle cells and neurons). However, due to the high myoglobin content of cardiac myocytes, which has been demonstrated to act as a scavenger of bioactive NO (Flogel et al. 2001), NO diffusion distances in the heart and within cardiac myocytes themselves are likely to be limited, as has been described for NOS2-derived NO (Wunderlich et al. 2003). Hence, compartmentalisation of NO synthesis within the same subcellular compartment as its downstream targets is likely to play a critical role in NOS signalling. In the case of the NOS-sGC-cGMP signalling pathway, sGC has been traditionally considered to be a cytoplasmic enzyme; however, more recently, a subpopulation of sGC has been localised to the plasma membrane both in noncardiac cells (Zabel et al. 2002) and cardiac myocytes (Agullo et al. 2005), where it is located in caveola-enriched lipid rafts, which not only puts sGC in the location of NO synthesis by NOS3 and NOS1 but also appears to protect it from oxidation (Tsai et al. 2012).

11.3.2 Soluble Guanylate Cyclase

Catalytically active sGC is a heterodimer consisting of α and β subunits (Kamisaki et al. 1986). In the cardiovascular system, the $\alpha_1\beta_1$ heterodimer of sGC is the main isoform responsible for NO-induced cGMP synthesis. Activation of sGC by NO is dependent on a haem group, which is orientated between the two subunits principally through coordination with his-105 in the N-terminal part of the β_1 subunit. Mutation of this residue prevents haem binding and results in a sGC that retains basal activity but is unresponsive to NO (Wedel et al. 1994). Early in vitro studies suggested that binding of NO to this haem group was sufficient to induce up to 200-fold activation of sGC (Humbert et al. 1990). However, in the past few years, a more complex picture of sGC activation has emerged, whereby NO binding to the haem group induces stable and moderate (approximately tenfold increase) activation of sGC, and that maximal NO activation of sGC is dependent on subsequent non-haem-NO interaction (Cary et al. 2005). This non-haem-NO interaction has very recently been demonstrated to be mediated by cysteine(s) and has been proposed to involve the formation of thionitroxide ($RSNO^{\bullet -}$—a short-lived chemical species formed by the direct nucleophilic addition of a thiol to NO) (Fernhoff et al. 2009). Numerous other factors may also influence sGC activity and expression, including cellular metabolic status (ATP acts as an allosteric inhibitor of sGC, so reductions in $[ATP]_i$ may lead to the amplification of NO-cGMP signals (Cary et al. 2005; Ruiz-Stewart et al. 2004)), cellular redox status (for review, see Munzel et al. 2005) and cGMP-dependent protein kinase I (cGKI)-mediated phosphorylation of Ser64 on the α_1 subunit (which inhibits sGC activity, thereby providing a negative feedback mechanism limiting sGC responsiveness to NO) (Zhou et al. 2008).

11.3.3 Cardiac cGMP Effectors

In cardiac myocytes, cGMP exerts its effects through the modulation of phosphodiesterase (PDE) activity (which breaks down cyclic nucleotides—see Sect. 11.3.3.2) and the activation of cGK signalling.[2]

11.3.3.1 Cardiac cGKI

Two cGK genes have been identified in mammalian cells encoding cGKI (including α and β splice variants, which differ in their N-terminal leucine zipper (LZ) interaction domain) and membrane-bound cGKII. Cardiac myocytes express cGKI; by immunoblotting, the α-isoform is the most highly expressed in crude cardiac extracts (Geiselhoringer et al. 2004). Functional cGKI is a homodimer with each subunit being composed of three domains: an N-terminal domain that contains an autoinhibitory region and a LZ motif (that mediates homodimerisation and interaction with scaffolding and/or target proteins), a regulatory domain containing cGMP binding sites and a C-terminal catalytic domain that phosphorylates target proteins as well as catalysing autophosphorylation to increase cGMP affinity. Binding of cGMP to the regulatory domain induces a conformational change releasing the suppression of cGKI activity by the autoinhibitory region (Francis et al. 2010). It has recently been demonstrated that oxidation of cGKIα, leading to disulphide bond formation between Cys42 residues of cGKIα monomers, can result in cGMP-independent activation of cGKIα; this has been shown to be biologically relevant in the role of H_2O_2 as an endothelium-derived hyperpolarising factor (EDHF) in the coronary vasculature (Burgoyne et al. 2007) and in sepsis-associated hypotension (Rudyk et al. 2013), but whether it is relevant in cardiac myocyte biology remains to be determined.

11.3.3.2 Cardiac PDEs: cGMP Catabolism and Compartmentalisation

PDEs are a large group of enzymes, categorised into 11 different families (based on structure, kinetics and substrate specificity) that are important downstream targets for cyclic nucleotide signals. Although principally involved in the breakdown of cyclic nucleotides, thereby playing an important role in the localisation of cAMP and cGMP signals to specific subcellular compartments, they also provide a mechanism for crosstalk between cGMP and cAMP signalling pathways. The cGMP-hydrolysing PDEs will be briefly discussed here. For information on cardiac PDEs, compartmentalisation and crosstalk between cyclic nucleotides by PDEs, the reader is referred to the following comprehensive reviews (Fischmeister et al. 2006; Omori and Korera 2007).

[2] cGMP can also activate cGMP-sensitive cyclic nucleotide-gated (CNG) cation channels and hyperpolarisation-activated cyclic nucleotide-gated (HCN) cation channels, the latter of which are found in cardiac pacemaker cells. However, to date, there is very little data in the literature concerning cGMP-dependent regulation of cardiac pacemaker activity (Herring et al (2001). NO-cGMP pathway increases the hyperpolarisation-activated current, I(f), and heart rate during adrenergic stimulation (Cardiovasc Res 52(3):446–453).

cGMP-hydrolysing PDEs

The PDEs that have been implicated in the cGMP catabolism in cardiac myocytes are PDE1, PDE2, PDE3, PDE5 and, more recently, PDE9.[3] PDE1 and PDE2 have dual-substrate specificity for cAMP and cGMP, whereas PDE5 is highly specific for cGMP. A fourth PDE family member, PDE3, although nominally designated as having dual-substrate specificity, is mainly described as a cGMP-inhibited cAMP PDE, as its catalytic rate for cAMP degradation is much greater than that for cGMP degradation (Zaccolo and Movsesian 2007). Nevertheless, PDE3 hydrolytic activity has been shown to contribute significantly to cGMP hydrolysis in both normal and failing human heart (Vandeput et al. 2007, 2009), and recent evidence has implicated PDE3 as the major cGMP-PDE responsible for the control of basal cytosolic $[cGMP]_i$ in adult mouse cardiac myocytes (Götz et al. 2013).

The most widely studied cardiac PDE is PDE5, mainly due to the extensive body of research into the therapeutic potential of PDE5 inhibition for the treatment of heart failure (Guazzi et al. 2011a, b; Zhang et al. 2010). Although only expressed at very low levels in cardiac myocytes in the normal heart (Takimoto 2005), PDE5 expression, and hence cGMP catabolism, is increased in both animal models of heart failure (Lu et al. 2010) and human heart failure (Pokreisz et al. 2009; Vandenwijngaert et al. 2013). Immunocytochemistry in isolated cardiac myocytes has suggested that, although expressed throughout the cell, PDE5 expression is localised to the Z-band striations; this localisation to Z-bands is dependent on the presence of a functional NOS3-sGC signalling complex, as it is absent in myocytes lacking NOS3 or in which NOS3 has been chronically inhibited by administration of L-NAME in drinking water (Takimoto 2005) and can be restored in L-NAME-treated animals by concurrent treatment with the NO-independent sGC activator BAY 41-8543 (Nagayama et al. 2008). Structurally, PDE5 consists of an N-terminal regulatory domain, containing two GAF domains (A and B), and a C-terminal catalytic domain. Binding of cGMP to the GAF-A domain of PDE5 (Zoraghi et al. 2005) induces a conformational change in PDE5 and activation of the catalytic domain. Phosphorylation of an N-terminal Ser (Ser92 and Ser102 of bovine and human PDE5, respectively) by cGK (and/or PKA) stabilises/maintains cGMP catabolism through increasing the affinity of cGMP binding to the GAF-A domain (Corbin et al. 2000), thus providing a negative feedback loop whereby cGK can limit elevations in cGMP through activation of cGMP catabolism. Indeed, functional evidence supporting this paradigm has recently been obtained in adult rat cardiac myocytes whereby cGK was demonstrated to activate PDE5, thereby limiting sub-sarcolemmal elevations in cGMP induced through the activation of sGC by exogenous NO (Castro et al. 2010).

The PDE1 family is unique amongst PDEs in that it is regulated by Ca^{2+}/CaM; similar to other Ca^{2+}-/CaM-dependent enzymes, Ca^{2+}/CaM binding activates the catalytic

[3] In addition to PDEs, $[cGMP]_i$ may be also be regulated by the export through the multidrug resistance (MDR) transporters (members of the ATP-binding cassette protein family), encoded by the ABCC genes. Several isoforms were previously identified in the cardiovascular system, MDR5 in the human heart [320] and MDR4 and MDR5 in vascular smooth muscle cells—in the latter, MDR4 was specifically shown to mediate the export of cGMP and influence its intracellular concentration, to the same extent as non-specific PDE inhibition with IBMX [321]. Whether MDR4 is expressed in human cardiac myocytes and similarly mediates cGMP efflux remains an open question.

domain through the release of an autoinhibitory domain (Sonnenburg et al. 1995); PKA-mediated phosphorylation of PDE1 reduces Ca^{2+}-/CaM-binding affinity, thereby inhibiting PDE1-mediated cyclic nucleotide hydrolysis (Sharma and Wang 1985). Three subfamilies of PDE1 have been identified (PDE1A, PDE1B and PDE1C)—both PDE1A and PDE1C have been identified in human hearts, whereas only PDE1A has been identified in rodent hearts (Miller et al. 2009; Vandeput et al. 2007). In human cardiac myocytes, PDE1C has been demonstrated to localise to both the Z and M lines and to be a major contributor to cGMP breakdown in both the normal and failing human heart (Vandeput et al. 2007, 2009). Recently, cardiac myocyte PDE1 has been shown to be upregulated in several rodent heart failure models (Miller et al. 2009).

PDE2, sometimes referred to as a cGMP-activated cAMP PDE (Zaccolo and Movsesian 2007), is, in fact, a dual-substrate specificity PDE, activated by cGMP binding to two N-terminal GAF domains. PDE2 is expressed in cardiac myocytes; both fractionation experiments and immunocytochemistry in neonatal rat cardiac myocytes have shown that it is targeted to the plasma membrane, where it is enriched at cell-cell junctions and Z-lines (Mongillo et al. 2006).

PDE9A, also a cGMP-specific PDE, was more recently identified in rodent and human hearts, where, contrary to PDE5, it modulates cardiac remodelling independently of NOS, suggesting that this isoform specifically hydrolyses cGMP pools generated by natriuretic peptides (Lee et al. 2015).

Compartmentalisation of cGMP Signals and Crosstalk Between cGMP and cAMP Signals

PDEs are downstream targets for cGMP signals and play an important role in the compartmentalisation of cGMP signals generated by both sGC and particulate guanylate cyclase (pGC (GC-A or NPRA), the plasma membrane receptor for natriuretic peptides) as well as being involved in crosstalk between cGMP and cAMP signalling pathways, as illustrated in Figs. 11.3 and 11.4.

The use of recombinant cyclic nucleotide-gated (CNG) channels as cyclic nucleotide sensors has allowed sub-sarcolemmal changes in [cGMP] to be measured in isolated cardiac myocytes and has implicated PDE2 and PDE5 in the compartmentalisation of cGMP signals. In adult rat cardiac myocytes, application of exogenous NO was demonstrated to result in small increases in the sarcolemmal CNG current (I_{CNG}), whereas activation of the plasma membrane pGC by application of ANP or BNP resulted in much greater increases in I_{CNG}. In both cases, application of the non-specific PDE inhibitor, IBMX, significantly elevated I_{CNG}, indicating the PDEs play a role in limiting cGMP signals downstream of both pGC and sGC. Subsequently, using selective inhibitors for PDE2 (Bay 607550) and PDE5 (sildenafil), it was demonstrated that PDE5 and, to a lesser extent, PDE2 are involved in the compartmentalisation of sGC cGMP signals, whereas PDE2 alone limits the spread of cGMP signals downstream of pGC (Castro et al. 2006). The disparity between the functional effects of ANP and PDE5 inhibition on β-AR-mediated increases in cardiac contractility provides further evidence supporting the role of PDE5 in the compartmentalisation of sGC signals but not pGC signals—PDE5 inhibition with sildenafil inhibits β-AR-mediated increases in cardiac contractility, whereas activation of pGC by ANP has no effect, despite causing much greater elevations

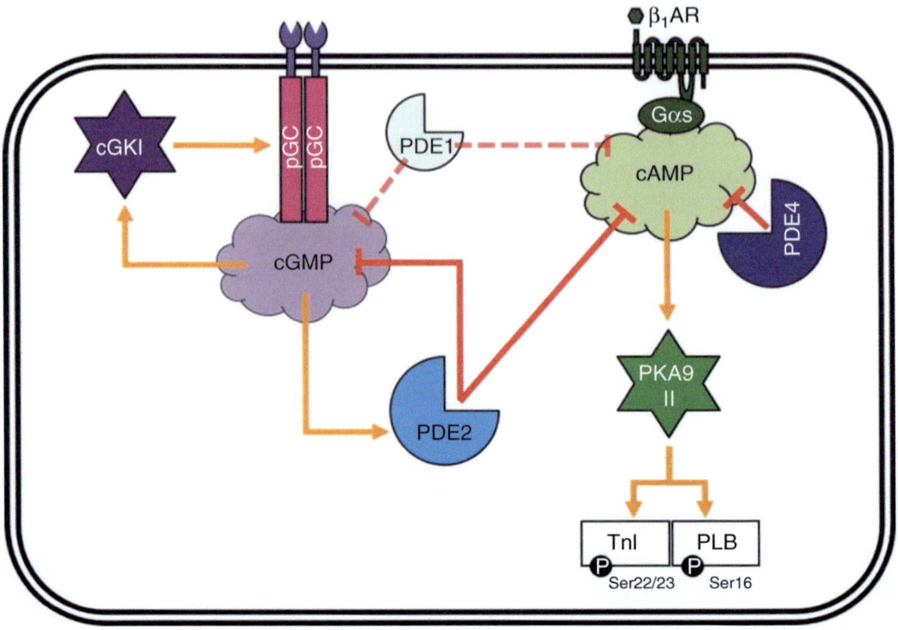

Fig. 11.3 Compartmentalisation of pGC-derived cGMP signals and crosstalk between cGMP and cAMP signals; critical role of PDE2. The cGMP-activated PDE2 plays a key role in the compartmentalisation of cGMP signals downstream of pGC. Given its dual-substrate specificity, it also plays an important role in the compartmentalisation of cAMP signals, thereby providing a mechanism whereby elevations in cGMP downstream of pGC can inhibit cAMP-dependent responses downstream of β-AR stimulation

in total myocardial $[cGMP]_i$ (Takimoto 2005; Tokimoto et al. 2007). Further evidence for a NOS3-sGC-PDE5 signalosome is provided by the observations that the effects of PDE5 inhibition on β-AR-mediated increases in cardiac contractility are absent in adult ventricular myocytes from NOS3 KO mice (and L-NAME-treated WT mice) and in cells in which sGC has been pharmacologically inhibited (with ODQ) (Takimoto 2005; Wang et al. 2009). More recently, the β_3AR and cGKI have been demonstrated to be important upstream and downstream mediators underlying the effects of PDE5 inhibition on β-AR-mediated increases in cardiac contractility, the former coupling to NOS3 activation and the latter decreasing myofilament Ca^{2+} sensitivity by promoting troponin I phosphorylation (Lee et al. 2010).

In addition to being involved in the hydrolysis and compartmentalisation of cGMP signals, PDE2 acts as an integration point for crosstalk between cAMP and cGMP signals. β-AR-mediated regulation of cardiac contractility is dependent on the compartmentalisation of cAMP-PKA signals downstream of β-AR activation; in neonatal rat ventricular myocytes (NRVMs), β-AR stimulation results in greater activation of cAMP synthesis in the PKA Type II compartment (as opposed to the PKA Type I compartment[4]), leading to the phosphorylation of downstream targets

[4]cAMP synthesis in PKA Type I compartment is activated by agonists such as prostaglandin E1, glucagon-like peptide 1 and glucagon.

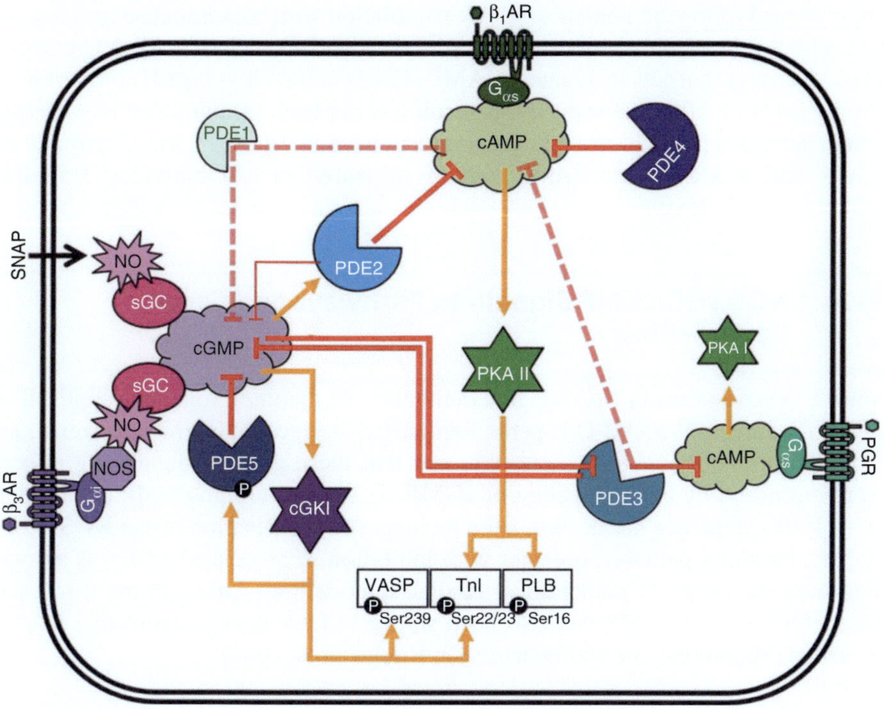

Fig. 11.4 Compartmentalisation of sGC-derived cGMP signals and crosstalk between cGMP and cAMP signals; roles of PDE5, PDE2 and PDE3. The compartmentalisation of sGC-derived cGMP signals is dependent on the cGMP-specific PDE5, which, through a signalosome with the β_3AR, NOS, sGC and cGKI, is involved in the regulation of the β-AR-stimulated contractile responses. PDE2, although playing a minor role in the compartmentalisation of sGC-derived cGMP signals, acts to inhibit cAMP-dependent responses downstream of β-AR stimulation. In addition, through cGMP-mediated inhibition of cAMP hydrolysis by PDE3, sGC-derived cGMP can also potentiate cAMP-dependent signalling, which, in human cardiac myocytes, may influence the β-AR-stimulated contractile response

involved in the regulation of cardiac contractility such as phospholamban (PLB) and troponin I (TnI) (Di Benedetto et al. 2008). Subsequently, elevations in $[cGMP]_i$ have been demonstrated to modify these cAMP signals downstream of β-AR activation; in the presence of the NO donor SNAP, the rises in cAMP signals in the PKA Types I and II compartments are inverted, with an enhanced cAMP response in the PKA Type I compartment and an attenuated response in the Type II compartment; on the other hand, application of ANP selectively reduces the β-AR-mediated cAMP response in the PKA Type II compartment. This effect of elevations in $[cGMP]_i$ to attenuate the cAMP signal in the PKA Type II compartment is due to the activation of cAMP hydrolysis by PDE2, whilst the enhanced cAMP response in the PKA Type I compartment is due to the cGMP-dependent inhibition of cAMP hydrolysis by PDE3. The functional significance of cGMP-dependent activation of cAMP hydrolysis is demonstrated by the fact that selective inhibition of PDE2 abolishes the ability of exogenous NO to inhibit β-AR-mediated increases in cardiac contractility (Stangherlin et al. 2011). Furthermore, in NRVMs, enhanced cAMP signals are observed following non-specific AR stimulation with noradrenaline in the presence of β_3AR blockade, demonstrating tightly compartmentalised β_3AR-NO-cGMP-PDE2 pathway that acts to attenuate cAMP signals in the PKA Type II compartment (Mongillo et al. 2006). However, this result has not been recapitulated in ARVMs, where selective β_3AR blockade is reported to have no effect on the magnitude of β_1AR- and β_2AR-specific cAMP signals measured at cell crests or T-tubules (Nikolaev et al. 2010).

11.4 NOS-sGC-cGMP Signalling Pathways and Cardiac Remodelling

Since in vitro studies in NRVMs first demonstrated that the exogenous application of NO (S-nitroso-N-acetyl-D,L-penicillamine (SNAP)) could inhibit noradrenaline-induced cardiac myocyte hypertrophy, and that these growth suppressing effects were mimicked by the application of cGMP analogue 8-bromo-cGMP (Calderone et al. 1998), evidence has accumulated to suggest that activation of the NOS-sGC-cGMP signalling pathway, together with inhibition of associated PDEs, is associated with inhibition of pathological cardiac remodelling. Although not discussed here, pGC-derived cGMP has also been shown to have a cardioprotective role as reviewed (Hammond and Balligand 2012; Ritchie et al. 2009).

11.4.1 Cardiac NOS Isoforms and Cardiac Remodelling

In order to Investate the roles of NOS-derived NO in influencing cardiac remodelling, numerous KO and TG models have been utilised and are discussed further below. However, there are several important limitations that must be considered when assessing the phenotypes of these animals. Firstly, in global KO models, lack of NOS in cell types other than cardiac myocytes will also influence the phenotype; for example, in NOS3 KO mice, lack of NOS3 in endothelial cells results in hypertension, and in NOS2 KO mice, inflammatory reactions are dampened as a consequence of lack of NOS2 in inflammatory cells. Secondly, in TG models, the subcellular localisation of the overexpressed NOS may differ from endogenous NOS and therefore influence the phenotype. Lastly, in TG models, the level of overexpression of the transgene may result in substrate and/or cofactor availability becoming limiting, leading to NOS uncoupling, and the production of $O_2^{\bullet-}$ as opposed to NO. Furthermore, there is evidence in the literature to suggest that the levels of NO produced may influence downstream signalling events; in cancer cells, it has been demonstrated that low, medium and high concentrations of exogenous NO activate ERK phosphorylation, hypoxia-inducible factor 1α (HIF-1α) accumulation and p53 phosphorylation, respectively (Thomas et al. 2004).

The evidence in the literature largely points towards NOS3 signalling playing a protective role against cardiac remodelling, and it is largely presumed that the NO produced by cardiac myocyte NOS3 acts indirectly through activation of sGC (Brunner et al. 2003; Takimoto 2005; Wang et al. 2009). Following MI, NOS3 KO mice show an enhanced cardiac myocyte hypertrophy in non-ischaemic myocardium, which is associated with a greater deterioration in cardiac function and the development of dilated cardiomyopathy (Scherrer-Crosbie et al. 2001) (regardless of whether arterial blood pressure is normalised or not). Similarly, providing that NOS3 remains in its dimeric form in WT mice, NOS3 KO mice develop greater cardiac hypertrophy and fibrosis and show greater deterioration in cardiac function following TAC (Buys et al. 2007; Ichinose et al. 2004). However, if the pressure-overload stimulus is harsh enough to result in NOS3 uncoupling, leading to increased oxidative stress, then a greater cardiac hypertrophy and deterioration in cardiac function are seen in WT mice rather than NOS3 KO mice (Takimoto 2005). Conversely, mice with cardiac-specific overexpression of NOS3 (with a fivefold increase in NOS3 protein expression versus WT, correctly targeted to caveolae) show an improved outcome post-MI, associated with less cardiac myocyte hypertrophy in the non-infarcted myocardium (Janssens et al. 2004).

Concerning NOS1, there is strong evidence from studies in NOS1 KO mice that NOS1-derived NO plays a protective role against cardiac remodelling. Lack of NOS1 has been demonstrated to exacerbate LV remodelling and functional deterioration following MI (Dawson et al. 2005; Saraiva et al. 2005) and is associated with increased cardiac oxidative stress due to enhanced xanthine oxidoreductase activity in NOS1 KO myocytes (Khan et al. 2004). However, on the whole, the opposite phenotype is not observed in cardiac-specific NOS1 TG mice, perhaps reflecting imbalances between NOS1 and cofactor levels, and the in vitro observation that NOS1 has the greatest propensity of the NOS isoforms to become uncoupled (Alderton et al. 2001). Two different cardiac-specific NOS1 transgenic mice have been described in the literature, with slightly different phenotypes; in one model, greater cardiac hypertrophy and impaired cardiac contractility are observed in TG mice in the absence of cardiac stress (Burkard et al. 2007), whereas in a second model, no differences in cardiac morphology and function are detected at baseline, but following pressure overload, TG mice show greater hypertrophy (HW:BW and cardiac myocyte) without upregulation of molecular markers of cardiac hypertrophy, together with better preservation of cardiac function and Ca^{2+} cycling, indicative of adaptive, as opposed to maladaptive, cardiac remodelling. The different phenotypes between these two NOS1 TG models are most likely due to small differences in subcellular localisation of the overexpressed NOS1; in the former model, NOS1 was demonstrated to interact with SERCA and the LTCC but not with the RyR, whereas in the latter model, NOS1 interaction with Cav-3 and RyR was observed. Interestingly, although $[cGMP]_i$ has been demonstrated to be elevated in NOS1 TG mice (Burkard et al. 2007), the effects of NOS1-derived NO on EC coupling and RyR function have been shown to be mediated directly via S-nitrosylation (Gonzalez et al. 2007; Wang et al. 2008; Zhang et al. 2008) leading to the speculation that NOS1 may predominantly signal via nitrosylating agents, whereas NOS3 may predominantly signal via cGMP (Ziolo 2008), although definitive evidence is lacking, especially in the case of NOS1-derived NO and cardiac remodelling.

The role of NOS2 in cardiac remodelling is not clearly resolved, other than to say it does not appear to protect against cardiac remodelling. Lack of NOS2 has either been observed to have no effect on cardiac hypertrophy (Hataishi et al. 2006; Sun et al. 2009) or to exacerbate cardiac remodelling and transition to heart failure following pressure overload (Zhang et al. 2007); the differences in phenotypes are most likely attributable to the harshness of the stimuli used and the corresponding inflammatory system involvement in the remodelling response in wild-type animals. Likewise, different phenotypes have been reported in cardiac-specific NOS2 TG mice; in one model, conditional overexpression of NOS2 in cardiac myocytes was associated with an increased cardiac hypertrophy and sudden death (Mungrue et al. 2002), whereas in a second model, non-conditional model minimal differences in cardiac morphology and function were detected, and no increased mortality was observed (Heger et al. 2002), a phenotype dependent on the NO scavenging properties of myoglobin (Wunderlich et al. 2003). These conflicting results have been the subject of intense debate (Godecke and Schrader 2004; Mungrue et al. 2003) and most likely reflect mechanistic differences in the TG mouse model used.

11.4.2 sGC and Cardiac Remodelling

To date, most of the evidence in the literature infers that NOS-mediated inhibition of cardiac remodelling is mediated indirectly through the activation of sGC (as opposed to directly through S-nitrosylation, although there is ample evidence that this plays a key role in the cardioprotective effect of NOS signalling in the setting of ischaemia-reperfusion (Sun and Murphy 2010) and in the balancing of cardiac performance with oxygen supply (Haldar and Stamler 2013), so the paucity of data suggesting a role in cardiac remodelling is somewhat anomalous). However, at present, there is only limited data from pharmacological and genetic studies directly manipulating sGC activity to reinforce this hypothesis, although the data from inhibition of cGMP-hydrolysing PDEs provides convincing evidence for an anti-remodelling role of cGMP (discussed in Sect. 11.4.3).

sGC can be activated independently of NO by pharmacological agents such as cinaciguat (BAY 58–2667) (Stasch et al. 2002) and BAY 41-2272 (Stasch et al. 2001). In vitro studies in NRVMs have demonstrated that cinaciguat and BAY 41-2272 can inhibit the development of cardiac myocyte hypertrophy (Irvine et al. 2012); likewise in in vivo models of hypertensive cardiac disease, both agents attenuate cardiac hypertrophic remodelling, but, given their blood pressure-lowering effects, it is not possible to differentiate the extent to which their anti-hypertrophic effects are due to direct action on cardiac myocytes themselves, versus effects secondary to reductions in cardiac preload and afterload (Boerrigter et al. 2007; Zanfolin et al. 2006). Cardiac remodelling studies in mice with global deletion of β_1 subunit of sGC (which leads to total loss of sGC enzyme activity) are precluded as mice die before reaching maturity due to gastrointestinal dysmotility (Friebe et al. 2007). Mouse models lacking the α_1 subunit of sGC have been independently described by two different groups (Buys et al. 2009; Mergia et al. 2006); interestingly, genetic background appears to influence whether these KO mice show gender-specific hypertension, as is the case for sGCα_1 KO on 129S6 background (Vermeersch et al. 2007), or are normotensive/show very mild hypertension, as observed in sGCα_1 KO on C57BL/6 background (Buys et al. 2009; Mergia et al. 2006), highlighting some of the problems associated with TG mouse models (Cook et al. 2009). Echocardiographic assessment of LV dimension and function in 10–14-week-old mice shows no differences between wild-type mice and sGCα_1 KO mice on C57BL/6 background (Buys et al. 2009), although cardiac myocytes isolated from these KO mice do show impaired sarcomere shortening at baseline (Cawley et al. 2011). The negative inotropic effect of β_3AR stimulation, as well as of exogenous NONOates on top of isoproterenol, was also abrogated in cardiac myocytes from these sGCα_1 KO mice, supporting the existence of a functional β_3AR/NO/sGC pathway counteracting the effect of β_1-β_2AR stimulation (Cawley et al. 2011). However, there is no published data on the response of these mice to cardiac stress. Furthermore, there are no published reports describing the effects of cardiac-specific knockout or overexpression of sGC.

11.4.3 PDE5 Inhibition and Cardiac Remodelling

Increasing amounts of evidence suggest that, with the development of heart failure, cGMP hydrolysis becomes elevated; heart failure has been associated with elevations in PDE5 expression (Lu et al. 2010; Nagendran et al. 2007) and, more recently, PDE1 expression (Miller et al. 2009). Over the past 5 years, numerous studies have provided evidence that the inhibition of cGMP hydrolysis, thereby raising $[cGMP]_i$, has a beneficial effect against cardiac remodelling (Nagayama et al. 2009a, b; Takimoto 2005; Zhang et al. 2010; Hassan and Ketat 2005), including in human heart failure (Guazzi et al. 2011a, b). In mouse models, PDE5 inhibition using sildenafil has been demonstrated to attenuate cardiac remodelling induced by pressure overload following TAC (Takimoto 2005) and, more importantly, both to reverse mild pre-existing TAC-induced cardiac remodelling (when sildenafil treatment is initiated 1 week post-TAC (Takimoto 2005)) and to prevent further advancement of cardiac remodelling and deterioration of cardiac function in cases of more advanced, pre-existing disease (when sildenafil treatment is initiated 3 weeks post-TAC (Nagayama et al. 2009a, b)). Intriguingly, sildenafil treatment has been demonstrated to be more effective in preventing cardiac hypertrophy when harsher stimuli are applied, as its beneficial effects are dependent of the inhibition of pathological stress signalling pathways (such as calcineurin and ERK), which are only activated when the hypertrophic stimulus reaches sufficient magnitude (Nagayama et al. 2009a, b). Whether sildenafil treatment specifically elevates cardiac myocyte [cGMP]i through inhibition of PDE5, and whether PDE5 inhibition truly represents a therapeutic target in the treatment of human heart failure, has been questioned, as differences between myocardial cGMP-hydrolysing activities have been highlighted between normal and failing human and mouse ventricular samples (Vandeput et al. 2009). In mouse myocardial extracts, cGMP hydrolysis has been attributed to PDE1 and PDE5 (with an increased contribution from PDE5 in heart failure samples), whereas in human myocardial extracts, the vast majority of cGMP hydrolytic activity was attributable to PDE1 and PDE3; in these human samples, sildenafil was demonstrated to have no effect on cGMP hydrolysis at low concentrations (at which it is specific for PDE5) but to only inhibit cGMP hydrolysis at 100x higher concentrations (at which it is capable of inhibiting PDE1). In the case of the murine target for sildenafil, a recently published study has demonstrated that, in a mouse model in which PDE5 is conditionally overexpressed in cardiac myocytes, the re-suppression of PDE5 expression (therefore mimicking the effects of sildenafil) reverses TAC-induced cardiac remodelling (Zhang et al. 2010), thereby providing evidence that in mouse cardiac myocytes at least, PDE5 inhibition does account for the effects of sildenafil; however, whether this is true in the human heart remains to be established. Given the effectiveness of sildenafil to reverse adverse cardiac remodelling and restore cardiac function in animal models, sildenafil treatment has been tested as an adjunct therapy in human heart failure patients; despite promising results from small clinical studies (Guazzi et al. 2011a, b; Lewis et al. 2007), the recently reported results from the large, multicentre RELAX trial[5] (Redfield et al. 2012) are somewhat underwhelming, with no improvements in exercise capacity or clinical status (including indices of left ventricular hypertrophy) seen over placebo in patients with heart failure with preserved ejection fraction (HFpEF[6]) (Redfield et al. 2013). However, the degree of left ventricular hypertrophy in the patient cohort was less severe than in some of the smaller studies that yielded promising results (Guazzi et al. 2011a, b), leading the authors to hypothesise that remodelling response in patient cohort was not great enough to allow PDE5 inhibition to exert beneficial effects, akin to what has been observed in the mouse, where the anti-hypertrophic efficacy of PDE5 inhibition is dependent on severity of pressure overload (Nagayama et al. 2009a, b). An alternative explanation for the lack of beneficial effect of sildenafil in this patient population may reflect the fact that myocardial cGMP levels and cGKI activity have been reported to be decreased in patients with HFpEF (compared with those with heart failure with reduced ejection fraction (HFrEF) or aortic stenosis), in the face of unchanged PDE5 protein levels (van Heerebeek et al. 2012); thus depressed cGMP-CGKI signalling in this setting most likely reflects deficiencies upstream of PDE5. Another multicentre trial has recently been initiated examining the effects of chronic PDE5 inhibition with tadalafil on clinical outcomes in patients with left ventricular systolic dysfunction and secondary pulmonary hypertension (**P**hosphodiesterase Type 5 **I**nhibition with **T**adalafil **CH**anges Outcomes in **H**eart **F**ailure (PITCH-HF); NCT01910389 (New England Research Institutes 2013)), and results are expected in April 2017.

11.4.4 β_3ARs and Cardiac Remodelling

We used a cardiac-specific transgenic model expressing low levels of the human β3AR under the alpha-MHC promoter to study the effect of β3AR on cardiac remodelling (Belge et al. 2014). β3AR attenuated cardiac myocyte hypertrophy in response to continuous or repetitive infusion of isoproterenol or angiotensin II, with a reduction of morphometric indexes (LV/TL ex vivo, echocardiography in vivo) as well as

[5] RELAX—PhosphodiesteRasE-5 Inhibition to Improve CLinical Status and EXercise Capacity in Diastolic Heart Failure

[6] Also commonly referred to as diastolic heart failure

reduced myocyte transverse area (by histological analysis), in parallel with reduced re-expression of the foetal gene programme typically accompanying hypertrophy. Hypertrophy was similarly reduced in rat neonatal cardiac myocytes with adenoviral expression of the human β3AR in response to three different agonists (phenylephrine, endothelin-1, isoproterenol). This anti-hypertrophic effect was NO dependent, as NOS inhibition abrogated the protection in vitro and in vivo. Mechanistically, we found co-localisation of β3AR with both eNOS and nNOS in caveolae, as well as coupling of β3AR to cGMP production using FRET sensors; β3AR expression inhibited the transcriptional activity of nuclear factor of activated T-cells (NFAT), and the anti-hypertrophic response was abrogated by cGKI inhibition (Belge et al. 2014).

Notably, β3AR expression also attenuated myocardial interstitial fibrosis in response to isoproterenol and angiotensin II infusions, suggesting paracrine effects initiated by β3AR signalling in cardiac myocytes (which exclusively expressed the receptor in this transgenic model). Likewise, preferential β3AR agonists (e.g. BRL37344) protected mice submitted to transaortic constriction (TAC) from adverse remodelling, with decreased hypertrophy and preserved LV function (Niu et al. 2012). This protection was lost in NOS1 KO mice. More recently, β3AR were implicated in the protection afforded by β1AR blockade (Trappanese et al. 2015). In a dog model of heart failure due to volume overload (mitral regurgitation), metoprolol promoted the upregulation of β3AR expression, its interaction with and activation of nNOS, followed by downstream cGMP production in specific membrane microdomains. Mechanistically, metoprolol protected sGC from oxidation, suggesting that some of the cardioprotective effects of β1AR blockade may result from enhanced expression and coupling of β3AR to NO/cGMP in the remodelling heart.

11.4.5 Mechanisms of cGMP Modulation of Cardiac Remodelling

Since studies in isolated cardiac myocytes demonstrated that the activation of cGK-dependent signalling pathways downstream of NO is involved in the negative inotropic effect of exogenous NO (Layland 2002) and that the anti-hypertrophic effects of exogenous NO are significantly enhanced by adenoviral overexpression of cGK1Iβ (Wollert et al. 2002), most of the beneficial effects of elevations in $[cGMP]_i$ in cardiac myocytes have been attributable to the activation of cGK signalling. However, until recently, direct evidence for this paradigm has been lacking, as genetic absence of cGKI (cGKI KO) is associated with premature mortality due to severe gastrointestinal abnormalities (Pfeifer et al. 1998). In an attempt to circumvent this problem, so-called rescue mice have been developed, in which smooth muscle-specific expression of cGKIβ (the main smooth muscle cGKI isoform) is restored (“knocked in”) on the cGKI KO background (Weber et al. 2007). Data produced in these mice has questioned the proposed cardioprotective role of cGKI (Lukowski et al. 2010) as, following chronic β-AR stimulation or pressure overload (induced by TAC), no differences in cardiac hypertrophy were observed between WT and rescue mice. However, the authors provide no evidence that cGK was in fact activated by the cardiac stresses in their WT mice, and others have questioned whether the stimuli applied were harsh enough to activate the pathological hypertrophic signalling pathways that cGKI has been proposed to inhibit (see below) (Kass and Takimoto 2010). More recent data, produced in two alternative mouse models deficient in cardiac cGKI function, has provided evidence that cGKI does play a protective role in the setting of cardiac remodelling. Firstly, mice with cardiac myocyte-specific deletion of cGKI develop dilated cardiomyopathy (with marked deterioration in cardiac function) following angiotensin II infusion or pressure overload (Frantz et al. 2013). Secondly, cGKIα LZ mutant mice (mice harbouring a knock-in mutation in cGKIα that disrupts the tertiary structure of the N-terminal LZ region, thereby preventing binding with proteins whose binding is dependent on the LZ, but retain cGKIα kinase activity (Michael et al. 2008)) show greater cardiac remodelling, functional deterioration and mortality following pressure overload, and furthermore, the protective effect of sildenafil treatment was lost in these mice (Blanton et al. 2012). Cellular targets implicated in cGMP-cGKI inhibition of cardiac remodelling are discussed further below and illustrated in Fig. 11.5.

11.4.6 cGMP-cGK-Dependent Modulation of Pathological Ca^{2+} Signals

The major mechanism whereby the cGMP-cGK signalling pathway modulates cardiac remodelling appears to be through attenuating the activation of pathological Ca^{2+} signals that lead to the activation of the calcineurin-nuclear factor of activated T-cell (NFAT) pathway and Ca^{2+}-calmodulin-activated protein kinase II (CamKII) implicated in cardiac remodelling (Ling et al. 2009; Molkentin et al. 1998).

The inhibition of calcineurin-NFAT signalling by elevations in $[cGMP]_i$ was first reported in isolated cardiac myocytes, whereby exogenous NO/cGMP was shown to attenuate αAR-mediated increases in NFAT transcriptional activity—an effect enhanced by adenoviral mediated cGK overexpression (Fiedler et al. 2002). Subsequently, inhibition of calcineurin activation was demonstrated to partially account for the anti-hypertrophic effect of PDE5 inhibition (both in vitro and in vivo) (Takimoto 2005). Conversely, the enhanced cardiac remodelling in pGC KO mice was shown to be associated with and dependent upon enhanced activation of the calcineurin-NFAT signalling pathway, (Kilic et al. 2005, 2007) as demonstrated by the regression of cardiac hypertrophy upon pharmacological inhibition of calcineurin (with FK506) (Tokudome et al. 2005).

In addition to inhibiting calcineurin activation, the anti-remodelling effects of cGMP have also been associated with the decreased activation of another Ca^{2+}-dependent signalling pathway—CaMKII. In mice lacking CnAβ (which have greatly reduced calcineurin expression and activity and correspondingly show an attenuated cardiac hypertrophy following pressure overload or chronic neurohormone stimulation (Bueno et al. 2002)), anti-hypertrophic effects of PDE5 inhibition are still observed, indicative of cGMP acting independently of the calcineurin pathway; pressure-overload-induced activation of CaMKII was demonstrated to be inhibited by sildenafil in both WT and CnAβ KO mice (Hsu et al. 2009). Further evidence supporting cGMP signals attenuating CaMKII activation comes from pGC

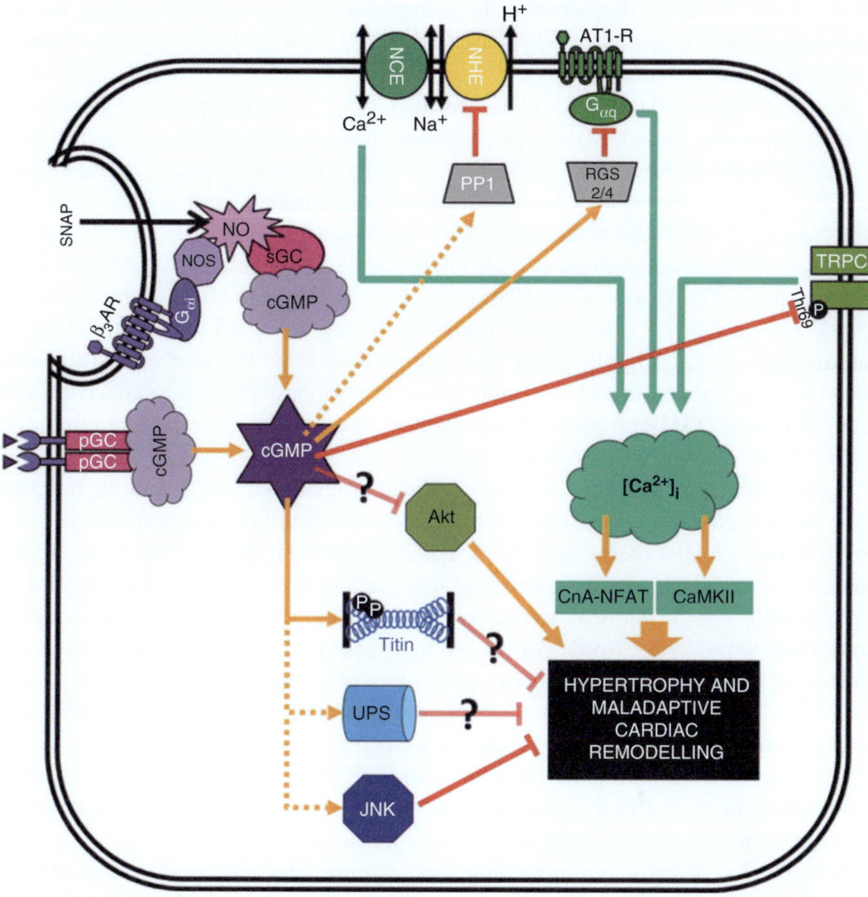

Fig. 11.5 cGKI targets implicated in the anti-remodelling effects of cGMP signalling. Multiple targets have been implicated in cGKI-mediated inhibition of pathological remodelling, including attenuation of Ca^{2+} signals (through the activation of RGS proteins to inhibit G_{aq} signalling, indirect inhibition of NHE activity via activation of protein phosphatase I (PPI) and inhibition of TRPC-mediated Ca^{2+} entry) that lead to activation of calcineurin-NFAT and CaMKII signalling associated with maladaptive remodelling. cGKI may also inhibit remodelling through attenuation of pathological Akt signalling and activation of JNK signalling. Lastly, cGKI phosphorylates titin and activates the UPS, both of which are key integration and regulatory nodes for cardiac signalling pathways

KO mice, in which the observed hypertrophic phenotype is associated with the upregulation of CaMKII expression, increased CaMKII autophosphorylation and phosphorylation of CaMKII downstream targets (Kilic et al. 2005, 2007; Kirchhof et al. 2004; Klaiber et al. 2010).

Given that elevations in $[cGMP]_i$ are ineffective in preventing increases in NFAT transcriptional activity in isolated cardiac myocytes overexpressing a constitutively active calcineurin mutant (Takimoto 2005; Fiedler et al. 2002) and that inhibition of two different Ca^{2+}-dependent signalling pathways underlies the anti-hypertrophic effect of the cGMP-cGK pathway, the search for cGK targets accounting for its inhibition of both calcineurin-NFAT signalling and CaMKII signalling has focused on upstream Ca^{2+} handling.

The first suggested target for cGK was the LTCC; in phenylephrine (PE)-treated cardiac myocytes, NO/cGMP suppress single-channel open probability and reduce $[Ca^{2+}]$ transients (Fiedler et al. 2002); the inhibitory effects of exogenous NO/cGMP on LTCC activity are enhanced in cardiac myocytes isolated from TG mice overexpressing cGKI (Schroder et al. 2003). Ser1928 of the α_{1C} subunit and Ser496 of the β_2 subunit have been identified as cGK phosphorylation targets in cardiac myocytes (Yang et al. 2007). However, whether Ca^{2+} entry through the LTCC, which plays a crucial role in cardiac myocyte EC coupling, is also involved in the activation of pathological calcineurin signalling is a contentious issue (Houser and Molkentin 2008); hence, although the LTCC may be an important cGK target underlying the acute effects of cGMP on cardiac contractility, whether it is an important target for remodelling remains to be established.

A second target involved, albeit indirectly, in intracellular Ca^{2+} handling, that has been implicated in the anti-hypertrophic effects of cGMP is the sarcolemmal Na^+/H^+ exchanger (NHE) (Kilic et al. 2005, 2007; Perez et al. 2007; Yeves et al. 2010). The activity of NHE, which imports Na^+ into cardiac myocytes, is closely linked to Ca^{2+} handling via the Na^+/Ca^{2+} exchanger, which, when $[Na^+]$ is elevated, may operate in reverse mode to extrude Na^+, in exchange for Ca^{2+}, thereby leading to elevations in $[Ca^{2+}]_i$. Recently, a genetic model has provided evidence that increased NHE activity is able to generate Ca^{2+} signals that induce cardiac hypertrophy and failure (Nakamura et al. 2008). Evidence for cGMP-dependent inhibition of NHE activity comes from both pGC KO mice and PDE5 inhibition studies. The cardiac hypertrophy in pGC KO mice has been associated with increased NHE1 activity, and chronic treatment of these mice with cariporide, an NHE inhibitor, is associated with regression of cardiac remodelling; interestingly, the anti-hypertrophic effects of NHE inhibition in this instance were associated with decreased activation of the CaMKII pathway, without attenuating the activation of the calcineurin pathway (Kilic et al. 2005). Conversely, enhanced $[cGMP]_i$, as a result of PDE5 inhibition, has been associated with decreased NHE activity, an effect attenuated by pharmacological inhibition of cGK with KT5823 (Perez et al. 2007). More recently, this cGK-dependent inhibition of NHE activity by elevations in [cGMP]i has been suggested to occur indirectly, via a cGK-dependent activation of protein phosphatase 1, which then dephosphorylates critical residues in the C-terminal tail of NHE1, thereby reducing its activity (Yeves et al. 2010).

In the past few years, canonical transient receptor potential (TRPC) channels (a family of non-selective cation channels) have emerged as important regulators of pathological cardiac remodelling (Eder and Molkentin 2011). Specifically, TRPC3 and TRPC6 have been identified as specific initiators and regulators of Ca^{2+} signals leading to calcineurin activation in cardiac myocytes (Bush et al. 2006; Wu et al. 2010), and, pertinent to this discussion, in noncardiac cells, both have been demonstrated to be regulated by cGK. To date, most work has focused on TRPC6 as a cGK target in cardiac myocytes. In isolated cardiac myocytes, both ANP and PDE5 inhibition with sildenafil have been demonstrated to increase the phosphorylation of TRPC6 on Thr69, (which is associated with the suppression of TRPC6 channel current), via cGK-dependent mechanisms (Koitabashi et al. 2009; Nishida et al. 2010). Genetic absence of TRPC3 and TRPC6 (Klaiber et al. 2010), knockdown of TRPC3 and TRPC6 expression and pharmacological inhibition of TRPC channels with BTP2 (Kinoshita et al. 2010) abolish the inhibitory effects of ANP on elevations and oscillations in $[Ca^{2+}]_i$ as a result of $G_{\alpha q}$ stimulation, indicating the significance of TRPC channels as a target for the anti-hypertrophic effects of cGMP. Furthermore, in pGC KO mice (which show elevated TRPC6 expression), pharmacological inhibition of TRPC channels (with BTP2) is associated with a regression of cardiac myocyte hypertrophy (HW:BW, myocyte size and molecular markers or hypertrophy), and conversely, overexpression of TRPC6 on a pGC KO background is associated with an exaggerated cardiac hypertrophy (Kinoshita et al. 2010).

Another target which has been demonstrated to account, at least in part, for the anti-hypertrophic effects of cGMP, and could account for decreased Ca^{2+} signalling and reduced calcineurin activation, is the "regulator of G-protein signalling" (RGS) family of proteins (Klaiber et al. 2010; Takimoto et al. 2009; Tokudome et al. 2008). RGS proteins play an important role in turning off GPCR signalling by acting as GTPase-activating proteins, thereby accelerating the rate of GTP hydrolysis by G_α and subsequent reconstitution of the GDP-bound heterotrimeric G-protein complex and terminating phospholipase-C-mediated IP_3 synthesis and subsequent IP_3-mediated Ca^{2+} release from intracellular stores. In cardiac myocytes, cGK-mediated phosphorylation and activation of RGS2 (Takimoto et al. 2009) and RGS4 (Tokudome et al. 2008) have been shown to contribute towards the anti-hypertrophic effects cGMP. In RGS2 KO mice, PDE5 inhibition with sildenafil treatment does not attenuate the development of cardiac hypertrophy or the activation of calcineurin signalling pathways following pressure overload despite similar increases in cGK activity as seen in WT mice (Takimoto et al. 2009). Furthermore, in myocytes isolated from RGS2 KO mice, ANP is unable to inhibit elevations in $[Ca^{2+}]_i$ following $G_{\alpha q}$ stimulation (Klaiber et al. 2010); similarly, overexpression of a dominant-negative form of RGS4 in isolated cardiac myocytes is associated with an attenuation of the inhibitory effects of ANP on IP_3 production, cardiac myocyte hypertrophy and NppA expression following $G_{\alpha q}$ stimulation (Tokudome et al. 2008). Accordingly, the expression and phosphorylation of RGS4 are attenuated in pGC KO mice, and furthermore, adverse cardiac remodelling observed in these mice can be rescued by cardiac-specific overexpression of RGS4 (Tokudome et al. 2008).

11.4.6.1 Other Mechanisms of cGMP-cGKI Modulation of Pathological Remodelling

In addition to inhibiting pathological Ca^{2+} signalling pathways, the anti-hypertrophic effects of cGMP have been associated with the inhibition of Akt-GSK3β signalling (Takimoto 2005; Kilic et al. 2005; Hsu et al. 2009) (by undefined mechanisms acting upstream of Akt itself, as PDE5 inhibition does not prevent cardiac hypertrophy in TG mice with cardiac-specific overexpression of a constitutively active form of Akt (Takimoto 2005)) and, more recently, cGKI-dependent activation of Jnk signalling (Blanton et al. 2012). cGKI-dependent signalling has been shown to target the elastic I-band region of the giant sarcomeric protein titin (Kruger and Linke 2011) where it promotes the phosphorylation of residues in the N2-Bus and N2-A regions. This decreases titin band stiffness promoting diastolic relaxation and also, given both regions are involved in protein-protein interactions, is implicated in the integration and co-ordination of signalling pathways controlling cardiac remodelling (Kruger and Linke 2011). More recently, cGKI has been demonstrated to stimulate the proteolytic activity of ubiquitin proteasome system (UPS) (Ranek et al. 2013) which is implicated in the regulation of cardiac hypertrophic signalling (including JNK and calcineurin pathways) (Portbury et al. 2012).

Conclusion

All the evidence reviewed above highlights the importance of subcellular confinement of components of the NOS/cGMP pathway for signalling specificity and proper regulation. This may provide some explanation for the apparent discrepancy between phenotypes of transgenic models that are confounded by excessive expression or mislocalisation of specific signalling elements, all of which should be examined for accurate interpretation. Likewise, the effects of exogenous NO supply on cardiac myocytes are expected to significantly differ from and may not faithfully reproduce those of endogenous NO generation from compartmentalised NOS isoforms. In this regard, any therapeutic exploitation of the protective effects of NO/cGMP against adverse cardiac remodelling, as reviewed above, would probably better be based on activation of surface receptors coupled to specific downstream NO-dependent effectors. Based on preclinical data (and emerging clinical evidence), the β3AR may fulfil such requirements. β3AR's expression in human myocardium, its resistance to desensitisation and upregulation in diseased hearts would make it an attractive target for pharmacological modulation of cardiac remodelling. This is currently being tested in a prospective, randomised, placebo-controlled multicentric European trial (www.beta3lvh.eu) in patients with structural cardiac remodelling (stage B, AHA) at risk of developing heart failure with preserved ejection fraction. This and other trials with β3AR agonists will tell whether the paradigm fulfils its promises.

Compliance with Ethical Standards

Conflict of Interest Statement The authors declare that they have no conflict of interest.

References

Agullo L et al (2005) Membrane association of nitric oxide-sensitive guanylyl cyclase in cardiomyocytes. Cardiovasc Res 68:65–74

Alderton WK, Cooper CE, Knowles RG (2001) Nitric oxide synthases: structure, function and inhibition. Biochem J 357:593–615

Angelone T, Filice E, Quintieri AM, Imbrogno S, Recchia A, Pulerà E, Mannarino C, Pellegrino D, Cerra MC (2008) Beta3-adrenoceptors modulate left ventricular relaxation in the rat heart via the NO-cGMP-PKG pathway. Acta Physiol (Oxf). 193(3):229–239.

Ashley EA, Sears CE, Bryant SM, Watkins HC, Casadei B (2002) Cardiac nitric oxide synthase 1 regulates basal and β-adrenergic contractility in murine ventricular myocytes. Circulation 105:3011–3016

Balligand JL, Feron O, Dessy C (2009) eNOS activation by physical forces: from short-term regulation of contraction to chronic remodeling of cardiovascular tissues. Physiol Rev 89:481–534

Balligand JL, Kelly RA, Marsden PA, Smith TW, Michel T (1993a) Control of cardiac muscle cell function by an endogenous nitric oxide signaling system. Proc Natl Acad Sci U S A 90(1):347–351

Balligand JL, Ungureanu D, Kelly RA, Kobzik L, Pimental D, Michel T, Smith TW (1993b) Abnormal contractile function due to induction of nitric oxide synthesis in rat cardiac myocytes follows exposure to activated macrophage-conditioned medium. J Clin Invest. 91(5): 2314–2319

Balligand JL et al (1994) Cytokine-inducible nitric oxide synthase (iNOS) expression in cardiac myocytes. Characterisation and regulation of iNOS expression and detection of iNOS activity in single cardiac myocytes in vitro. J Biol Chem 269:27580–27588

Balligand JL et al (1995) Nitric oxide-dependent parasympathetic signaling is due to activation of constitutive endothelial (type III) nitric oxide synthase in cardiac myocytes. J Biol Chem 270:14582–14586

Barouch LA et al (2002) Nitric oxide regulates the heart by spatial confinement of nitric oxide synthase isoforms. Nature 416:337–339

Belge C, Massion PB, Pelat M, Balligand JL (2005) Nitric oxide and the heart: update on new paradigms. Ann N Y Acad Sci 1047:173–182

Belge C, et al (2014) Enhanced expression of beta3-adrenoceptors in cardiac myocytes attenuates neurohormone-induced hypertrophic remodeling through nitric oxide synthase. Circulation 129:451–462

Bendall JK et al (2004) Role of myocardial neuronal nitric oxide synthase-derived nitric oxide in β-adrenergic hyporesponsiveness after myocardial infarction-induced heart failure in rat. Circulation 110:2368–2375

Blanton RM et al (2012) Protein kinase G Iα inhibits pressure overload-induced cardiac remodeling and is required for the cardioprotective effect of sildenafil in vivo. J Am Heart Assoc 1:e003731

Boerrigter G et al (2007) Targeting heme-oxidized soluble guanylate cyclase in experimental heart failure. Hypertension 49:1128–1133

Brunner F et al (2003) Attenuation of myocardial ischemia/reperfusion injury in mice with myocyte-specific overexpression of endothelial nitric oxide synthase. Cardiovasc Res 57:55–62

Bueno OF et al (2002) Impaired cardiac hypertrophic response in Calcineurin Aβ -deficient mice. Proc Natl Acad Sci U S A 99:4586–4591

Burgoyne JR et al (2007) Cysteine redox sensor in PKGIα enables oxidant-induced activation. Science 317:1393–1397

Burkard N et al (2007) Conditional neuronal nitric oxide synthase overexpression impairs myocardial contractility. Circ Res 100:e32–e44

Bush EW et al (2006) Canonical transient receptor potential channels promote cardiomyocyte hypertrophy through activation of calcineurin signaling. J Biol Chem 281:33487–33496

Buys ES et al (2007) Cardiomyocyte-restricted restoration of nitric oxide synthase 3 attenuates left ventricular remodeling after chronic pressure overload. Am J Physiol Heart Circ Physiol 293:H620–H627

Buys ES et al (2009) sGC$\alpha_1\beta_1$ attenuates cardiac dysfunction and mortality in murine inflammatory shock models. Am J Phys Heart Circ Phys 297:H654–H663

Calderone A, Thaik CM, Takahashi N, Chang DL, Colucci WS (1998) Nitric oxide, atrial natriuretic peptide, and cyclic GMP inhibit the growth-promoting effects of norepinephrine in cardiac myocytes and fibroblasts. J Clin Invest 101:812–818

Calvert JW et al (2011) Exercise Protects Against Myocardial Ischemia-Reperfusion Injury via Stimulation of β_3-Adrenergic Receptors and Increased Nitric Oxide Signaling: Role of Nitrite and Nitrosothiols. Circ Res 108:1448–1458

Cary SP, Winger JA, Marletta MA (2005) Tonic and acute nitric oxide signaling through soluble guanylate cyclase is mediated by nonheme nitric oxide, ATP, and GTP. Proc Natl Acad Sci U S A 102:13064–13069
Castro LR, Schittl J, Fischmeister R (2010) Feedback control through cGMP-dependent protein kinase contributes to differential regulation and compartmentation of cGMP in rat cardiac myocytes. Circ Res 107:1232–1240
Castro LR, Verde I, Cooper DM, Fischmeister R (2006) Cyclic guanosine monophosphate compartmentation in rat cardiac myocytes. Circulation 113:2221–2228
Cawley SM et al (2011) sGCα1 mediates the negative inotropic effects of NO in cardiac myocytes independent of changes in calcium handling. Am J Physiol Heart Circ Physiol 2:2
Chen CA et al (2010) S-glutathionylation uncouples eNOS and regulates its cellular and vascular function. Nature 468:1115–1118
Cook SA, Clerk A, Sugden PH (2009) Are transgenic mice the 'alkahest' to understanding myocardial hypertrophy and failure? J Mol Cell Cardiol 46:118–129
Corbin JD, Turko IV, Beasley A, Francis SH (2000) Phosphorylation of phosphodiesterase-5 by cyclic nucleotide-dependent protein kinase alters its catalytic and allosteric cGMP-binding activities. Eur J Biochem 267:2760–2767
Dawson D et al (2005) nNOS gene deletion exacerbates pathological left ventricular remodeling and functional deterioration after myocardial infarction. Circulation 112:3729–3737
Devic E, Xiang Y, Gould D, Kobilka B (2001) β-adrenergic receptor subtype-specific signaling in cardiac myocytes from β_1 and β_2 adrenoceptor knockout mice. Mol Pharmacol 60:577–583
Di Benedetto G et al (2008) Protein kinase A type I and type II define distinct intracellular signaling compartments. Circ Res 103:836–844
Eder P, Molkentin JD (2011) TRPC Channels As Effectors of Cardiac Hypertrophy. Circ Res 108:265–272
Emorine LJ, Marullo S, Briend-Sutren MM, Patey G, Tate K, Delavier-Klutchko C, Strosberg AD (1989) Molecular characterization of the human beta 3-adrenergic receptor. Science. 245(4922):1118–1121
Fernhoff NB, Derbyshire ER, Marletta MA (2009) A nitric oxide/cysteine interaction mediates the activation of soluble guanylate cyclase. Proc Natl Acad Sci U S A 106:21602–21607
Feron O et al (1996) Endothelial nitric oxide synthase targeting to caveolae. Specific interactions with caveolin isoforms in cardiac myocytes and endothelial cells. J Biol Chem 271: 22810–22814
Fiedler B et al (2002) Inhibition of calcineurin-NFAT hypertrophy signaling by cGMP-dependent protein kinase type I in cardiac myocytes. Proc Natl Acad Sci U S A 99:11363–11368
Fischmeister R et al (2006) Compartmentation of cyclic nucleotide signaling in the heart: the role of cyclic nucleotide phosphodiesterases. Circ Res 99:816–828
Flogel U, Merx MW, Godecke A, Decking UK, Schrader J (2001) Myoglobin: A scavenger of bioactive NO. Proc Natl Acad Sci U S A 98:735–740
Förstermann U, Sessa WC (2012) Nitric oxide synthases: regulation and function. Eur Heart J 33:829–837
Francis SH, Busch JL, Corbin JD, Sibley D (2010) cGMP-dependent protein kinases and cGMP phosphodiesterases in nitric oxide and cGMP action. Pharmacol Rev 62:525–563
Frantz S et al (2013) Stress-dependent dilated cardiomyopathy in mice with cardiomyocyte-restricted inactivation of cyclic GMP-dependent protein kinase I. Eur Heart J 34:1233–1244
Friebe A, Mergia E, Dangel O, Lange A, Koesling D (2007) Fatal gastrointestinal obstruction and hypertension in mice lacking nitric oxide-sensitive guanylyl cyclase. Proc Natl Acad Sci U S A 104:7699–7704
Fujimura T et al (1999) Expression and possible functional role of the β_3-adrenoceptor in human and rat detrusor muscle. J Urol 161:680–685
Geiselhoringer A, Gaisa M, Hofmann F, Schlossmann J (2004) Distribution of IRAG and cGKI-isoforms in murine tissues. FEBS Lett 575:19–22
Godecke A, Schrader J (2004) The Janus faces of NO? Circ Res 94:e55

Gonzalez DR, Beigi F, Treuer AV, Hare JM (2007) Deficient ryanodine receptor S-nitrosylation increases sarcoplasmic reticulum calcium leak and arrhythmogenesis in cardiomyocytes. Proc Natl Acad Sci U S A 104:20612–20617

Götz, K., et al. (2013) PDE3 regulates cGMP levels and cGMP/cAMP cross-talk in adult mouse ventricular cardiomyocytes. Emmy Noether Group of the DFG, Department of Cardiology and Pneumology, European Heart Research, Institute Göttingen, Georg August University Medical Center, University of Göttingen, D-37075 Göttingen, Germany.

Gauthier C, Tavernier G, Charpentier F, Langin D, Le Marec H (1996) Functional beta3-adrenoceptor in the human heart. J Clin Invest. 98(2):556–562

Gauthier C, Leblais V, Kobzik L, Trochu JN, Khandoudi N, Bril A, Balligand JL, Le Marec H (1998) The negative inotropic effect of beta3-adrenoceptor stimulation is mediated by activation of a nitric oxide synthase pathway in human ventricle. J Clin Invest. 102(7):1377–1384

Guazzi M, Vicenzi M, Arena R, Guazzi MD (2011a) PDE5 inhibition with Sildenafil improves left ventricular diastolic function, cardiac geometry, and clinical status in patients with stable systolic heart failure: results of a 1-year, prospective, randomized, placebo-controlled study. Circ Heart Fail 4:8–17

Guazzi M, Vicenzi M, Arena R, Guazzi MD (2011b) Pulmonary hypertension in heart failure with preserved ejection fraction: a target of phosphodiesterase-5 inhibition in a 1-year study. Circulation 124:164–174

Haldar SM, Stamler JS (2013) S-nitrosylation: integrator of cardiovascular performance and oxygen delivery. J Clin Invest 123:101–110

Hammond J, Balligand J-L (2012) Nitric oxide synthase and cyclic GMP signaling in cardiac myocytes: From contractility to remodeling. J Mol Cell Cardiol 52:330–340

Hare JM, Loh E, Creager MA, Colucci WS (1995) Nitric oxide inhibits the positive inotropic response to β-adrenergic stimulation in humans with left ventricular dysfunction. Circulation 92:2198–2203

Hassan MA, Ketat AF (2005) Sildenafil citrate increases myocardial cGMP content in rat heart, decreases its hypertrophic response to isoproterenol and decreases myocardial leak of creatine kinase and troponin T. BMC Pharmacol 5:10

Hataishi R et al (2006) Nitric oxide synthase 2 and pressure-overload-induced left ventricular remodelling in mice. Exp Physiol 91:633–639

van Heerebeek L et al (2012) Low Myocardial Protein Kinase G Activity in Heart Failure With Preserved Ejection Fraction. Circulation 126:830–839

Heger J et al (2002) Cardiac-specific overexpression of inducible nitric oxide synthase does not result in severe cardiac dysfunction. Circ Res 90:93–99

Houser SR, Molkentin JD (2008) Does contractile Ca^{2+} control calcineurin-NFAT signaling and pathological hypertrophy in cardiac myocytes? Sci Signal 24(1):pe31

Hsu S et al (2009) Phosphodiesterase 5 inhibition blocks pressure overload-induced cardiac hypertrophy independent of the calcineurin pathway. Cardiovasc Res 81:301–309

Humbert P et al (1990) Purification of soluble guanylyl cyclase from bovine lung by a new immunoaffinity chromatographic method. Eur J Biochem 190:273–278

Ichinose F et al (2004) Pressure overload-induced LV hypertrophy and dysfunction in mice are exacerbated by congenital NOS3 deficiency. Am J Physiol Heart Circ Physiol 286:H1070–H1075

Idigo WO et al (2012) Regulation of endothelial nitric-oxide synthase (NOS) S-glutathionylation by neuronal NOS: evidence of a functional interaction between myocardial constitutive NOS isoforms. J Biol Chem 287:43665–43673

Irvine JC et al (2012) The soluble guanylyl cyclase activator Bay 58-2667 selectively limits cardiomyocyte hypertrophy. PLoS One 7:e44481

Janssens S et al (2004) Cardiomyocyte-specific overexpression of nitric oxide synthase 3 improves left ventricular performance and reduces compensatory hypertrophy after myocardial infarction. Circ Res 94:1256–1262

Kamisaki Y et al (1986) Soluble guanylate cyclase from rat lung exists as a heterodimer. J Biol Chem 261:7236–7241

Kass DA, Takimoto E (2010) Regulation and role of myocyte cyclic GMP-dependent protein kinase-1. Proc Natl Acad Sci U S A 107(24):E98; author reply E99

Khan SA et al (2004) Neuronal nitric oxide synthase negatively regulates xanthine oxidoreductase inhibition of cardiac excitation-contraction coupling. Proc Natl Acad Sci U S A 101: 15944–15948

Kilic A, Bubikat A, Gassner B, Baba HA, Kuhn M (2007) Local actions of atrial natriuretic peptide counteract angiotensin II stimulated cardiac remodeling. Endocrinology 148:4162–4169

Kilic A et al (2005) Enhanced activity of the myocardial Na^+/H^+ exchanger NHE-1 contributes to cardiac remodeling in atrial natriuretic peptide receptor-deficient mice. Circulation 112:2307–2317

Kinoshita H et al (2010) Inhibition of TRPC6 channel activity contributes to the antihypertrophic effects of natriuretic peptides-guanylyl cyclase-A signaling in the heart. Circ Res 106:1849–1860

Kirchhof P et al (2004) Ventricular arrhythmias, increased cardiac calmodulin kinase II expression, and altered repolarisation kinetics in ANP receptor deficient mice. J Mol Cell Cardiol 36:691–700

Klaiber M et al (2010) Novel insights into the mechanisms mediating the local antihypertrophic effects of cardiac atrial natriuretic peptide: role of cGMP-dependent protein kinase and RGS2. Basic Res Cardiol 30:30

Kohr MJ et al (2011) Characterisation of potential S-nitrosylation sites in the myocardium. Am J Physiol Heart Circ Physiol 300:H1327–H1335

Koitabashi N et al (2009) Cyclic GMP/PKG-dependent inhibition of TRPC6 channel activity and expression negatively regulates cardiomyocyte NFAT activation Novel mechanism of cardiac stress modulation by PDE5 inhibition. J Mol Cell Cardiol 48(4):713–724

Kruger M, Linke WA (2011) The giant protein titin: a regulatory node that integrates myocyte signaling pathways. J Biol Chem 286:9905–9912

Landmesser U et al (2003) Oxidation of tetrahydrobiopterin leads to uncoupling of endothelial cell nitric oxide synthase in hypertension. J Clin Invest 111:1201–1209

Layland J, Li JM, Shah AM (2002) Role of cyclic GMP-dependent protein kinase in the contractile response to exogenous nitric oxide in rat cardiac myocytes. J Physiol 540(Pt 2):457–467

Lee DI et al (2010) PDE5A suppression of acute β-adrenergic activation requires modulation of myocyte β_3 signaling coupled to PKG-mediated troponin I phosphorylation. Basic Res Cardiol 105:337–347

Lee DI et al (2015) Phosphodiesterase 9A controls nitric oxide-independent cGMP and hypertrophic heart disease. Nature 519:472–476

Lewis GD et al (2007) Sildenafil improves exercise capacity and quality of life in patients with systolic heart failure and secondary pulmonary hypertension. Circulation 116:1555–1562

Lima B, Forrester MT, Hess DT, Stamler JS (2010) S-nitrosylation in cardiovascular signaling. Circ Res 106:633–646

Lindquist JM, Fredriksson JM, Rehnmark S, Cannon B, Nedergaard J (2000) β_3- and α_1-adrenergic Erk1/2 activation is Src- but not G_i-mediated in Brown adipocytes. J Biol Chem 275:22670–22677

Ling H et al (2009) Requirement for Ca^{2+}/calmodulin-dependent kinase II in the transition from pressure overload-induced cardiac hypertrophy to heart failure in mice. J Clin Invest 119:1230–1240

Lu Z et al (2010) Oxidative stress regulates left ventricular PDE5 expression in the failing heart. Circulation 121:1474–1483

Lukowski R et al (2010) Cardiac hypertrophy is not amplified by deletion of cGMP-dependent protein kinase I in cardiomyocytes. Proc Natl Acad Sci U S A 8:8

Massion PB, Feron O, Dessy C, Balligand JL (2003) Nitric oxide and cardiac function: ten years after, and continuing. Circ Res 93:388–398

Mergia E, Friebe A, Dangel O, Russwurm M, Koesling D (2006) Spare guanylyl cyclase NO receptors ensure high NO sensitivity in the vascular system. J Clin Invest 116:1731–1737

Michael SK et al (2008) High blood pressure arising from a defect in vascular function. Proc Natl Acad Sci 105:6702–6707

Miller CL et al (2009) Role of Ca^{2+}/calmodulin-stimulated cyclic nucleotide phosphodiesterase 1 in mediating cardiomyocyte hypertrophy. Circ Res 105:956–964

Mohamed TMA et al (2011) Plasma Membrane Calcium Pump (PMCA4)-Neuronal Nitric-oxide Synthase Complex Regulates Cardiac Contractility through Modulation of a Compartmentalized Cyclic Nucleotide Microdomain. J Biol Chem 286:41520–41529

Molkentin JD et al (1998) A calcineurin-dependent transcriptional pathway for cardiac hypertrophy. Cell 93:215–228

Mongillo M et al (2006) Compartmentalized phosphodiesterase-2 activity blunts β-adrenergic cardiac inotropy via an NO/cGMP-dependent pathway. Circ Res 98:226–234

Mungrue IN, Stewart DJ, Husain M (2003) The Janus faces of iNOS. Circ Res 93:e74

Mungrue IN et al (2002) Cardiomyocyte overexpression of iNOS in mice results in peroxynitrite generation, heart block, and sudden death. J Clin Invest 109:735–743

Munzel T, Daiber A, Ullrich V, Mulsch A (2005) Vascular consequences of endothelial nitric oxide synthase uncoupling for the activity and expression of the soluble guanylyl cyclase and the cGMP-dependent protein kinase. Arterioscler Thromb Vasc Biol 25:1551–1557

Nagayama T, Zhang M, Hsu S, Takimoto E, Kass DA (2008) Sustained soluble guanylate cyclase stimulation offsets nitric-oxide synthase inhibition to restore acute cardiac modulation by sildenafil. J Pharmacol Exp Ther 326:380–387

Nagayama T et al (2009a) Pressure-overload magnitude-dependence of the anti-hypertrophic efficacy of PDE5A inhibition. J Mol Cell Cardiol 46:560–567

Nagayama T et al (2009b) Sildenafil stops progressive chamber, cellular, and molecular remodeling and improves calcium handling and function in hearts with pre-existing advanced hypertrophy caused by pressure overload. J Am Coll Cardiol 53:207–215

Nagendran J et al (2007) Phosphodiesterase type 5 is highly expressed in the hypertrophied human right ventricle, and acute inhibition of phosphodiesterase type 5 improves contractility. Circulation 116:238–248

Nakamura TY, Iwata Y, Arai Y, Komamura K, Wakabayashi S (2008) Activation of Na^+/H^+ exchanger 1 is sufficient to generate Ca^{2+} signals that induce cardiac hypertrophy and heart failure. Circ Res 103:891–899

New England Research Institutes (2013) Phosphodiesterase Type 5 inhibition with Tadalafil changes outcomes in heart failure (PITCH-HF). Vol. 2014. National Library of Medicine (US), Bethesda (MD).

Nikolaev VO et al (2010) β_2-adrenergic receptor redistribution in heart failure changes cAMP compartmentation. Science 327:1653–1657

Nishida M et al (2010) Phosphorylation of TRPC6 channels at Thr69 is required for anti-hypertrophic effects of phosphodiesterase 5 inhibition. J Biol Chem 285:13244–13253

Niu X et al (2012) Cardioprotective Effect of β_3 Adrenergic Receptor Agonism—Role of Neuronal Nitric Oxide Synthase. J Am Coll Cardiol 59:1979–1987

Oceandy D et al (2007) Neuronal Nitric Oxide Synthase Signaling in the Heart Is Regulated by the Sarcolemmal Calcium Pump 4b. Circulation 115:483–492

Omori K, Kotera J (2007) Overview of PDEs and their regulation. Circ Res 100:309–327

Pacher P, Beckman JS, Liaudet L (2007) Nitric oxide and peroxynitrite in health and disease. Physiol Rev 87:315–424

Palmer RM, Ferrige AG, Moncada S (1987) Nitric oxide release accounts for the biological activity of endothelium-derived relaxing factor. Nature 327:524–526

Perez NG et al (2007) Phosphodiesterase 5A inhibition induces Na^+/H^+ exchanger blockade and protection against myocardial infarction. Hypertension 49:1095–1103

Pfeifer A et al (1998) Defective smooth muscle regulation in cGMP kinase I-deficient mice. EMBO J 17:3045–3051

Pokreisz P et al (2009) Ventricular phosphodiesterase-5 expression is increased in patients with advanced heart failure and contributes to adverse ventricular remodeling after myocardial infarction in mice. Circulation 119:408–416

Portbury AL, Ronnebaum SM, Zungu M, Patterson C, Willis MS (2012) Back to your heart: ubiquitin proteasome system-regulated signal transduction. J Mol Cell Cardiol 52:526–537

Pou S, Keaton L, Surichamorn W, Rosen GM (1999) Mechanism of Superoxide Generation by Neuronal Nitric-oxide Synthase. J Biol Chem 274:9573–9580

Pou S, Pou WS, Bredt DS, Snyder SH, Rosen GM (1992) Generation of superoxide by purified brain nitric oxide synthase. J Biol Chem 267:24173–24176

Ranek MJ, Terpstra EJ, Li J, Kass DA, Wang X (2013) Protein Kinase G Positively Regulates Proteasome-Mediated Degradation of Misfolded Proteins. Circulation 128(4):365–376

Redfield MM et al (2012) PhosphodiesteRasE-5 Inhibition to Improve CLinical Status and EXercise Capacity in Diastolic Heart Failure (RELAX) trial: rationale and design. Circulation: Heart Failure 5:653–659

Redfield MM et al (2013) Effect of phosphodiesterase-5 inhibition on exercise capacity and clinical status in heart failure with preserved ejection fraction: a randomized clinical trial. JAMA 309:1268–1277

Ritchie RH et al (2009) Exploiting cGMP-based therapies for the prevention of left ventricular hypertrophy: NO* and beyond. Pharmacol Ther 124:279–300

Robidoux J et al (2006) Maximal β_3-adrenergic regulation of lipolysis involves Src and epidermal growth factor receptor-dependent ERK1/2 activation. J Biol Chem 281:37794–37802

Rudyk O et al (2013) Protein kinase G oxidation is a major cause of injury during sepsis. Proc Natl Acad Sci U S A 110:9909–9913

Ruiz-Stewart I et al (2004) Guanylyl cyclase is an ATP sensor coupling nitric oxide signaling to cell metabolism. Proc Natl Acad Sci U S A 101:37–42

Saraiva RM et al (2005) Deficiency of neuronal nitric oxide synthase increases mortality and cardiac remodeling after myocardial infarction: role of nitroso-redox equilibrium. Circulation 112:3415–3422

Scherrer-Crosbie M et al (2001) Endothelial nitric oxide synthase limits left ventricular remodeling after myocardial infarction in mice. Circulation 104:1286–1291

Schroder F et al (2003) Single L-type Ca^{2+} channel regulation by cGMP-dependent protein kinase type I in adult cardiomyocytes from PKG I transgenic mice. Cardiovasc Res 60:268–277

Sears CE, Ashley EA, Casadei B (2004) Nitric oxide control of cardiac function: is neuronal nitric oxide synthase a key component? Philos Trans R Soc Lond Ser B Biol Sci 359:1021–1044

Seddon M, Shah AM, Casadei B (2007) Cardiomyocytes as effectors of nitric oxide signalling. Cardiovasc Res 75:315–326

Sharma RK, Wang JH (1985) Differential regulation of bovine brain calmodulin-dependent cyclic nucleotide phosphodiesterase isoenzymes by cyclic AMP-dependent protein kinase and calmodulin-dependent phosphatase. Proc Natl Acad Sci U S A 82:2603–2607

Soeder KJ, Snedden SK, Cao W, Della Rocca GJ, Daniel KW, Luttrell LM, Collins S (1999) The beta3-adrenergic receptor activates mitogen-activated protein kinase in adipocytes through a Gi-dependent mechanism. J Biol Chem. 23;274(17):12017–12022

Sonnenburg WK et al (1995) Identification of inhibitory and calmodulin-binding domains of the PDE1A1 and PDE1A2 calmodulin-stimulated cyclic nucleotide phosphodiesterases. J Biol Chem 270:30989–31000

Stangherlin A et al (2011) cGMP Signals Modulate cAMP Levels in a Compartment-Specific Manner to Regulate Catecholamine-Dependent Signaling in Cardiac Myocytes. Circ Res 17:17

Stasch JP et al (2001) NO-independent regulatory site on soluble guanylate cyclase. Nature 410:212–215

Stasch JP et al (2002) NO- and haem-independent activation of soluble guanylyl cyclase: molecular basis and cardiovascular implications of a new pharmacological principle. Br J Pharmacol 136:773–783

Strosberg AD (1997) Structure and function of the beta 3-adrenergic receptor. Annu Rev Pharmacol Toxicol 37:421–450

Sun J, Morgan M, Shen RF, Steenbergen C, Murphy E (2007) Preconditioning results in S-nitrosylation of proteins involved in regulation of mitochondrial energetics and calcium transport. Circ Res 101:1155–1163

Sun J, Murphy E (2010) Protein S-nitrosylation and cardioprotection. Circ Res 106:285–296
Sun Y et al (2009) Deletion of inducible nitric oxide synthase provides cardioprotection in mice with 2-kidney, 1-clip hypertension. Hypertension 53:49–56
Takimoto E, Champion HC, Belardi D, Moslehi J, Mongillo M, Mergia E, Montrose DC, Isoda T, Aufiero K, Zaccolo M, Dostmann WR, Smith CJ, Kass DA (2005) cGMP catabolism by phosphodiesterase 5A regulates cardiac adrenergic stimulation by NOS3-dependent mechanism. Circ Res 96(1):100–109
Takimoto E et al (2007) Compartmentalisation of cardiac β-adrenergic inotropy modulation by phosphodiesterase type 5. Circulation 115:2159–2167
Takimoto E et al (2009) Regulator of G protein signaling 2 mediates cardiac compensation to pressure overload and antihypertrophic effects of PDE5 inhibition in mice. J Clin Invest 119:408–420
Thomas DD et al (2004) Hypoxic inducible factor 1α, extracellular signal-regulated kinase, and p53 are regulated by distinct threshold concentrations of nitric oxide. Proc Natl Acad Sci U S A 101:8894–8899
Tokudome T et al (2005) Calcineurin-nuclear factor of activated T cells pathway-dependent cardiac remodeling in mice deficient in guanylyl cyclase A, a receptor for atrial and brain natriuretic peptides. Circulation 111:3095–3104
Tokudome T et al (2008) Regulator of G-protein signaling subtype 4 mediates antihypertrophic effect of locally secreted natriuretic peptides in the heart. Circulation 117:2329–2339
Trappanese DM et al (2015) Chronic beta1-adrenergic blockade enhances myocardial beta3-adrenergic coupling with nitric oxide-cGMP signaling in a canine model of chronic volume overload: new insight into mechanisms of cardiac benefit with selective beta1-blocker therapy. Basic Res Cardiol 110:456–470
Tsai EJ et al (2012) Pressure-overload-induced subcellular relocalisation/oxidation of soluble guanylyl cyclase in the heart modulates enzyme stimulation. Circ Res 110:295–303
Vandenwijngaert S et al (2013) Increased cardiac myocyte PDE5 levels in human and murine pressure overload hypertrophy contribute to adverse LV remodeling. PLoS One 8:e58841
Vandeput F et al (2007) Cyclic nucleotide phosphodiesterase PDE1C1 in human cardiac myocytes. J Biol Chem 282:32749–32757
Vandeput F et al (2009) cGMP-hydrolytic activity and its inhibition by sildenafil in normal and failing human and mouse myocardium. J Pharmacol Exp Ther 330:884–891
Vaniotis G et al (2013) Regulation of cardiac nitric oxide signaling by nuclear β-adrenergic and endothelin receptors. J Mol Cell Cardiol 62:58–68
Varghese P et al (2000) β_3-adrenoceptor deficiency blocks nitric oxide-dependent inhibition of myocardial contractility. J Clin Invest 106:697–703
Vásquez-Vivar J et al (1998) Superoxide generation by endothelial nitric oxide synthase: The influence of cofactors. Proc Natl Acad Sci 95:9220–9225
Vermeersch P et al (2007) Soluble guanylate cyclase-α1 deficiency selectively inhibits the pulmonary vasodilator response to nitric oxide and increases the pulmonary vascular remodeling response to chronic hypoxia. Circulation 116:936–943
Wang H, Kohr MJ, Traynham CJ, Ziolo MT (2009) Phosphodiesterase 5 restricts NOS3/Soluble guanylate cyclase signaling to L-type Ca^{2+} current in cardiac myocytes. J Mol Cell Cardiol 47:304–314
Wang H et al (2008) Neuronal nitric oxide synthase signaling within cardiac myocytes targets phospholamban. Am J Physiol Cell Physiol 294:C1566–C1575
Watts VL et al (2013) Anti-hypertrophic and anti-oxidant effect of β_3-adrenergic stimulation in myocytes requires differential neuronal NOS phosphorylation. J Mol Cell Cardiol 62:8–17
Weber S et al (2007) Rescue of cGMP kinase I knockout mice by smooth muscle specific expression of either isozyme. Circ Res 101:1096–1103
Wedel B et al (1994) Mutation of His-105 in the β1 subunit yields a nitric oxide-insensitive form of soluble guanylyl cyclase. Proc Natl Acad Sci U S A 91:2592–2596
Wollert KC et al (2002) Gene transfer of cGMP-dependent protein kinase I enhances the antihypertrophic effects of nitric oxide in cardiomyocytes. Hypertension 39:87–92

Wright PT et al (2014) Caveolin-3 regulates compartmentation of cardiomyocyte β_2-adrenergic receptor-mediated cAMP signaling. J Mol Cell Cardiol 67:38–48

Wu X, Eder P, Chang B, Molkentin JD (2010) TRPC channels are necessary mediators of pathologic cardiac hypertrophy. Proc Natl Acad Sci U S A 107:7000–7005

Wunderlich C, Flogel U, Godecke A, Heger J, Schrader J (2003) Acute inhibition of myoglobin impairs contractility and energy state of iNOS-overexpressing hearts. Circ Res 92:1352–1358

Xia Y, Dawson VL, Dawson TM, Snyder SH, Zweier JL (1996) Nitric oxide synthase generates superoxide and nitric oxide in arginine-depleted cells leading to peroxynitrite-mediated cellular injury. Proc Natl Acad Sci U S A 93:6770–6774

Xu KY, Huso DL, Dawson TM, Bredt DS, Becker LC (1999) Nitric oxide synthase in cardiac sarcoplasmic reticulum. Proc Natl Acad Sci U S A 96:657–662

Yang L et al (2007) Protein kinase G phosphorylates $Ca_v1.2$ α_{1c} and β_2 subunits. Circ Res 101:465–474

Yeves AM et al (2010) Decreased activity of the Na^+/H^+ exchanger by phosphodiesterase 5A inhibition is attributed to an increase in protein phosphatase activity. Hypertension 56:690–695

Zabel U et al (2002) Calcium-dependent membrane association sensitizes soluble guanylyl cyclase to nitric oxide. Nat Cell Biol 4:307–311

Zaccolo M, Movsesian MA (2007) cAMP and cGMP signaling cross-talk: role of phosphodiesterases and implications for cardiac pathophysiology. Circ Res 100:1569–1578

Zanfolin M et al (2006) Protective effects of BAY 41-2272 (sGC stimulator) on hypertension, heart, and cardiomyocyte hypertrophy induced by chronic L-NAME treatment in rats. J Cardiovasc Pharmacol 47:391–395

Zhang ZS, Cheng HJ, Onishi K, Ohte N, Wannenburg T, Cheng CP (2005) Enhanced inhibition of L-type Ca2+ current by beta3-adrenergic stimulation in failing rat heart. J Pharmacol Exp Ther 315(3):1203–1211

Zhang P et al (2007) Inducible nitric oxide synthase deficiency protects the heart from systolic overload-induced ventricular hypertrophy and congestive heart failure. Circ Res 100:1089–1098

Zhang YH et al (2008) Reduced phospholamban phosphorylation is associated with impaired relaxation in left ventricular myocytes from neuronal NO synthase-deficient mice. Circ Res 102:242–249

Zhang M et al (2010) Myocardial remodeling is controlled by myocyte-targeted gene regulation of phosphodiesterase type 5. J Am Coll Cardiol 56:2021–2030

Zhou Z et al (2008) Protein kinase G phosphorylates soluble guanylyl cyclase on serine 64 and inhibits its activity. Arterioscler Thromb Vasc Biol 28:1803–1810

Ziolo MT (2008) The fork in the nitric oxide road: cyclic GMP or nitrosylation? Nitric Oxide 18:153–156

Ziolo MT, Kohr MJ, Wang H (2008) Nitric oxide signaling and the regulation of myocardial function. J Mol Cell Cardiol 45:625–632

Zoraghi R, Bessay EP, Corbin JD, Francis SH (2005) Structural and functional features in human PDE5A1 regulatory domain that provide for allosteric cGMP binding, dimerisation, and regulation. J Biol Chem 280:12051–12063

Zou MH, Shi C, Cohen RA (2002) Oxidation of the zinc-thiolate complex and uncoupling of endothelial nitric oxide synthase by peroxynitrite. J Clin Invest 109:817–826

Zweier JL, Chen CA, Druhan LJ (2011) S-glutathionylation reshapes our understanding of endothelial nitric oxide synthase uncoupling and nitric oxide/reactive oxygen species-mediated signaling. Antioxid Redox Signal 14:1769–1775

12 Compartmentation of Natriuretic Peptide Signalling in Cardiac Myocytes: Effects on Cardiac Contractility and Hypertrophy

Lise Román Moltzau, Silja Meier, Kjetil Wessel Andressen, and Finn Olav Levy

Contents

L.R. Moltzau • S. Meier • K.W. Andressen • F.O. Levy (✉)
Department of Pharmacology, Faculty of Medicine, University of Oslo and Oslo University Hospital, Oslo, Norway

Center for Heart Failure Research, Faculty of Medicine, University of Oslo and Oslo University Hospital, Oslo, Norway
e-mail: l.r.moltzau@medisin.uio.no; silja.meier@medisin.uio.no; k.w.andressen@medisin.uio.no; f.o.levy@medisin.uio.no

V. Nikolaev, M. Zaccolo (eds.), *Microdomains in the Cardiovascular System*, Cardiac and Vascular Biology 3, DOI 10.1007/978-3-319-54579-0_12

Abstract

Natriuretic peptides have received considerable attention in the recent years both as biomarkers for heart failure and as treatment for the disease. Atrial (ANP) and B-type natriuretic peptide (BNP) stimulate the natriuretic peptide receptor (NPR)-A, whereas C-type natriuretic peptide (CNP) stimulates the NPR-B. Both NPR-A and NPR-B are guanylyl cyclases which generate cGMP when stimulated. All three peptides bind the NPR-C which does not possess guanylyl cyclase (GC) activity. Evidence is accumulating that natriuretic peptides are involved in regulating contractility and hypertrophic responses through a complex system of signalling pathways, both cGMP dependent and independent, in different compartments in cardiomyocytes. This review discusses compartmentalisation of natriuretic peptide signalling in cardiomyocytes, with focus on contractile and antihypertrophic effects in healthy and diseased hearts.

12.1 Introduction

One of the first indications of compartmentation of intracellular signalling in the heart was the difference between the effects of prostaglandin E_1 (PGE_1) and isoproterenol (Iso). Whereas both increased cyclic adenosine 3',5'-monophosphate (cAMP), only Iso increased contractility (Hayes et al. 1979; Kaumann and Birnbaumer 1974). Later it was found that PGE_1 increased cAMP in the soluble fraction of cardiomyocyte homogenates, whereas Iso increased cAMP in both the soluble and particulate fractions (Buxton and Brunton 1983). Further, it was found that cAMP in the particulate fraction was important for regulation of contractility (Aass et al. 1988). The use of subcellularly localised fluorescence resonance energy transfer (FRET)-based sensors elegantly supported these earlier studies (Di Benedetto et al. 2008). Lately, evidence is accumulating that cyclic guanosine 3',5'-monophosphate (cGMP) signalling is also compartmented. The second messenger cGMP was discovered about 50 years ago (Ashman et al. 1963) and can act through the activation of protein kinase G (PKG), phosphodiesterases (PDEs) and cyclic nucleotide-gated channels (CNGC) (Francis et al. 2010). Cyclic GMP signalling through PKG and PDEs is most studied. Stimuli that activate guanylyl cyclases (GCs) increase the formation of cGMP from guanosine triphosphate (GTP). Of the two families of GCs that produce cGMP in the cells, soluble GC (sGC) is cytosolic and activated by nitric oxide (NO), whereas natriuretic peptide receptors (NPRs) are membrane-bound or particulate GCs (pGCs; Potter 2011). This review will focus on compartmentation of both cGMP-dependent and cGMP-independent signalling by natriuretic peptides through NPRs.

12.2 Natriuretic Peptides and Their Receptors in Cardiomyocytes

Atrial natriuretic peptide (ANP) was discovered in the early 1980s (de Bold et al. 1981), followed by its family members brain natriuretic peptide (BNP) (Sudoh et al. 1988) and C-type natriuretic peptide (CNP) some years later (Sudoh et al. 1990).

In healthy hearts, ANP is mainly synthesised in atria, and very little ANP is found in the ventricle tissue. However, during heart failure (HF), the ventricle is an important source of circulating ANP (Nishikimi et al. 2011; Yasue et al. 1994). BNP is secreted from the left ventricle both in normal and failing hearts (Yasue et al. 1994). Both ANP and BNP are increased in heart failure and are thus used as biomarkers as their expression is correlated with the severity of the disease. BNP (or more specifically NT-proBNP) is routinely measured for both clinical diagnosis and management of HF (Dickstein et al. 2008; Hunt et al. 1994). CNP has been shown to be produced in endothelial cells (Suga et al. 1992b), but studies also report production in rat cardiac fibroblasts (Horio et al. 2003) and neonatal cardiomyocytes (Del Ry et al. 2011). Early studies failed to detect CNP in normal and failing human hearts (Takahashi et al. 1992), but a more recent report found increased production and increased plasma levels of CNP in heart failure patients (Del Ry et al. 2005).

ANP, BNP and CNP differentially bind to the three different natriuretic peptide receptors, NPR-A, NPR-B and NPR-C. NPR-A is activated by ANP and BNP (Suga et al. 1992a), whereas CNP activates NPR-B (Koller et al. 1991; Suga et al. 1992a). All three natriuretic peptides bind to NPR-C (Suga et al. 1992a). NPR-A is also referred to as GC-A or NPR1, NPR-B as GC-B or NPR2 and NPR-C as the clearance receptor or NPR3 (Potter et al. 2006). NPR-A and NPR-B possess GC activity increasing cGMP when stimulated, whereas NPR-C is devoid of GC activity (Leitman et al. 1986; Takayanagi et al. 1987). Although the main function of NPR-C is suggested to be the removal of natriuretic peptides from the circulation, studies have also shown signalling functions for this receptor (Rose and Giles 2008). Natriuretic peptides can either bind to NPR-A or NPR-B and increase cGMP or they can bind NPR-C and signal through this receptor without directly increasing cGMP. In general, studies involving ANP or BNP (NPR-A/NPR-C agonists) far exceed those performed with CNP (NPR-B/NPR-C agonist) and even less have been performed to compare the effects and signalling after stimulation with the different natriuretic peptides.

Messenger RNA of the three receptors is found in rat cardiomyocytes, suggesting that NPR-A, NPR-B and NPR-C are all expressed (Lin et al. 1995). Whether NPR-A is downregulated in heart failure has been discussed (Cabiati et al. 2010; Kuhn 2003). However, several studies show indications of reduced NPR-A activity and density in different heart failure preparations (Dickey et al. 2007, 2011; Singh et al. 2006). Whereas NPR-A stimulation (ANP) resulted in higher pGC activity than after NPR-B stimulation (CNP) in normal mouse hearts, the NPR-A effect was reduced in heart failure below the preserved NPR-B effect (Dickey et al. 2007). Particulate GC activity was found in normal human cardiomyocytes after both NPR-A and NPR-B stimulation (Dickey et al. 2011). However, the pGC activity of both receptors seemed to be low in failing cardiomyocytes as both ANP and CNP caused a small nominal, but not significant, increase in pGC activity (Dickey et al. 2011). Further, Singh et al. (2006) showed that the density of NPR-A was significantly downregulated in patients with ischaemic heart disease.

Natriuretic peptides (NP) have lately received much attention for potential treatment of heart failure as natriuretic peptides cause beneficial haemodynamic effects such as vasodilation, natriuresis and diuresis both in normal subjects (Holmes et al.

1993) and in HF patients (Abraham et al. 1998; Marcus et al. 1996; Mills et al. 1999). Further, they are shown to be antihypertrophic (Potter et al. 2009). Nesiritide (synthetic BNP) showed beneficial effects in clinical trials of heart failure (Colucci et al. 2000). However, a later study did not support routine use of nesiritide in patients with acute heart failure due to lack of effect on mortality and rehospitalisation (O'Connor et al. 2011). Recently, a chimeric peptide of CNP and dendroaspis NP (DNP; CD-NP, cenderitide) with dual specificity for both NPR-A and NPR-B (Dickey et al. 2008) was proposed for heart failure treatment. Cenderitide has later showed promising results (Lee et al. 2009), but information from ongoing clinical trials is limited as of yet.[1] Recently, the PARADIGM-HF clinical trial on heart failure patients with reduced ejection fraction showed beneficial effects of a combination of the angiotensin II receptor antagonist valsartan and the neprilysin (NEP) inhibitor sacubitril, thus targeting also the breakdown of natriuretic peptides, compared to the angiotensin converting enzyme (ACE) inhibitor enalapril (McMurray et al. 2014). Valsartan/sacubitril (Entresto®) is now in clinical trial (PARAGON-HF) for heart failure with preserved ejection fraction (HFpEF).[2] Thus, recent evidence may indicate a beneficial role of natriuretic peptides in heart failure, presumably based on their beneficial vascular effects. However, the direct effects of natriuretic peptides in the heart are less investigated.

12.3 Compartmentation of Natriuretic Peptide-Induced Effects

12.3.1 Signalling of Natriuretic Peptides Causing Contractile Effects

Effects on cardiac contractility may involve positive or negative inotropic effects (altered contractility) or positive lusitropic effects (enhanced relaxation). Studies investigating the effects of different natriuretic peptides on cardiac contractility have been performed in various species and preparations, and the divergent results complicate conclusions both in normal (Beaulieu et al. 1997; Brusq et al. 1999; Hirose et al. 1998; Lainchbury et al. 2000; Nir et al. 2001; Ohte et al. 1999; Pierkes et al. 2002; Sodi et al. 2008; Su et al. 2005a; Tajima et al. 1998; Vaxelaire et al. 1989; Wollert et al. 2003; Zhang et al. 2005) and failing hearts (Afzal et al. 2011; Lainchbury et al. 2000; Moltzau et al. 2013, 2014a, 2014b; Ohte et al. 1999; Qvigstad et al. 2010). Table 12.1 summarises these various results obtained. The reasons for the observed differences in inotropic effects are not entirely clear. However, it is likely that compartmentation of natriuretic peptide-induced effects is at least part of an explanation.

[1] https://clinicaltrials.gov/ct2/show/NCT00839007, https://clinicaltrials.gov/ct2/show/NCT00620308, https://clinicaltrials.gov/ct2/show/NCT02071602, https://clinicaltrials.gov/ct2/show/NCT01750905, https://clinicaltrials.gov/ct2/show/nct02359227

[2] https://clinicaltrials.gov/ct2/show/NCT01920711

Table 12.1 Overview of the inotropic responses to natriuretic peptides in different preparations

Preparation	Species	Model	NPR-A (ANP/BNP)	NPR-B (CNP)	References
Cardiomyocytes					
Ventricular	Mouse	Normal		NIR	Su et al. (2005a), Zhang et al. (2007)
Normal	Mouse	Normal	No IR (ANP)		Takimoto et al. (2007)
LV	Mouse	Normal		PIR	Frantz et al. (2013)
Ventricular	Rabbit	Normal	NIR	NIR	Zhang et al. (2005)
LV	Rat	Normal	NIR (BNP)		Sodi et al. (2008)
Neonatal, ventricular	Rat	Normal		NIR	Nir et al. (2001)
Embryonic ventricular cells	Chick	Normal	NIR (ANP)		Vaxelaire et al. (1989)
Ventricular, normal and Hyp (renal hypertension)	Rabbit	Normal vs Hyp		Normal: NIR Hyp: No IR	Moalem et al. (2006)
LV, normal and Hyp (aortic banding)	Rat	Normal vs Hyp	Normal: NIR Hyp: No IR		Tajima et al. (1998)
Ventricular, normal and Hyp (aortic banding)	Mouse	Normal vs Hyp	Normal: NIR (BNP) Hyp: Attenuated NIR (BNP)	Normal: NIR Hyp: Attenuated NIR	Su et al. (2005b), Tan et al. (2010)
Normal and mild Hyp (TAC model)	Mouse	Normal vs Hyp	Normal: No IR Hyp: PIR (ANP, in the presence of β_1- and β_2-AR signalling)		Perera et al. (2015)
Multicellular preparations					
Papillary muscle; LV	Rat	Normal	No IR (ANP)	NIR	Brusq et al. (1999)
Muscle strips; LV, normal and HF (coronary artery ligation post-infarction model)	Rat	Normal vs HF	No IR	Normal and HF: NIR (alone) Increased PIR (in the presence of cAMP signalling)	Afzal et al. (2011), Meier et al. (2015), Moltzau et al. (2013), Moltzau et al. (2014a), Moltzau et al. (2014b), Qvigstad et al. (2010)

(continued)

Table 12.1 (continued)

Preparation	Species	Model	NPR-A (ANP/ BNP)	NPR-B (CNP)	References
RA and LV preparations	Dog	Normal	No IR in atria or ventricle	PIR in atria and ventricle	Beaulieu et al. (1997), Hirose et al. (1998)
Whole heart					
Whole heart; normal and HF (pacing induced)	Dog	Normal vs HF	Normal: NIR (ANP) HF: NIR (ANP)		Ohte et al. (1999)
Whole heart	Mouse	Normal	No IR (ANP)	Initial PIR, then NIR	Pierkes et al. (2002)
Whole heart	Mouse	Normal		PIR	Wollert et al. (2003)
Whole heart; normal and HF (pacing induced)	Dog	Normal vs HF	Normal: Mild PIR HF: No IR	Normal: Mild PIR HF: No IR	Lainchbury et al. (2000)

CM cardiomyocytes, *HF* heart failure, *Hyp* hypertrophy, *IR* inotropic response, *LV* left ventricle, *NIR* negative inotropic response, *NP* natriuretic peptides, *PIR* positive inotropic response, *RA* right atrium, *TAC* transverse aortic constriction

In rat papillary muscles, stimulation with CNP increased phosphorylation of phospholamban (PLB) at Ser16 and troponin I (TnI) at Ser23/24 and caused a negative inotropic response and a positive lusitropic response. These effects were associated with increased sarcoplasmic/endoplasmic reticulum Ca^{2+}-ATPase (SERCA)2 activity (Brusq et al. 1999). This was later also shown in a rat model of heart failure where the lusitropic and negative inotropic response to CNP was explained by the concerted effects of PLB and TnI phosphorylation (Moltzau et al. 2013). The lusitropic and negative inotropic responses, with an initial positive inotropic response to CNP, have also been shown in working mouse heart associated with PLB Ser16 phosphorylation (Pierkes et al. 2002). The negative inotropic response to CNP was abolished in both a PLB-knockout (KO) model (Zhang et al. 2007) and a SERCA2a-KO model (Moltzau et al. 2013), which illustrates a role for SERCA2 in the negative inotropic response to CNP. The negative inotropic response to CNP is through PKG-dependent mechanisms as the effect of CNP was reduced in the presence of PKG inhibitors (Rp-8-[(4-chlorophenyl)thio]-cGMPS trimethylamine and KT5823) in normal mouse and rabbit (Su et al. 2005a; Zhang et al. 2005) and failing rat (Moltzau et al. 2013) ventricular cardiomyocytes. However, to complicate the picture, the positive inotropic response to CNP in mice seen in some studies was also suggested to be through increased PLB Ser16 phosphorylation and increased Ca^{2+} transients through PKG (Frantz et al. 2013; Wollert et al. 2003). Possible reasons for these opposing negative and positive inotropic responses to CNP will be discussed below. Whereas some studies have shown no effect on contractility by

ANP or BNP neither in normal nor heart failure animals (Afzal et al. 2011; Brusq et al. 1999; Moltzau et al. 2014b; Pierkes et al. 2002; Qvigstad et al. 2010), Zhang et al. (2005) showed negative inotropic responses to all three natriuretic peptides, in rabbit ventricular cardiomyocytes. Some studies show attenuation of natriuretic peptide-induced negative inotropic response in hypertrophic hearts (Moalem et al. 2006; Su et al. 2005b; Tan et al. 2010), whereas in a rat heart failure model, both the negative inotropic and positive lusitropic responses to CNP were increased compared to sham-operated rats (Moltzau et al. 2013). Further, one study showed no effect of natriuretic peptides on contractility in heart failure dogs as opposed to normal dogs (Lainchbury et al. 2000), whereas one study also showed an effect of ANP in heart failure (Ohte et al. 1999).

In addition to the proposed pathway through PKG affecting PLB and TnI phosphorylation and SERCA2 activity in normal (Brusq et al. 1999; Wollert et al. 2003) and failing hearts (Moltzau et al. 2013), several other pathways have been suggested to contribute to a possible effect of natriuretic peptides on contractility. ANP in atrial frog cardiomyocytes have been shown to decrease the L-type Ca^{2+} channel (LTCC) current ($I_{Ca,L}$) (Gisbert and Fischmeister 1988). A PKG-mediated effect on $I_{Ca,L}$ was shown with ANP in rabbit ventricular cardiomyocytes (Tohse et al. 1995) and BNP in rat left ventricle combined with a negative inotropic response (Sodi et al. 2008). Both BNP and CNP decreased $I_{Ca,L}$ in failing rat cardiomyocytes; however, only CNP caused negative inotropic response (Moltzau et al. 2014a). PKG was shown to inhibit both basal and cAMP-stimulated $I_{Ca,L}$ in ventricular cardiomyocytes from rat hearts (Mery et al. 1991; Sumii and Sperelakis 1995). In another study, ANP reduced contractility associated with the decrease in intracellular pH in normal rats. In the presence of 5-(N-ethyl-N-isopropyl)-amiloride, an inhibitor of the Na^+/H^+-exchanger (NHE, mainly NHE1 in the heart (Wakabayashi et al. 2013)), ANP did not cause effects on contractility nor pH, suggesting alteration in pH as the mechanism for ANP-induced negative inotropic response in these hearts (Tajima et al. 1998). Reduction in pH is known to reduce Ca^{2+} sensitivity of myofilaments (Fabiato and Fabiato 1978; Spitzer and Bridge 1992). In addition to causing negative inotropic responses through the PKG pathway, natriuretic peptides can influence cAMP signalling through cGMP-mediated effects on PDE2 and PDE3. This crosstalk between cGMP and cAMP and how it affects contractility will be further discussed below. Signalling pathways shown to be involved and possible pathways that can be involved in causing contractile effects of natriuretic peptides are illustrated in Fig. 12.1. Taken together, natriuretic peptides cause both negative and positive inotropic responses. The effects of natriuretic peptides on contractility are evidently complex. Whether the different results in a number of studies are due to specific changes in compartmentation, activation of different signalling pathways in different species, models of diseases or experimental setup is not clear. The possible mechanisms by which natriuretic peptides can cause both negative and positive inotropic responses will be discussed in more detail below.

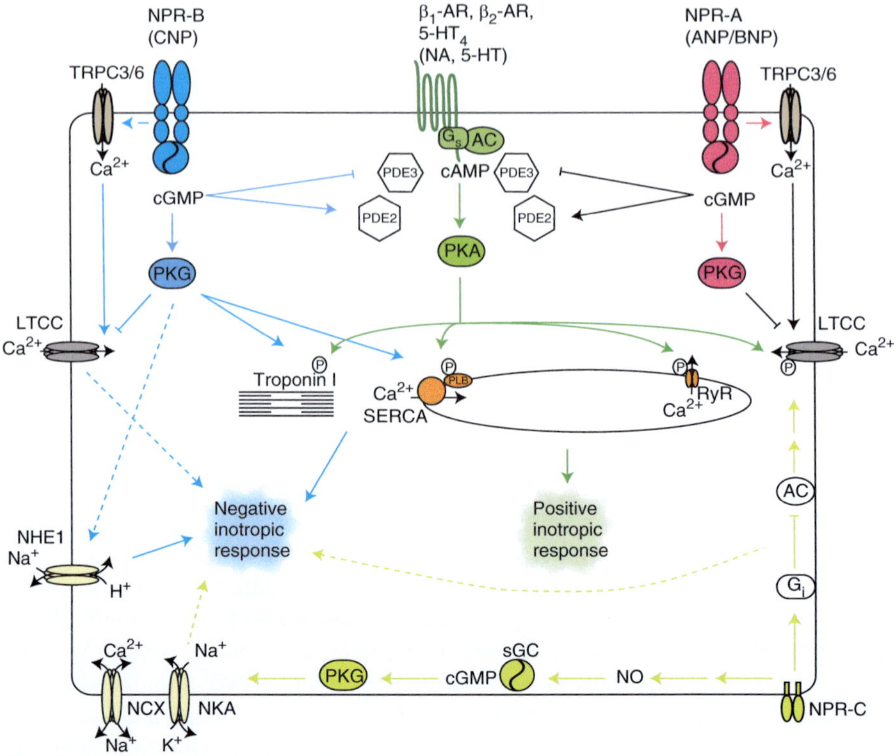

Fig. 12.1 Overview of natriuretic peptide signalling pathways implicated in the regulation of contractility in cardiomyocytes. Dashed lines are signalling pathways not yet elucidated. Abbreviations: *5-HT* serotonin, *LTCC* L-type calcium channel, *NA* noradrenaline, *NCX* Na^+/Ca^{2+}-exchanger, *NHE1* Na^+/H^+-exchanger, *NKA* Na^+/K^+-ATPase, *NO* nitric oxide, *PLB* phospholamban, *RyR* ryanodine receptor, *SERCA* sarco/endoplasmatic reticulum Ca^{2+}-ATPase, *TRPC3/6* transient receptor potential canonical channel 3/6. See text for details

12.3.2 Signalling of Natriuretic Peptides Causing Antihypertrophic Effects

It is well documented that natriuretic peptides have antihypertrophic effects in the heart (Calvieri et al. 2012). However, it is not clear which signalling pathways are the most important and if compartmentation is important for their effects. To maintain the focus of this review, only studies including the antihypertrophic effects of natriuretic peptides in cardiomyocytes will be discussed. Several studies show that stimulating NPR-A reduces hypertrophic responses and markers both alone and in the presence of agonists, stimulating receptors activating $G_{q/11}$, such as AngII (Hayashi et al. 2004; Laskowski et al. 2006; O'Tierney et al. 2010; Rosenkranz et al. 2003), phenylephrine (PE), endothelin (ET)-1 (Hayashi et al. 2004) and norepinephrine (NE) (Calderone et al. 1998). In contrast, G_s-stimulated hypertrophic responses do not seem to be affected by natriuretic peptides (Klaiber et al. 2010), indicating compartmentation of the signalling affecting G-protein-coupled receptor (GPCR) signalling. In agreement with this, an NPR-A KO model shows cardiac hypertrophy (Bubikat et al. 2005; Holtwick et al. 2003). Further, HS-142-1, an NPR-A/NPR-B antagonist, both reduces the antihypertrophic response to BNP (Chen et al. 2014) and causes hypertrophic responses in basal and PE-stimulated conditions in neonatal cardiomyocytes (Horio et al. 2000). Overexpression of NPR-A in mice seems to reduce cardiomyocyte size (Kishimoto et al. 2001). CU-NP, a chimeric peptide shown to stimulate both NPR-A and NPR-B, exhibited antihypertrophic effects (Kilic et al. 2010). The antihypertrophic response to natriuretic peptides does not seem to be unique for the NPR-A as also CNP was shown to be antihypertrophic (Izumiya et al. 2012; Rosenkranz et al. 2003; Tokudome et al. 2004). The similar antihypertrophic effects of ANP/BNP and CNP do not indicate different compartmentation of the antihypertrophic cGMP signal caused by NPR-A and NPR-B stimulation.

It appears that inhibition of the Ca^{2+}-calcineurin-NFAT pathway is responsible for the antihypertrophic effect of natriuretic peptides (Kilic et al. 2010; Tokudome et al. 2008). However, the link between natriuretic peptides and Ca^{2+} decrease has been discussed. Several studies show that the antihypertrophic effects of natriuretic peptides are cGMP dependent, and some also show that the mechanisms for their antihypertrophic effects are PKG dependent (Kinoshita et al. 2010; Klaiber et al. 2010; Laskowski et al. 2006; O'Tierney et al. 2010; Rosenkranz et al. 2003; Tokudome et al. 2008). Regulator of G-protein signalling (RGS)2 and RGS4 and Ca^{2+}-permeable transient receptor potential canonical (TRPC)3/6 channels are phosphorylated by PKG in close proximity of NPR-A and have been shown to be involved in the antihypertrophic actions of ANP (Kinoshita et al. 2010; Klaiber et al. 2011; Klaiber et al. 2010; Tokudome et al. 2008). Several additional mechanisms have been proposed, including decreased activation of Ca^{2+}/calmodulin-dependent kinase II (CaMKII) (Kilic et al. 2005; Tokudome et al. 2004), the Akt pathway (Kilic et al. 2005; O'Tierney et al. 2010), NHE1 (Kilic et al. 2010; Kilic et al. 2005), extracellular signal-regulated kinase (ERK) (Bubikat et al. 2005; Hayashi et al. 2004; O'Tierney et al. 2010; Tokudome et al. 2004), Smad pathway dependent on p38 mitogen-activated protein kinase (MAPK) (Chen et al. 2014), increased MAPK phosphatase 1 expression (Hayashi et al. 2004) and antioxidant effects (Laskowski et al. 2006). Whether all these data are results of the same mechanism through phosphorylation of RGS2/4 or whether natriuretic peptides can target different proteins in the same signalling pathway is not clear. An overview of the signalling pathways of natriuretic peptides regulating hypertrophic signalling is shown in Fig. 12.2. Even though there are not many studies addressing compartmentation of the cGMP signal after NPR-A and NPR-B in hypertrophy, there are indications that the antihypertrophic effect of ANP through NPR-A can be both cGMP dependent and cGMP independent. There is evidence that TRPC3/6 can be modified by ANP independently of cGMP by conformational changes in a complex between the NPR-A and TRPC3/6 (Klaiber et al. 2011). Cyclic GMP-independent effects of natriuretic peptides are further discussed below.

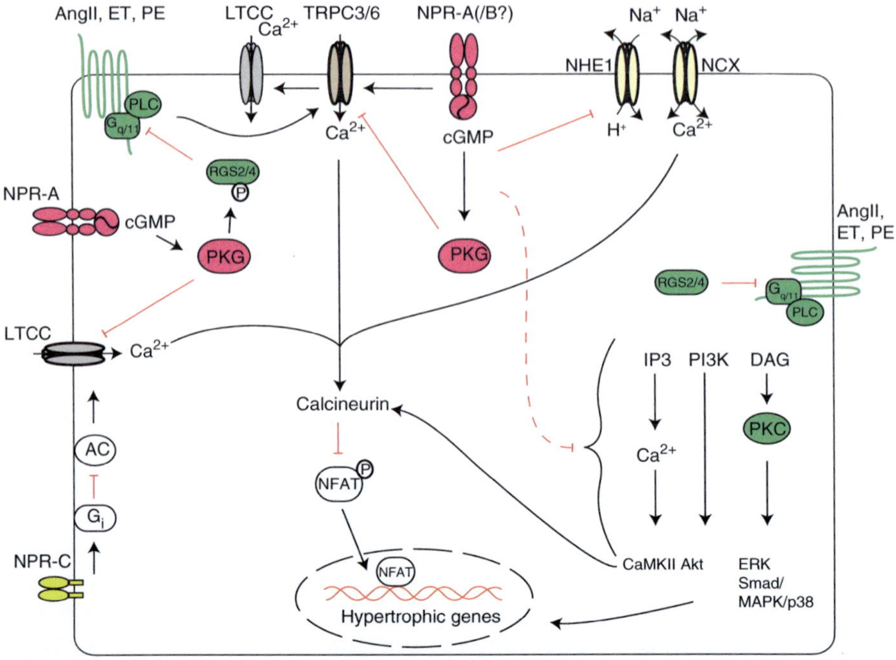

Fig. 12.2 Overview of natriuretic peptide signalling pathways implicated in the regulation of hypertrophic responses in cardiomyocytes. Dashed lines are signalling pathways not yet elucidated. NPR-A and possibly NPR-B reduce CaMKII, Akt, ERK and MAPK/p38 through Smad activity causing antihypertrophic effects; however, it is not known where in the signalling pathways cGMP interacts. Abbreviations: *Akt/PKB* protein kinase B, *AngII* angiotensin II, *CaMKII* Ca^{2+}/calmodulin-dependent protein kinase II, *ERK* extracellular signal-regulated kinase, *ET* endothelin, *DAG* diacylglycerol, *IP3* inositol trisphosphate, *LTCC* L-type calcium channel, *MAPK* mitogen-activated protein kinase, *NFAT* nuclear factor of activated Tcells, *NHE1* Na^+/H^+-exchanger, *NCX* Na^+/Ca^{2+}-exchanger, *RGS2/4* regulator of Gprotein signalling 2/4, *PE* phenylephrine, *PI3K* phosphoinositide 3-kinase, *PLC* phospholipase C, *TRPC3/6* transient receptor potential canonical channel 3/6. See text for details

12.3.3 Crosstalk Between Natriuretic Peptide-Induced cGMP and cAMP

The phosphodiesterases PDE2 and PDE3 break down cAMP and cGMP, but their activity can also be regulated by cGMP from natriuretic peptide stimulation. The interactions between PDEs, cGMP and cAMP could contribute to compartmentalise signalling from natriuretic peptides. Cyclic GMP competitively inhibits the cAMP hydrolytic activity of PDE3 due to the low V_{max} for hydrolysis of cGMP as compared to cAMP, whereas cGMP increases the activity of PDE2 by binding to one of its GAF (cGMP-binding PDEs, *Anabaena* adenylyl cyclase and *Escherichia coli* FhlA) domains (Omori and Kotera 2007). Most of the studies investigating PDE involvement in crosstalk between the cGMP and cAMP signalling pathways were performed either by activating the sGC pathway (Mery et al. 1993; Mongillo et al.

2006; Stangherlin et al. 2011; Wen et al. 2004) or by using cGMP (Fischmeister and Hartzell 1987; Hartzell and Fischmeister 1986; Ono and Trautwein 1991; Vandecasteele et al. 2001). However, in the recent years, more studies have also addressed natriuretic peptides. Wen et al. (2004) found that CNP increased both cGMP and cAMP efflux in rabbit atria. It was suggested that the increase in cAMP efflux was due to PDE3 inhibition by cGMP and a consequent increase in cAMP, as the effect was not seen in the presence of the PDE3 inhibitor milrinone. Supporting this, we and others later showed that CNP activating NPR-B, but not BNP activating NPR-A, increased cGMP causing inhibition of PDE3 and consequently increase of cAMP-mediated inotropic response in failing (Afzal et al. 2011; Qvigstad et al. 2010) and normal rat left ventricle (Meier et al. 2015) and normal left porcine atrium (Weninger et al. 2012). The involvement of PDE3 inhibition in the CNP-induced cAMP increase was confirmed in transgenic mouse cardiomyocytes expressing FRET-based cAMP sensors. In these mice, CNP increased Iso-induced cAMP in the absence but not in the presence of PDE3 inhibition (Götz et al. 2014). The lack of (or moderate) effect of BNP/ANP on cAMP-mediated signalling and inotropic response, despite similar global increase of cGMP as by CNP, indicates different compartmentation of the cGMP signal from the two receptors (NPR-A and NPR-B) (Afzal et al. 2011; Götz et al. 2014; Qvigstad et al. 2010).

In contrast to the lack of effect of BNP on cAMP-mediated inotropic response in our studies (Afzal et al. 2011; Moltzau et al. 2014a; Qvigstad et al. 2010), others have found that cGMP generated by NPR-A can inhibit PDE3 and activate PDE2. Stangherlin et al. (2011) showed in neonatal cardiomyocytes that ANP reduced Iso-induced cAMP through a PDE2-dependent pathway in a compartment presumed to be involved in regulation of contractility. Perera et al. (2015) showed that ANP caused an increased cardiomyocyte shortening in the presence of Iso (β_1- and β_2-adrenoceptor stimulation) in a mild transverse aortic constriction (TAC) model, most likely through cGMP inhibition of PDE3. This effect was not present in normal hearts. This was explained by a redistribution of PDE2 from the β_1-adrenoceptor compartment in normal to the β_2-adrenoceptor compartment in TAC myocytes. Furthermore, ANP induced cell shortening in the presence of β_2-adrenoceptor stimulation in normal, but not in TAC mice. As ANP increased Iso and not β_2-adrenoceptor-induced effects in diseased cells and hearts, a likely interpretation is that ANP augments β_1-adrenoceptor-stimulated contractility. However, direct evidence of this is still lacking. This was in line with a more dominant PDE2 activity in the β_2-adrenoceptor compartment in TAC (Perera et al. 2015). On the other hand, in failing rat hearts, NPR-A stimulation did not affect β_1-adrenoceptor or 5-HT_4-mediated signalling (Afzal et al. 2011; Qvigstad et al. 2010).

Natriuretic peptides via cGMP can decrease $I_{Ca,L}$ through two mechanisms, either directly via PKG (Sodi et al. 2008; Tohse et al. 1995) as discussed above or by activation of PDE2 as shown in frog ventricular (Fischmeister and Hartzell 1987) and human atrial cardiomyocytes (Vandecasteele et al. 2001). Although the cGMP-mediated $I_{Ca,L}$ reduction in rats seems to occur mainly through PKG activation (Mery et al. 1991), natriuretic peptides could also inhibit the $I_{Ca,L}$ by PDE2 activation, causing subsequent decrease in cAMP and reduced activation of LTCC by PKA. In line with this hypothesis, PDE2 inhibition with EHNA was found to reduce the sensitivity of the negative inotropic response to CNP (Moltzau et al. 2014b; Su et al. 2005a). This may be explained by increased cAMP and increased LTCC activation counteracting the PKG-mediated negative inotropic response due to reduced cGMP-mediated activation of PDE2 by CNP.

12.3.4 Regulation of cGMP Levels and cGMP-Dependent Effects by PDEs

Compared to the numerous studies of PDE regulation of cAMP signalling, relatively few studies have addressed the effects of PDEs on cGMP signalling. Castro et al. (2006) found PDE5 to regulate sGC- but not NPR-A-generated cGMP, whereas PDE2 regulated cGMP from both sources showing compartmentation between the sGC and the pGC pathways. Using a membrane-bound sensor derived from CNG channels, NPR-A-generated cGMP in rat cardiomyocytes was found to be mainly regulated by PDE2 (Castro et al. 2006). Supporting this observation, we showed that global cellular cGMP generated by both NPR-A and NPR-B was strongly regulated by PDE2 in failing rat cardiomyocytes (Moltzau et al. 2014a; Moltzau et al. 2014b). However, these findings seem to contradict a previous study where PDE2 inhibition increased ANP- but not CNP-induced cGMP in adult rat cardiomyocytes (Doyle et al. 2002). Further, Doyle et al. (2002) found PDE3 to be the main regulator of cGMP from NPR-B, whereas we found that PDE3 inhibition caused only a small global cGMP increase after both NPR-A and NPR-B stimulation in failing rat cardiomyocytes (Moltzau et al. 2014b). However, despite low global cGMP increase under PDE3 inhibition, cGMP might increase locally as a result of decreased degradation. Götz et al. (2014) later showed that PDE3 inhibition increased CNP-induced cGMP using a transgenic mouse model for visualisation of cGMP. We found that PDE3 inhibition with cilostamide increased the sensitivity of the negative inotropic response and positive lusitropic response to CNP (Moltzau et al. 2014b), which would be in agreement with a localised compartmented increase of cGMP activating the PKG-mediated pathway. In contrast to the effects seen with CNP, in a heart failure model, neither a negative inotropic response nor a lusitropic response to BNP was revealed in the presence of PDE2, PDE3 and PDE5 inhibition. Thus, regulation by these PDEs is not the main explanation why BNP and CNP signal in different compartments (Moltzau et al. 2014a).

As discussed above, a possible role of PDE5 in regulating sGC-generated cGMP was suggested (Castro et al. 2006; Takimoto et al. 2005). Cyclic GMP generated by the NO-sGC system is functionally separated from cGMP generated by ANP/NPR-A, as PDE5 inhibition and NO donors modulated Iso-induced contractility in mouse hearts, whereas ANP did not (Takimoto et al. 2007). Later, Fischmeister and co-workers further illustrated the compartmentation between the NO-sGC pathway and the NPR-A receptor by reporting that PKG activation caused a positive feedback mechanism on the NPR-A receptor, thus increasing its own activity, but a negative feedback mechanism on sGC-generated cGMP by activating PDE5 (Castro et al. 2010). The mechanism of the observed positive feedback mechanism is not yet fully established. Further support that cGMP is generated in different compartments by sGC and pGC was obtained by Afzal et al. (2011) in failing rat ventricle, where we found that activation of the NO pathway with SIN-1 caused a reduction, whereas CNP caused an increase in β_1-adrenoceptor-induced inotropic responses. This was presumably through activation of PKG- and cGMP-mediated inhibition of PDE3, respectively. However, to complicate this view, the functional separation between sGC and pGC seems to be dependent on the cGMP target affected. Both the NO pathway and CNP increased the inotropic response to 5-HT_4 serotonin receptor stimulation, in contrast to their divergent effects on the β_1-AR-induced inotropic response, even though both β_1-AR and 5-HT_4 receptors increase cAMP signalling (Afzal et al. 2011). PDE5 has mainly been linked to regulation of sGC-generated cGMP. However, Zhang et al. (2012) found a redistribution of PDE5 in a hypertrophic model and that PDE5 in hypertrophic cardiomyocytes can regulate NPR-A-induced cGMP. Recently, Lee et al. (2015) showed that PDE9A was upregulated in cardiomyocytes from a hypertrophic mouse model and in patients with dilated cardiomyopathy, HFpEF and aortic stenosis. Furthermore, PDE9A was important in regulating NPR-A-induced cGMP increase. PDE9A upregulation seemed to be a maladaptive mechanism in the diseased hearts (Lee et al. 2015), supporting the suggestion that cGMP generated by natriuretic peptides may be beneficial in failing hearts (Boerrigter et al. 2009).

12.3.5 cGMP-Independent Signalling of Natriuretic Peptides

In addition to activating cGMP-dependent pathways, NPR-A stimulation by ANP in mice has also been shown to activate a cGMP-independent pathway through TRPC3/6 and subsequent activation of the LTCC (Fig. 12.1. and Fig. 12.2.). Under normal conditions, both cGMP-dependent mechanisms through the PKG pathway reducing the Ca^{2+} current and cGMP-independent mechanisms through the TRPC3/6 pathway (increasing Ca^{2+}) seem to be active together, causing no net changes in Ca^{2+} in this model. However, when the PKG pathway was inhibited, an increased Ca^{2+} through TRPC3/6 was revealed (Klaiber et al. 2011). Further, the activation of TRPC3/6 seemed to be more active in a hypertrophic model (Klaiber et al. 2011).

ANP, BNP and CNP increase cGMP, and thus it is reasonable to interpret an increase of cGMP as an effect through NPR-A and NPR-B, respectively. However, NPR-C could also be involved as all three peptides can bind this receptor (Suga et al. 1992a). There are very few studies of potential signalling pathways activated by NPR-C in cardiomyocytes. In an early study, ANP was found to decrease adenylyl cyclase (AC) activity and cAMP in both atrial and ventricular neonatal rat cardiomyocytes (Anand-Srivastava and Cantin 1986), and this was in other cell types suggested to be caused by NPR-C-mediated signalling coupled to G_i (Anand-Srivastava et al. 1990; Murthy et al. 2000; Resink et al. 1988). In atrial bullfrog myocytes, CNP was shown to inhibit Iso-stimulated $I_{Ca,L}$, an effect suggested to be attributable to NPR-C signalling and decreased AC activity (Rose et al. 2003). Thus, natriuretic peptides can potentially cause inhibition of the LTCC through three different mechanisms: cGMP-dependent mechanisms through PDE2 or PKG activation, as discussed above, and a cGMP-independent mechanism through NPR-C. William et al. (2008) showed that ANP could stimulate the Na^+/K^+ ATPase through NPR-C in rabbit ventricular cardiomyocytes. This effect was abolished in the presence of inhibitors of the NO-sGC pathway; thus the effect of ANP through NPR-C was NO-dependent (William et al. 2008). Furthermore, it is shown that natriuretic peptides modulate the ATP-sensitive K^+ channels in cardiomyocytes. The mechanism for this seems to be complex as BNP and CNP decreased the open probability of the channels in line with the effect of a cGMP-analogue, whereas the NPR-C agonist cANF/cANP (4-23) increased the activity of the channel (Burley et al. 2014). Thus, the effect of natriuretic peptides may both increase and decrease the activity of the channel due to their ability to increase cGMP trough NPR-A and NPR-B or to activate the NPR-C.

12.4 Consequences of Compartmentation of Natriuretic Peptide Signalling

12.4.1 What Determines if Natriuretic Peptides Cause Negative or Positive Inotropic Responses?

It is challenging to interpret the effect of natriuretic peptides on contractility, as studies show both positive and negative inotropic responses of only one or all natriuretic peptides. There may be several reasons for these opposing effects. The fact that one hormone can activate (1) different cGMP-dependent signalling pathways through the same receptor, (2) both cGMP-dependent and cGMP-independent pathways through the same receptor or (3) different signalling pathways through stimulating several of the natriuretic peptide receptors illustrates the complexity of the signalling. This might explain why studies show opposing effects of natriuretic peptides and that under some conditions, one pathway is more active than another.

Cyclic AMP through PKA and cGMP through PKG affect several of the same targets. Both PKA and PKG phosphorylate PLB and TnI causing increased SERCA2 activity and Ca^{2+} desensitisation, respectively. Still, PKA activation is associated with a positive inotropic response, whereas PKG activation is associated with a negative inotropic response. One might speculate that the reason why cAMP-mediated receptor activation (i.e. β-adrenoceptors) gives a positive inotropic effect is PKA-mediated activation of other targets than those of PKG, mainly the LTCC, but also possibly the ryanodine receptor (RyR) (El-Armouche and Eschenhagen 2009) and phosphorylation of cardiac myosin-binding protein C (cMyBPC) (Colson et al. 2010). It is clear that if the cAMP-mediated pathway is activated, natriuretic peptides will increase this signalling through cGMP-mediated PDE3 inhibition (Afzal et al. 2011; Qvigstad et al. 2010; Weninger et al. 2012). Thus, if activators of cAMP signalling such as noradrenaline are not properly washed out or blocked during experimental conditions, natriuretic peptides could cause positive inotropic responses. In accordance with this, the negative inotropic response to natriuretic peptides has mostly been shown in isolated muscle strips or cardiomyocytes (Brusq et al. 1999; Moalem et al. 2006; Moltzau et al. 2013; Moltzau et al. 2014a; Nir et al. 2001; Sodi et al. 2008; Su et al. 2005a; Tajima et al. 1998; Tan et al. 2010; Vaxelaire et al. 1989; Zhang et al. 2005; Zhang et al. 2007), whereas the positive inotropic response is mostly shown in the whole heart or tissue (Beaulieu et al. 1997; Hirose et al. 1998; Lainchbury et al. 2000; Pierkes et al. 2002; Wollert et al. 2003), where endogenous cAMP stimulators are more likely to be present. Whether the PDE3-inhibitory effect of cGMP can regulate basal cAMP and contractility (in the absence of cAMP stimulation by GPCRs) is still not known.

It is well established that increased SERCA2 activity can give a lusitropic response through faster Ca^{2+} removal. Since β-adrenoceptor stimulation, which increases PLB phosphorylation resulting in increased SERCA2 activity, is known to give a positive inotropic response, it is easy to overlook the fact that activation of SERCA2 is in its nature a Ca^{2+} sequestrating and thus relaxing mechanism, which, when occurring in isolation, could potentially give a negative inotropic response in addition to the lusitropic response. Thus, in the absence of cAMP, natriuretic peptides activating PKG and subsequent PLB phosphorylation causing increased SERCA2 activity might cause negative inotropic response. CNP was shown to cause phosphorylation of TnI Ser23/24 in normal and failing rat ventricle (Brusq et al. 1999; Moltzau et al. 2013; Moltzau et al. 2014a; Moltzau et al. 2014b). However, could phosphorylation of TnI contribute to or even cause a negative inotropic or a positive lusitropic response alone? Decreased myofilament Ca^{2+} sensitivity is accompanied by increased off-rate of Ca^{2+} from TnC, which will contribute to faster reversal of the actin-myosin interaction and facilitate the increased Ca^{2+} sequestration caused by increased SERCA2 activity. In PLB-KO mice, with constantly elevated SERCA2 activity, CNP showed no negative inotropic response as opposed to control mice (Zhang et al. 2007). This indicates that the presence of constantly activated SERCA2 does not provide a sufficient condition for eliciting negative inotropic response by CNP-mediated TnI phosphorylation, thus indicating that SERCA2 should also be at an activity level which can be activated further (through removal of inhibition by PLB). It is also consistent with a hypothesis that TnI phosphorylation alone cannot cause the functional responses seen with CNP. We have proposed that phosphorylation of both PLB and TnI is needed for the negative inotropic response to natriuretic peptides observed in the absence of cAMP stimulators (Moltzau et al. 2013). Thus, if one of these targets cannot be phosphorylated, a negative inotropic response would not be detected. All natriuretic peptides have been shown to decrease $I_{Ca,L}$ (Gisbert and Fischmeister 1988; Moltzau et al. 2014a; Sodi et al. 2008; Tohse et al. 1995), which might contribute to a negative inotropic response. However, we observed that isolated decrease in $I_{Ca,L}$ was not sufficient for a negative inotropic response, as both BNP and CNP decreased $I_{Ca,L}$, but only CNP caused a negative inotropic response. In addition, CNP phosphorylated PLB and TnI, as opposed to BNP (Moltzau et al. 2014a). Thus, a combination of proteins phosphorylated seems to be essential for the effects. However, it is an interesting question whether TnI phosphorylation is really required for the negative inotropic response to CNP. An experimental approach to address this could be to test whether CNP would elicit a negative inotropic response in a TnI-KO or TnI-phosphorylation-deficient mouse, which to our knowledge has not been done. In summary, one or more of the pathways shown to have an effect on contractility (Fig. 12.1) might be more or less activated under different conditions (experimental setting, different species, animal models, etc.), causing different combinations of protein phosphorylation and thus the different inotropic responses: none, negative or positive.

12.4.2 Natriuretic Peptide Signalling—Beneficial or Not?

Natriuretic peptides are believed to be beneficial both in hypertrophy and heart failure (Nishikimi et al. 2006). However, natriuretic peptides activate several pathways both in healthy and diseased hearts, some believed to be beneficial and some detrimental. It might not be so easy to determine the net effect of simultaneously activating several signalling pathways. As previously mentioned CNP can increase cGMP causing inhibition of PDE3 and further enhance cAMP-mediated signalling. PDE3 seems to be the main phosphodiesterase controlling the inotropic response in the human ventricle (Afzal et al. 2008; Molenaar et al. 2013). Since inhibition of PDE3 has been proven to be detrimental in heart failure (Amsallem et al. 2005), the cGMP-mediated inhibition of PDE3 might also be detrimental. If so, a selective NPR-B antagonist could potentially be of use in heart failure (Bach et al. 2014). At the same time, CNP can activate the PKG pathway (Moltzau et al. 2013), believed to be beneficial in heart failure due to negative inotropic and antihypertrophic effects (Boerrigter et al. 2009). These pathways might simultaneously be active. However, the relative contribution of the two opposing pathways seems to be different in the absence and presence of GPCR-stimulated cAMP signalling. In the absence of cAMP signalling, the activation of the PKG pathway seems to be the dominant signalling pathway of CNP as the overall effect is a negative inotropic response (Moltzau et al. 2013; Moltzau et al. 2014a; Moltzau et al. 2014b). However, in the presence of cAMP signalling, the NPR-B-cGMP-PDE3 pathway seems to mediate the main response as the inhibitory component of CNP, presumably through PKG, does not overrun the sensitisation of β_1-AR- and 5-HT_4-induced inotropic response by CNP (Afzal et al. 2011; Qvigstad et al. 2010; Weninger et al. 2012). In a model of right ventricular hypertrophy, the PDE5 inhibitor sildenafil was shown to increase cGMP and seemed to preferentially increase contractility by a PDE3-dependent mechanism over a PKG-mediated mechanism. Furthermore, PKG1 activity was reduced in the hypertrophic right ventricle (Nagendran et al. 2007). The outcome of cGMP signalling in human heart failure patients would probably depend on the balance between PDE3 inhibition and PKG activation, which has not been sufficiently established.

The NPR-A agonist ANP has also been shown to activate pathways that will counteract each other. ANP increases cGMP-dependent activation of the PKG pathway and causes beneficial antihypertrophic effects by decreasing Ca^{2+} influx through the inactivation of TRPC3/6 channels. However, at the same time, ANP is suggested to activate the same channels by a direct interaction between TRPC3/6 and NPR-A. This cGMP-independent mechanism seems to be more pronounced in the hypertrophic hearts (Klaiber et al. 2011). As previously discussed, natriuretic peptides might also simultaneously activate both the NPR-A or NPR-B receptor and the NPR-C and cause opposing effects (Burley et al. 2014).

12.4.3 Compartmentation of the cGMP Signal Generated by NPR-A and NPR-B

The striking difference between the effects of NPR-A and NPR-B stimulation in some studies is intriguing. Several studies show that NPR-B can mediate effects on contractility both alone and in the presence of cAMP signalling, whereas NPR-A does not, despite both receptors generating cGMP (Afzal et al. 2011; Brusq et al. 1999; Moltzau et al. 2014a; Pierkes et al. 2002; Qvigstad et al. 2010). A more mechanistic study showed that CNP causes phosphorylation of PLB and TnI and increased SERCA2 activity leading to increased Ca^{2+}-transients, whereas BNP does not (Moltzau et al. 2014a). This discrepancy in compartmentation does not seem to be present for the antihypertrophic actions of natriuretic peptides as all three natriuretic peptides show antihypertrophic effects (Izumiya et al. 2012; Rosenkranz et al. 2003; Tokudome et al. 2004). An explanation for the different effects seen with BNP and CNP on contractility might be different localisation of the receptors or different binding proteins linked to the receptors causing different compartmentation of the signal. There are several studies showing the role of A-kinase anchoring proteins (AKAPs) in the cAMP signalling pathway, where AKAPs are responsible for localisation of receptors, effector proteins and PDEs to distinct compartments in the cell (Carnegie et al. 2009). However, less is known about potential G kinase anchoring proteins (GKAPs), anchoring proteins involved in targeting cGMP signalling to different localisations within the cells (Corradini et al. 2015; Francis et al. 2010). There are several known cGMP-dependent protein kinase (G kinase) interacting proteins (GKIPs) that might influence the localisation of PKG to different regions of the cell. However, GKIPs seem to be different from the AKAPs. Often, translocation and increased cGMP are needed for the association of a GKIP to PKG (Francis et al. 2010). In HEK293 cells, PKG translocated to the plasma membrane upon ANP-stimulation of the NPR-A receptor, and PKG activation increased the GC activity of the receptor (Airhart et al. 2003). This positive feedback mechanism of NPR-A and PKG was later also found in cardiomyocytes (Castro et al. 2010). The role of binding proteins, GKIPs and GKAPs in different effects and compartmentation of natriuretic peptides is largely unexplored.

12.5 Visualising cGMP After NPR Stimulation in Cardiomyocytes

As discussed above, in many studies NPR-A and NPR-B receptors both increase cGMP in cardiomyocytes, whereas the effect of this cGMP is rather different (Afzal et al. 2011; Brusq et al. 1999; Moltzau et al. 2014a; Pierkes et al. 2002; Qvigstad et al. 2010). This suggests that cGMP is increased in different subcellular compartments following NPR-A and NPR-B stimulation. Traditional biochemical techniques measuring global cellular levels of cGMP cannot discriminate between these subcellular microdomains. Early experiments, measuring subsarcolemmal cGMP through monitoring the current of a cyclic nucleotide-gated ion channel in adult cardiomyocytes discriminated between cGMP produced by sGC and NPR-A (Castro et al. 2006, 2010). In the field of cAMP, local concentrations of cAMP have been monitored with fluorescence (or Förster) resonance energy transfer (FRET)-based sensors. These sensors have been localised to various subcellular compartments. An important consideration to bear in mind when using these localised sensors is how the localisation is achieved. One option is to tag the sensor construct with a protein, presuming that the fused construct will localise similarly to the native protein. Another approach is to dock the sensor to a desired target protein by fusing the sensor to a protein interaction domain known to interact with the protein of interest. A third method, useful to target the sensor to a particular organelle, is to fuse the sensor to a localisation signal, e.g. a nuclear localisation signal (NLS). Depending on the method, the success and specificity of the localisation will depend on several factors, e.g. expression level of the sensor fusion proteins, the capacity of the desired protein interactions, the specificity of the signalling sequence, etc. Localised FRET-based cAMP sensors have been made for proteins such as regulatory subunits of PKA types I and II, plasma membrane, SERCA and mitochondria (Di Benedetto et al. 2008; Lefkimmiatis et al. 2013; Perera et al. 2015; Sprenger et al. 2015).

For cGMP, several FRET-based sensors have been developed with varying affinity for cGMP using different binding domains for cGMP, such as PKGI and the GAF domains of PDE2 or PDE5 (Honda et al. 2001; Niino et al. 2009; Nikolaev et al. 2006; Russwurm et al. 2007). Of these, the Cygnet sensor (PKGI) was fused to RI and RII regulatory subunits of PKA where ANP-stimulation increased cGMP more in the RII compartment in neonatal cardiomyocytes, and this cGMP increased PDE2 activity towards β-adrenoceptor-stimulated cAMP (Stangherlin et al. 2011). In adult cardiomyocytes, only the FRET-based sensor Red cGES-DE5 (GAF domain of PDE5) has high enough affinity (40 nM; Niino et al. 2009) to monitor local concentrations of cGMP (Götz et al. 2014). In transgenic mice expressing this sensor, CNP-stimulation increased cGMP levels significantly more than ANP-stimulation, and the contribution of different PDEs to this response was monitored (Götz et al. 2014). The localisation of this sensor to various subcellular compartments would be of interest to determine the local concentrations of cGMP following NPR-A and NPR-B stimulation and the importance of local PDEs to control this cGMP.

Conclusion

Several levels of functional evidence indicate different effects of NPR-A and NPR-B stimulation in the same cell, consistent with compartmented intracellular signalling. A better understanding of the basis for these differential effects and the compartmented signalling is on the way, using modern molecular methods including FRET-based sensors to study intracellular localisation of the second messenger cGMP and its downstream effects such as PKG activation. Given the recent interest in the role of natriuretic peptides in treatment of heart failure and their diverse effects in cardiovascular physiology and pathophysiology, there is reason to look forward to the future development in this area of research.

Compliance with Ethical Standards

Conflict of Interest Statement The authors declare that they have no conflict of interest.

References

Aass H, Skomedal T, Osnes JB (1988) Increase of cyclic AMP in subcellular fractions of rat heart muscle after beta-adrenergic stimulation: prenalterol and isoprenaline caused different distribution of bound cyclic AMP. J Mol Cell Cardiol 20:847–860

Abraham WT, Lowes BD, Ferguson DA, Odom J, Kim JK, Robertson AD, Bristow MR, Schrier RW (1998) Systemic hemodynamic, neurohormonal, and renal effects of a steady-state infusion of human brain natriuretic peptide in patients with hemodynamically decompensated heart failure. J Card Fail 4:37–44

Afzal F, Andressen KW, Mørk HK, Aronsen JM, Sjaastad I, Dahl CP, Skomedal T, Levy FO, Osnes JB, Qvigstad E (2008) 5-HT_4-elicited positive inotropic response is mediated by cAMP and regulated by PDE3 in failing rat and human cardiac ventricles. Br J Pharmacol 155:1005–1014. doi:10.1038/bjp.2008.339[pii]

Afzal F, Qvigstad E, Aronsen JM, Moltzau LR, Sjaastad I, Skomedal T, Osnes JB, Levy FO (2011) Agents increasing cyclic GMP amplify 5-HT_4-elicited positive inotropic response in failing rat cardiac ventricle. Naunyn Schmiedebergs Arch Pharmacol 384:543–553. doi:10.1007/s00210-011-0670-6 [doi]

Airhart N, Yang YF, Roberts CT Jr, Silberbach M (2003) Atrial natriuretic peptide induces natriuretic peptide receptor-cGMP-dependent protein kinase interaction. J Biol Chem 278:38693–38698. doi:10.1074/jbc.M304098200 [doi]; M304098200 [pii]

Amsallem E, Kasparian C, Haddour G, Boissel JP, Nony P (2005) Phosphodiesterase III inhibitors for heart failure. Cochrane Database Syst Rev CD002230. doi:10.1002/14651858.CD002230.pub2 [doi]

Anand-Srivastava MB, Cantin M (1986) Atrial natriuretic factor receptors are negatively coupled to adenylate cyclase in cultured atrial and ventricular cardiocytes. Biochem Biophys Res Commun 138:427–436. doi:0006-291X(86)90299-8 [pii]

Anand-Srivastava MB, Sairam MR, Cantin M (1990) Ring-deleted analogs of atrial natriuretic factor inhibit adenylate cyclase/cAMP system. Possible coupling of clearance atrial natriuretic factor receptors to adenylate cyclase/cAMP signal transduction system. J Biol Chem 265:8566–8572

Ashman DF, Lipton R, Melicow MM, Price TD (1963) Isolation of adenosine 3', 5'-monophosphate and guanosine 3', 5'-monophosphate from rat urine. Biochem Biophys Res Commun 11:330–334

Bach T, Bergholtz S, Riise J, Qvigstad E, Skomedal T, Osnes JB, Levy FO (2014) Identification of small molecule NPR-B antagonists by high throughput screening—potential use in heart failure. Naunyn Schmiedebergs Arch Pharmacol 387:5–14. doi:10.1007/s00210-013-0940-6 [doi]

Beaulieu P, Cardinal R, Page P, Francoeur F, Tremblay J, Lambert C (1997) Positive chronotropic and inotropic effects of C-type natriuretic peptide in dogs. Am J Phys 273:H1933–H1940

Boerrigter G, Lapp H, Burnett JC (2009) Modulation of cGMP in heart failure: a new therapeutic paradigm. Handb Exp Pharmacol 191:485–506. doi:10.1007/978-3-540-68964-5_21 [doi]

Brusq JM, Mayoux E, Guigui L, Kirilovsky J (1999) Effects of C-type natriuretic peptide on rat cardiac contractility. Br J Pharmacol 128:206–212. doi:10.1038/sj.bjp.0702766 [doi]

Bubikat A, De Windt LJ, Zetsche B, Fabritz L, Sickler H, Eckardt D, Godecke A, Baba HA, Kuhn M (2005) Local atrial natriuretic peptide signaling prevents hypertensive cardiac hypertrophy in endothelial nitric-oxide synthase-deficient mice. J Biol Chem 280:21594–21599. doi:10.1074/jbc.M501103200

Burley DS, Cox CD, Zhang J, Wann KT, Baxter GF (2014) Natriuretic peptides modulate ATP-sensitive K^+ channels in rat ventricular cardiomyocytes. Basic Res Cardiol 109:402. doi:10.1007/s00395-014-0402-4

Buxton IL, Brunton LL (1983) Compartments of cyclic AMP and protein kinase in mammalian cardiomyocytes. J Biol Chem 258:10233–10239

Cabiati M, Campan M, Caselli C, Prescimone T, Giannessi D, Del Ry S (2010) Sequencing and cardiac expression of natriuretic peptide receptors A and C in normal and heart failure pigs. Regul Pept 162:12–17. doi:10.1016/j.regpep.2010.02.004

Calderone A, Thaik CM, Takahashi N, Chang DL, Colucci WS (1998) Nitric oxide, atrial natriuretic peptide, and cyclic GMP inhibit the growth-promoting effects of norepinephrine in cardiac myocytes and fibroblasts. J Clin Invest 101:812–818. doi:10.1172/jci119883

Calvieri C, Rubattu S, Volpe M (2012) Molecular mechanisms underlying cardiac antihypertrophic and antifibrotic effects of natriuretic peptides. J Mol Med (Berl) 90:5–13. doi:10.1007/s00109-011-0801-z [doi]

Carnegie GK, Means CK, Scott JD (2009) A-kinase anchoring proteins: from protein complexes to physiology and disease. IUBMB Life 61:394–406. doi:10.1002/iub.168 [doi]

Castro LR, Schittl J, Fischmeister R (2010) Feedback control through cGMP-dependent protein kinase contributes to differential regulation and compartmentation of cGMP in rat cardiac myocytes. Circ Res 107:1232–1240. doi:CIRCRESAHA.110.226712 [pii]; 10.1161/CIRCRESAHA.110.226712 [doi]

Castro LR, Verde I, Cooper DM, Fischmeister R (2006) Cyclic guanosine monophosphate compartmentation in rat cardiac myocytes. Circulation 113:2221–2228. doi:CIRCULATIONAHA.105.599241 [pii]; 10.1161/CIRCULATIONAHA.105.599241 [doi]

Chen Y, Yao F, Chen S, Huang H, Wu L, He J, Dong Y (2014) Endogenous BNP attenuates cardiomyocyte hypertrophy induced by Ang II via p38 MAPK/Smad signaling. Die Pharmazie 69:833–837

Colson BA, Locher MR, Bekyarova T, Patel JR, Fitzsimons DP, Irving TC, Moss RL (2010) Differential roles of regulatory light chain and myosin binding protein-C phosphorylations in the modulation of cardiac force development. J Physiol 588:981–993. doi:jphysiol.2009.183897 [pii]; 10.1113/jphysiol.2009.183897 [doi]

Colucci WS, Elkayam U, Horton DP, Abraham WT, Bourge RC, Johnson AD, Wagoner LE, Givertz MM, Liang CS, Neibaur M, Haught WH, LeJemtel TH, Nesiritide Study Group. (2000) Intravenous nesiritide, a natriuretic peptide, in the treatment of decompensated congestive heart failure. N Engl J Med 343:246–253. doi:10.1056/NEJM200007273430403 [doi]

Corradini E, Burgers PP, Plank M, Heck AJ, Scholten A (2015) Huntingtin-associated protein 1 (HAP1) is a cGMP-dependent kinase anchoring protein (GKAP) specific for the cGMP-dependent protein kinase Ibeta isoform. J Biol Chem 290:7887–7896. doi:10.1074/jbc.M114.622613

de Bold AJ, Borenstein HB, Veress AT, Sonnenberg H (1981) A rapid and potent natriuretic response to intravenous injection of atrial myocardial extract in rats. Life Sci 28:89–94

Del Ry S, Cabiati M, Vozzi F, Battolla B, Caselli C, Forini F, Segnani C, Prescimone T, Giannessi D, Mattii L (2011) Expression of C-type natriuretic peptide and its receptor NPR-B in cardiomyocytes. Peptides 32:1713–1718. doi:S0196-9781(11)00239-7 [pii]; 10.1016/j.peptides.2011.06.014 [doi]

Del Ry S, Passino C, Maltinti M, Emdin M, Giannessi D (2005) C-type natriuretic peptide plasma levels increase in patients with chronic heart failure as a function of clinical severity. Eur J Heart Fail 7:1145–1148. doi:S1388-9842(04)00378-2 [pii]; 10.1016/j.ejheart.2004.12.009 [doi]

Di Benedetto G, Zoccarato A, Lissandron V, Terrin A, Li X, Houslay MD, Baillie GS, Zaccolo M (2008) Protein kinase A type I and type II define distinct intracellular signaling compartments. Circ Res 103:836–844. doi:CIRCRESAHA.108.174813 [pii]; 10.1161/CIRCRESAHA.108.174813 [doi]

Dickey DM, Burnett JC Jr, Potter LR (2008) Novel bifunctional natriuretic peptides as potential therapeutics. J Biol Chem 283:35003–35009. doi:M804538200 [pii]; 10.1074/jbc.M804538200 [doi]

Dickey DM, Dries DL, Margulies KB, Potter LR (2011) Guanylyl cyclase (GC)-A and GC-B activities in ventricles and cardiomyocytes from failed and non-failed human hearts: GC-A is inactive in the failed cardiomyocyte. J Mol Cell Cardiol. doi:S0022-2828(11)00473-1 [pii]; 10.1016/j.yjmcc.2011.11.007 [doi]

Dickey DM, Flora DR, Bryan PM, Xu X, Chen Y, Potter LR (2007) Differential regulation of membrane guanylyl cyclases in congestive heart failure: natriuretic peptide receptor (NPR)-B, Not NPR-A, is the predominant natriuretic peptide receptor in the failing heart. Endocrinology 148:3518–3522. doi:en.2007-0081 [pii]; 10.1210/en.2007-0081 [doi]

Dickstein K, Cohen-Solal A, Filippatos G, McMurray JJ, Ponikowski P, Poole-Wilson PA, Stromberg A, van Veldhuisen DJ, Atar D, Hoes AW, Keren A, Mebazaa A, Nieminen M, Priori SG, Swedberg K (2008) ESC Guidelines for the diagnosis and treatment of acute and chronic heart failure 2008: the Task Force for the Diagnosis and Treatment of Acute and Chronic Heart Failure 2008 of the European Society of Cardiology. Developed in collaboration with the Heart Failure Association of the ESC (HFA) and endorsed by the European Society of Intensive Care Medicine (ESICM). Eur Heart J 29:2388–2442.

Doyle DD, Upshaw-Earley J, Bell EL, Palfrey HC (2002) Natriuretic peptide receptor-B in adult rat ventricle is predominantly confined to the nonmyocyte population. Am J Physiol Heart Circ Physiol 282:H2117–H2123. doi:10.1152/ajpheart.00988.2001 [doi]

El-Armouche A, Eschenhagen T (2009) Beta-adrenergic stimulation and myocardial function in the failing heart. Heart Fail Rev 14:225–241. doi:10.1007/s10741-008-9132-8 [doi]

Fabiato A, Fabiato F (1978) Effects of pH on the myofilaments and the sarcoplasmic reticulum of skinned cells from cardiac and skeletal muscles. J Physiol 276:233–255

Fischmeister R, Hartzell HC (1987) Cyclic guanosine 3',5'-monophosphate regulates the calcium current in single cells from frog ventricle. J Physiol 387:453–472

Francis SH, Busch JL, Corbin JD, Sibley D (2010) cGMP-dependent protein kinases and cGMP phosphodiesterases in nitric oxide and cGMP action. Pharmacol Rev 62:525–563. doi:62/3/525 [pii]; 10.1124/pr.110.002907 [doi]

Frantz S, Klaiber M, Baba HA, Oberwinkler H, Volker K, Gabetaner B, Bayer B, Abebetaer M, Schuh K, Feil R, Hofmann F, Kuhn M (2013) Stress-dependent dilated cardiomyopathy in mice with cardiomyocyte-restricted inactivation of cyclic GMP-dependent protein kinase I. Eur Heart J 34:1233–1244. doi:ehr445 [pii]; 10.1093/eurheartj/ehr445 [doi]

Gisbert MP, Fischmeister R (1988) Atrial natriuretic factor regulates the calcium current in frog isolated cardiac cells. Circ Res 62:660–667

Götz KR, Sprenger JU, Perera RK, Steinbrecher JH, Lehnart SE, Kuhn M, Gorelik J, Balligand JL, Nikolaev VO (2014) Transgenic mice for real-time visualization of cGMP in intact adult cardiomyocytes. Circ Res 114:1235–1245. doi:10.1161/circresaha.114.302437

Hartzell HC, Fischmeister R (1986) Opposite effects of cyclic GMP and cyclic AMP on Ca^{2+} current in single heart cells. Nature 323:273–275. doi:10.1038/323273a0 [doi]

Hayashi D, Kudoh S, Shiojima I, Zou Y, Harada K, Shimoyama M, Imai Y, Monzen K, Yamazaki T, Yazaki Y, Nagai R, Komuro I (2004) Atrial natriuretic peptide inhibits cardiomyocyte hypertrophy through mitogen-activated protein kinase phosphatase-1. Biochem Biophys Res Commun 322:310–319. doi:10.1016/j.bbrc.2004.07.119

Hayes JS, Brunton LL, Brown JH, Reese JB, Mayer SE (1979) Hormonally specific expression of cardiac protein kinase activity. Proc Natl Acad Sci U S A 76:1570–1574

Hirose M, Furukawa Y, Kurogouchi F, Nakajima K, Miyashita Y, Chiba S (1998) C-type natriuretic peptide increases myocardial contractility and sinus rate mediated by guanylyl cyclase-linked natriuretic peptide receptors in isolated, blood-perfused dog heart preparations. J Pharmacol Exp Ther 286:70–76

Holmes SJ, Espiner EA, Richards AM, Yandle TG, Frampton C (1993) Renal, endocrine, and hemodynamic effects of human brain natriuretic peptide in normal man. J Clin Endocrinol Metab 76:91–96

Holtwick R, van Eickels M, Skryabin BV, Baba HA, Bubikat A, Begrow F, Schneider MD, Garbers DL, Kuhn M (2003) Pressure-independent cardiac hypertrophy in mice with cardiomyocyte-restricted inactivation of the atrial natriuretic peptide receptor guanylyl cyclase-A. J Clin Invest 111:1399–1407. doi:10.1172/jci17061

Honda A, Adams SR, Sawyer CL, Lev-Ram V, Tsien RY, Dostmann WR (2001) Spatiotemporal dynamics of guanosine 3',5'-cyclic monophosphate revealed by a genetically encoded, fluorescent indicator. Proc Natl Acad Sci U S A 98:2437–2442. doi:10.1073/pnas.051631298

Horio T, Nishikimi T, Yoshihara F, Matsuo H, Takishita S, Kangawa K (2000) Inhibitory regulation of hypertrophy by endogenous atrial natriuretic peptide in cultured cardiac myocytes. Hypertension 35:19–24

Horio T, Tokudome T, Maki T, Yoshihara F, Suga S, Nishikimi T, Kojima M, Kawano Y, Kangawa K (2003) Gene expression, secretion, and autocrine action of C-type natriuretic peptide in cultured adult rat cardiac fibroblasts. Endocrinology 144:2279–2284

Hunt PJ, Richards AM, Espiner EA, Nicholls MG, Yandle TG (1994) Bioactivity and metabolism of C-type natriuretic peptide in normal man. J Clin Endocrinol Metab 78:1428–1435

Izumiya Y, Araki S, Usuku H, Rokutanda T, Hanatani S, Ogawa H (2012) Chronic C-Type natriuretic peptide infusion attenuates angiotensin II-Induced myocardial superoxide production and cardiac Remodeling. Int J Vasc Med 2012:246058. doi:10.1155/2012/246058

Kaumann AJ, Birnbaumer L (1974) Prostaglandin E1 action on sinus pacemaker and adenylyl cyclase in kitten myocardium. Nature 251:515–517

Kilic A, Rajapurohitam V, Sandberg SM, Zeidan A, Hunter JC, Said Faruq N, Lee CY, Burnett JC Jr, Karmazyn M (2010) A novel chimeric natriuretic peptide reduces cardiomyocyte hypertrophy through the NHE-1-calcineurin pathway. Cardiovasc Res 88:434–442. doi:10.1093/cvr/cvq254

Kilic A, Velic A, De Windt LJ, Fabritz L, Voss M, Mitko D, Zwiener M, Baba HA, van Eickels M, Schlatter E, Kuhn M (2005) Enhanced activity of the myocardial Na^+/H^+ exchanger NHE-1 contributes to cardiac remodeling in atrial natriuretic peptide receptor-deficient mice. Circulation 112:2307–2317. doi:10.1161/circulationaha.105.542209

Kinoshita H, Kuwahara K, Nishida M, Jian Z, Rong X, Kiyonaka S, Kuwabara Y, Kurose H, Inoue R, Mori Y, Li Y, Nakagawa Y, Usami S, Fujiwara M, Yamada Y, Minami T, Ueshima K, Nakao K (2010) Inhibition of TRPC6 channel activity contributes to the antihypertrophic effects of natriuretic peptides-guanylyl cyclase-A signaling in the heart. Circ Res 106:1849–1860. doi:10.1161/circresaha.109.208314

Kishimoto I, Rossi K, Garbers DL (2001) A genetic model provides evidence that the receptor for atrial natriuretic peptide (guanylyl cyclase-A) inhibits cardiac ventricular myocyte hypertrophy. Proc Natl Acad Sci U S A 98:2703–2706. doi:10.1073/pnas.051625598

Klaiber M, Dankworth B, Kruse M, Hartmann M, Nikolaev VO, Yang RB, Volker K, Gassner B, Oberwinkler H, Feil R, Freichel M, Groschner K, Skryabin BV, Frantz S, Birnbaumer L, Pongs O, Kuhn M (2011) A cardiac pathway of cyclic GMP-independent signaling of guanylyl cyclase A, the receptor for atrial natriuretic peptide. Proc Natl Acad Sci U S A 108:18500–18505. doi:1103300108 [pii]; 10.1073/pnas.1103300108 [doi]

Klaiber M, Kruse M, Volker K, Schroter J, Feil R, Freichel M, Gerling A, Feil S, Dietrich A, Londono JE, Baba HA, Abramowitz J, Birnbaumer L, Penninger JM, Pongs O, Kuhn M (2010) Novel insights into the mechanisms mediating the local antihypertrophic effects of cardiac atrial natriuretic peptide: role of cGMP-dependent protein kinase and RGS2. Basic Res Cardiol 105:583–595. doi:10.1007/s00395-010-0098-z

Koller KJ, Lowe DG, Bennett GL, Minamino N, Kangawa K, Matsuo H, Goeddel DV (1991) Selective activation of the B natriuretic peptide receptor by C-type natriuretic peptide (CNP). Science 252:120–123

Kuhn M (2003) Structure, regulation, and function of mammalian membrane guanylyl cyclase receptors, with a focus on guanylyl cyclase-A. Circ Res 93:700–709. doi:10.1161/01.RES.0000094745.28948.4D [doi];93/8/700 [pii]

Lainchbury JG, Burnett JC Jr, Meyer D, Redfield MM (2000) Effects of natriuretic peptides on load and myocardial function in normal and heart failure dogs. Am J Physiol Heart Circ Physiol 278:H33–H40

Laskowski A, Woodman OL, Cao AH, Drummond GR, Marshall T, Kaye DM, Ritchie RH (2006) Antioxidant actions contribute to the antihypertrophic effects of atrial natriuretic peptide in neonatal rat cardiomyocytes. Cardiovasc Res 72:112–123. doi:10.1016/j.cardiores.2006.07.006

Lee CY, Chen HH, Lisy O, Swan S, Cannon C, Lieu HD, Burnett JC Jr. (2009) Pharmacodynamics of a novel designer natriuretic peptide, CD-NP, in a first-in-human clinical trial in healthy subjects. J Clin Pharmacol 49:668–673. doi:0091270009336233 [pii]; 10.1177/0091270009336233 [doi]

Lee DI, Zhu G, Sasaki T, Cho GS, Hamdani N, Holewinski R, Jo SH, Danner T, Zhang M, Rainer PP, Bedja D, Kirk JA, Ranek MJ, Dostmann WR, Kwon C, Margulies KB, Van Eyk JE, Paulus WJ, Takimoto E, Kass DA (2015) Phosphodiesterase 9A controls nitric-oxide-independent cGMP and hypertrophic heart disease. Nature 519:472–476. doi:10.1038/nature14332

Lefkimmiatis K, Leronni D, Hofer AM (2013) The inner and outer compartments of mitochondria are sites of distinct cAMP/PKA signaling dynamics. J Cell Biol 202:453–462. doi:10.1083/jcb.201303159

Leitman DC, Andresen JW, Kuno T, Kamisaki Y, Chang JK, Murad F (1986) Identification of multiple binding sites for atrial natriuretic factor by affinity cross-linking in cultured endothelial cells. J Biol Chem 261:11650–11655

Lin X, Hanze J, Heese F, Sodmann R, Lang RE (1995) Gene expression of natriuretic peptide receptors in myocardial cells. Circ Res 77:750–758

Marcus LS, Hart D, Packer M, Yushak M, Medina N, Danziger RS, Heitjan DF, Katz SD (1996) Hemodynamic and renal excretory effects of human brain natriuretic peptide infusion in patients with congestive heart failure. A double-blind, placebo-controlled, randomized crossover trial. Circulation 94:3184–3189

McMurray JJ, Packer M, Desai AS, Gong J, Lefkowitz MP, Rizkala AR, Rouleau JL, Shi VC, Solomon SD, Swedberg K, Zile MR, Investigators P-H, Committees (2014) Angiotensin-neprilysin inhibition versus enalapril in heart failure. N Engl J Med 371:993–1004. doi:10.1056/NEJMoa1409077

Meier S, Andressen KW, Aronsen JM, Sjaastad I, Skomedal T, Osnes J-B, Qvigstad E, Levy FO, Moltzau LR (2015) Enhancement of cAMP-mediated inotropic responses by CNP is regulated differently by PDE2 in normal and failing hearts. BMC Pharmacol Toxicol 16:A66. doi:10.1186/2050-6511-16-s1-a66

Mery PF, Lohmann SM, Walter U, Fischmeister R (1991) Ca^{2+} current is regulated by cyclic GMP-dependent protein kinase in mammalian cardiac myocytes. Proc Natl Acad Sci U S A 88:1197–1201

Mery PF, Pavoine C, Belhassen L, Pecker F, Fischmeister R (1993) Nitric oxide regulates cardiac Ca^{2+} current. Involvement of cGMP-inhibited and cGMP-stimulated phosphodiesterases through guanylyl cyclase activation. J Biol Chem 268:26286–26295

Mills RM, LeJemtel TH, Horton DP, Liang C, Lang R, Silver MA, Lui C, Chatterjee K (1999) Sustained hemodynamic effects of an infusion of nesiritide (human b-type natriuretic peptide) in heart failure: a randomized, double-blind, placebo-controlled clinical trial. Natrecor Study Group. J Am Coll Cardiol 34:155–162

Moalem J, Davidov T, Zhang Q, Grover GJ, Weiss HR, Scholz PM (2006) Negative inotropic effects of C-type natriuretic peptide are attenuated in hypertrophied ventricular myocytes associated with reduced cyclic GMP production. J Surg Res 135:38–44. doi:S0022-4804(06)00012-6 [pii]; 10.1016/j.jss.2006.01.012 [doi]

Molenaar P, Christ T, Hussain RI, Engel A, Berk E, Gillette KT, Chen L, Galindo-Tovar A, Krobert KA, Ravens U, Levy FO, Kaumann AJ (2013) PDE3, but not PDE4, reduces β_1-and β_2-adrenoceptor-mediated inotropic and lusitropic effects in failing ventricle from metoprolol-treated patients. Br J Pharmacol 169:528–538. doi:10.1111/bph.12167

Moltzau LR, Aronsen JM, Meier S, Nguyen CH, Hougen K, Ørstavik O, Sjaastad I, Christensen G, Skomedal T, Osnes JB, Levy FO, Qvigstad E (2013) SERCA2 activity is involved in the CNP-mediated functional responses in failing rat myocardium. Br J Pharmacol 170:366–379. doi:10.1111/bph.12282 [doi]

Moltzau LR, Aronsen JM, Meier S, Skogestad J, Ørstavik O, Lothe GB, Sjaastad I, Skomedal T, Osnes JB, Levy FO, Qvigstad E (2014a) Different compartmentation of responses to brain natriuretic peptide and C-type natriuretic peptide in failing rat ventricle. J Pharmacol Exp Ther 350:681–690. doi:jpet.114.214882 [pii]; 10.1124/jpet.114.214882 [doi]

Moltzau LR, Meier S, Aronsen JM, Afzal F, Sjaastad I, Skomedal T, Osnes JB, Levy FO, Qvigstad E (2014b) Differential regulation of C-type natriuretic peptide-induced cGMP and functional responses by PDE2 and PDE3 in failing myocardium. Naunyn Schmiedebergs Arch Pharmacol. doi:10.1007/s00210-013-0953-1 [doi]

Mongillo M, Tocchetti CG, Terrin A, Lissandron V, Cheung YF, Dostmann WR, Pozzan T, Kass DA, Paolocci N, Houslay MD, Zaccolo M (2006) Compartmentalized phosphodiesterase-2 activity

blunts beta-adrenergic cardiac inotropy via an NO/cGMP-dependent pathway. Circ Res 98:226–234. doi:01.RES.0000200178.34179.93 [pii]; 10.1161/01.RES.0000200178.34179.93 [doi]
Murthy KS, Teng BQ, Zhou H, Jin JG, Grider JR, Makhlouf GM (2000) G_{i-1}/G_{i-2}-dependent signaling by single-transmembrane natriuretic peptide clearance receptor. Am J Physiol Gastrointest Liver Physiol 278:G974–G980
Nagendran J, Archer SL, Soliman D, Gurtu V, Moudgil R, Haromy A, St AC, Webster L, Rebeyka IM, Ross DB, Light PE, Dyck JR, Michelakis ED (2007) Phosphodiesterase type 5 is highly expressed in the hypertrophied human right ventricle, and acute inhibition of phosphodiesterase type 5 improves contractility. Circulation 116:238–248. doi:CIRCULATIONAHA.106.655266 [pii]; 10.1161/CIRCULATIONAHA.106.655266 [doi]
Niino Y, Hotta K, Oka K (2009) Simultaneous live cell imaging using dual FRET sensors with a single excitation light. PLoS One 4:e6036. doi:10.1371/journal.pone.0006036
Nikolaev VO, Gambaryan S, Lohse MJ (2006) Fluorescent sensors for rapid monitoring of intracellular cGMP. Nat Methods 3:23–25. doi:10.1038/nmeth816
Nir A, Zhang DF, Fixler R, Burnett JC Jr, Eilam Y, Hasin Y (2001) C-type natriuretic peptide has a negative inotropic effect on cardiac myocytes. Eur J Pharmacol 412:195–201
Nishikimi T, Kuwahara K, Nakao K (2011) Current biochemistry, molecular biology, and clinical relevance of natriuretic peptides. J Cardiol 57:131–140. doi:S0914-5087(11)00003-7 [pii]; 10.1016/j.jjcc.2011.01.002 [doi]
Nishikimi T, Maeda N, Matsuoka H (2006) The role of natriuretic peptides in cardioprotection. Cardiovasc Res 69:318–328. doi:10.1016/j.cardiores.2005.10.001
O'Connor CM, Starling RC, Hernandez AF, Armstrong PW, Dickstein K, Hasselblad V, Heizer GM, Komajda M, Massie BM, McMurray JJ, Nieminen MS, Reist CJ, Rouleau JL, Swedberg K, Adams KF Jr, Anker SD, Atar D, Battler A, Botero R, Bohidar NR, Butler J, Clausell N, Corbalan R, Costanzo MR, Dahlstrom U, Deckelbaum LI, Diaz R, Dunlap ME, Ezekowitz JA, Feldman D, Felker GM, Fonarow GC, Gennevois D, Gottlieb SS, Hill JA, Hollander JE, Howlett JG, Hudson MP, Kociol RD, Krum H, Laucevicius A, Levy WC, Mendez GF, Metra M, Mittal S, Oh BH, Pereira NL, Ponikowski P, Tang WH, Tanomsup S, Teerlink JR, Triposkiadis F, Troughton RW, Voors AA, Whellan DJ, Zannad F, Califf RM (2011) Effect of nesiritide in patients with acute decompensated heart failure. N Engl J Med 365:32–43. doi:10.1056/NEJMoa1100171
O'Tierney PF, Chattergoon NN, Louey S, Giraud GD, Thornburg KL (2010) Atrial natriuretic peptide inhibits angiotensin II-stimulated proliferation in fetal cardiomyocytes. J Physiol 588:2879–2889. doi:10.1113/jphysiol.2010.191098
Ohte N, Cheng CP, Suzuki M, Little WC (1999) Effects of atrial natriuretic peptide on left ventricular performance in conscious dogs before and after pacing-induced heart failure. J Pharmacol Exp Ther 291:589–595
Omori K, Kotera J (2007) Overview of PDEs and their regulation. Circ Res 100:309–327. doi:100/3/309 [pii]; 10.1161/01.RES.0000256354.95791.f1 [doi]
Ono K, Trautwein W (1991) Potentiation by cyclic GMP of beta-adrenergic effect on Ca^{2+} current in guinea-pig ventricular cells. J Physiol 443:387–404
Perera RK, Sprenger JU, Steinbrecher JH, Hubscher D, Lehnart SE, Abesser M, Schuh K, El-Armouche A, Nikolaev VO (2015) Microdomain switch of cGMP-regulated phosphodiesterases leads to ANP-induced augmentation of beta-adrenoceptor-stimulated contractility in early cardiac hypertrophy. Circ Res 116:1304–1311. doi:10.1161/circresaha.116.306082
Pierkes M, Gambaryan S, Boknik P, Lohmann SM, Schmitz W, Potthast R, Holtwick R, Kuhn M (2002) Increased effects of C-type natriuretic peptide on cardiac ventricular contractility and relaxation in guanylyl cyclase A-deficient mice. Cardiovasc Res 53:852–861
Potter LR (2011) Guanylyl cyclase structure, function and regulation. Cell Signal 23:1921–1926. doi:S0898-6568(11)00280-4 [pii]; 10.1016/j.cellsig.2011.09.001 [doi]
Potter LR, Abbey-Hosch S, Dickey DM (2006) Natriuretic peptides, their receptors, and cyclic guanosine monophosphate-dependent signaling functions. Endocr Rev 27:47–72. doi:er.2005-0014 [pii]; 10.1210/er.2005-0014 [doi]
Potter LR, Yoder AR, Flora DR, Antos LK, Dickey DM (2009) Natriuretic peptides: their structures, receptors, physiologic functions and therapeutic applications. Handb Exp Pharmacol 191:341–366. doi:10.1007/978-3-540-68964-5_15 [doi]

Qvigstad E, Moltzau LR, Aronsen JM, Nguyen CH, Hougen K, Sjaastad I, Levy FO, Skomedal T, Osnes JB (2010) Natriuretic peptides increase beta$_1$-adrenoceptor signalling in failing hearts through phosphodiesterase 3 inhibition. Cardiovasc Res 85:763–772. doi:cvp364 [pii]; 10.1093/cvr/cvp364 [doi]

Resink TJ, Panchenko MP, Tkachuk VA, Buhler FR (1988) Involvement of N$_i$ protein in the functional coupling of the atrial natriuretic factor (ANF) receptor to adenylate cyclase in rat lung plasma membranes. Eur J Biochem 174:531–535

Rose RA, Giles WR (2008) Natriuretic peptide C receptor signalling in the heart and vasculature. J Physiol 586:353–366. doi:jphysiol.2007.144253 [pii]; 10.1113/jphysiol.2007.144253 [doi]

Rose RA, Lomax AE, Giles WR (2003) Inhibition of L-type Ca^{2+} current by C-type natriuretic peptide in bullfrog atrial myocytes: an NPR-C-mediated effect. Am J Physiol Heart Circ Physiol 285:H2454–H2462. doi:10.1152/ajpheart.00388.2003 [doi];00388.2003 [pii]

Rosenkranz AC, Woods RL, Dusting GJ, Ritchie RH (2003) Antihypertrophic actions of the natriuretic peptides in adult rat cardiomyocytes: importance of cyclic GMP. Cardiovasc Res 57:515–522

Russwurm M, Mullershausen F, Friebe A, Jager R, Russwurm C, Koesling D (2007) Design of fluorescence resonance energy transfer (FRET)-based cGMP indicators: a systematic approach. Biochem J 407:69–77. doi:10.1042/bj20070348

Singh G, Kuc RE, Maguire JJ, Fidock M, Davenport AP (2006) Novel snake venom ligand dendroaspis natriuretic peptide is selective for natriuretic peptide receptor-A in human heart: downregulation of natriuretic peptide receptor-A in heart failure. Circ Res 99:183–190. doi:10.1161/01.RES.0000232322.06633.d3

Sodi R, Dubuis E, Shenkin A, Hart G (2008) B-type natriuretic peptide (BNP) attenuates the L-type calcium current and regulates ventricular myocyte function. Regul Pept 151:95–105. doi:S0167-0115(08)00113-4 [pii]; 10.1016/j.regpep.2008.06.006 [doi]

Spitzer KW, Bridge JH (1992) Relationship between intracellular pH and tension development in resting ventricular muscle and myocytes. Am J Phys 262:C316–C327

Sprenger JU, Perera RK, Steinbrecher JH, Lehnart SE, Maier LS, Hasenfuss G, Nikolaev VO (2015) In vivo model with targeted cAMP biosensor reveals changes in receptor-microdomain communication in cardiac disease. Nat Commun 6:6965. doi:10.1038/ncomms7965

Stangherlin A, Gesellchen F, Zoccarato A, Terrin A, Fields LA, Berrera M, Surdo NC, Craig MA, Smith G, Hamilton G, Zaccolo M (2011) cGMP signals modulate cAMP levels in a compartment-specific manner to regulate catecholamine-dependent signaling in cardiac myocytes. Circ Res 108:929–939. doi:CIRCRESAHA.110.230698 [pii]; 10.1161/CIRCRESAHA.110.230698 [doi]

Su J, Scholz PM, Weiss HR (2005a) Differential effects of cGMP produced by soluble and particulate guanylyl cyclase on mouse ventricular myocytes. Exp Biol Med (Maywood) 230:242–250. doi:230/4/242 [pii]

Su J, Zhang Q, Moalem J, Tse J, Scholz PM, Weiss HR (2005b) Functional effects of C-type natriuretic peptide and nitric oxide are attenuated in hypertrophic myocytes from pressure-overloaded mouse hearts. Am J Physiol Heart Circ Physiol 288:H1367–H1373. doi:00880.2004 [pii]; 10.1152/ajpheart.00880.2004 [doi]

Sudoh T, Kangawa K, Minamino N, Matsuo H (1988) A new natriuretic peptide in porcine brain. Nature 332:78–81. doi:10.1038/332078a0 [doi]

Sudoh T, Minamino N, Kangawa K, Matsuo H (1990) C-type natriuretic peptide (CNP): a new member of natriuretic peptide family identified in porcine brain. Biochem Biophys Res Commun 168:863–870

Suga S, Nakao K, Hosoda K, Mukoyama M, Ogawa Y, Shirakami G, Arai H, Saito Y, Kambayashi Y, Inouye K (1992a) Receptor selectivity of natriuretic peptide family, atrial natriuretic peptide, brain natriuretic peptide, and C-type natriuretic peptide. Endocrinology 130:229–239

Suga S, Nakao K, Itoh H, Komatsu Y, Ogawa Y, Hama N, Imura H (1992b) Endothelial production of C-type natriuretic peptide and its marked augmentation by transforming growth factor-beta. Possible existence of "vascular natriuretic peptide system". J Clin Invest 90:1145–1149. doi:10.1172/JCI115933 [doi]

Sumii K, Sperelakis N (1995) cGMP-dependent protein kinase regulation of the L-type Ca^{2+} current in rat ventricular myocytes. Circ Res 77:803–812

Tajima M, Bartunek J, Weinberg EO, Ito N, Lorell BH (1998) Atrial natriuretic peptide has different effects on contractility and intracellular pH in normal and hypertrophied myocytes from pressure-overloaded hearts. Circulation 98:2760–2764

Takahashi T, Allen PD, Izumo S (1992) Expression of A-, B-, and C-type natriuretic peptide genes in failing and developing human ventricles. Correlation with expression of the Ca^{2+}-ATPase gene. Circ Res 71:9–17

Takayanagi R, Snajdar RM, Imada T, Tamura M, Pandey KN, Misono KS, Inagami T (1987) Purification and characterization of two types of atrial natriuretic factor receptors from bovine adrenal cortex: guanylate cyclase-linked and cyclase-free receptors. Biochem Biophys Res Commun 144:244–250. doi:S0006-291X(87)80502-8 [pii]

Takimoto E, Belardi D, Tocchetti CG, Vahebi S, Cormaci G, Ketner EA, Moens AL, Champion HC, Kass DA (2007) Compartmentalization of cardiac beta-adrenergic inotropy modulation by phosphodiesterase type 5. Circulation 115:2159–2167. doi:CIRCULATIONAHA.106.643536 [pii]; 10.1161/CIRCULATIONAHA.106.643536 [doi]

Takimoto E, Champion HC, Belardi D, Moslehi J, Mongillo M, Mergia E, Montrose DC, Isoda T, Aufiero K, Zaccolo M, Dostmann WR, Smith CJ, Kass DA (2005) cGMP catabolism by phosphodiesterase 5A regulates cardiac adrenergic stimulation by NOS3-dependent mechanism. Circ Res 96:100–109. doi:01.RES.0000152262.22968.72 [pii]; 10.1161/01.RES.0000152262.22968.72 [doi]

Tan T, Scholz PM, Weiss HR (2010) Hypoxia inducible factor-1 improves the negative functional effects of natriuretic peptide and nitric oxide signaling in hypertrophic cardiac myocytes. Life Sci 87:9–16. doi:S0024-3205(10)00199-2 [pii]; 10.1016/j.lfs.2010.05.002 [doi]

Tohse N, Nakaya H, Takeda Y, Kanno M (1995) Cyclic GMP-mediated inhibition of L-type Ca^{2+} channel activity by human natriuretic peptide in rabbit heart cells. Br J Pharmacol 114:1076–1082

Tokudome T, Horio T, Soeki T, Mori K, Kishimoto I, Suga S, Yoshihara F, Kawano Y, Kohno M, Kangawa K (2004) Inhibitory effect of C-type natriuretic peptide (CNP) on cultured cardiac myocyte hypertrophy: interference between CNP and endothelin-1 signaling pathways. Endocrinology 145:2131–2140. doi:10.1210/en.2003-1260

Tokudome T, Kishimoto I, Horio T, Arai Y, Schwenke DO, Hino J, Okano I, Kawano Y, Kohno M, Miyazato M, Nakao K, Kangawa K (2008) Regulator of G-protein signaling subtype 4 mediates antihypertrophic effect of locally secreted natriuretic peptides in the heart. Circulation 117:2329–2339. doi:10.1161/circulationaha.107.732990

Vandecasteele G, Verde I, Rucker-Martin C, Donzeau-Gouge P, Fischmeister R (2001) Cyclic GMP regulation of the L-type Ca^{2+} channel current in human atrial myocytes. J Physiol 533:329–340

Vaxelaire JF, Laurent S, Lacolley P, Briand V, Schmitt H, Michel JB (1989) Atrial natriuretic peptide decreases contractility of cultured chick ventricular cells. Life Sci 45:41–48

Wakabayashi S, Hisamitsu T, Nakamura TY (2013) Regulation of the cardiac Na^+/H^+ exchanger in health and disease. J Mol Cell Cardiol 61:68–76. doi:10.1016/j.yjmcc.2013.02.007

Wen JF, Cui X, Jin JY, Kim SM, Kim SZ, Kim SH, Lee HS, Cho KW (2004) High and low gain switches for regulation of cAMP efflux concentration: distinct roles for particulate GC- and soluble GC-cGMP-PDE3 signaling in rabbit atria. Circ Res 94:936–943. doi:10.1161/01.RES.0000123826.70125.4D [doi];01.RES.0000123826.70125.4D [pii]

Weninger S, De Maeyer JH, Lefebvre RA (2012) Study of the regulation of the inotropic response to 5-HT_4 receptor activation via phosphodiesterases and its cross-talk with C-type natriuretic peptide in porcine left atrium. Naunyn Schmiedeberg Arch Pharmacol 385:565–577. doi:10.1007/s00210-012-0746-y

William M, Hamilton EJ, Garcia A, Bundgaard H, Chia KK, Figtree GA, Rasmussen HH (2008) Natriuretic peptides stimulate the cardiac sodium pump via NPR-C-coupled NOS activation. Am J Physiol Cell Physiol 294:C1067–C1073. doi:00243.2007 [pii]; 10.1152/ajpcell.00243.2007 [doi]

Wollert KC, Yurukova S, Kilic A, Begrow F, Fiedler B, Gambaryan S, Walter U, Lohmann SM, Kuhn M (2003) Increased effects of C-type natriuretic peptide on contractility and calcium regulation in murine hearts overexpressing cyclic GMP-dependent protein kinase I. Br J Pharmacol 140:1227–1236. doi:10.1038/sj.bjp.0705567 [doi];sj.bjp.0705567 [pii]

Yasue H, Yoshimura M, Sumida H, Kikuta K, Kugiyama K, Jougasaki M, Ogawa H, Okumura K, Mukoyama M, Nakao K (1994) Localization and mechanism of secretion of B-type natriuretic peptide in comparison with those of A-type natriuretic peptide in normal subjects and patients with heart failure. Circulation 90:195–203

Zhang Q, Moalem J, Tse J, Scholz PM, Weiss HR (2005) Effects of natriuretic peptides on ventricular myocyte contraction and role of cyclic GMP signaling. Eur J Pharmacol 510:209–215. doi:S0014-2999(05)00061-0 [pii]; 10.1016/j.ejphar.2005.01.031 [doi]

Zhang Q, Scholz PM, Pilzak A, Su J, Weiss HR (2007) Role of phospholamban in cyclic GMP mediated signaling in cardiac myocytes. Cell Physiol Biochem 20:157–166. doi:000104163 [pii]; 10.1159/000104163 [doi]

Zhang M, Takimoto E, Lee DI, Santos CX, Nakamura T, Hsu S, Jiang A, Nagayama T, Bedja D, Yuan Y, Eaton P, Shah AM, Kass DA (2012) Pathological cardiac hypertrophy alters intracellular targeting of phosphodiesterase type 5 from nitric oxide synthase-3 to natriuretic peptide signaling. Circulation 126:942–951. doi:10.1161/circulationaha.112.090977

Cyclic GMP/Protein Kinase Localized Signaling and Disease Implications

13

Brian Leei Lin, David A. Kass, and Dong I. Lee

Contents

Abstract

Cyclic guanosine 3′,5′-monophosphate (cGMP) and its downstream target, protein kinase G (PKG or cGK), play central roles in cellular regulation and are important to cardiovascular homeostasis and disease pathophysiology. Cyclic GMP is synthesized via either nitric oxide (NO) or natriuretic peptide (NP) stimulation pathways, each coupled to corresponding cyclases, and catabolized by select members of the phosphodiesterase superfamily. Growing evidence now

B.L. Lin • D.I. Lee, Ph.D. (✉)
Division of Cardiology, Department of Medicine,
The Johns Hopkins University Medical Institutions,
720 Rutland Avenue, ROSS 858, Baltimore, MD 21205, USA
e-mail: dlee120@jhmi.edu

D.A. Kass, M.D. (✉)
Division of Cardiology, Department of Medicine,
The Johns Hopkins University Medical Institutions,
720 Rutland Avenue, ROSS 858, Baltimore, MD 21205, USA

Department of Pharmacology and Molecular Sciences,
Johns Hopkins University, Baltimore, MD 21205, USA
e-mail: dkass@jhmi.edu

V. Nikolaev, M. Zaccolo (eds.), *Microdomains in the Cardiovascular System*,
Cardiac and Vascular Biology 3, DOI 10.1007/978-3-319-54579-0_13

supports control of cGMP and PKG in distinct microdomains within the myocyte, which results in differential downstream targeting. This regional control stems from distinct localization of the relevant signaling components and their capacity to translocate in the cell under both physiological and pathophysiological conditions to further impact the net response. This chapter discusses current understanding of microdomain regulation of the cGMP/PKG pathway, as this information is important to optimally leverage their effects for the treatment of cardiovascular disease.

13.1 Introduction

Cyclic GMP and consequent PKG signaling is initiated by nitric oxide (NO) or natriuretic peptide (NP) stimulation of their respective cyclases and is terminated by cGMP hydrolysis by members of the superfamily of phosphodiesterases (PDEs) (Kokkonen and Kass 2016; Kuhn 2016; Derbyshire and Marletta 2012). NO stimulates soluble guanylyl cyclase (sGC), while natriuretic peptide (NP) activates a receptor-bound guanylyl cyclase (GC-A or GC-B) (Fig. 13.1). Differences in their cellular location play a key role in compartmentalizing the signaling. Despite its

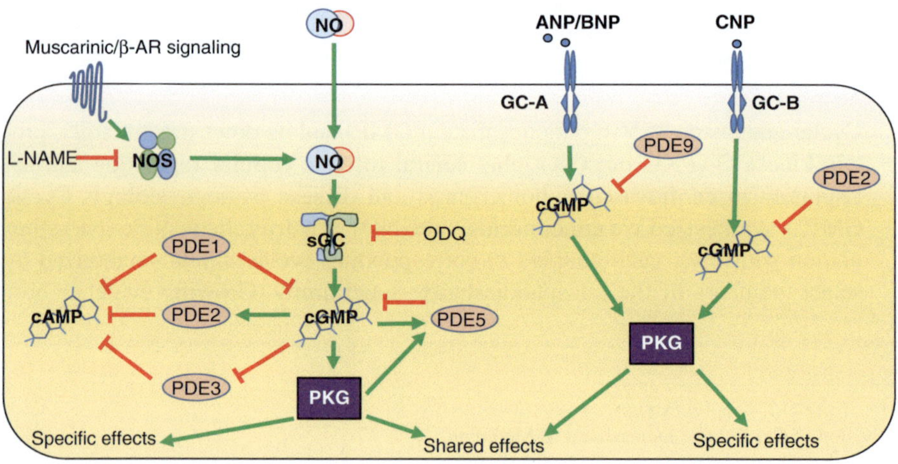

Fig. 13.1 Local regulation of cGMP/PKG. Cyclic GMP is synthesized by either nitric oxide (NO)-soluble guanylate cyclase (sGC) or natriuretic peptide (NP) receptor guanylate cyclase (GC-A, GC-B) stimulation and then further regulated by select members of the phosphodiesterase (PDE) family to degrade it to GMP. Cyclic GMP in turn activates protein kinase G (PKG), and this targets a number of shared downstream proteins independent of the upstream activation pathway, as well as specific proteins that do depend upon the pathway. PDE1-3, PDE5, and PDE9 are also depicted on the scheme, and their regulation of or by cGMP or cAMP is shown. Cyclic GMP binds to both PDE2 and PDE5, activating the former to hydrolyze cAMP and the latter to hydrolyze cGMP. PDE5 activation also involves its phosphorylation by PKG. *ANP*, *BNP*, *CNP* A-type, B-type, and C-type natriuretic peptide; *β-AR* β-adrenergic receptor; *cGMP* cyclic guanosine 3′,5′-monophosphate; *GC-A*, *GC-B* guanylyl cyclase-A, guanylyl cyclase-B; *NO* nitric oxide; *NOS* nitric oxide-synthase; *PDE* phosphodiesterase; *PKG* protein kinase G; *sGC* soluble guanylate cyclase

name, sGC is found in both cytosolic and membrane compartments, the latter including both caveolin-3 (Cav3)-enriched microdomains and those lacking Cav3 (Liu et al. 2013; Tsai et al. 2012). NP receptors reside in the plasma membrane, though their precise subdomain localization remains uncertain (Kuhn 2016). Once synthesized, cGMP binds to regulatory domains on PKG, with PKG1α being the most prominent isoform in the heart (Geiselhoringer et al. 2004), and this activates the kinase. Cyclic GMP hydrolysis is achieved by PDEs that convert cGMP to 5′-GMP. At least five different cGMP-targeting PDEs have been identified to play this role in cardiac myocytes, though their relative importance continues to be elucidated and appears to vary depending upon underlying conditions (Kokkonen and Kass 2016). These PDEs also have different cyclic nucleotide specificity, and intracellular localization which means different targeted cGMP pools, further localizing cGMP regulation. For example, PDE5A and PDE9A localize to Z-disk to impact NO-activated cGMP, or T-tubules to impact NP-derived cGMP pools, respectively (Lee et al. 2015). Intracellular localization of many of these signaling components is also dynamic and can be influenced by physiological and pathological stimuli. As PKG activation generally counters pathological cardiac stress, this pathway appears particularly important in conditions of heart disease. For example, under excess cardiac stress or toxic insult, cGMP/PKG regulates Ca^{2+} homeostasis, mitochondrial function and its permeability transition pore, G-protein coupled signaling, hormone- and mechanosensitive cation channels, protein quality control, sarcomere function, and other factors (Inserte and Garcia-Dorado 2015). Some of these are more impacted by NO-activated pathways and others by NP signaling, imparting another layer of differential control in normal and diseased hearts.

Recent discoveries have expanded therapeutic approaches to stimulate PKG signaling. Initially, NO donors, organo-nitrates, and natriuretic peptides were the primary means available. With the discovery of PDE5A, selective inhibitors were added and have become clinical pharmaceuticals (Kass 2012). New tools include direct activators and stimulators of sGC, inhibitors of NP proteolysis by neprilysin, and inhibitors of PDE9A (Mcmurray et al. 2014; Gheorghiade et al. 2013; Lee et al. 2015). Rather than being redundant methods to stimulate cGMP/PKG, each approach confers selective effects based on the signaling compartment and costimulation that is present. In the following sections, we discuss the localized regulation of this pathway, including stimulation of cGMP synthesis, its hydrolysis by PDEs, relocalization effects and protein interactions, and the influence of oxidant stress.

13.2 Modulation of cGMP by NO or NP Pathways

Soluble GC is a heterodimer containing an α_1 or α_2 subunit and heme-binding β_1 or β_2 subunit, with the $\alpha_1\beta_1$ combination being the most common isoform (Buys et al. 2009). The N-terminus of the β_1 subunit serves as the primary NO sensor, triggering synthesis of cGMP from GTP. NO synthesis is itself compartmentalized by regional expression of one of three NO synthase isoforms: neuronal (nNOS or NOS1), inducible (iNOS or NOS2), and endothelial (eNOS or NOS3) (Forstermann and Sessa 2012). NOS1 and NOS3 are constitutively expressed in the cardiomyocyte, while

NOS2 is expressed in response to inflammatory signaling (Seddon et al. 2007). NOS3 colocalizes with caveolin-3, nuclear membrane, β-adrenergic receptors (β-AR), and L-type Ca^{2+} channels (LTCC), where synthesized NO negatively regulates contractile regulation and excitation-contraction coupling (Simon et al. 2014; Mongillo et al. 2006; Balligand 2013). NOS1 localizes to the sarcoplasmic reticulum (SR) membrane where it coprecipitates with xanthine oxidoreductase (XOR) and the ryanodine receptor (RyR2). It generates NO that facilitates Ca^{2+} uptake and release largely by cGMP-independent mechanisms (Zhang et al. 2014b; Sears et al. 2003). This localization provides negative regulation of XOR as mice lacking NOS1 have excess XOR-dependent superoxide generation (Kinugawa et al. 2005). NOS1 can relocalize to the plasma membrane which is observed following ischemia/reperfusion (I/R) injury. Here it colocalizes with caveolin-3 and promotes S-nitrosylation of the LTCC (Sun et al. 2006). This may provide anti-arrhythmic effects by impeding LTCC oxidation (Yue et al. 2015), but may also be detrimental by leaving XOR unattended, impairing SR calcium cycling by ROS-dependent changes (Wang et al. 2010).

The second cGMP synthetic pathway couples with the natriuretic peptide receptor, which upon binding to NP stimulates endogenous GC activity. NPs are a family of hormones secreted by the heart in response to stretch (Kuhn 2016). There are three NPs: A-type, B-type, and C-type (ANP, BNP, and CNP, respectively) (Kuhn 2016). ANP and BNP were first referred to as atrial and brain natriuretic peptide, respectively, based on presumed tissue of origin, but both are now known to be expressed by cardiac muscle. C-type NP is produced by vascular tissue, though myocyte synthesis has also been reported (Del Ry 2013; Del Ry et al. 2011).

ANP and BNP both stimulate the NPR-A type receptor coupled to GC-A, whereas CNP stimulates NPR-B coupled to GC-B (Fig. 13.1). Both receptors and associated cyclase form macromolecular complexes with GC activity localized to the inner plasma membrane (Kuhn 2016). Both GC-A and GC-B are expressed in cardiac myocytes, vascular smooth muscle and endothelial cells, and fibroblasts. Exogenous CNP suppresses fibrosis in vitro, whereas ANP and BNP have a more limited impact (Horio et al. 2003). Interestingly, despite low expression in cardiomyocytes, CNP and NPR-B also appear necessary for regulating hypertrophic responses in rat hearts, as transgenic rats that overexpressed a dominant negative GC-B mutant develop spontaneous cardiac hypertrophy (Langenickel et al. 2006). Conversely, cardiac-specific CNP overexpression in mice with myocardial infarction develop less hypertrophy (Wang et al. 2007).

The natriuretic peptide receptor-C (NPR-C) lacks guanylyl cyclase activity (Koller and Goeddel 1992; Anand-Srivastava 2005) and instead mediates signaling via G_i-proteins. It also provides a clearance receptor, where NP binding leads to peptide degradation. NPR-C binds all three NPs and is abundantly expressed in atria myocytes, fibroblasts, vascular smooth muscle and adipocytes. Though NPR-C does not directly enhance cGMP synthesis, it has been shown to activate G_i-proteins and eNOS signaling (Li et al. 2014).

The impact of sGC versus GC-A or GC-B activation, each stimulating PKG, can overlap, but there are also distinct differences indicating compartmentation. In adult rat ventricular myocytes (ARVM), the local pools of sGC-derived cGMP are present in the cytosol, whereas GC-A-derived cGMP is found at the plasma membrane

(Castro et al. 2006). In frog ventricular myocytes, sGC stimulation blocks LTCC currents, whereas GC-A stimulation by ANP does not (Mery et al. 1993; Gisbert and Fischmeister 1988). In mammalian cells, sGC stimulation has minimal impact on the LTCC, but rather desensitizes the myofilaments to calcium by phosphorylation of troponin-I to reduce contraction (Su et al. 2005; Layland et al. 2002; Lee et al. 2010). This response is not observed with GC-A stimulation (Takimoto et al. 2007).

A major challenge to determining compartmentalization of cGMP/PKG signaling in the intact cardiac myocyte has been the lack of a PKG kinase activity sensor and low cGMP concentrations that in turn require very high sensitivity probes. Unlike the Förster resonance energy transfer (FRET)-based biosensors to visualize adenosine 3′,5′-monophosphate (cAMP) (Adams et al. 1991; Stangherlin and Zaccolo 2012; Dipilato et al. 2004; Stangherlin et al. 2011), cGMP-FRET sensors (Honda et al. 2001; Nikolaev et al. 2006; Gotz et al. 2014; Perera et al. 2015) have struggled to provide adequate signal/noise to detect cGMP in adult myocytes, particularly when generated via NO-sGC stimulation. Some other sensors have been most successful in detecting cGMP localization regulated by NP stimuli (Gotz et al. 2014; Stangherlin et al. 2011; Krawutschke et al. 2015; Perera et al. 2015). Another approach is a cGMP fluorescent biosensor called FlincG (Nausch et al. 2008) which does not involve FRET. FlincG has rapid kinetics and reasonable sensitivity to low cGMP concentration (1–10 nM) and has been used in intact neonatal myocytes (Lee et al. 2015). However, this has not been useful in adult myocytes likely due to inadequate signal-to-noise. Efforts to improve on these sensors continue.

13.3 Modulation of cGMP/PKG by Localized PDEs in the Heart

In addition to localizing a cGMP/PKG signal by compartmentation of the relevant agonists and cyclases, cGMP is further constrained by a localized distribution of selective PDEs (Francis et al. 2011; Kim and Kass 2016). Among these, seven are known to be expressed in the heart: PDE1–5, 8, and 9. PDE1, 2, and 3 hydrolyze both cAMP and cGMP, PDE4 and PDE8 are cAMP specific, and PDE5 and PDE9 are cGMP specific.

PDE1 was the first identified dual cAMP/cGMP esterase and is activated by Ca^{2+}/calmodulin (CaM) (Sonnenburg et al. 1998). It is expressed as three isoforms, PDE1A, 1B, and 1C, which vary in substrate selectivity. PDE1A and 1B are more selective for cGMP, whereas PDE1C has similar selectivity for both cGMP and cAMP. Mice and rats express primarily PDE1A, whereas larger mammals and humans express largely PDE1C (Johnson et al. 2012; Vandeput et al. 2007). PDE1A inhibition attenuates hypertrophy and fibrosis in small rodents, and activation of PKG is considered important though increased cAMP signaling also plays a role (Miller et al. 2009, 2011). PDE1C-deficient mice also show reduced TGF-β-mediated fibrosis and hypertrophy (Knight et al. 2016), and this appears due to increased cAMP not cGMP. The nature of this compartmentalized cAMP and its upstream agonists remains under study, though cell imaging found PDE1C localized at M-band and Z-disk locations (Vandeput et al. 2007).

PDE2A is another dual-substrate esterase and is found in the cytosol, plasma membrane, sarcoplasmic reticulum, Golgi, and nuclear membrane (Conti and Beavo 2007). These different localizations of PDE2A relate in part to N-terminus splice variants and affect regulation of specific compartmentalized cGMP and cAMP. A unique feature of PDE2A is its activation of cAMP hydrolysis upon cGMP binding to regulatory GAF domains in the N-terminus. This positions PDE2A as a cAMP/cGMP crosstalk PDE (Stangherlin et al. 2011). An example of such regulation is its role in countering β-adrenergic agonism by β_3-AR-dependent stimulation of NO-cGMP synthesis (Mongillo et al. 2006). PDE2A appears upregulated in the failing heart, and this is coupled to reduced β-adrenergic stimulated signaling (Mehel et al. 2013). The pathophysiological role of PDE2A appears more complex, as studies found PDE2A inhibition suppresses hypertrophy (Zoccarato et al. 2015; Bubb et al. 2014). PDE2A promotes hypertrophic signaling by stimulating local PKA activation to phosphorylate nuclear factor of activated T cells (NFAT), inhibiting its nuclear translocation and hypertrophy activation (Zoccarato et al. 2015). PDE2A inhibition has also been demonstrated to promote cGMP-dependent suppression of right ventricular hypertrophy (Bubb et al. 2014). Yet mice with myocyte-targeted PDE2A3 (human variant) overexpression display less arrhythmia and improved contractile function after myocardial infarction (Vettel et al. 2016) related to reduced β-AR agonism and consequent cAMP signaling. Upregulation of PDE2A3 also activates myofibroblast formation by reducing cAMP signaling (Vettel et al. 2014) and is reported to underlie coronary vascular dysfunction from increased endothelial permeability due to NP stimulation (Chen et al. 2016). Thus, PDE2A3 modulation of function is altered depending on cell type and under pathophysiological conditions. Whether PDE2A1 or PDE2A2 regulate function differently than PDE2A3 has yet to be determined.

PDE3 primarily regulates cAMP (Liu and Maurice 1998), but can be inhibited by cGMP competitive binding in the catalytic site, providing another crosstalk mechanism between cAMP and cGMP (Zaccolo and Movsesian 2007). Two primary subfamilies, PDE3A and 3B, are expressed, with additional splice variants (e.g., PDE3A1, A2, A3) that confer different myocardial effects. PDE3A1 associates with sarcoplasmic reticulum calcium ATPase (SERCa2a), phospholamban, and AKAP18 forming a protein complex that regulates Ca^{2+} uptake and contractility (Ahmad et al. 2015). PDE3A1 incorporation is activated by PKA, providing a negative feedback loop to calcium uptake. PDE3 isoform hydrolysis of cAMP is inhibited by low concentrations of cGMP, which confers a positive inotropic effect to the latter (Vila-Petroff et al. 1999). The microdomains where PDE3A variants reside can also be altered by disease, thereby impacting cGMP/cAMP crosstalk. Perera et al. reported that in early adaptive cardiac hypertrophy, PDE2A relocalizes from β_1-AR-associated non-caveolin-enriched microdomains to β_2-AR-containing caveolae domains (Perera et al. 2015). In contrast, PDE3A that is normally colocalized with β_2-AR exhibits reduced expression in this microdomain. The result is enhanced ANP augmentation of β-adrenergic activation, where cAMP coupled to the β_2-AR declines, but cAMP coupled with β_1-AR increases.

PDE5A was the first cGMP-selective PDE identified, and its therapeutic inhibition is widely used to treat erectile dysfunction and pulmonary hypertension. In cardiomyocytes, PDE5A preferentially targets cGMP generated via NO-sGC rather than NP-GC-A (Kokkonen and Kass 2016). PDE5A normally localizes to the sarcomere Z-disk, but this distribution can become diffuse in myocytes from dilated failing or late-stage hypertrophic hearts and conditions in which NOS3 is genetically deleted or pharmacologically inhibited (Senzaki et al. 2001; Zhang et al. 2008; Nagayama et al. 2008). Restoring NOS activity or directly stimulating sGC reverses this relocalization. Like NO, PDE5 inhibition suppresses β-adrenergic contractile stimulation, an effect that requires PKG phosphorylation of troponin-I (Lee et al. 2010). Selective targeting of PDE5 to NO rather than ANP-stimulated cGMP synthesis has been confirmed using cGMP fluorescent reporter proteins (Lee et al. 2015) and cGMP-sensitive ion channels (Castro et al. 2006). Functional selectivity was revealed in several studies showing that PDE5-inhibition suppresses chronic LV dysfunction and pressure overload-induced hypertrophy (Takimoto et al. 2005, 2007; Kukreja et al. 2005; Lee et al. 2010).

PDE9A is the most cGMP-selective degrading member of the PDE superfamily. It was first cloned in 1998 (Fisher et al. 1998) and is expressed in the brain, gut, kidney, and heart. Recently, protein expression and activity has been revealed in the cardiomyocytes and revealed to selectively target cGMP generated by the NP-signaling pathway (Lee et al. 2015). Unlike PDE5A, PDE9A localizes to T-tubule membranes, and its inhibition attenuates cardiac hypertrophy induced by pressure overload regardless of whether NOS is concomitantly inhibited or not. PDE9 expression is substantially increased in diseased hearts in both humans with heart failure and in pressure-overloaded mice. While both PDE5 and PDE9 target cGMP and regulate PKG, the downstream protein targets have some overlap, but are not identical. Furthermore, there are different transcriptional signatures with selective inhibition of each as well (Lee et al. 2015).

13.4 Modifying Intracellular Localization and Targeting by cGMP/PKG

When PKG1α is stimulated by cGMP in both vascular smooth muscle and cardiac myocytes, it rapidly migrates from a normal diffuse cytosolic distribution to one intensified at the outer plasma membrane (Takimoto et al. 2009; Tang et al. 2003). This migration is thought to play an important role in enhancing kinase targeting to membrane resident proteins, enhancing their suppression. The mechanism for this translocation remains unknown, though it does not occur if a mutant PKG1α lacking the N-terminus coiled-coil (leucine zipper) binding motif is expressed (Blanton et al. 2012). The movement of PKG upon activation differs from PKA which generally remains stationary in microdomains associated with its binding with A-kinase anchoring protein complexes (Logue and Scott 2010).

There are several proteins known to be targeted by PKG that co-migrate to the plasma membrane upon stress stimulation. These include members of the regulator of G-protein signaling (RGS) proteins, RGS2 and RGS4, and the transient receptor potential canonical channel—TRPC6 (Fig. 13.2). Both RGS proteins contain leucine zipper motifs that interact with PKG1α. Subsequent phosphorylation and

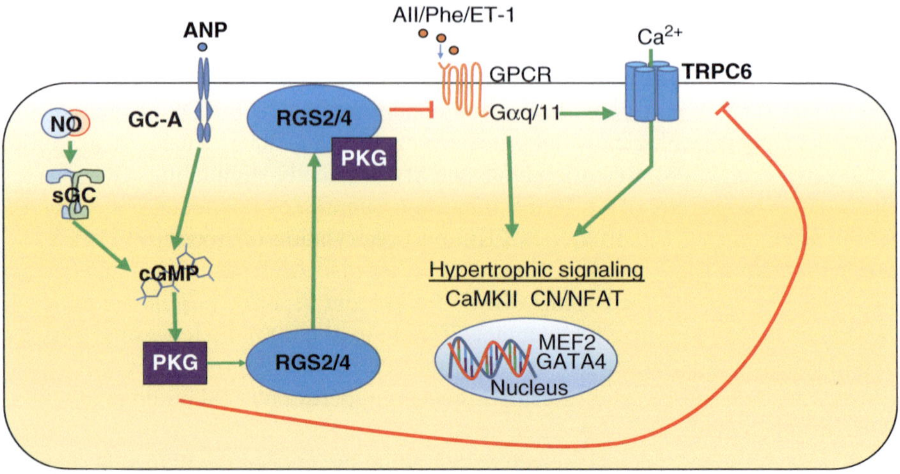

Fig. 13.2 PKG activation and translocation regulates RGS2/4 and TRPC6-mediated hypertrophic signaling. PKG activation counters hypertrophic stimulation in cardiac myocytes in part by its activation and co-migration with regulator of G-protein signaling proteins RGS2 and RGS4. Their translocation to the plasma membrane results in the suppression of Gαq/11-coupled agonist stimulation. PKG also phosphorylates and co-migrates with transient receptor potential canonical channel type 6 (TRPC6), inhibiting Ca^{2+} influx coupled to Gαq/11 stimulation. This Ca^{2+} plays an important role in activating calcineurin and consequently nuclear factor of activated T cells (NFAT). PKG activation suppresses this signaling as well. *Ang II* angiotensin II, *CaMKII* Ca^{2+}/calmodulin-dependent protein kinase II, *CN* calcineurin, *ET-1* endothelin-1, *GATA4* GATA-binding protein 4, *GPCR* G-protein coupled receptor, *MEF2* myocyte enhancer factor 2, *NO* nitric oxide, *Phe* phenylephrine. Other abbreviations as in Fig. 13.1

activation of the RGS protein by PKG1α and co-migration to the plasma membrane results in the dissociation of GTP from the activated Gαq/11 protein, suppressing agonist-receptor activity (Zhang and Mende 2011; Heximer et al. 1997; Tamirisa et al. 1999; Roy et al. 2003). RGS4 is upregulated in the failing human myocardium (Mittmann et al. 2002), where it primarily localizes to cardiomyocytes (Tokudome et al. 2008). This may confer an adaptation to Gαq-hyperactivation by neurohormones such as norepinephrine, angiotensin II, and endothelin. RGS4 stimulation plays an important role in NP-mediated suppression of myocardial hypertrophy (Tokudome et al. 2008). A similar impact is observed with RGS2 (Takimoto et al. 2009), which is activated by NO- and NP-dependent PKG (Klaiber et al. 2010). RGS2 is a more selective inhibitor of Gαq signaling and potent suppressor of associated hypertrophic signaling (Hao et al. 2006; Zhang et al. 2006). RGS2 knockout mice exhibit exacerbated cardiac phenotype in response to pressure overload (Takimoto et al. 2009), and this is rescued by co-inhibiting phospholipase-C activity, consistent with its targeting Gαq-coupled signaling.

Another localization-dependent PKG target is TRPC6, a nonselective, non-voltage-gated cation channel that conducts principally calcium (Dietrich and Gudermann 2014). It resides in different microdomains within the cell but is translocated to the plasma membrane upon activation (Cayouette et al. 2004; Chaudhuri et al. 2016). TRPC1, 3, and 6 have been best studied in the heart and are each linked to pathological cardiac hypertrophy and fibrosis (Bush et al. 2006; Kuwahara et al. 2006; Nakayama et al. 2006; Onohara et al. 2006). The calcium current transmitted through the channels activates calcineurin- and calmodulin-dependent protein kinase II (CaMKII), which couple to transcriptional regulators mediated as hypertrophic and pro-fibrotic gene program (Klaiber et al. 2010; Nakamura et al. 2015; Shi et al. 2013). All three channel proteins can be modified by PKG, with specific N-terminus residues being identified and proven functional for TRPC3 and TRPC6 (Takahashi et al. 2008; Koitabashi et al. 2010; Zhang et al. 2014a). Phosphorylation reduces channel conductance and suppresses activation of NFAT and CaMKII in cells stimulated by pro-hypertrophic agonists (Nakayama et al. 2006; Wu et al. 2010; Makarewich et al. 2014; Seo et al. 2014b; Koitabashi et al. 2010; Nakamura et al. 2015). PKG1α modification does not relocate TRPC6 from the plasma membrane to the cell interior in HEK cells (Takahashi et al. 2008), suggesting its suppression of activity involves direct regulation. Phosphorylation of TRPC6 may impact other channel proteins as well, by acting as a poison peptide in the commonly hetero-tetrameric channel.

PKG1α also plays an important role in myocyte mechanosensing, and this requires TRPC6 targeting (Seo et al. 2014a). Upon an increase in cardiac muscle (or myocyte) length and systolic load, force rises immediately due to length-dependent calcium activation, but no calcium transient change is observed. However, a subsequent and slower force rise follows, called stress-stimulated contractility or slow force response, and this is calcium dependent and attributed to mechanically activated cation channels (Seo et al. 2014a; Kockskamper et al. 2008; Cingolani et al. 2013). Intriguingly, activation of PKG virtually eliminates stress-stimulated contractility (Seo et al. 2014a). The effect from PKG activation requires TRPC6, since genetic deletion of the cation channel protein prevents the PKG effect. The precise localization of TRPC6 channels in the outer membrane remains unknown, though some have suggest the channels reside in part in caveolin-3-enriched microdomains (Makarewich et al. 2014). Disruption of the plasma membrane in cells and muscle lacking dystrophin augments mechano-stimulation via TRPC6 in particular and likely contributes to disease pathobiology. However, this effect is also suppressible by PKG activation.

PKG1α also phosphorylates proteins in the proteasome (Ranek et al. 2013, 2014) to enhance protein quality control, phospholamban (Mattiazzi et al. 2005) and phospholemman (Madhani et al. 2010) to regulate calcium cycling, and titin (Kruger et al. 2009), myosin-binding protein C (Thoonen et al. 2015), and troponin-I to modulate sarcomere function. However, whether these targets are differentially impacted by the signaling source of cGMP and/or PDE regulating its localized concentration remains to be determined.

13.5 Role of Oxidative Stress in Altering cGMP/PKG Signaling

Given the prominent role of NO to the cGMP signaling pathway, it is perhaps not surprising that excess production of reactive oxygen species can impact this signaling cascade. However, this impact is now recognized to extend far beyond the chemical interaction of NO and superoxide and formation of nitrosative species, but to

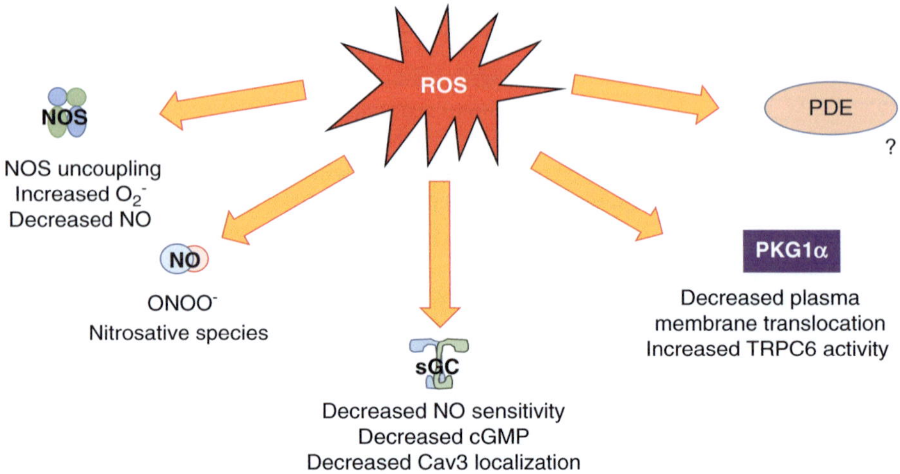

Fig. 13.3 Mechanisms of ROS disruption/modulation of cGMP/PKG signaling. Reactive oxygen species (ROS) interferes with the normal NO-cGMP-PKG signaling cascade in myocytes by modifying the proteins and/or messengers. ROS uncouples NOS resulting in decreased NO and increased superoxide anion synthesis by the enzyme. NO chemically interacts with oxidative species to form nitrosative species, such as peroxynitrite ($ONOO^-$), limiting the normal signaling associated with NO—including its activation of sGC. ROS modifies sGC (oxidation) which reduces its responsiveness to NO and this cGMP synthesis. This in part involves a decline in sGC-caveolin-3 (Cav3) microdomain localization augmenting sGC to oxidation. ROS can oxidize PKG1α to form a dimer between cysteine 42 residues on the N-terminus homomonomers. When sustained for an hour or more, this suppresses PKG1α translocation to the plasma membrane and impacts its negative modulation of TRPC6. The result is enhanced hypertrophic signaling. Abbreviations as in earlier figures

involve posttranslational modification and associated localization of key components of the signaling pathway (Fig. 13.3). Oxidants alter NOS at its heme-containing core, deplete the cell of the obligate NOS cofactor tetrahydrobiopterin (BH4), and result in NOS3 S-glutathionylation. All three changes lead to what is termed enzyme uncoupling, wherein superoxide rather than NO is generated from L-arginine catabolism (Karbach et al. 2014). Oxidation of sGC also occurs at its heme-core resulting in depressed NO responsiveness and thereby cGMP synthesis. Lastly, oxidation of PKG1α involves formation of a disulfide bond between homodimer cysteine-42 residues downstream of the protein-binding domain. This alters intracellular localization (Prysyazhna and Eaton 2015) as well as intrinsic kinase activity (Burgoyne et al. 2007).

In vascular cells, calcium-induced NOS3 activity results in the membrane translocation of sGC that potentiates the synthesis of cGMP (Zabel et al. 2002). Both cytosolic and membrane localization of sGC are observed in cardiomyocytes (Tsai et al. 2012; Liu et al. 2013), and this likely plays a role in its redox regulation. At rest, the sGC β_1-subunit is found in caveolin-3-enriched membranes with a smaller component in heavier membrane subfractions. The location impacts NO sensitivity, which is 3× greater if sGC is in caveolin-3-enriched domains related to differences in sGC oxidation. The caveolae microdomain serves as a reductive estuary where sGC is most potent for generating cGMP, and migration from this domain by sustained pressure overload (mice) or volume overload (dogs) may contribute to depressed NO-sGC-cGMP. Caveolin-3 knockout mice also have greater sGC oxidation and blunted sGC sensitivity to NO, suggesting a functional requirement for a caveolin-3/sGC compartment for optimal NO-stimulated cGMP signaling.

Oxidation on PKG1α at its C42 residue increases its kinase activity in the absence of cGMP in vitro, and this regulates small resistance artery tone and blood pressure in vivo. (Prysyazhna et al. 2012; Stubbert et al. 2014). The C42 residue needed for disulfide formation is specific to the PKG1α isoform, the most prominent in both vascular smooth muscle and cardiac myocytes, and is reactive to H_2O_2, nitrocysteine, and H_2S (Burgoyne and Eaton 2009; Stubbert et al. 2014). Mice expressing a "redox-dead" PKG1α C42S knock-in (C42S-KI) develop mild spontaneous hypertension and show reduced vasodilation with nitroglycerin and sepsis (Prysyazhna et al. 2012; Rudyk et al. 2012, 2013). This supports a role of PKG1α oxidation to blood pressure regulation and hypotensive responses with oxidative stress. H_2O_2-induced PKG1α oxidation induces trafficking from cytoplasm to the plasma membrane in smooth muscle (Dou et al. 2012; Zhang et al. 2012), where it activates large-conductance Ca^{2+}-activated K^+ channels (BK_{Ca}), resulting in H_2O_2-induced vasodilation (Zhang et al. 2012).

In cardiomyocytes, studies have shown PKG1α oxidation also impacts its intracellular localization, but rather than sending it to the plasma membrane, sustained oxidation moves the kinase to the cytosol (Nakamura et al. 2015). C42S-KI mice are actually protected against chronic pressure overload, and this is accompanied by a sustained relocalization of the mutant PKG1α to the plasma membrane. This appears to improve PKG1α suppression of TRPC6 by direct phosphorylation. Myocytes expressing both the redox-dead PKG1α and mutant TRPC6 that cannot be phosphorylated by PKG1α (T70A, S322A) lose the benefit from the PKG1α C42S mutation. This is not likely the sole differentially targeted protein related to PKG1α redox, and recent studies have revealed a number of these differences (Scotcher et al. 2016). The regulating sources of ROS and associated antioxidants for PKG1α oxidation remain unknown. However, studies have demonstrated that therapy which attenuates PKG1α oxidation also attenuates cardiac injury (Prysyazhna et al. 2016).

Conclusion

In textbooks and even recent reviews, the majority of diagrams depicting the cGMP/PKG signaling system show multiple inputs converging on a common cellular cGMP and PKG pool. The diagram will depict placement of the proteins as static and downstream targets as determined simply by PKG1α activation. We now know that this is incorrect and that instead there are multiple microdomains regulating selective cGMP and associated PKG signaling. Moreover, these domains are mutable due to posttranslational modifications such as oxidation, intracellular translocation, and changes in the net components of a local signalosome. We also know that cardiac stress remodels this cascade, that this applies to sustained stress-induced myocardial disease, and that it impacts the capacity of the PKG1α activation to protect the heart as well as determine which therapeutic agents are likely to be most effective. Far more remains unknown. The protein partners responsible for PKG1α migration are not yet resolved. How oxidation moves the kinase to and from the plasma membrane is unknown. What different kinase targets are facilitated and if this impacts the efficacy of PDE modulation remains to be determined. Solving these questions should help develop more precise leveraging of the pathway as a treatment for heart disease.

Compliance with Ethical Standards

Conflict of Interest Statements B. L. L. and D. I. L. declare that they have no conflicts of interest to declare. D. A. K. has consulting relationships with Intra-Cellular Therapies Inc., Ironwood Inc., and Theravance Inc. and has received research funding from Pfizer and Boehringer Ingelheim.

References

Adams SR, Harootunian AT, Buechler YJ, Taylor SS, Tsien RY (1991) Fluorescence ratio imaging of cyclic AMP in single cells. Nature 349:694–697

Ahmad F, Shen W, Vandeput F, Szabo-Fresnais N, Krall J, Degerman E, Goetz F, Klussmann E, Movsesian M, Manganiello V (2015) Regulation of sarcoplasmic reticulum Ca2+ ATPase 2 (SERCA2) activity by phosphodiesterase 3A (PDE3A) in human myocardium: phosphorylation-dependent interaction of PDE3A1 with SERCA2. J Biol Chem 290:6763–6776

Anand-Srivastava MB (2005) Natriuretic peptide receptor-C signaling and regulation. Peptides 26:1044–1059

Balligand JL (2013) Beta3-adrenoreceptors in cardiovascular diseases: new roles for an “old” receptor. Curr Drug Deliv 10:64–66

Blanton RM, Takimoto E, Lane AM, Aronovitz M, Piotrowski R, Karas RH, Kass DA, Mendelsohn ME (2012) Protein kinase G Ialpha inhibits pressure overload-induced cardiac remodeling and is required for the cardioprotective effect of sildenafil in vivo. J Am Heart Assoc 1: e003731

Bubb KJ, Trinder SL, Baliga RS, Patel J, Clapp LH, Macallister RJ, Hobbs AJ (2014) Inhibition of phosphodiesterase 2 augments cGMP and cAMP signaling to ameliorate pulmonary hypertension. Circulation 130:496–507

Burgoyne JR, Eaton P (2009) Transnitrosylating nitric oxide species directly activate type I protein kinase A, providing a novel adenylate cyclase-independent cross-talk to beta-adrenergic-like signaling. J Biol Chem 284:29260–29268

Burgoyne JR, Madhani M, Cuello F, Charles RL, Brennan JP, Schroder E, Browning DD, Eaton P (2007) Cysteine redox sensor in PKGIa enables oxidant-induced activation. Science 317:1393–1397

Bush EW, Hood DB, Papst PJ, Chapo JA, Minobe W, Bristow MR, Olson EN, Mckinsey TA (2006) Canonical transient receptor potential channels promote cardiomyocyte hypertrophy through activation of calcineurin signaling. J Biol Chem 281:33487–33496

Buys ES, Cauwels A, Raher MJ, Passeri JJ, Hobai I, Cawley SM, Rauwerdink KM, Thibault H, Sips PY, Thoonen R, Scherrer-Crosbie M, Ichinose F, Brouckaert P, Bloch KD (2009) sGC(alpha)1(beta)1 attenuates cardiac dysfunction and mortality in murine inflammatory shock models. Am J Physiol Heart Circ Physiol 297:H654–H663

Castro LR, Verde I, Cooper DM, Fischmeister R (2006) Cyclic guanosine monophosphate compartmentation in rat cardiac myocytes. Circulation 113:2221–2228

Cayouette S, Lussier MP, Mathieu EL, Bousquet SM, Boulay G (2004) Exocytotic insertion of TRPC6 channel into the plasma membrane upon Gq protein-coupled receptor activation. J Biol Chem 279:7241–7246

Chaudhuri P, Rosenbaum MA, Sinharoy P, Damron DS, Birnbaumer L, Graham LM (2016) Membrane translocation of TRPC6 channels and endothelial migration are regulated by Calmodulin and PI3 kinase activation. Proc Natl Acad Sci U S A 113:2110–2115

Chen W, Spitzl A, Mathes D, Nikolaev VO, Werner F, Weirather J, Spiranec K, Rock K, Fischer JW, Kammerer U, Stegner D, Baba HA, Hofmann U, Frantz S, Kuhn M (2016) Endothelial actions of ANP enhance myocardial inflammatory infiltration in the early phase after acute infarction. Circ Res 119:237–248

Cingolani HE, Perez NG, Cingolani OH, Ennis IL (2013) The Anrep effect: 100 years later. Am J Physiol Heart Circ Physiol 304:H175–H182

Conti M, Beavo J (2007) Biochemistry and physiology of cyclic nucleotide phosphodiesterases: essential components in cyclic nucleotide signaling. Annu Rev Biochem 76:481–511

Del Ry S (2013) C-type natriuretic peptide: a new cardiac mediator. Peptides 40:93–98

Del Ry S, Cabiati M, Vozzi F, Battolla B, Caselli C, Forini F, Segnani C, Prescimone T, Giannessi D, Mattii L (2011) Expression of C-type natriuretic peptide and its receptor NPR-B in cardiomyocytes. Peptides 32:1713–1718

Derbyshire ER, Marletta MA (2012) Structure and regulation of soluble guanylate cyclase. Annu Rev Biochem 81:533–559

Dietrich A, Gudermann T (2014) TRPC6: physiological function and pathophysiological relevance. Handb Exp Pharmacol 222:157–188

Dipilato LM, Cheng X, Zhang J (2004) Fluorescent indicators of cAMP and Epac activation reveal differential dynamics of cAMP signaling within discrete subcellular compartments. Proc Natl Acad Sci U S A 101:16513–16518

Dou D, Zheng X, Liu J, Xu X, Ye L, Gao Y (2012) Hydrogen peroxide enhances vasodilatation by increasing dimerization of cGMP-dependent protein kinase type Ialpha. Circ J 76:1792–1798

Fisher DA, Smith JF, Pillar JS, St Denis SH, Cheng JB (1998) Isolation and characterization of PDE9A, a novel human cGMP-specific phosphodiesterase. J Biol Chem 273:15559–15564

Forstermann U, Sessa WC (2012) Nitric oxide synthases: regulation and function. Eur Heart J 33:829–837. 837a–837d

Francis SH, Blount MA, Corbin JD (2011) Mammalian cyclic nucleotide phosphodiesterases: molecular mechanisms and physiological functions. Physiol Rev 91:651–690

Geiselhoringer A, Gaisa M, Hofmann F, Schlossmann J (2004) Distribution of Iraq and cGKI-isoforms in murine tissues. FEBS Lett 575:19–22

Gheorghiade M, Marti CN, Sabbah HN, Roessig L, Greene SJ, Bohm M, Burnett JC, Campia U, Cleland JG, Collins SP, Fonarow GC, Levy PD, Metra M, Pitt B, Ponikowski P, Sato N, Voors AA, Stasch JP, Butler J, Academic Research Team in Heart F (2013) Soluble guanylate cyclase: a potential therapeutic target for heart failure. Heart Fail Rev 18:123–134

Gisbert MP, Fischmeister R (1988) Atrial natriuretic factor regulates the calcium current in frog isolated cardiac cells. Circ Res 62:660–667

Gotz KR, Sprenger JU, Perera RK, Steinbrecher JH, Lehnart SE, Kuhn M, Gorelik J, Balligand JL, Nikolaev VO (2014) Transgenic mice for real-time visualization of cGMP in intact adult cardiomyocytes. Circ Res 114:1235–1245

Hao J, Michalek C, Zhang W, Zhu M, Xu X, Mende U (2006) Regulation of cardiomyocyte signaling by RGS proteins: differential selectivity towards G proteins and susceptibility to regulation. J Mol Cell Cardiol 41:51–61

Heximer SP, Watson N, Linder ME, Blumer KJ, Hepler JR (1997) RGS2/G0S8 is a selective inhibitor of Gqalpha function. Proc Natl Acad Sci U S A 94:14389–14393

Honda A, Adams SR, Sawyer CL, Lev-Ram V, Tsien RY, Dostmann WR (2001) Spatiotemporal dynamics of guanosine 3',5'-cyclic monophosphate revealed by a genetically encoded, fluorescent indicator. Proc Natl Acad Sci U S A 98:2437–2442

Horio T, Tokudome T, Maki T, Yoshihara F, Suga S, Nishikimi T, Kojima M, Kawano Y, Kangawa K (2003) Gene expression, secretion, and autocrine action of C-type natriuretic peptide in cultured adult rat cardiac fibroblasts. Endocrinology 144:2279–2284

Inserte J, Garcia-Dorado D (2015) The cGMP/PKG pathway as a common mediator of cardioprotection: translatability and mechanism. Br J Pharmacol 172:1996–2009

Johnson WB, Katugampola S, Able S, Napier C, Harding SE (2012) Profiling of cAMP and cGMP phosphodiesterases in isolated ventricular cardiomyocytes from human hearts: comparison with rat and Guinea pig. Life Sci 90:328–336

Karbach S, Wenzel P, Waisman A, Munzel T, Daiber A (2014) eNOS uncoupling in cardiovascular diseases—the role of oxidative stress and inflammation. Curr Pharm Des 20:3579–3594

Kass DA (2012) Cardiac role of cyclic-GMP hydrolyzing phosphodiesterase type 5: from experimental models to clinical trials. Curr Heart Fail Rep 9:192–199

Kim GE, Kass DA (2016) Cardiac phosphodiesterases and their modulation for treating heart disease. In: Handbook of experimental pharmacology. Springer, Berlin

Kinugawa S, Huang H, Wang Z, Kaminski PM, Wolin MS, Hintze TH (2005) A defect of neuronal nitric oxide synthase increases xanthine oxidase-derived superoxide anion and attenuates the control of myocardial oxygen consumption by nitric oxide derived from endothelial nitric oxide synthase. Circ Res 96:355–362

Klaiber M, Kruse M, Volker K, Schroter J, Feil R, Freichel M, Gerling A, Feil S, Dietrich A, Londono JE, Baba HA, Abramowitz J, Birnbaumer L, Penninger JM, Pongs O, Kuhn M (2010) Novel insights into the mechanisms mediating the local antihypertrophic effects of cardiac atrial natriuretic peptide: role of cGMP-dependent protein kinase and RGS2. Basic Res Cardiol 105:583–595

Knight WE, Chen S, Zhang Y, Oikawa M, Wu M, Zhou Q, Miller CL, Cai Y, Mickelsen DM, Moravec C, Small EM, Abe J, Yan C (2016) PDE1C deficiency antagonizes pathological cardiac remodeling and dysfunction. Proc Natl Acad Sci U S A 113(45):E7116–E7125

Kockskamper J, Von Lewinski D, Khafaga M, Elgner A, Grimm M, Eschenhagen T, Gottlieb PA, Sachs F, Pieske B (2008) The slow force response to stretch in atrial and ventricular myocardium from human heart: functional relevance and subcellular mechanisms. Prog Biophys Mol Biol 97:250–267

Koitabashi N, Aiba T, Hesketh GG, Rowell J, Zhang M, Takimoto E, Tomaselli GF, Kass DA (2010) Cyclic GMP/PKG-dependent inhibition of TRPC6 channel activity and expression negatively regulates cardiomyocyte NFAT activation novel mechanism of cardiac stress modulation by PDE5 inhibition. J Mol Cell Cardiol 48:713–724

Kokkonen K, Kass DA (2016) Nanodomain regulation of cardiac cyclic nucleotide signaling by phosphodiesterases. Annu Rev Pharmacol Toxicol 57:455–479

Koller KJ, Goeddel DV (1992) Molecular biology of the natriuretic peptides and their receptors. Circulation 86:1081–1088

Krawutschke C, Koesling D, Russwurm M (2015) Cyclic GMP in vascular relaxation: export is of similar importance as degradation. Arterioscler Thromb Vasc Biol 35:2011–2019

Kruger M, Kotter S, Grutzner A, Lang P, Andresen C, Redfield MM, Butt E, Dos Remedios CG, Linke WA (2009) Protein kinase G modulates human myocardial passive stiffness by phosphorylation of the titin springs. Circ Res 104:87–94

Kuhn M (2016) Molecular physiology of membrane guanylyl cyclase receptors. Physiol Rev 96:751–804

Kukreja RC, Salloum F, Das A, Ockaili R, Yin C, Bremer YA, Fisher PW, Wittkamp M, Hawkins J, Chou E, Kukreja AK, Wang X, Marwaha VR, Xi L (2005) Pharmacological preconditioning with sildenafil: basic mechanisms and clinical implications. Vasc Pharmacol 42:219–232

Kuwahara K, Wang Y, Mcanally J, Richardson JA, Bassel-Duby R, Hill JA, Olson EN (2006) TRPC6 fulfills a calcineurin signaling circuit during pathologic cardiac remodeling. J Clin Invest 116:3114–3126

Langenickel TH, Buttgereit J, Pagel-Langenickel I, Lindner M, Monti J, Beuerlein K, Al-Saadi N, Plehm R, Popova E, Tank J, Dietz R, Willenbrock R, Bader M (2006) Cardiac hypertrophy in transgenic rats expressing a dominant-negative mutant of the natriuretic peptide receptor B. Proc Natl Acad Sci U S A 103:4735–4740

Layland J, Li JM, Shah AM (2002) Role of cyclic GMP-dependent protein kinase in the contractile response to exogenous nitric oxide in rat cardiac myocytes. J Physiol 540:457–467

Lee DI, Vahebi S, Tocchetti CG, Barouch LA, Solaro RJ, Takimoto E, Kass DA (2010) PDE5A suppression of acute beta-adrenergic activation requires modulation of myocyte beta-3 signaling coupled to PKG-mediated troponin I phosphorylation. Basic Res Cardiol 105:337–347

Lee DI, Zhu G, Sasaki T, Cho GS, Hamdani N, Holewinski R, Jo SH, Danner T, Zhang M, Rainer PP, Bedja D, Kirk JA, Ranek MJ, Dostmann WR, Kwon C, Margulies KB, Van Eyk JE, Paulus

WJ, Takimoto E, Kass DA (2015) Phosphodiesterase 9A controls nitric-oxide-independent cGMP and hypertrophic heart disease. Nature 519:472–476
Li Y, Sarkar O, Brochu M, Anand-Srivastava MB (2014) Natriuretic peptide receptor-C attenuates hypertension in spontaneously hypertensive rats: role of nitroxidative stress and Gi proteins. Hypertension 63:846–855
Liu H, Maurice DH (1998) Expression of cyclic GMP-inhibited phosphodiesterases 3A and 3B (PDE3A and PDE3B) in rat tissues: differential subcellular localization and regulated expression by cyclic AMP. Br J Pharmacol 125:1501–1510
Liu Y, Dillon AR, Tillson M, Makarewich C, Nguyen V, Dell'italia L, Sabri AK, Rizzo V, Tsai EJ (2013) Volume overload induces differential spatiotemporal regulation of myocardial soluble guanylyl cyclase in eccentric hypertrophy and heart failure. J Mol Cell Cardiol 60:72–83
Logue JS, Scott JD (2010) Organizing signal transduction through A-kinase anchoring proteins (AKAPs). FEBS J 277:4370–4375
Madhani M, Hall AR, Cuello F, Charles RL, Burgoyne JR, Fuller W, Hobbs AJ, Shattock MJ, Eaton P (2010) Phospholemman Ser69 phosphorylation contributes to sildenafil-induced cardioprotection against reperfusion injury. Am J Physiol Heart Circ Physiol 299:H827–H836
Makarewich CA, Zhang H, Davis J, Correll RN, Trappanese DM, Hoffman NE, Troupes CD, Berretta RM, Kubo H, Madesh M, Chen X, Gao E, Molkentin JD, Houser SR (2014) Transient receptor potential channels contribute to pathological structural and functional remodeling after myocardial infarction. Circ Res 115:567–580
Mattiazzi A, Mundina-Weilenmann C, Guoxiang C, Vittone L, Kranias E (2005) Role of phospholamban phosphorylation on Thr17 in cardiac physiological and pathological conditions. Cardiovasc Res 68:366–375
Mcmurray JJ, Packer M, Desai AS, Gong J, Lefkowitz MP, Rizkala AR, Rouleau JL, Shi VC, Solomon SD, Swedberg K, Zile MR, PARADIGM-HF Investigators & Committees (2014) Angiotensin-neprilysin inhibition versus enalapril in heart failure. N Engl J Med 371:993–1004
Mehel H, Emons J, Vettel C, Wittkopper K, Seppelt D, Dewenter M, Lutz S, Sossalla S, Maier LS, Lechene P, Leroy J, Lefebvre F, Varin A, Eschenhagen T, Nattel S, Dobrev D, Zimmermann WH, Nikolaev VO, Vandecasteele G, Fischmeister R, El-Armouche A (2013) Phosphodiesterase-2 is up-regulated in human failing hearts and blunts beta-adrenergic responses in cardiomyocytes. J Am Coll Cardiol 62:1596–1606
Mery PF, Pavoine C, Belhassen L, Pecker F, Fischmeister R (1993) Nitric oxide regulates cardiac Ca2+ current. Involvement of cGMP-inhibited and cGMP-stimulated phosphodiesterases through guanylyl cyclase activation. J Biol Chem 268:26286–26295
Miller CL, Oikawa M, Cai Y, Wojtovich AP, Nagel DJ, Xu X, Xu H, Florio V, Rybalkin SD, Beavo JA, Chen YF, Li JD, Blaxall BC, Abe J, Yan C (2009) Role of Ca2+/calmodulin-stimulated cyclic nucleotide phosphodiesterase 1 in mediating cardiomyocyte hypertrophy. Circ Res 105:956–964
Miller CL, Cai Y, Oikawa M, Thomas T, Dostmann WR, Zaccolo M, Fujiwara K, Yan C (2011) Cyclic nucleotide phosphodiesterase 1A: a key regulator of cardiac fibroblast activation and extracellular matrix remodeling in the heart. Basic Res Cardiol 106:1023–1039
Mittmann C, Chung CH, Hoppner G, Michalek C, Nose M, Schuler C, Schuh A, Eschenhagen T, Weil J, Pieske B, Hirt S, Wieland T (2002) Expression of ten RGS proteins in human myocardium: functional characterization of an upregulation of RGS4 in heart failure. Cardiovasc Res 55:778–786
Mongillo M, Tocchetti CG, Terrin A, Lissandron V, Cheung YF, Dostmann WR, Pozzan T, Kass DA, Paolocci N, Houslay MD, Zaccolo M (2006) Compartmentalized phosphodiesterase-2 activity blunts beta-adrenergic cardiac inotropy via an NO/cGMP-dependent pathway. Circ Res 98:226–234
Nagayama T, Zhang M, Hsu S, Takimoto E, Kass DA (2008) Sustained soluble guanylate cyclase stimulation offsets nitric-oxide synthase inhibition to restore acute cardiac modulation by sildenafil. J Pharmacol Exp Ther 326:380–387
Nakamura T, Ranek MJ, Lee DI, Shalkey Hahn V, Kim C, Eaton P, Kass DA (2015) Prevention of PKG1alpha oxidation augments cardioprotection in the stressed heart. J Clin Invest 125:2468–2472

Nakayama H, Wilkin BJ, Bodi I, Molkentin JD (2006) Calcineurin-dependent cardiomyopathy is activated by TRPC in the adult mouse heart. FASEB J 20:1660–1670

Nausch LW, Ledoux J, Bonev AD, Nelson MT, Dostmann WR (2008) Differential patterning of cGMP in vascular smooth muscle cells revealed by single GFP-linked biosensors. Proc Natl Acad Sci U S A 105:365–370

Nikolaev VO, Gambaryan S, Lohse MJ (2006) Fluorescent sensors for rapid monitoring of intracellular cGMP. Nat Methods 3:23–25

Onohara N, Nishida M, Inoue R, Kobayashi H, Sumimoto H, Sato Y, Mori Y, Nagao T, Kurose H (2006) TRPC3 and TRPC6 are essential for angiotensin II-induced cardiac hypertrophy. EMBO J 25:5305–5316

Perera RK, Sprenger JU, Steinbrecher JH, Hubscher D, Lehnart SE, Abesser M, Schuh K, El-Armouche A, Nikolaev VO (2015) Microdomain switch of cGMP-regulated phosphodiesterases leads to ANP-induced augmentation of beta-adrenoceptor-stimulated contractility in early cardiac hypertrophy. Circ Res 116:1304–1311

Prysyazhna O, Eaton P (2015) Redox regulation of cGMP-dependent protein kinase Ialpha in the cardiovascular system. Front Pharmacol 6:139

Prysyazhna O, Rudyk O, Eaton P (2012) Single atom substitution in mouse protein kinase G eliminates oxidant sensing to cause hypertension. Nat Med 18:286–290

Prysyazhna O, Burgoyne JR, Scotcher J, Grover S, Kass D, Eaton P (2016) Phosphodiesterase 5 inhibition limits doxorubicin-induced heart failure by attenuating protein kinase G Ialpha oxidation. J Biol Chem 291:17427–17436

Ranek MJ, Terpstra EJ, Li J, Kass DA, Wang X (2013) Protein kinase G positively regulates proteasome-mediated degradation of misfolded proteins. Circulation 128:365–376

Ranek MJ, Kost CK Jr, Hu C, Martin DS, Wang X (2014) Muscarinic 2 receptors modulate cardiac proteasome function in a protein kinase G-dependent manner. J Mol Cell Cardiol 69:43–51

Roy AA, Lemberg KE, Chidiac P (2003) Recruitment of RGS2 and RGS4 to the plasma membrane by G proteins and receptors reflects functional interactions. Mol Pharmacol 64:587–593

Rudyk O, Prysyazhna O, Burgoyne JR, Eaton P (2012) Nitroglycerin fails to lower blood pressure in redox-dead Cys42Ser PKG1alpha knock-in mouse. Circulation 126:287–295

Rudyk O, Phinikaridou A, Prysyazhna O, Burgoyne JR, Botnar RM, Eaton P (2013) Protein kinase G oxidation is a major cause of injury during sepsis. Proc Natl Acad Sci U S A 110: 9909–9913

Scotcher J, Prysyazhna O, Boguslavskyi A, Kistamas K, Hadgraft N, Martin ED, Worthington J, Rudyk O, Rodriguez Cutillas P, Cuello F, Shattock MJ, Marber MS, Conte MR, Greenstein A, Greensmith DJ, Venetucci L, Timms JF, Eaton P (2016) Disulfide-activated protein kinase G Ialpha regulates cardiac diastolic relaxation and fine-tunes the Frank-Starling response. Nat Commun 7:13187

Sears CE, Bryant SM, Ashley EA, Lygate CA, Rakovic S, Wallis HL, Neubauer S, Terrar DA, Casadei B (2003) Cardiac neuronal nitric oxide synthase isoform regulates myocardial contraction and calcium handling. Circ Res 92:e52–e59

Seddon M, Shah AM, Casadei B (2007) Cardiomyocytes as effectors of nitric oxide signalling. Cardiovasc Res 75:315–326

Senzaki H, Smith CJ, Juang GJ, Isoda T, Mayer SP, Ohler A, Paolocci N, Tomaselli GF, Hare JM, Kass DA (2001) Cardiac phosphodiesterase 5 (cGMP-specific) modulates beta-adrenergic signaling in vivo and is down-regulated in heart failure. FASEB J 15:1718–1726

Seo K, Rainer PP, Lee DI, Hao S, Bedja D, Birnbaumer L, Cingolani OH, Kass DA (2014a) Hyperactive adverse mechanical stress responses in dystrophic heart are coupled to transient receptor potential canonical 6 and blocked by cGMP-protein kinase G modulation. Circ Res 114:823–832

Seo K, Rainer PP, Shalkey Hahn V, Lee DI, Jo SH, Andersen A, Liu T, Xu X, Willette RN, Lepore JJ, Marino JP Jr, Birnbaumer L, Schnackenberg CG, Kass DA (2014b) Combined TRPC3 and TRPC6 blockade by selective small-molecule or genetic deletion inhibits pathological cardiac hypertrophy. Proc Natl Acad Sci U S A 111:1551–1556

Shi J, Geshi N, Takahashi S, Kiyonaka S, Ichikawa J, Hu Y, Mori Y, Ito Y, Inoue R (2013) Molecular determinants for cardiovascular TRPC6 channel regulation by Ca2+/calmodulin-dependent kinase II. J Physiol 591:2851–2866

Simon JN, Duglan D, Casadei B, Carnicer R (2014) Nitric oxide synthase regulation of cardiac excitation-contraction coupling in health and disease. J Mol Cell Cardiol 73:80–91

Sonnenburg WK, Rybalkin SD, Bornfeldt KE, Kwak KS, Rybalkina IG, Beavo JA (1998) Identification, quantitation, and cellular localization of PDE1 calmodulin-stimulated cyclic nucleotide phosphodiesterases. Methods 14:3–19

Stangherlin A, Zaccolo M (2012) cGMP-cAMP interplay in cardiac myocytes: a local affair with far-reaching consequences for heart function. Biochem Soc Trans 40:11–14

Stangherlin A, Gesellchen F, Zoccarato A, Terrin A, Fields LA, Berrera M, Surdo NC, Craig MA, Smith G, Hamilton G, Zaccolo M (2011) cGMP signals modulate cAMP levels in a compartment-specific manner to regulate catecholamine-dependent signaling in cardiac myocytes. Circ Res 108:929–939

Stubbert D, Prysyazhna O, Rudyk O, Scotcher J, Burgoyne JR, Eaton P (2014) Protein kinase G Ialpha oxidation paradoxically underlies blood pressure lowering by the reductant hydrogen sulfide. Hypertension 64:1344–1351

Su J, Scholz PM, Weiss HR (2005) Differential effects of cGMP produced by soluble and particulate guanylyl cyclase on mouse ventricular myocytes. Exp Biol Med (Maywood) 230:242–250

Sun J, Picht E, Ginsburg KS, Bers DM, Steenbergen C, Murphy E (2006) Hypercontractile female hearts exhibit increased S-nitrosylation of the L-type Ca2+ channel alpha1 subunit and reduced ischemia/reperfusion injury. Circ Res 98:403–411

Takahashi S, Lin H, Geshi N, Mori Y, Kawarabayashi Y, Takami N, Mori MX, Honda A, Inoue R (2008) Nitric oxide-cGMP-protein kinase G pathway negatively regulates vascular transient receptor potential channel TRPC6. J Physiol 586:4209–4223

Takimoto E, Champion HC, Li M, Belardi D, Ren S, Rodriguez ER, Bedja D, Gabrielson KL, Wang Y, Kass DA (2005) Chronic inhibition of cyclic GMP phosphodiesterase 5A prevents and reverses cardiac hypertrophy. Nat Med 11:214–222

Takimoto E, Belardi D, Tocchetti CG, Vahebi S, Cormaci G, Ketner EA, Moens AL, Champion HC, Kass DA (2007) Compartmentalization of cardiac beta-adrenergic inotropy modulation by phosphodiesterase type 5. Circulation 115:2159–2167

Takimoto E, Koitabashi N, Hsu S, Ketner EA, Zhang M, Nagayama T, Bedja D, Gabrielson KL, Blanton R, Siderovski DP, Mendelsohn ME, Kass DA (2009) Regulator of G protein signaling 2 mediates cardiac compensation to pressure overload and antihypertrophic effects of PDE5 inhibition in mice. J Clin Invest 119:408–420

Tamirisa P, Blumer KJ, Muslin AJ (1999) RGS4 inhibits G-protein signaling in cardiomyocytes. Circulation 99:441–447

Tang KM, Wang GR, Lu P, Karas RH, Aronovitz M, Heximer SP, Kaltenbronn KM, Blumer KJ, Siderovski DP, Zhu Y, Mendelsohn ME (2003) Regulator of G-protein signaling-2 mediates vascular smooth muscle relaxation and blood pressure. Nat Med 9:1506–1512

Thoonen R, Giovanni S, Govindan S, Lee DI, Wang GR, Calamaras TD, Takimoto E, Kass DA, Sadayappan S, Blanton RM (2015) Molecular screen identifies cardiac myosin-binding protein-C as a protein kinase G-Ialpha substrate. Circ Heart Fail 8:1115–1122

Tokudome T, Kishimoto I, Horio T, Arai Y, Schwenke DO, Hino J, Okano I, Kawano Y, Kohno M, Miyazato M, Nakao K, Kangawa K (2008) Regulator of G-protein signaling subtype 4 mediates antihypertrophic effect of locally secreted natriuretic peptides in the heart. Circulation 117:2329–2339

Tsai EJ, Liu Y, Koitabashi N, Bedja D, Danner T, Jasmin JF, Lisanti MP, Friebe A, Takimoto E, Kass DA (2012) Pressure-overload-induced subcellular relocalization/oxidation of soluble guanylyl cyclase in the heart modulates enzyme stimulation. Circ Res 110:295–303

Vandeput F, Wolda SL, Krall J, Hambleton R, Uher L, Mccaw KN, Radwanski PB, Florio V, Movsesian MA (2007) Cyclic nucleotide phosphodiesterase PDE1C1 in human cardiac myocytes. J Biol Chem 282:32749–32757

Vettel C, Lammle S, Ewens S, Cervirgen C, Emons J, Ongherth A, Dewenter M, Lindner D, Westermann D, Nikolaev VO, Lutz S, Zimmermann WH, El-Armouche A (2014) PDE2-mediated cAMP hydrolysis accelerates cardiac fibroblast to myofibroblast conversion and is antagonized by exogenous activation of cGMP signaling pathways. Am J Physiol Heart Circ Physiol 306:H1246–H1252

Vettel C, Lindner M, Dewenter M, Lorenz K, Schanbacher C, Riedel M, Lammle S, Meinecke S, Mason FE, Sossalla S, Geerts A, Hoffmann M, Wunder F, Brunner FJ, Wieland T, Mehel H, Karam S, Lechene P, Leroy J, Vandecasteele G, Wagner M, Fischmeister R, El-Armouche A (2016) Phosphodiesterase 2 protects against catecholamine-induced arrhythmia and preserves contractile function after myocardial infarction. Circ Res 120(1):120–132

Vila-Petroff MG, Younes A, Egan J, Lakatta EG, Sollott SJ (1999) Activation of distinct cAMP-dependent and cGMP-dependent pathways by nitric oxide in cardiac myocytes. Circ Res 84:1020–1031

Wang Y, De Waard MC, Sterner-Kock A, Stepan H, Schultheiss HP, Duncker DJ, Walther T (2007) Cardiomyocyte-restricted over-expression of C-type natriuretic peptide prevents cardiac hypertrophy induced by myocardial infarction in mice. Eur J Heart Fail 9:548–557

Wang H, Viatchenko-Karpinski S, Sun J, Gyorke I, Benkusky NA, Kohr MJ, Valdivia HH, Murphy E, Gyorke S, Ziolo MT (2010) Regulation of myocyte contraction via neuronal nitric oxide synthase: role of ryanodine receptor S-nitrosylation. J Physiol 588:2905–2917

Wu X, Eder P, Chang B, Molkentin JD (2010) TRPC channels are necessary mediators of pathologic cardiac hypertrophy. Proc Natl Acad Sci U S A 107:7000–7005

Yue ZJ, Xu PT, Jiao B, Chang H, Song Z, Xie MJ, Yu ZB (2015) Nitric oxide protects L-type calcium channel of cardiomyocyte during long-term isoproterenol stimulation in tail-suspended rats. Biomed Res Int 2015:780814

Zabel U, Kleinschnitz C, Oh P, Nedvetsky P, Smolenski A, Muller H, Kronich P, Kugler P, Walter U, Schnitzer JE, Schmidt HH (2002) Calcium-dependent membrane association sensitizes soluble guanylyl cyclase to nitric oxide. Nat Cell Biol 4:307–311

Zaccolo M, Movsesian MA (2007) cAMP and cGMP signaling cross-talk: role of phosphodiesterases and implications for cardiac pathophysiology. Circ Res 100:1569–1578

Zhang P, Mende U (2011) Regulators of G-protein signaling in the heart and their potential as therapeutic targets. Circ Res 109:320–333

Zhang W, Anger T, Su J, Hao J, Xu X, Zhu M, Gach A, Cui L, Liao R, Mende U (2006) Selective loss of fine tuning of Gq/11 signaling by RGS2 protein exacerbates cardiomyocyte hypertrophy. J Biol Chem 281:5811–5820

Zhang M, Koitabashi N, Nagayama T, Rambaran R, Feng N, Takimoto E, Koenke T, O'Rourke B, Champion HC, Crow MT, Kass DA (2008) Expression, activity, and pro-hypertrophic effects of PDE5A in cardiac myocytes. Cell Signal 20:2231–2236

Zhang DX, Borbouse L, Gebremedhin D, Mendoza SA, Zinkevich NS, Li R, Gutterman DD (2012) H_2O_2-induced dilation in human coronary arterioles: role of protein kinase G dimerization and large-conductance Ca^{2+}-activated K+ channel activation. Circ Res 110:471–480

Zhang P, Ma Y, Wang Y, Ma X, Huang Y, Li RA, Wan S, Yao X (2014a) Nitric oxide and protein kinase G act on TRPC1 to inhibit 11,12-EET-induced vascular relaxation. Cardiovasc Res 104:138–146

Zhang YH, Jin CZ, Jang JH, Wang Y (2014b) Molecular mechanisms of neuronal nitric oxide synthase in cardiac function and pathophysiology. J Physiol 592:3189–3200

Zoccarato A, Surdo NC, Aronsen JM, Fields LA, Mancuso L, Dodoni G, Stangherlin A, Livie C, Jiang H, Sin YY, Gesellchen F, Terrin A, Baillie GS, Nicklin SA, Graham D, Szabo-Fresnais N, Krall J, Vandeput F, Movsesian M, Furlan L, Corsetti V, Hamilton G, Lefkimmiatis K, Sjaastad I, Zaccolo M (2015) Cardiac hypertrophy is inhibited by a local pool of cAMP regulated by phosphodiesterase 2. Circ Res 117:707–719

Part II

Calcium Microdomains

Distribution and Regulation of L-Type Ca^{2+} Channels in Cardiomyocyte Microdomains

14

Alexey V. Glukhov, Anamika Bhargava, and Julia Gorelik

Contents

A.V. Glukhov
Department of Medicine, University of Wisconsin-Madison School of Medicine and Public Health, Madison, WI 53705, USA
e-mail: aglukhov@medicine.wisc.edu

A. Bhargava
Department of Biotechnology, Indian Institute of Technology Hyderabad, Kandi, Sangareddy, Telangana 502285, India
e-mail: abhargava@iith.ac.in

J. Gorelik (✉)
Department of Cardiovascular Sciences, Imperial Centre for Translational and Experimental Medicine, National Heart and Lung Institute, Imperial College, London W120NN, UK
e-mail: j.gorelik@imperial.ac.uk

V. Nikolaev, M. Zaccolo (eds.), *Microdomains in the Cardiovascular System*, Cardiac and Vascular Biology 3, DOI 10.1007/978-3-319-54579-0_14

Abstract

Cardiac excitation involves action potential generation by individual cells and its conduction from cell to cell through intercellular gap junctions. Excitation of the cellular membrane results in opening of the voltage-gated L-type Ca^{2+} channels, which allow a small amount of Ca^{2+} to enter the cell. This triggers the release of a much greater amount of Ca^{2+} from the intracellular Ca^{2+} store, the sarcoplasmic reticulum, and gives rise to the systolic Ca^{2+} transient and contraction. These processes are highly regulated by the autonomic nervous system, which ensures the acute and reliable contractile function of the heart and the short-term modulation of this function upon changes in heart rate or workload. Recently, it became evident that discrete clusters of L-type Ca^{2+} channels exist in the sarcolemma, where they form an interacting network with regulatory proteins and receptors. It allows the specificity, reliability, and accuracy of autonomic modulation of the excitation-contraction processes by a variety of neurohormonal pathways. Disruption in subcellular targeting of calcium channels and associated signaling pathways may contribute to the pathophysiology of a variety of cardiac diseases including heart failure and certain arrhythmias. This chapter reviews the emerging understanding of microdomain-specific distribution, functioning, regulation, and remodeling of L-type Ca^{2+} channels in atrial and ventricular myocytes and their contributions to the cellular signaling and cardiac pathology.

14.1 Introduction

Execution of many cellular processes in terms of specificity and reliability depends on calcium signals that are tightly regulated in space and time and organized in microdomains. One of the important regulators of calcium level in cardiac myocytes is voltage-dependent L-type Ca^{2+} channels (LTCCs), which are essential to numerous cellular processes including excitability, excitation-contraction coupling (EC coupling), hormone secretion, and regulation of gene expression. The influx of Ca^{2+} through LTCCs (L-type Ca^{2+} current, $I_{Ca,L}$) is tightly controlled and compartmentalized within the cardiac myocytes to accommodate such diverse functions. It has long been recognized that discrete clusters of LTCCs exist along the sarcolemma, and studies in recent years have greatly extended our understanding of how specific subcellular localization impacts their function and regulation (Balijepalli et al. 2006; Balijepalli and Kamp 2008; Bryant et al. 2014; Nichols et al. 2010). Importantly, LTCCs form a protein-protein interacting network with various structural and regulatory proteins working together as part of a macromolecular signaling complex (Best and Kamp 2012; Cerrone and Delmar 2014; Willoughby and Cooper 2007). This protein-protein interaction is tightly controlled and regulated by numerous signaling pathways such as adrenergic, muscarinic, adenosine, prostaglandin, angiotensin, etc. The diversity of cellular responses achieved with a limited pool of second messengers is made possible through the organization of essential signaling components into microdomains such as specific membrane-based structures as T-tubules (Kamp and Hell 2000), lipid rafts/caveolae (Balijepalli et al. 2006; Pani and Singh 2009), and costameres (Hong et al. 2014). In addition to these physical compartments, scaffolding proteins (e.g., A-kinase-anchoring proteins (AKAPs), caveolins, and focal adhesion kinase) (Balijepalli et al. 2006; Balijepalli and Kamp 2008; Carnegie et al. 2009; Gray et al. 1997; Nichols et al. 2010) facilitate the formation of multi-protein complexes which create further segregation and refinement in intracellular Ca^{2+} signaling. During pathological remodeling, when cell structural integrity is altered, the protein-protein interaction is lost (Nattel et al. 2007). Disruption of normal subcellular targeting of LTCCs and associated signaling proteins may contribute to the pathophysiology of a variety of cardiac diseases including heart failure and certain arrhythmias (Schaper et al. 2002).

14.2 Microdomain-Specific Distribution and Regulation of LTCCs in Ventricular Myocytes

One of the earliest evidence that LTCCs distribute in clusters/microdomains was reported by Takagishi and colleagues (Takagishi et al. 1997). With the use of confocal microscopy and label-fracture replicas, they suggested that, in cardiac myocytes, calcium channels are organized predominantly in the form of clusters in the peripheral plasma membrane. Label-fracture replica is a technique that allows cytochemical mapping of membrane surfaces. Cell surfaces labeled with an electron-dense marker are freeze-fractured, and fractured faces are then replicated by platinum/carbon evaporation. This reveals surface distribution of the label coincident with the platinum/carbon replica of the fracture face. A similar organization of calcium channels in cardiomyocytes has also been revealed by confocal microscopy where they are present in the transverse-tubule (T-tubules), deep periodic invaginations of the sarcolemma (Carl et al. 1995). It was hypothesized that LTCC clusters may be present adjacent to junctional sarcoplasmic reticulum to trigger calcium-induced calcium release (CICR). More recently a number of specific signaling microdomains in cardiac myocytes have been identified; these include T-tubules, in particular, junctions with the sarcoplasmic reticulum (SR) in dyads, and domains outside dyads such as lipid rafts, caveolae, and nucleus (Balijepalli and Kamp 2008). These microdomains are thought to be not only involved in regulating contraction but also other signaling events. A number of LTCC subpopulations are identified in cardiomyocytes that associate with different microdomains. This differential association with different microdomains allows these different subpopulations to exhibit unique regulation and functional roles (Best and Kamp 2012).

14.2.1 T-Tubule (Dyads)

In mammalian cardiac myocytes, LTCCs have been localized to both surface and T-tubular sarcolemma (Carl et al. 1995; Takagishi et al. 1997). T-tubule network represents a complex system of interconnected membrane structures continuous with the extracellular space. T-tubules extend deep into the cardiac myocytes at Z-lines which bring the sarcolemma in close proximity with junctional SR (jSR) (Carl et al. 1995; Takagishi et al. 1997). Coupling of LTCCs present within the T-tubular membrane with jSR throughout the cell facilitates synchronous Ca^{2+} release upon depolarization stimulus during EC coupling (Gathercole et al. 2000; Scriven et al. 2010). The major subpopulation of LTCCs which participates in EC coupling is present predominantly within the T-tubule network. In earlier experiments, when the T-tubule system was depleted from the surface membrane following osmotic shock, $I_{Ca,L}$ decreased by as much as 75–80% without altering SR Ca^{2+} load (Kawai et al. 1999). Similar observations have been reported in numerous follow-up studies (Bers 2002). This indicates that LTCCs which reside in the non-T-tubular microdomains play a particularly important role in SR Ca^{2+} loading (Kawai et al. 1999).

Direct evidence of localization of LTCCs within the T-tubule microdomains came from the study of Bhargava and colleagues (Bhargava et al. 2013) who used the novel state-of-the-art technique of super-resolution scanning patch clamp to record functional LTCCs from T-tubular and crest sarcolemma of ventricular myocytes (Fig. 14.1). LTCC current was more frequently recorded from T-tubule microdomains than in the rest of sarcolemma (crest), where LTCC current was rarely detected. Therefore,

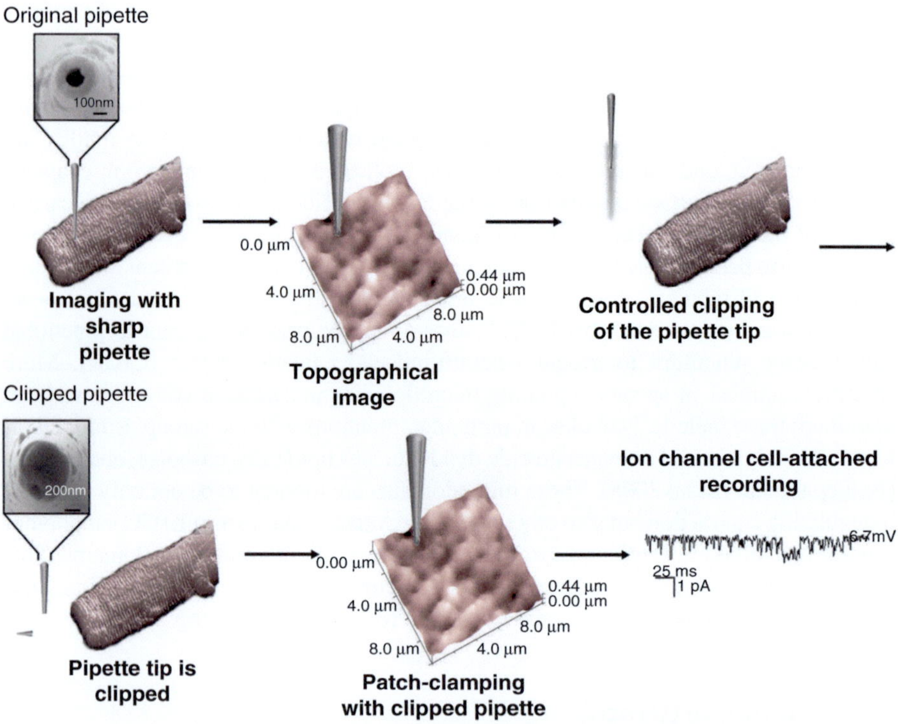

Fig. 14.1 Super-resolution scanning patch-clamp method. First, a topographical image of a cardiomyocyte is generated by the SICM with a 100 nm nanopipette. Then the inner tip diameter of the nanopipette is widened to ~350 nm by controlled clipping to increase the area of attachment. For that, the pipette is navigated to an area free of cardiomyocytes, the fall rate is increased, and the pipette is allowed to impact on the surface. As a result, the pipette breaks, and, owing the conical shape of the pipette, its tip diameter increases resulting in a stepwise increase of the pipette current as its resistance drops. After clipping, the pipette is lowered to a specific location (T-tubule or crest) and a gigaseal is established. Single ion channels are recorded in cell-attached mode. Insets show the pipette tip size before and after clipping. Reproduced with permission from Bhargava et al. (2013)

multiple evidences show the majority of functional LTCCs are present within the T-tubules where they participate in forming complexes with the SR (dyads). LTCCs located outside dyads (extradyadic) do not seem to contribute directly to EC coupling but they may participate in other cellular processes (Makarewich et al. 2012). Importantly, these extradyadic channels are not randomly distributed throughout the sarcolemma, but they are also present in specialized microdomains with unique biochemical composition where they associate with important signaling molecules (Balijepalli and Kamp 2008; Dart 2010; Takagishi et al. 1997). A subset of extradyadic LTCCs localize to cholesterol and sphingolipid-enriched regions of the plasma membrane known as lipid microdomains or "rafts" as described below.

14.2.2 Lipid Rafts/Caveolae

Lipid rafts represent areas of lipid bilayer that are more rigid and ordered than the more fluid bulk of the bilayer. It has been proposed that multiple types of rafts can exist based on the differences in lipid and protein composition (Pike 2004), but strong evidence in support of this statement is still lacking (Edidin 2003; Munro 2003). Over the years, rafts have attracted attention because of their ability to selectively aggregate interacting signaling molecules, and thus they may be key components in spatial organization of cell processes/signaling pathways (Patel et al. 2008; Simons and Toomre 2000). By providing such spatial organization, lipid rafts not only facilitate intramolecular cross talk but also expedite the signal relay. These cell processes include signal transduction pathways, apoptosis, viral infections, cell adhesion and migration, synaptic transmission, organization of the cytoskeleton, plasma membrane protein sorting, etc. (Balijepalli et al. 2006; Razani et al. 2002; Zhang et al. 2011). Many of these cell processes involve release of Ca^{2+}, and therefore a lot of studies have focused on Ca^{2+} regulation by lipid rafts (Pani and Singh 2009).

A number of ion channels, including LTCCs, are present in the lipid raft microdomains. The lipid composition of the membrane in which the ion channels are embedded influences the channel function both directly and indirectly. When a channel protein undergoes conformational change, there is a local disturbance in the surrounding bilayer, and thus the overall energetic cost of a channel transition between different states not only includes channel activation energy but also the energy associated with membrane deformation (Andersen and Koeppe 2007; Edidin 2003). This implies that any change in the lipid environment could alter the channel activity. Evidence of direct modulation of ion channels by lipids comes from the specific interaction between channel proteins and sphingolipids that regulate voltage sensing and/or channel gating (Epshtein et al. 2009; Romanenko et al. 2002). There is also strong evidence that cholesterol directly modulates LTCCs. Exposure of smooth muscle cells to cholesterol-enriched liposomes leads to a gradual increase in LTCC current and a positive shift in voltage dependence of inactivation (Sen et al. 1992).

Ion channels present in the lipid microdomains are also modulated by association to other proteins in the microdomains. One such association of LTCCs with other proteins exists in caveolae of cardiac myocytes where LTCCs form part of macromolecular signaling complex. Caveolae (Latin for "little caves," singular = caveola) are raft-like domains that form morphologically distinct structures of flask-shaped invaginations of the plasma membrane or vesicles adjacent to it (Palade and Bruns 1968; Simionescu et al. 1975). Caveolae contain the scaffold protein caveolin, which enables the rafts to adopt the flask-like shape. Caveolae appear in a variety of cell types and are thought to compartmentalize numerous signaling processes (Boulware et al. 2007; Razani et al. 2002; Song et al. 1996). There are three known isoforms of caveolin proteins, caveolin 1 (Cav1), caveolin 2 (Cav2), and caveolin 3 (Cav3) (Razani et al. 2002), oligomers of which line the membrane of caveolae. Caveolins also serve as markers to help sort appropriate proteins to caveolae rather than non-caveolar lipid rafts. In that respect it is notable that in cardiac myocytes, LTTCs co-localize with Cav3 (the major isoform of caveolin present in the heart) in immunochemical confocal microscopy and immunogold electron microscopy studies, and indeed LTCCs are found in caveolae in cardiac myocytes (Balijepalli et al. 2006; Cavalli et al. 2007; Shibata et al. 2006).

14.2.3 Nucleus

Apart from cardiac myocyte sarcolemma, nuclear localization was confirmed for a fragment of LTCC molecule, namely, its carboxyl terminus. Western blot analysis showed that nuclear extracts of cardiomyocytes appear to contain cleaved LTCC carboxyl terminus, and GFP-tagged LTCC carboxyl terminus showed nuclear localization (Schroder et al. 2009). This is the most unusual finding but it suggests that the carboxyl terminus of LTCC may be involved in regulating gene transcription (Schroder et al. 2009).

14.2.4 ß-Adrenergic Regulation of the LTCC Subpopulation in Caveolar Microdomains

Several neuronal and hormonal signaling pathways regulate LTCCs located in different microdomains (Balijepalli et al. 2006; Balijepalli and Kamp 2008; Carl et al. 1995; Takagishi et al. 1997). This neurohormonal regulation is fundamental to the ability of the heart to adapt to stress via alteration of heart rate and contractility. Most prominent of these regulations is the β-adrenergic regulation of LTCCs. There are three subtypes of β-adrenergic receptors (ARs: β_1, β_2, and β_3) in the cardiac myocytes, and each has its own distinct set of effector proteins and downstream signals. In cardiac myocytes, the subpopulation of LTCCs within the caveolae is present as a macromolecular complex made of β_2AR, adenylyl cyclase (AC), $G_{\alpha s}$, $G_{\alpha i}$, and protein phosphatase 2A (PP2A) (Balijepalli et al. 2006). Disruption of caveolae with either methyl-ß-cyclodextrin (MßCD) or small interfering RNA directed against Cav3 abolishes β_2AR stimulation of LTCCs. However, this does not affect $ß_1$AR stimulation of LTCC current (Balijepalli et al. 2006). This means that there is a specific coupling of LTCCs with β_2ARs within caveolae and also suggests that LTCCs coupled to $ß_1$AR signaling may be located outside of caveolar microdomains. Supporting this, Ca^{2+}/calmodulin-dependent protein kinase II (CaMKII), which is known to mediate apoptosis upon $ß_1$AR activation of LTCC current, resides in the non-caveolar domains in cardiac myocytes (Chesley et al. 2000; George and Pitt 2006; Zhu et al. 2003).

Activation of ßARs appears to alter cardiac LTCC current via a dual mechanism. LTCC is phosphorylated by protein kinase A (PKA) at serine 1928 which increases the open channel probability of the LTCC and therefore results in an enhancement of the whole-cell $I_{Ca,L}$ (McDonald et al. 1994; Osterrieder et al. 1982). Also, superfusion with isoproterenol (nonselective βAR agonist) results in a biphasic response of LTCC current where low concentrations of isoproterenol trigger a fast increase of LTCC current. This fast mechanism acts via $G_{\alpha S}$ protein; $G_{\alpha S}$-coupled receptors activate AC, which results in the production of cAMP. The subcellular localization of cAMP thus produced differs depending upon the receptor involved. For example, stimulation of βARs using isoproterenol leads to cAMP accumulation within cytosolic and caveolar compartments, whereas activation of another $G_{\alpha S}$-coupled receptor, the E-type prostaglandin receptor 4, results only in cytoplasmic cAMP production. This can be easily monitored by the use of novel microdomain-targeted Foster resonance energy transfer (FRET)-based biosensors (Perera and Nikolaev 2013; Sprenger and Nikolaev 2013). Therefore, it can be assumed that restricted production of cAMP within the caveolar compartment upon β_2AR activation is responsible for the signal to stay local in the vicinity of caveolar microdomain.

14.2.5 Heart Failure-Associated Remodeling of Cardiomyocyte Microdomains and LTCCs

In heart failure (HF), extensive remodeling of the myocardium takes place which affects all components of the cardiac myocytes; in particular, some sarcomeres are lost (Fig. 14.2a–f). For a long time, the importance of LTCC dysregulation in HF was ignored due to the fact that no changes in the $I_{Ca,L}$ or density of LTCCs were observed in failing hearts using standard whole-cell patch-clamp technique (Kamp and He 2002). Whole-cell currents are determined by the number of functional channels, the probability of a channel being open (open probability, P_o), and the current through a single open channel. Thus, the observation that the peak $I_{Ca,L}$ is similar in failing and non-failing myocytes using the whole-cell patch-clamp technique may be misleading as it doesn't take into account that the changes in channel density, gating, and single-channel current when they happen in different directions may cancel each other out. As described before (Sect. 14.2.2), gating and other channel kinetics may be affected by the microdomains in which the channels reside.

In both hypertrophied and failing hearts, LTCC current loses its ability to trigger CICR. Because LTCC density and SR Ca^{2+} release channels are unaffected, it seems that there is a change in the relation between these two proteins as CICR is largely dependent on the proximity of these two proteins. This defective CICR is restored by ßAR stimulation in hypertrophied but not in failing hearts suggesting that failure of compensatory mechanisms can lead to progression from hypertrophy to HF (Gomez et al. 1997).

During remodeling of the myocardium, the geometry and protein composition of various subcellular compartments alter. An extensive proteomics study revealed that 30 proteins are specifically up- or downregulated in membrane microdomains in

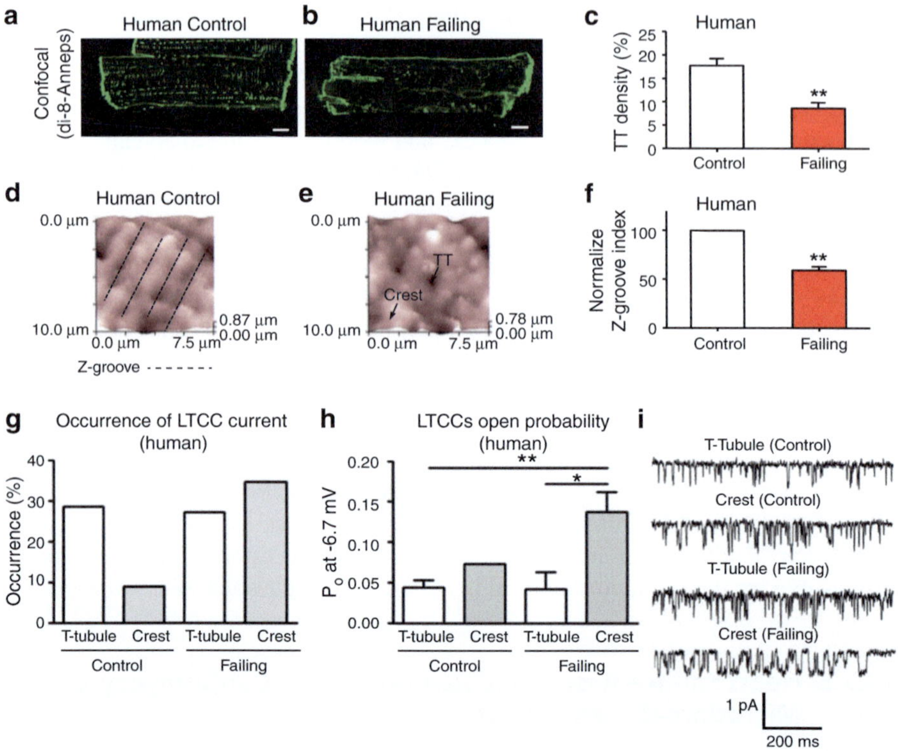

Fig. 14.2 Remodeling of T-tubule microdomains and consequences for LTCCs in HF. Example confocal images of human control (**a**) and failing (**b**) cardiomyocytes showing membranes stained with di-8-ANNEPS, scale bar 10 μm. (**c**) T-tubule density in control and failing cells. Human control $n = 20$, failing $n = 8$. $^{**}P < 0.01$. (**d**) SICM scan from a 10 μm × 10 μm portion of cell membrane showing regular undulations, indicating spatially alternating T-tubule invaginations and surface membrane "crests" in human control. (**e**) These regular structures are absent in human failing cardiomyocytes. (**f**) Z-groove index in human failing cells normalized to control average value. Human control $n = 30$, failing $n = 59$; top. $^{**}P < 0.01$. (**g**) Chances of obtaining a LTCC current (% occurrence) in human control and failing cells. $^{**}P < 0.01$. (**h**) P_o during a voltage step to V_{max} (to −6.7 mV) in human control and failing cells. Human control, T-tubule $n = 6$, crest $n = 1$; failing, T-tubule $n = 6$, crest $n = 9$. $^{*}P < 0.05$, $^{**}P < 0.01$. (**i**) Representative single-channel traces at −6.7 mV showing LTCC activity in human cardiomyocytes. Reproduced with permission from Sanchez-Alonso et al. (2016)

human HF (Banfi et al. 2006) suggesting an extensive microdomain disturbance. If different microdomains are disturbed in HF, it is plausible that LTCC function is affected which in turn will affect the downstream signaling. Remodeling of different microdomains and their effect on LTCC regulation in HF is described below.

14.2.5.1 T-Tubular/Dyadic Microdomains in Cardiovascular Disease and Consequences for LTCCs

During structural remodeling in HF, both T-tubule density and regularity reduce (Louch et al. 2004; Lyon et al. 2009; Sanchez-Alonso et al. 2016) (Fig. 14.2a–f). T-tubule remodeling may alter the geometry of the dyadic cleft or it may potentially disrupt macromolecular signaling complexes located within these microdomains (Gomez et al. 1997; Heinzel et al. 2008, 2011). Because a major population of cardiac LTCCs reside in T-tubular microdomains, it is conceivable that disruption of the latter would influence the function of LTCCs in T-tubular microdomains. In accordance, there is a significant reduction in the density of LTCCs in T-tubular microdomains during HF which has been shown using biochemical methods such as membranes fractionation (Balijepalli et al. 2003; He et al. 2001). Thereafter, recent development of techniques such as "super-resolution scanning patch clamp" allowed scientists to measure activity of functional LTCCs in various microdomains. Such direct recording of LTCC current either from T-tubules or crest (sarcolemmal) microdomains in normal and failing hearts shows that a significant percentage of functional LTCCs are lost in T-tubular microdomains in HF (Sanchez-Alonso et al. 2016). A further interesting finding is that more LTCC current is present in the crest (sarcolemma) where it is rarely present in control (Sanchez-Alonso et al. 2016) (Fig. 14.2g–i). This observation leads to an indication that perhaps LTCCs are not universally lost but redistributed. This direct evidence of loss of LTCCs from T-tubular microdomains and appearance in noncanonical microdomains (crests) is in fact an answer to various open questions that existed. The redistribution of LTCCs in cardiomyocyte microdomains possibly explains the nonexistence of differences under normal and failing conditions in the whole-cell $I_{Ca,L}$ or $I_{Ca,L}$ density (recorded by patch clamp) as this technique cannot differentiate between microdomain-specific currents. This shifting of microdomains also can explain the impaired CICR as the crest LTCCs are no longer in close proximity to sarcolemmal release channels required for CICR.

14.2.5.2 Caveolar Microdomains in Cardiovascular Disease and Consequences for LTCCs

It is clear from the above discussion that caveolar microdomains are prominent in cardiac myocytes. The importance of these microdomains for cardiovascular physiology is better understood by the phenotypes of mice lacking caveolin genes (genes encoding scaffolding proteins that line caveolae). An interesting feature of caveolin proteins is that their expression is decreased with age, and studies on caveolin knock out (KO) mice support the hypothesis that the loss of caveolin protein causes an aged phenotype (Kawabe et al. 2001; Ratajczak et al. 2003). This suggests caveolins as potential therapeutic targets in the treatment of age-related disorders such as cardiovascular disease.

Among the known three isoforms of caveolin, Cav3 is the predominant caveolin in cardiac myocytes. However, Cav1 KO mice also develop a severe cardiomyopathy leading to significantly shortened life-span (Park et al. 2003). More importantly there are structural abnormalities in the hearts of Cav1 KO mice. Cav1 KO hearts have significantly enlarged ventricular chambers, abnormal ventricular wall thickness, hypertrophy, and decreased contractility (Cohen et al. 2003; Zhao et al. 2002). Indeed, cardiac myocytes from Cav3 KO mice completely lack caveolae, and these mice develop a progressive cardiomyopathy similarly to Cav1 KO mice. Genetic abnormalities such as mutations in caveolin proteins can also cause cardiomyopathies. For example, mutations found in Cav3 are linked to familial hypertrophic cardiomyopathy and inherited arrhythmogenic syndrome (Hayashi et al. 2004).

Similar to HF, T-tubules in the Cav3 KO mice are disorganized (described in Sect. 14.2.5.1), and therefore one may assume that will have consequences for LTCCs distribution and function comparable to the disorganization of T-tubule network during remodeling in HF (Sect. 14.2.5.1). Interestingly, Cav3 expression level is decreased in the failing human heart (Feiner et al. 2011). On the other hand, overexpression of Cav3 attenuates HF (Andersen and Koeppe 2007; Horikawa et al. 2011; Tsutsumi et al. 2008). Thus, Cav3 may be a potential target for therapies to ameliorate failing heart.

The population of LTCCs in cardiac myocytes which resides in caveolae is shown to specifically couple to β_2ARs, which reside nearby in the same caveolae (Balijepalli et al. 2006) (Fig. 14.3). Disruption of caveolae results in the disruption of localized coupling between β_2ARs and LTCCs. This is shown in cardiac myocytes either treated with MβCD to deplete cholesterol (and therefore deplete caveolae) or treated with siRNA to knock Cav3 down. After either treatment LTCCs no

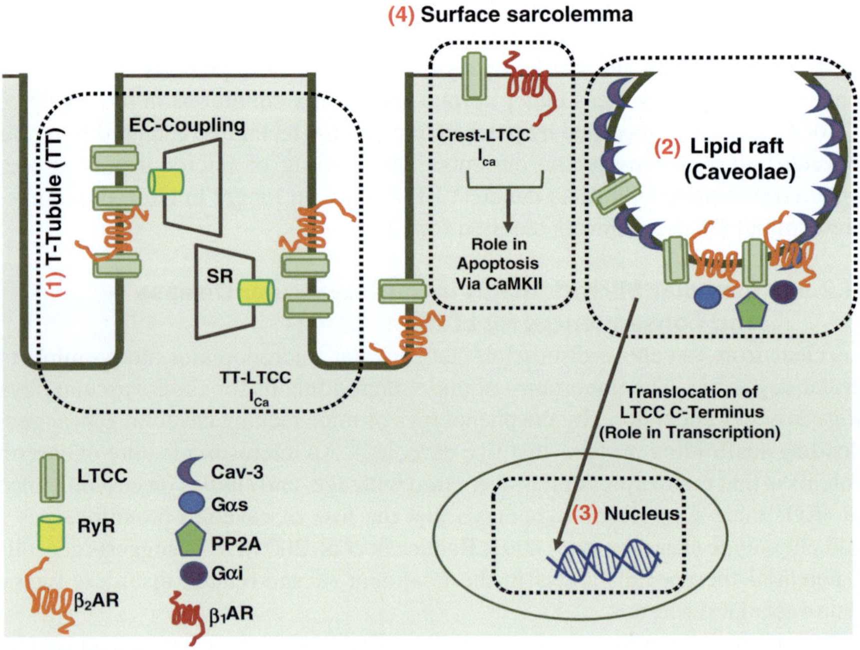

Fig. 14.3 LTCC microdomains in cardiomyocytes. (*1*) T-tubule microdomain where LTCCs are involved in EC Coupling. Single-channel LTCC current (I_{Ca}) shows low open probability. Predominant ßARs are $ß_2$ARs. (*2*) Specialized lipid rafts called caveolae house LTCCs together with $ß_2$AR, caveolin-3 (Cav-3), $G_{\alpha i}$, $G_{\alpha s}$, PP2A. Caveolar LTCCs do not contribute to EC coupling but are a site for localized cAMP generation upon stimulation of cardiomyocytes by $ß_2$ARs which are involved in many cellular processes. (*3*) C-terminus of LTCCs can translocate to nucleus and can act as a transcription factor. (*4*) Sarcolemmal (crest) LTCCs also have low open probability and mainly couple with $ß_1$ARs. $ß_1$AR activation of LTCC current activates CaMKII which mediates apoptosis in non-caveolar microdomains

longer increase their current upon activation of β_2ARs (Balijepalli et al. 2006). However, caveolar disorganization does not affect β_1AR of LTCCs (Balijepalli et al. 2006). This has relevance to age-related heart diseases where caveolae are reduced or destroyed.

14.2.5.3 ß-Adrenergic Microdomains in Cardiovascular Disease and Consequences for LTCCs

As described before, βARs exist as three subtypes, β_1, β_2, and β_3. The former two are important in the regulation of EC coupling of myocardium (Woo and Xiao 2012). Stimulation of βAR results in the activation of the G_s-AC-cAMP-PKA signaling cascade. In cardiac myocytes, the phosphorylation of PKA substrates including phospholamban, LTCC, ryanodine receptor, and other proteins results in the Ca^{2+} transient and eventually contraction. In HF, density, localization, and coupling of βAR subtypes with downstream signals are disturbed (Bristow et al. 1982, 1986; He et al. 2005; Lohse et al. 2003). Comprehensive evidence points out opposing functional roles of β_1- and β_2ARs in regulating myocyte viability and myocardial remodeling (Woo and Xiao 2012). In contrast with the β_2AR stimulation which shows a cardioprotective effect, prolonged β_1AR stimulation has a detrimental effect on myocardium. Normally, β_1ARs stimulate PKA-mediated phosphorylation of a number of key proteins (as described above), whereas no such effect is seen with β_2AR stimulation (Woo and Xiao 2012). However, β_2AR stimulation leads to LTCC phosphorylation and thereby enhancement of LTCC current with smaller inotropic and lusitropic effects. During HF, β_1ARs are downregulated at both the mRNA and protein levels (Bristow et al. 1986), and their density at the plasma membrane is reduced by 50%, whereas β_2ARs remains unchanged (Bristow et al. 1986). As a consequence, the β_1/β_2ARs ratio changes from 80:20 in healthy hearts to 60:40 in failing hearts, which indicates a prominent role β_2AR signaling in HF.

These βAR-subtype-specific differences are attributed to cAMP compartmentation in cardiac myocytes. Both physical and functional compartmentation happens as a result of separation that exists between compartments in cardiac myocytes, e. g,. T-tubules and caveolae. Experiments using myocytes isolated from hearts of transgenic mice that express a FRET sensor for cAMP (Nikolaev et al. 2006b) revealed that the β_1AR-mediated cAMP signal propagates throughout large parts of the cell, whereas the β_2AR signal is locally confined in T-tubules (Nikolaev et al. 2006a). By combining FRET-based cAMP imaging with scanning ion conductance microscopy (SICM/FRET method), this research is taken to the next level. With this combination, functional localization of βARs has been revealed for the first time. With this method, surface topography image of a cardiac myocyte is obtained with nanoscale resolution using SICM which reveals features such as z-grooves, T-tubules, and the crests (Nikolaev et al. 2010). Next, the nanopipette, the sensitive probe used for SICM, is placed above a feature of interest, e.g., a T-tubule. Last, an agonist or an antagonist of βARs is applied selectively to this defined morphological microdomain by extruding the solution containing the said substance from the nanopipette. It was shown with this method that selective stimulation of β_1ARs in both T-tubule and crest regions results in robust cAMP synthesis. In contrast,

β_2AR-selective stimulation results in cAMP signals only in the T-tubules, but not in other regions of the sarcolemma. This indicates that β_2ARs have a preferential location and function in T-tubules.

The combined SICM/FRET technique revealed that in HF, the distribution of β_1- and β_2ARs along the sarcolemma is disturbed. Widespread membrane β_1AR localization remains unchanged but β_2ARs lose their exclusive location in the T-tubules and are found also in the membrane areas between T-tubules (crests) (Nikolaev et al. 2010). Interestingly, following the β_2ARs redistribution, their subcellular cAMP gradients are also remodeled, as cAMP signals from β_2AR stimulation start to diffuse throughout the entire cytosol, similarly to the β_1AR-mediated cAMP signals.

Also, in the failing heart, the selective downregulation of β_1AR is associated with an upregulation of G_i and an enhanced β_2AR-G_i signaling (Woo and Xiao 2012). β_2AR levels though remain unchanged, β_2AR coupling efficiency to G_s is reduced leading to G_s-mediated responses such as cAMP production and positive inotropic effect. Enhanced β_2AR signaling through G_i strongly blunts the increase in LTCC current also by β_1AR stimulation. This kind of blunted adrenergic regulation of LTCC current is observed in animal models of HF and in human HF (Kamp and He 2002). In another aspect of this story, stimulation of LTCCs by β_1- and β_2ARs is mediated by different isoforms of ACs which show distinct compartmentation due to their interaction with Cav3 (Timofeyev et al. 2013). Up- or downregulation of these ACs in HF can cause disturbances in the regulation of LTCCs by βARs.

14.3 Microdomain-Specific Distribution and Regulation of LTCCs in Atrial Myocytes

14.3.1 T-Tubule Structure in Atrial versus Ventricular Myocytes

In contrast to ventricular myocytes, the role of T-tubules in atria is less clear. Atrial myocytes have long been perceived as having no or very few T-tubules (Brette et al. 2002; Smyrnias et al. 2010; Tidball et al. 1991). However, recent experimental evidence demonstrates that atrial myocytes from certain species, such as sheep (Dibb et al. 2009), cows, horses, humans (Richards et al. 2011), and even rodents (Dibb et al. 2013; Kirk et al. 2003; Smyrnias et al. 2010), do possess T-tubules. Generally, atrial T-tubules are sparse and less regular when compared with those in ventricular myocytes, assessed both in situ (Wei et al. 2010) and in vitro (Lyon et al. 2009; Smyrnias et al. 2010). Using the peroxidase labeling method, Forssmann and Girardier revealed that there are two types of muscle cells in the right atrium of the rat (Forssmann and Girardier 1970). They found that in most atrial myocytes, T-tubule system was either missing or poorly developed; however, some atrial myocytes exhibited a highly developed network of T-tubules. Later, Kirk et al. isolated cardiomyocytes separately from the right and left rat atria and found that about half of the atrial myocytes possesses an irregular transverse-axial tubular system (Kirk et al. 2003). The authors demonstrated that cells with T-tubules had a larger mean diameter than cells without T-tubules and were more common in the left atrium.

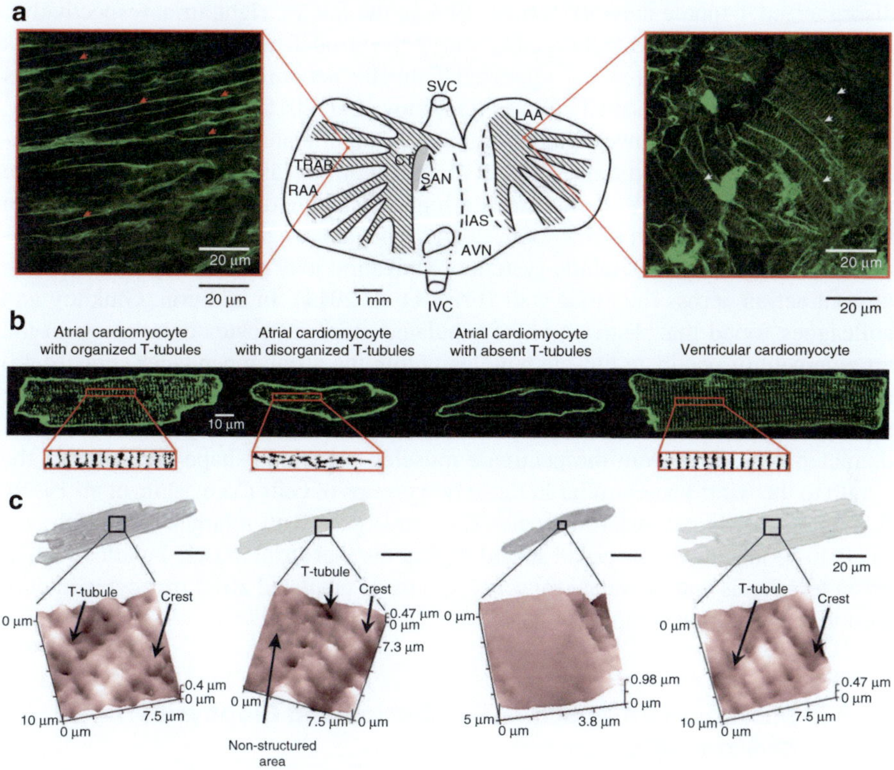

Fig. 14.4 Spatial heterogeneity of the atrial T-tubular system: in situ and in vitro measurements. (**a**) In situ confocal imaging of T-tubules (TTs) in intact rat atrial preparation stained with WGA. In the middle, the schematic outlines of the isolated rat atria preparation showing the main anatomical features. The enlarged images from the endocardium of the *right* (RAA) and *left* (LAA) atrial appendages demonstrate typical atrial myocytes with organized TTs (*white arrows*), disorganized TTs (*red arrows*), or mixture of both types. *SVC* and *IVC* superior and inferior vena cava, *SAN* sinoatrial node, *CT* crista terminalis, *TRAB* trabeculae, *AVN* atrioventricular node, *IAS* interatrial septum. (**b**) Di-8-ANEPPS membrane staining showing a T-tubule network in ventricular myocytes and in atrial myocytes with organized, disorganized, and absent T-tubules. Below the confocal images, enlarged areas of 40 μm × 5 μm are shown that were binarized and used in T-tubule density and regularity measurements. (**c**) Correlation between surface structure and cell size. Optical images and topography scans (zoomed areas) of a ventricular myocyte and atrial myocytes with various degrees of organization of surface structures are shown. T-tubules, crests, and non-structured areas are indicated by arrows. Note that the cell shown in the right most panel does not possess any organized surfaces structures. Reproduced with permission from Glukhov et al. (2015a)

Further studies identified three groups of atrial myocytes: untubulated, tubulated with disorganized T-tubules, and organized tubulated atrial cells (Frisk et al. 2014; Glukhov et al. 2015a) (Fig. 14.4). Untubulated myocytes amount up to 30–40% of total atrial cells. The majority of atrial myocytes with organized T-tubules were isolated from the left atria (40 vs. 2% in the left vs. right atria, respectively). Conversely, myocytes isolated from the right atria were found more likely to have disorganized T-tubule network (26 vs. 59% in the left vs. right atria, respectively). The authors also showed that organization of the atrial T-tubular network correlates with cell width: cells showing organized T-tubular networks were larger than cells with disorganized or absent T-tubules (Glukhov et al. 2015a; Kirk et al. 2003).

Importantly, all the investigators highlighted significant anatomical heterogeneity of T-tubule system in atria. Frisk et al. used both isolated atrial cells and tissue from rat and pig hearts to demonstrate a higher T-tubule density in the epicardium than in the endocardium (Frisk et al. 2014). The authors proposed that such transmural heterogeneity of T-tubule system organization may promote synchronization of contraction across the atrial wall (Frisk et al. 2014). In addition, Glukhov and colleagues found that distribution of tubulated atrial myocytes correlate with the arrangement of pectinate muscle bundles within the atrial appendages (Fig. 14.4a) (Glukhov et al. 2015a). Similar results have been reported in rabbit atria where myocytes isolated from the crista terminalis were of significantly larger mean cell diameter than those from the pectinate muscles, while the shape (the ratio of the length to the width) was similar in those two groups of cells (Yamashita et al. 1995). Assuming that larger atrial myocytes (i.e., myocytes with a larger cell width) are likely to be tabulated, one could speculate that distinct atrial muscle bundles such as crista terminalis and pectinate muscles consist of tabulated atrial myocytes in contrast to the atria free wall.

14.3.2 Microdomain-Specific Distribution and Biophysics of Atrial LTCC

Anatomical variation of T-tubule system among atrial myocytes has been proposed to underlie the heterogeneity of the whole-cell $I_{Ca,L}$ measured within the atria. In healthy dogs, $I_{Ca,L}$ was found to be the largest in crista terminalis cells, intermediate in cells from the right atrial appendage and pectinate muscles, and the smallest in atrioventricular ring cells (Feng et al. 1998). In contrast, in rabbit right atrial myocytes isolated from different areas, whole-cell clamp recordings showed no anatomical variation in the density of $I_{Ca,L}$ (Yamashita et al. 1995). At the same time, Frisk and colleagues used both isolated atrial cells and tissue to demonstrate that in pig and rat atria there was a high variability in the distribution of T-tubules and $I_{Ca,L}$ among cells, with a steep dependence of $I_{Ca,L}$ on atrial myocyte capacitance and T-tubule density (Frisk et al. 2014). The authors observed a higher T-tubule density in the epicardium than endocardium which could form the transmural gradient in $I_{Ca,L}$ density and thus may promote synchronization of contraction across the atrial wall.

It has been proposed that LTCCs distribute in the membrane of atrial myocytes differently as compared with ventricular myocytes due to lack of a regular T-tubule system. Atrial whole-cell $I_{Ca,L}$ has a greater contribution of extratubular channels as compared with ventricular myocytes, and substantial $I_{Ca,L}$ has been measured in atrial myocytes after T-tubules were experimentally destroyed (Frisk et al. 2014). This highlights a unique distribution of atrial LTCCs. Indeed, immunostaining for LTCCs demonstrated in atrial myocytes variable signal, in agreement with variable T-tubule organization (Fig. 14.5a) (Frisk et al. 2014). Although many cells did not

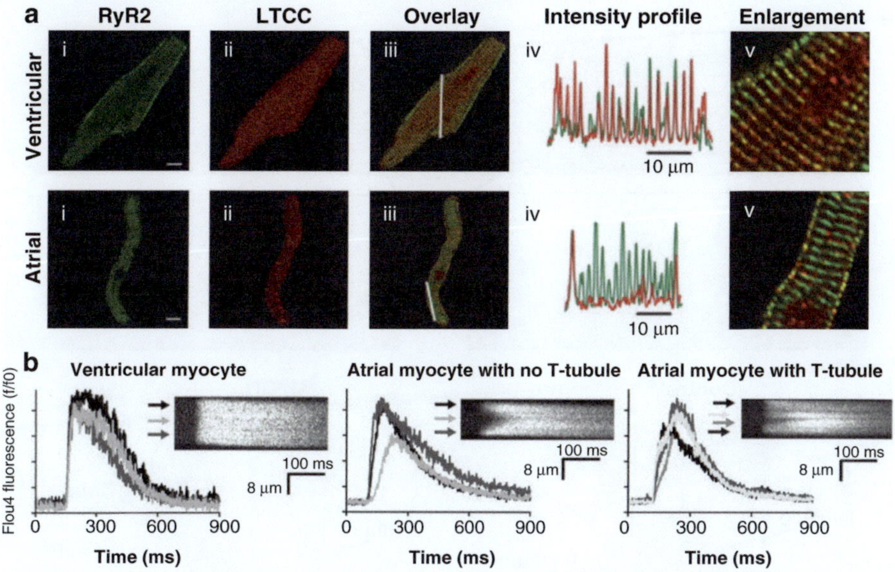

Fig. 14.5 Distribution and co-localization of ryanodine receptors (RyR) and L-type Ca^{2+} channels (LTCCs) determines a unique atrial Ca^{2+} signaling. (**a**) Localization of RyRs and LTCCs in ventricular and atrial myocytes. The images depict single ventricular (*top*) or atrial (*bottom*) myocytes immunostained for type 2 RyRs (*i*) or LTCCs (*ii*). The scale bars represent 10 μm. The intensity profiles (*iv*) were obtained by measuring the intensity of fluorescent antibody labeling across the cellular regions depicted by the white lines in the overlay images (*iii*). The white bars in the cell images represent 10 μm. Reproduced with permission from Smyrnias et al. (2010). (**b**) Ca^{2+} transients in ventricular and atrial myocytes with or without T-tubules. Panels depict the temporal and spatial properties of single depolarization-evoked Ca^{2+} transients in control ventricular myocyte (*left*), atrial myocyte with a "U"-shaped Ca^{2+} signal (*middle*), and atrial myocyte with an internal initiation site (*right*). The traces were obtained by sampling fluo-4 intensity along the line-scan images, as depicted by the correspondingly colored arrows. Reproduced with permission from Smyrnias et al. (2010)

exhibit significant LTCC labeling in the cell interior, a subset showed significant staining, and a minority exhibited LTCCs with a regular, striated organization (Bootman et al. 2006; Chen-Izu et al. 2006). Similar patterns of immunostaining were also observed by others (Schulson et al. 2011; Smyrnias et al. 2010).

The abovementioned studies lacked information on the functionality of LTCC proteins within a subcellular domain. Recent methodological advances have made it possible to image the topography of a live cardiomyocyte and to study clustering of functional ion channels from a specific microdomain (Bhargava et al. 2013). In a recent study by Glukhov and colleagues, microdomain-specific localization and functioning of atrial LTCCs was characterized (Glukhov et al. 2015a). The authors found that in both rat and human atrial myocytes, LTCCs distribute equally in T-tubules and crest areas of the sarcolemma (Fig. 14.6c), whereas, in ventricular myocytes, LTCCs primarily cluster in T-tubules. To link the extratubular LTCC to caveolae structures, the authors used two different approaches: treatment with MβCD and direct LTCC inhibition in Cav3-containig membranes using a specific peptide

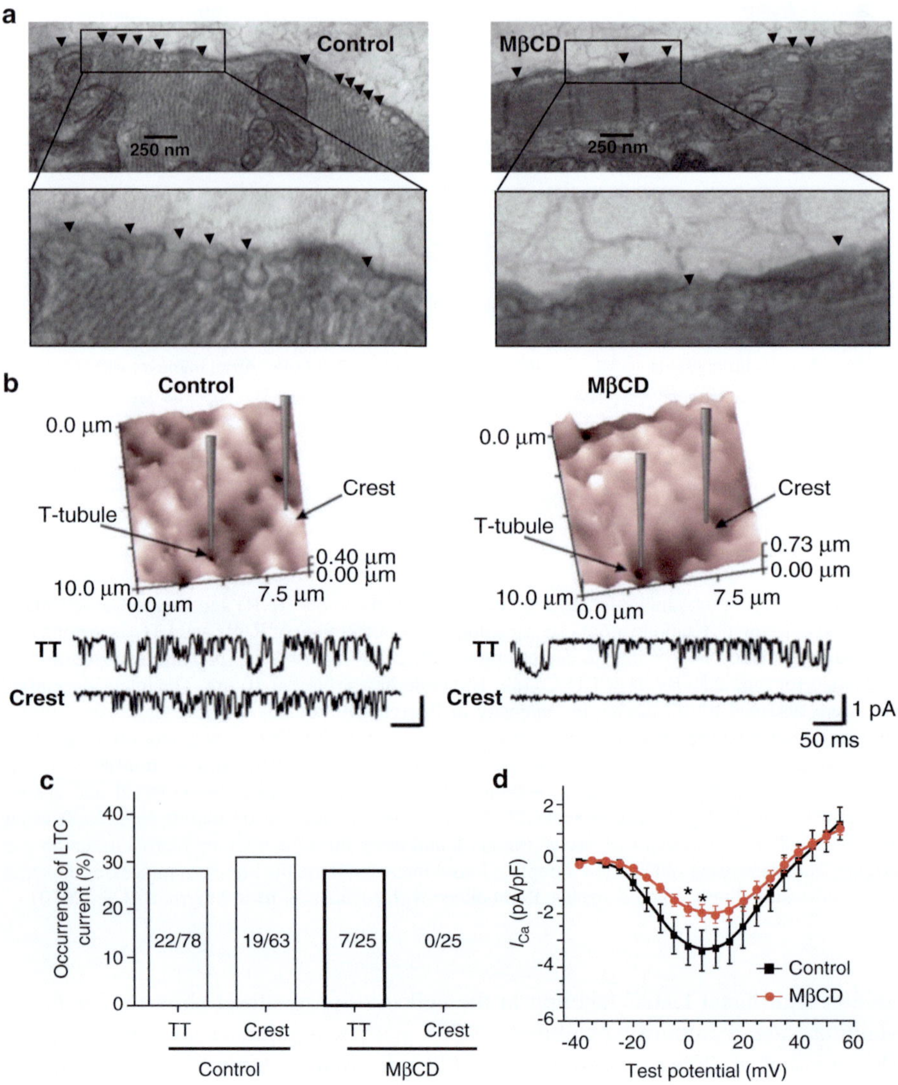

Fig. 14.6 Caveolae as a source of extratubular LTCCs. Cholesterol depletion removes caveolae and abolishes the occurrence of extratubular LTCCs decreasing whole-cell $I_{Ca,L}$. (**a**) Ultrastructural changes in rat atrial myocytes after methyl-β-cyclodextrin incubation (MßCD). (**b**) Typical 10 μm × 10 μm topographic scans of control (*left*) and MßCD-treated (*right*) rat atrial myocytes. Below are single-channel recordings obtained from the T-tubule (TT) and the crest of sarcolemma (crest). (**c**) Percentage of LTCC current occurrence in the T-tubules and crests. (**d**) Whole-cell $I_{Ca,L}$ density before and after MßCD treatment. Reproduced with permission from Glukhov et al. (2015a)

inhibitor Rem, a member of the GRK GTPase family (Makarewich et al. 2012). Incubation of atrial myocytes with MβCD resulted in ≈60% depletion of caveolae and complete elimination of extratubular LTCCs, with no effect on the channels located in T-tubules (Fig. 14.6a–c). Associated with LTCC removal from the crest of sarcolemma, MβCD significantly decreased the whole-cell $I_{Ca,L}$ density by ≈30% (Fig. 14.6d). Similarly, the caveolin-3-targeted LTCC-blocking agent, Rem peptide, a member of the RGK GTPase family, significantly decreased the occurrence of functional LTCC on the crest of atrial myocyte sarcolemma (Glukhov et al. 2015a). Altogether those results indicate that in atrial myocytes, extratubular LTCCs are localized in caveolae structures and may contribute to up to 30% of the atrial whole-cell $I_{Ca,L}$.

Importantly, Glukhov et al. have shown the microdomain-specific heterogeneity in LTCC biophysical properties. In rat, but not in human atrial myocytes, LTCCs found in the crest had ≈40% lower amplitudes measured at negative voltages than channels recorded in the T-tubules (Glukhov et al. 2015a). The authors demonstrated that this is due to a significantly higher probability of extratubular channels to open at low-amplitude sub-conductance states that at high-conductance states which is different from the channels located in T-tubules. In previous studies, open probability of different sub-conductance states was suggested to be determined by a phosphorylation status of the channel (Cloues and Sather 2000; Gondo et al. 1998) which could be significantly regulated by a microdomain-specific microenvironment (Stangherlin and Zaccolo 2012).

In contrast, in human atrial myocytes isolated from patients with no history of atrial fibrillation or HF in anamnesis, both extratubular channels and LTCCs located in T-tubules had similar unitary amplitudes but different open probabilities (P_o). The LTCCs located in T-tubules had almost twice higher open probability than extratubular channels which could be attributed to their differential structure (Foell et al. 2004; Mangoni et al. 2003). In rat atrial myocytes, extratubular LTCCs demonstrate very heterogeneous distribution of their open probability, where two distinct clusters of channels could be appreciated: one with a lower P_o when compared with the P_o for T-tubular LTCCs (like extratubular LTTCs in human atrial myocytes) and another with a higher P_o which was four times bigger than the average P_o of the first subpopulation and twice bigger than the P_o of T-tubular LTCCs. Extratubular channels with distinct P_o thus might be linked to caveolar (i.e., in association with Cav3) and non-caveolar lipid rafts (Best and Kamp 2012).

Multiple Ca^{2+} channel regulatory β-subunit ($Ca_v\beta$) isoforms are known to differentially regulate the functional properties and membrane trafficking of LTCCs. Foell et al. showed that in human and canine left ventricle myocardium, differential subcellular localization of β-subunit of LTCCs with $Ca_v\beta_{1b}$, $Cav\beta_2$, and $Ca_v\beta_3$ is being predominantly localized on the T-tubule sarcolemma, whereas $Ca_v\beta_{1a}$ and $Ca_v\beta_4$ are more prevalent in the surface sarcolemma (Foell et al. 2004). Importantly, β-subunit isoforms differentially modulate LTCCs biophysics, including open probability, because of regulatory effects within the channel protein complex (Foell et al. 2004). Hullin et al. have shown that co-expression of $Ca_v1.2$ with $Ca_v\beta_{2a}$ results in a significantly higher open probability of single LTCC when it is compared with $Ca_v1.2$ alone or co-expressed with $Ca_v\beta_{3a}$ (Hullin et al. 2003). Though the subcellular localization of $Ca_v\beta$ isoforms in atrial myocytes remains unknown, their compartmentalized expression may underlie the microdomain-specific heterogeneity in biophysical properties of atrial LTCCs. In addition, some anchoring proteins, such as AKAP150 (Cheng et al. 2011) and ankyrin-B (Le Scouarnec et al. 2008), may regulate spatial distribution and activity of LTCCs in atrial myocytes.

14.3.3 LTCCs and Unique Atrial Ca^{2+} Signaling

It has been proposed that due to lack of a regular T-tubular system, a distinct spatial distribution of LTCCs with regard to their coupling to Ca^{2+} release channels on the SR, the ryanodine receptors (RyR2s) (Fig. 14.5a) may underlie the unique Ca^{2+} signaling observed in atrial myocytes (Dobrev et al. 2009; Trafford et al. 2013). Different authors linked the organization of atrial T-tubule network to the degree of synchronization of subcellular Ca^{2+} transient releases from the SR. The sparsity of the T-tubule network in atrial myocytes leads to a Ca^{2+} transient that initiates at the periphery of the cell and propagates toward the cell center (Fig. 14.5b) (Bootman et al. 2006; Mackenzie et al. 2001; Woo et al. 2003), reminiscent of the spatiotemporal properties of detubulated ventricular myocytes (Brette et al. 2005) and cardiac Purkinje cells lacking T-tubules. Using confocal imaging of atrial myocytes double-labeled with the subcellular Ca^{2+} indicator fluo-4 and surface membrane indicator Di-8-ANEPPS, Kirk and colleagues showed that cells without T-tubules had U-shaped transients that started at the cell periphery, and cells with T-tubules had W-shaped transients that began simultaneously at the cell periphery and the T-tubules (Kirk et al. 2003). When Ca^{2+} release occurs only at the cell periphery and Ca^{2+} spreads by passive diffusion, the time-to-target curve is concave up, forming a U-shaped transient (Fig. 14.5b) (Kirk et al. 2003; Smyrnias et al. 2010).

The extent to which extratubular channels participate in Ca^{2+} transient formation and ECC remains uncertain and may be different for atrial and ventricular channels. In feline ventricular myocytes, a caveolae-targeted LTCC inhibition by Rem blocked a small fraction of $I_{Ca,L}$ without affecting Ca^{2+} transients or reducing contractility (Makarewich et al. 2012). In contrast, the study by Kirk et al. which demonstrated that atrial SR Ca^{2+} releases are located at the T-tubules as well as the peripheral sarcolemmal membrane highlights the importance of extratubular LTCC in initiation and synchronization of atrial Ca^{2+} transients (Kirk et al. 2003). Immunolabeling studies of isolated atrial myocytes demonstrated that the pattern of RyR localization in atrial myocytes has some similarity to that observed in ventricular cells, in that most RyRs lie within regularly spaced transverse striations corresponding to the positions of the Z-tubules (Fig. 14.5a). However, there is the expression of additional RyR clusters around the periphery of the atrial cells (Carl et al. 1995; Hatem et al. 1997; Mackenzie et al. 2001). The majority of $Ca_v1.2$ clusters (~60%) were tightly linked to the RyR2, suggesting that $Ca_v1.2$-mediated Ca^{2+} influx is the primary mechanism of ECC in atria (Schulson et al. 2011). Based on their coupling with LTCCs, junctional (i.e., coupled with $Ca_v1.2$) and non-junctional clusters of RyRs may differ in their probability of initiating SR Ca^{2+} release, with some having been designated as "eager" sites, while others require more time to be activated (Mackenzie et al. 2001; Trafford et al. 2013).

To facilitate the diffusion-driven Ca^{2+} propagation in detubulated atrial myocytes, Dobrev et al. (Dobrev et al. 2009) proposed that atrial RyRs are hypersensitive to cytosolic $[Ca^{2+}]_i$, probably due to either a higher SR Ca^{2+} content in atrial versus ventricular myocytes (Walden et al. 2009) or elevated phosphorylation of atrial RyRs. This means that a local Ca^{2+} release from the SR, or diffusive Ca^{2+} wave, is more likely to trigger another local Ca^{2+} release. It thus may improve the synchrony of the atrial $[Ca^{2+}]_i$ transients when T-tubules are disorganized or absent. In addition, different sensitivity of junctional and non-junctional RyRs to cytosolic Ca^{2+} has been recently proposed to contribute to the unique spatial patterning of Ca^{2+} signals during atrial myocyte ECC.

14.3.4 LTCC Remodeling in HF and Atrial Fibrillation

14.3.4.1 Atrial Fibrillation

Chronic atrial fibrillation (AF) involves important structural and electrical changes including significant downregulation of $I_{Ca,L}$ that results in shortened action potential duration and reduced atrial contractility (Brundel et al. 2001; Christ et al. 2004). Though it was initially suggested that such decrease in $I_{Ca,L}$ is transcriptionally mediated through downregulation of the pore-forming α_{1C} subunit (Brundel et al. 2001; Klein et al. 2003), recent studies in humans have shown that it is not the case, at both mRNA and protein levels (Christ et al. 2004; Schotten et al. 2003). Instead, the reduced $I_{Ca,L}$ has been proposed to be associated with the impaired basal phosphorylation of LTCCs due to increased protein phosphatase activity (Christ et al. 2004). These results, however, contradict data from single-channel measurements, where increased open probability of single LTCCs was found in AF patients (Klein et al. 2003).

This remained discrepant until the recent study by Balycheva et al. where several important components of $I_{Ca,L}$ downregulation have been revealed (Balycheva et al. 2014). First, the authors stated that a significant T-tubule system degradation found in both animal models of atrial fibrillation (Lenaerts et al. 2009; Wakili et al. 2010) and in AF patients (Balycheva et al. 2014) would contribute to the $I_{Ca,L}$ decrease by reducing the contribution of the T-tubular component of the whole-cell current. In addition, they found a dramatic reduction in a number of functional LTCCs available on the surface membrane which would subsequently reduce the contribution of the extratubular component of the $I_{Ca,L}$. The authors have shown that the enhanced activity of both tubular and the remaining extratubular channels, characterized by their increased open probability and availability, do not compensate for the whole-cell $I_{Ca,L}$ decrease (Balycheva et al. 2014).

Importantly, Balycheva and colleagues demonstrated that these changes were accompanied by unaffected $Ca_v1.2$ mRNA expression, while $Ca_v\beta_2$ mRNA was found to be downregulated (Balycheva et al. 2014). The authors concluded that the reduction in a number of functional LTCCs available on the membrane of AF cardiomyocytes, along with unchanged $Ca_v1.2$ expression level, may indicate an increase in a number of nonfunctional "silent" channels. Indeed, isoproterenol could significantly increase

$I_{Ca,L}$ (Christ et al. 2004) recruiting "silent" LTCCs (Balycheva et al. 2014). Together with the rightward shift of the $I_{Ca,L}$ current-voltage curve in AF, these may indicate phosphorylation-dependent mechanisms underlying a decreased number of functional LTCC available for activation during depolarization. Indeed, Christ et al. reported a decreased LTCC phosphorylation in AF patients which was associated with the increased type 2A (but not type 1) phosphatase protein expression and higher phosphatase activity in AF (Christ et al. 2004). At the same time, Klein et al. linked the increased single LTCC activity in AF patients to their hyperphosphorylation due to the reduced activity of both PP2A and PP1 phosphatases (Klein et al. 2003). The remaining paradox that two opposite mechanisms associated with enhanced LTCCs activity and decreased channels' availability (i.e., increased number of silent channels) both rely on channels' phosphorylation requires further investigation.

In addition to phosphorylation defects, structural remodeling may affect LTCC trafficking disturbing a proper localization of functional LTCCs (Schotten et al. 2003; Shaw and Colecraft 2013). This can include downregulation of $Ca_v\beta$ auxiliary subunits (Schotten et al. 2003), changes in A-kinase-anchoring protein (AKAPs) expression profile (Johnson et al. 2012; Kamp and Hell 2000), or enhanced channel degradation (Brundel et al. 2002).

14.3.4.2 Heart Failure

Decrease in atrial $I_{Ca,L}$ in HF has been shown in both animal models (Clarke et al. 2015) and patients with congestive HF (Ouadid et al. 1995). Similar to AF remodeling, T-tubule degradation observed in atrial myocytes in HF (Caldwell et al. 2014; Dibb et al. 2009; Glukhov et al. 2015a; Wakili et al. 2010) would decrease the T-tubule component of the $I_{Ca,L}$. However, an additional mechanism responsible for the reduction in $I_{Ca,L}$ in failing atria has been recently reported and linked to the reduction in the amplitude of the remaining tubular LTCCs. Despite the decreased $I_{Ca,L}$, an increase in SR Ca^{2+} load (caffeine-induced $[Ca^{2+}]_i$ release) has been observed in failing atria (Clarke et al. 2015; Yeh et al. 2008). In addition to the increased SR Ca^{2+} loading, a significant reduction in calsequestrin expression has been found in failing atria (Yeh et al. 2008), and this has been linked to increase in SR Ca^{2+} leak and atrial arrhythmogenesis, perhaps as a result of decreased SR Ca^{2+} buffering (Glukhov et al. 2015b). Both SR Ca^{2+} leak and elevated diastolic $[Ca^{2+}]_i$ may affect T-tubular LTCCs, causing a reduction in their amplitude, either through Ca^{2+}-dependent inactivation or dephosphorylation. Therefore, a disruption in the delicate balance of dynamic interactions between dyadic LTCCs and their microenvironment may alter Ca^{2+} signaling and can lead to pathological changes in cellular physiology.

Conclusion

In cardiac myocytes, LTCCs and associated regulatory receptors are spatially compartmentalized to multiple distinct subcellular microdomains, and that may impact upon their function and regulation. Importantly, many of these proteins form an interacting network where they work together as a part of a macromolecular signaling complex. A number of important LTCC-interacting networks have been identified in cardiac myocytes that associate with unique macromolecular signaling complexes and scaffolding proteins, which enables spatiotemporal modulation of cellular electrophysiology and contraction. These include complexes located in T-tubules, lipid rafts/caveolae, costameres, and intercalated disks where they are associated with different structural proteins. Such organization allows the specificity, reliability, and accuracy of autonomic modulation of excitation-contraction processes by a variety of neurohormonal pathways either via direct interaction or by second messengers through different G-protein-coupled receptors. In this respect, microdomain-specific localization of certain G-protein coupled receptors as well as subcellular compartmentation of a variety of second messengers and kinases should be also acknowledged (Harvey and Calaghan 2012; Insel et al. 2005; Stangherlin and Zaccolo 2012; Zhang and Mende 2011). In conclusion, all of the above considerations demonstrate the importance and the complexity of the microdomain-specific modulation and remodeling of LTCC biophysical properties. This extends beyond the classical concept of electrical remodeling in cardiac disease, according to which dysfunction can be explained by straightforward increases or decreases in protein expression alone. Instead, the emerging evidence stresses that alterations of spatial compartmentation of ion channels and receptors could be crucial for pathology, in addition to classically appreciated changes in protein expression and posttranslational modifications. Thus, a better understanding of the various subcellular macromolecular signaling complexes may enable new therapeutic approaches for predicting and ameliorating the risk of sudden cardiac death and malignant arrhythmias in patients with cardiac diseases.

Compliance with Ethical Standards

Conflict of Interest Statement The authors declare that they have no conflict of interest.

References

Andersen OS, Koeppe RE (2007) Bilayer thickness and membrane protein function: an energetic perspective. Annu Rev Biophys Biomol Struct 36:107–130

Balijepalli RC, Kamp TJ (2008) Caveolae, ion channels and cardiac arrhythmias. Prog Biophys Mol Biol 98:149–160

Balijepalli RC, Lokuta AJ, Maertz NA et al (2003) Depletion of T-tubules and specific subcellular changes in sarcolemmal proteins in tachycardia-induced heart failure. Cardiovasc Res 59:67–77

Balijepalli RC, Foell JD, Hall DD et al (2006) Localization of cardiac L-type Ca^{2+} channels to a caveolar macromolecular signaling complex is required for beta(2)-adrenergic regulation. Proc Natl Acad Sci U S A 103:7500–7505

Balycheva M, Glukhov A, Schobesberger S et al (2014) Increased open probability of L-type calcium channels localized in T-tubules in patients with chronic atrial fibrillation: role of channel subunits. Circulation 130:A18709

Banfi C, Brioschi M, Wait R et al (2006) Proteomic analysis of membrane microdomains derived from both failing and non-failing human hearts. Proteomics 6:1976–1988

Bers DM (2002) Cardiac excitation-contraction coupling. Nature 415:198–205

Best JM, Kamp TJ (2012) Different subcellular populations of L-type Ca^{2+} channels exhibit unique regulation and functional roles in cardiomyocytes. J Mol Cell Cardiol 52:376–387

Bhargava A, Lin X, Novak P et al (2013) Super-resolution scanning patch clamp reveals clustering of functional ion channels in adult ventricular myocyte. Circ Res 112:1112–1120
Bootman MD, Higazi DR, Coombes S et al (2006) Calcium signalling during excitation-contraction coupling in mammalian atrial myocytes. J Cell Sci 119:3915–3925
Boulware MI, Kordasiewicz H, Mermelstein PG (2007) Caveolin proteins are essential for distinct effects of membrane estrogen receptors in neurons. J Neurosci 27:9941–9950
Brette F, Komukai K, Orchard CH (2002) Validation of formamide as a detubulation agent in isolated rat cardiac cells. Am J Physiol Heart Circ Physiol 283:H1720–H1728
Brette F, Despa S, Bers DM et al (2005) Spatiotemporal characteristics of SR Ca^{2+} uptake and release in detubulated rat ventricular myocytes. J Mol Cell Cardiol 39:804–812
Bristow MR, Ginsburg R, Minobe W et al (1982) Decreased catecholamine sensitivity and beta-adrenergic-receptor density in failing human hearts. N Engl J Med 307:205–211
Bristow MR, Ginsburg R, Umans V et al (1986) Beta 1- and beta 2-adrenergic-receptor subpopulations in nonfailing and failing human ventricular myocardium: coupling of both receptor subtypes to muscle contraction and selective beta 1-receptor down-regulation in heart failure. Circ Res 59:297–309
Brundel BJ, Van Gelder IC, Henning RH et al (2001) Ion channel remodeling is related to intraoperative atrial effective refractory periods in patients with paroxysmal and persistent atrial fibrillation. Circulation 103:684–690
Brundel BJ, Ausma J, van Gelder IC et al (2002) Activation of proteolysis by calpains and structural changes in human paroxysmal and persistent atrial fibrillation. Cardiovasc Res 54: 380–389
Bryant S, Kimura TE, Kong CH et al (2014) Stimulation of I_{Ca} by basal PKA activity is facilitated by caveolin-3 in cardiac ventricular myocytes. J Mol Cell Cardiol 68:47–55
Caldwell JL, Smith CE, Taylor RF et al (2014) Dependence of cardiac transverse tubules on the BAR domain protein amphiphysin II (BIN-1). Circ Res 115:986–996
Carl SL, Felix K, Caswell AH et al (1995) Immunolocalization of sarcolemmal dihydropyridine receptor and sarcoplasmic reticular triadin and ryanodine receptor in rabbit ventricle and atrium. J Cell Biol 129:673–682
Carnegie GK, Means CK, Scott JD (2009) A-kinase anchoring proteins: from protein complexes to physiology and disease. IUBMB Life 61:394–406
Cavalli A, Eghbali M, Minosyan TY et al (2007) Localization of sarcolemmal proteins to lipid rafts in the myocardium. Cell Calcium 42:313–322
Cerrone M, Delmar M (2014) Desmosomes and the sodium channel complex: implications for arrhythmogenic cardiomyopathy and Brugada syndrome. Trends Cardiovasc Med 24: 184–190
Cheng EP, Yuan C, Navedo MF et al (2011) Restoration of normal L-type Ca^{2+} channel function during Timothy syndrome by ablation of an anchoring protein. Circ Res 109:255–261
Chen-Izu Y, McCulle SL, Ward CW et al (2006) Three-dimensional distribution of ryanodine receptor clusters in cardiac myocytes. Biophys J 91:1–13
Chesley A, Lundberg MS, Asai T et al (2000) The beta(2)-adrenergic receptor delivers an anti-apoptotic signal to cardiac myocytes through G_i-dependent coupling to phosphatidylinositol 3'-kinase. Circ Res 87:1172–1179
Christ T, Boknik P, Wohrl S et al (2004) L-type Ca^{2+} current downregulation in chronic human atrial fibrillation is associated with increased activity of protein phosphatases. Circulation 110:2651–2657
Clarke JD, Caldwell JL, Horn MA et al (2015) Perturbed atrial calcium handling in an ovine model of heart failure: potential roles for reductions in the L-type calcium current. J Mol Cell Cardiol 79:169–179
Cloues RK, Sather WA (2000) Permeant ion binding affinity in subconductance states of an L-type Ca^{2+} channel expressed in *Xenopus laevis* oocytes. J Physiol 524(Pt 1):19–36
Cohen AW, Park DS, Woodman SE et al (2003) Caveolin-1 null mice develop cardiac hypertrophy with hyperactivation of p42/44 MAP kinase in cardiac fibroblasts. Am J Physiol Cell Physiol 284:C457–C474

Dart C (2010) Lipid microdomains and the regulation of ion channel function. J Physiol 588:3169–3178

Dibb KM, Clarke JD, Horn MA et al (2009) Characterization of an extensive transverse tubular network in sheep atrial myocytes and its depletion in heart failure. Circ Heart Fail 2:482–489

Dibb KM, Clarke JD, Eisner DA et al (2013) A functional role for transverse (t-) tubules in the atria. J Mol Cell Cardiol 58:84–91

Dobrev D, Teos LY, Lederer WJ (2009) Unique atrial myocyte Ca^{2+} signaling. J Mol Cell Cardiol 46:448–451

Edidin M (2003) The state of lipid rafts: from model membranes to cells. Annu Rev Biophys Biomol Struct 32:257–283

Epshtein Y, Chopra AP, Rosenhouse-Dantsker A et al (2009) Identification of a C-terminus domain critical for the sensitivity of $K_{ir}2.1$ to cholesterol. Proc Natl Acad Sci U S A 106:8055–8060

Feiner EC, Chung P, Jasmin JF et al (2011) Left ventricular dysfunction in murine models of heart failure and in failing human heart is associated with a selective decrease in the expression of caveolin-3. J Card Fail 17:253–263

Feng J, Yue L, Wang Z et al (1998) Ionic mechanisms of regional action potential heterogeneity in the canine right atrium. Circ Res 83:541–551

Foell JD, Balijepalli RC, Delisle BP et al (2004) Molecular heterogeneity of calcium channel beta-subunits in canine and human heart: evidence for differential subcellular localization. Physiol Genomics 17:183–200

Forssmann WG, Girardier L (1970) A study of the T system in rat heart. J Cell Biol 44:1–19

Frisk M, Koivumaki JT, Norseng PA et al (2014) Variable t-tubule organization and Ca^{2+} homeostasis across the atria. Am J Physiol Heart Circ Physiol 307:H609–H620

Gathercole DV, Colling DJ, Skepper JN et al (2000) Immunogold-labeled L-type calcium channels are clustered in the surface plasma membrane overlying junctional sarcoplasmic reticulum in Guinea-pig myocytes-implications for excitation-contraction coupling in cardiac muscle. J Mol Cell Cardiol 32:1981–1994

George MS, Pitt GS (2006) The real estate of cardiac signaling: location, location, location. Proc Natl Acad Sci U S A 103:7535–7536

Glukhov AV, Balycheva M, Sanchez-Alonso JL et al (2015a) Direct evidence for microdomain-specific localization and remodeling of functional L-type calcium channels in rat and human atrial myocytes. Circulation 132:2372–2384

Glukhov AV, Kalyanasundaram A, Lou Q et al (2015b) Calsequestrin 2 deletion causes sinoatrial node dysfunction and atrial arrhythmias associated with altered sarcoplasmic reticulum calcium cycling and degenerative fibrosis within the mouse atrial pacemaker complex. Eur Heart J 36:686–697

Gomez AM, Valdivia HH, Cheng H et al (1997) Defective excitation-contraction coupling in experimental cardiac hypertrophy and heart failure. Science 276:800–806

Gondo N, Ono K, Mannen K et al (1998) Four conductance levels of cloned cardiac L-type Ca^{2+} channel alpha1 and alpha1/beta subunits. FEBS Lett 423:86–92

Gray PC, Tibbs VC, Catterall WA et al (1997) Identification of a 15-kDa cAMP-dependent protein kinase-anchoring protein associated with skeletal muscle L-type calcium channels. J Biol Chem 272:6297–6302

Harvey RD, Calaghan SC (2012) Caveolae create local signalling domains through their distinct protein content, lipid profile and morphology. J Mol Cell Cardiol 52:366–375

Hatem SN, Benardeau A, Rucker-Martin C et al (1997) Different compartments of sarcoplasmic reticulum participate in the excitation-contraction coupling process in human atrial myocytes. Circ Res 80:345–353

Hayashi T, Arimura T, Ueda K et al (2004) Identification and functional analysis of a caveolin-3 mutation associated with familial hypertrophic cardiomyopathy. Biochem Biophys Res Commun 313:178–184

He J, Conklin MW, Foell JD et al (2001) Reduction in density of transverse tubules and L-type Ca^{2+} channels in canine tachycardia-induced heart failure. Cardiovasc Res 49:298–307

He JQ, Balijepalli RC, Haworth RA et al (2005) Crosstalk of beta-adrenergic receptor subtypes through Gi blunts beta-adrenergic stimulation of L-type Ca^{2+} channels in canine heart failure. Circ Res 97:566–573

Heinzel FR, Bito V, Biesmans L et al (2008) Remodeling of T-tubules and reduced synchrony of Ca^{2+} release in myocytes from chronically ischemic myocardium. Circ Res 102: 338–346

Heinzel FR, MacQuaide N, Biesmans L et al (2011) Dyssynchrony of Ca^{2+} release from the sarcoplasmic reticulum as subcellular mechanism of cardiac contractile dysfunction. J Mol Cell Cardiol 50:390–400

Hong T, Yang H, Zhang SS et al (2014) Cardiac BIN1 folds T-tubule membrane, controlling ion flux and limiting arrhythmia. Nat Med 20:624–632

Horikawa YT, Panneerselvam M, Kawaraguchi Y et al (2011) Cardiac-specific overexpression of caveolin-3 attenuates cardiac hypertrophy and increases natriuretic peptide expression and signaling. J Am Coll Cardiol 57:2273–2283

Hullin R, Khan IF, Wirtz S et al (2003) Cardiac L-type calcium channel beta-subunits expressed in human heart have differential effects on single channel characteristics. J Biol Chem 278:21623–21630

Insel PA, Head BP, Ostrom RS et al (2005) Caveolae and lipid rafts: G protein-coupled receptor signaling microdomains in cardiac myocytes. Ann N Y Acad Sci 1047:166–172

Johnson KR, Nicodemus-Johnson J, Carnegie GK et al (2012) Molecular evolution of A-kinase anchoring protein (AKAP)-7: implications in comparative PKA compartmentalization. BMC Evol Biol 12:125

Kamp TJ, He JQ (2002) L-type Ca^{2+} channels gaining respect in heart failure. Circ Res 91:451–453

Kamp TJ, Hell JW (2000) Regulation of cardiac L-type calcium channels by protein kinase a and protein kinase C. Circ Res 87:1095–1102

Kawabe JI, Grant BS, Yamamoto M et al (2001) Changes in caveolin subtype protein expression in aging rat organs. Mol Cell Endocrinol 176:91–95

Kawai M, Hussain M, Orchard CH (1999) Excitation-contraction coupling in rat ventricular myocytes after formamide-induced detubulation. Am J Phys 277:H603–H609

Kirk MM, Izu LT, Chen-Izu Y et al (2003) Role of the transverse-axial tubule system in generating calcium sparks and calcium transients in rat atrial myocytes. J Physiol 547:441–451

Klein G, Schroder F, Vogler D et al (2003) Increased open probability of single cardiac L-type calcium channels in patients with chronic atrial fibrillation. Role of phosphatase 2A. Cardiovasc Res 59:37–45

Le Scouarnec S, Bhasin N, Vieyres C et al (2008) Dysfunction in ankyrin-B-dependent ion channel and transporter targeting causes human sinus node disease. Proc Natl Acad Sci U S A 105: 15617–15622

Lenaerts I, Bito V, Heinzel FR et al (2009) Ultrastructural and functional remodeling of the coupling between Ca^{2+} influx and sarcoplasmic reticulum Ca^{2+} release in right atrial myocytes from experimental persistent atrial fibrillation. Circ Res 105:876–885

Lohse MJ, Engelhardt S, Eschenhagen T (2003) What is the role of beta-adrenergic signaling in heart failure? Circ Res 93:896–906

Louch WE, Bito V, Heinzel FR et al (2004) Reduced synchrony of Ca^{2+} release with loss of T-tubules-a comparison to Ca^{2+} release in human failing cardiomyocytes. Cardiovasc Res 62:63–73

Lyon AR, MacLeod KT, Zhang Y et al (2009) Loss of T-tubules and other changes to surface topography in ventricular myocytes from failing human and rat heart. Proc Natl Acad Sci U S A 106:6854–6859

Mackenzie L, Bootman MD, Berridge MJ et al (2001) Predetermined recruitment of calcium release sites underlies excitation-contraction coupling in rat atrial myocytes. J Physiol 530: 417–429

Makarewich CA, Correll RN, Gao H et al (2012) A caveolae-targeted L-type Ca^{2+} channel antagonist inhibits hypertrophic signaling without reducing cardiac contractility. Circ Res 110: 669–674

Mangoni ME, Couette B, Bourinet E et al (2003) Functional role of L-type $Ca_v1.3$ Ca^{2+} channels in cardiac pacemaker activity. Proc Natl Acad Sci U S A 100:5543–5548
McDonald TF, Pelzer S, Trautwein W et al (1994) Regulation and modulation of calcium channels in cardiac, skeletal, and smooth muscle cells. Physiol Rev 74:365–507
Munro S (2003) Lipid rafts: elusive or illusive? Cell 115:377–388
Nattel S, Maguy A, Le Bouter S et al (2007) Arrhythmogenic ion-channel remodeling in the heart: heart failure, myocardial infarction, and atrial fibrillation. Physiol Rev 87:425–456
Nichols CB, Rossow CF, Navedo MF et al (2010) Sympathetic stimulation of adult cardiomyocytes requires association of AKAP5 with a subpopulation of L-type calcium channels. Circ Res 107:747–756
Nikolaev VO, Bunemann M, Schmitteckert E et al (2006a) Cyclic AMP imaging in adult cardiac myocytes reveals far-reaching beta1-adrenergic but locally confined beta2-adrenergic receptor-mediated signaling. Circ Res 99:1084–1091
Nikolaev VO, Gambaryan S, Lohse MJ (2006b) Fluorescent sensors for rapid monitoring of intracellular cGMP. Nat Methods 3:23–25
Nikolaev VO, Moshkov A, Lyon AR et al (2010) Beta2-adrenergic receptor redistribution in heart failure changes cAMP compartmentation. Science 327:1653–1657
Osterrieder W, Brum G, Hescheler J et al (1982) Injection of subunits of cyclic AMP-dependent protein kinase into cardiac myocytes modulates Ca^{2+} current. Nature 298:576–578
Ouadid H, Albat B, Nargeot J (1995) Calcium currents in diseased human cardiac cells. J Cardiovasc Pharmacol 25:282–291
Palade GE, Bruns RR (1968) Structural modulations of plasmalemmal vesicles. J Cell Biol 37:633–649
Pani B, Singh BB (2009) Lipid rafts/caveolae as microdomains of calcium signaling. Cell Calcium 45:625–633
Park DS, Cohen AW, Frank PG et al (2003) Caveolin-1 null (−/−) mice show dramatic reductions in life span. Biochemistry 42:15124–15131
Patel HH, Murray F, Insel PA (2008) Caveolae as organizers of pharmacologically relevant signal transduction molecules. Annu Rev Pharmacol Toxicol 48:359–391
Perera RK, Nikolaev VO (2013) Compartmentation of cAMP signalling in cardiomyocytes in health and disease. Acta Physiol (Oxf) 207:650–662
Pike LJ (2004) Lipid rafts: heterogeneity on the high seas. Biochem J 378:281–292
Ratajczak P, Damy T, Heymes C et al (2003) Caveolin-1 and -3 dissociations from caveolae to cytosol in the heart during aging and after myocardial infarction in rat. Cardiovasc Res 57: 358–369
Razani B, Woodman SE, Lisanti MP (2002) Caveolae: from cell biology to animal physiology. Pharmacol Rev 54:431–467
Richards MA, Clarke JD, Saravanan P et al (2011) Transverse tubules are a common feature in large mammalian atrial myocytes including human. Am J Physiol Heart Circ Physiol 301: H1996–H2005
Romanenko VG, Rothblat GH, Levitan I (2002) Modulation of endothelial inward-rectifier K+ current by optical isomers of cholesterol. Biophys J 83:3211–3222
Sanchez-Alonso JL, Bhargava A, O'Hara T et al (2016) Microdomain-specific modulation of L-type calcium channels leads to triggered ventricular arrhythmia in heart failure. Circ Res 119:944–955
Schaper J, Kostin S, Hein S et al (2002) Structural remodelling in heart failure. Exp Clin Cardiol 7:64–68
Schotten U, Haase H, Frechen D et al (2003) The L-type Ca^{2+}-channel subunits alpha1C and beta2 are not downregulated in atrial myocardium of patients with chronic atrial fibrillation. J Mol Cell Cardiol 35:437–443
Schroder E, Byse M, Satin J (2009) L-type calcium channel C terminus autoregulates transcription. Circ Res 104:1373–1381
Schulson MN, Scriven DR, Fletcher P et al (2011) Couplons in rat atria form distinct subgroups defined by their molecular partners. J Cell Sci 124:1167–1174

Scriven DR, Asghari P, Schulson MN et al (2010) Analysis of $Ca_v1.2$ and ryanodine receptor clusters in rat ventricular myocytes. Biophys J 99:3923–3929

Sen L, Bialecki RA, Smith E et al (1992) Cholesterol increases the L-type voltage-sensitive calcium channel current in arterial smooth muscle cells. Circ Res 71:1008–1014

Shaw RM, Colecraft HM (2013) L-type calcium channel targeting and local signalling in cardiac myocytes. Cardiovasc Res 98:177–186

Shibata EF, Brown TL, Washburn ZW et al (2006) Autonomic regulation of voltage-gated cardiac ion channels. J Cardiovasc Electrophysiol 17(Suppl 1):S34–S42

Simionescu M, Simionescu N, Palade GE (1975) Segmental differentiations of cell junctions in the vascular endothelium. The microvasculature. J Cell Biol 67:863–885

Simons K, Toomre D (2000) Lipid rafts and signal transduction. Nat Rev Mol Cell Biol 1:31–39

Smyrnias I, Mair W, Harzheim D et al (2010) Comparison of the T-tubule system in adult rat ventricular and atrial myocytes, and its role in excitation-contraction coupling and inotropic stimulation. Cell Calcium 47:210–223

Song KS, Scherer PE, Tang Z et al (1996) Expression of caveolin-3 in skeletal, cardiac, and smooth muscle cells. Caveolin-3 is a component of the sarcolemma and co-fractionates with dystrophin and dystrophin-associated glycoproteins. J Biol Chem 271:15160–15165

Sprenger JU, Nikolaev VO (2013) Biophysical techniques for detection of cAMP and cGMP in living cells. Int J Mol Sci 14:8025–8046

Stangherlin A, Zaccolo M (2012) Phosphodiesterases and subcellular compartmentalized cAMP signaling in the cardiovascular system. Am J Physiol Heart Circ Physiol 302:H379–H390

Takagishi Y, Rothery S, Issberner J et al (1997) Spatial distribution of dihydropyridine receptors in the plasma membrane of Guinea pig cardiac myocytes investigated by correlative confocal microscopy and label-fracture electron microscopy. J Electron Microsc 46:165–170

Tidball JG, Cederdahl JE, Bers DM (1991) Quantitative analysis of regional variability in the distribution of transverse tubules in rabbit myocardium. Cell Tissue Res 264:293–298

Timofeyev V, Myers RE, Kim HJ et al (2013) Adenylyl cyclase subtype-specific compartmentalization: differential regulation of L-type Ca^{2+} current in ventricular myocytes. Circ Res 112: 1567–1576

Trafford AW, Clarke JD, Richards MA et al (2013) Calcium signalling microdomains and the t-tubular system in atrial mycoytes: potential roles in cardiac disease and arrhythmias. Cardiovasc Res 98:192–203

Tsutsumi YM, Horikawa YT, Jennings MM et al (2008) Cardiac-specific overexpression of caveolin-3 induces endogenous cardiac protection by mimicking ischemic preconditioning. Circulation 118:1979–1988

Wakili R, Yeh YH, Yan Qi X et al (2010) Multiple potential molecular contributors to atrial hypocontractility caused by atrial tachycardia remodeling in dogs. Circ Arrhythm Electrophysiol 3:530–541

Walden AP, Dibb KM, Trafford AW (2009) Differences in intracellular calcium homeostasis between atrial and ventricular myocytes. J Mol Cell Cardiol 46:463–473

Wei S, Guo A, Chen B et al (2010) T-tubule remodeling during transition from hypertrophy to heart failure. Circ Res 107:520–531

Willoughby D, Cooper DM (2007) Organization and Ca^{2+} regulation of adenylyl cyclases in cAMP microdomains. Physiol Rev 87:965–1010

Woo AY, Xiao RP (2012) Beta-adrenergic receptor subtype signaling in heart: from bench to bedside. Acta Pharmacol Sin 33:335–341

Woo SH, Cleemann L, Morad M (2003) Spatiotemporal characteristics of junctional and nonjunctional focal Ca^{2+} release in rat atrial myocytes. Circ Res 92:e1–11

Yamashita T, Nakajima T, Hazama H et al (1995) Regional differences in transient outward current density and inhomogeneities of repolarization in rabbit right atrium. Circulation 92: 3061–3069

Yeh YH, Wakili R, Qi XY et al (2008) Calcium-handling abnormalities underlying atrial arrhythmogenesis and contractile dysfunction in dogs with congestive heart failure. Circ Arrhythm Electrophysiol 1:93–102

Zhang P, Mende U (2011) Regulators of G-protein signaling in the heart and their potential as therapeutic targets. Circ Res 109:320–333

Zhang Q, Timofeyev V, Qiu H et al (2011) Expression and roles of $Ca_v1.3$ (alpha1D) L-type Ca^{2+} channel in atrioventricular node automaticity. J Mol Cell Cardiol 50:194–202

Zhao YY, Liu Y, Stan RV et al (2002) Defects in caveolin-1 cause dilated cardiomyopathy and pulmonary hypertension in knockout mice. Proc Natl Acad Sci U S A 99:11375–11380

Zhu WZ, Wang SQ, Chakir K et al (2003) Linkage of beta1-adrenergic stimulation to apoptotic heart cell death through protein kinase A-independent activation of Ca^{2+}/calmodulin kinase II. J Clin Invest 111:617–625

The Role of Local Ca^{2+} Release for Ca^{2+} Alternans and SR-Ca^{2+} Leak

15

Karin P. Hammer and Lars S. Maier

Contents

K.P. Hammer (✉) • L.S. Maier (✉)
University Hospital Regensburg, Internal Medicine II, Cardiology,
Franz-Josef-Strauss-Allee 11, D-93053 Regensburg, Germany
e-mail: karin.hammer@ukr.de; lars.maier@ukr.de

V. Nikolaev, M. Zaccolo (eds.), *Microdomains in the Cardiovascular System*,
Cardiac and Vascular Biology 3, DOI 10.1007/978-3-319-54579-0_15

Abstract

Cardiac Ca^{2+} is the second messenger transducing electrical signals into mechanical responses during excitation-contraction coupling (ECC). The unique morphology of ventricular myocytes plays a pivotal role during ECC and assures the synchronous Ca^{2+} release into the cytosol and the orchestrated interplay of the channels and modulators involved. Alterations of one or more of the key players will cause instabilities during Ca^{2+} cycling that can have detrimental effects on the function of the myocyte. Local Ca^{2+} release is the underlying mechanism during ECC, and single spontaneous release events occur rarely under healthy conditions but increase during disease progression. This shift in the fine-tuned release machinery can propel toward more severe arrhythmogenic behavior. Increased SR leak can set the basis for Ca^{2+} alternans in single myocytes among other factors. Alternans as arrhythmogenic factors mechanistically link cardiac mechanical dysfunction and sudden cardiac death. On a cellular level, alternans can be seen early on during disease progression. Here we want to highlight the role of localized Ca^{2+} release for the development and maintenance of alternans in single myocytes and the intact heart.

Abbreviations

$[Ca^{2+}]$	Ca^{2+} concentration
$[Ca^{2+}]_{cleft}$	Ca^{2+} concentration in the cleft
$[Ca^{2+}]_i$	Intracellular Ca^{2+} concentration
$[Ca^{2+}]_{mito}$	Mitochondrial Ca^{2+} concentration
$[Ca^{2+}]_{SR}$	SR-Ca^{2+} concentration
AP	Action potential
BCL	Basic cycle length
CaMKII	Ca^{2+}/calmodulin-dependent protein kinase II
CaT	Ca^{2+} transient
CICR	Ca^{2+}-induced Ca^{2+} release
DI	Diastolic interval
ECC	Excitation-contraction coupling
ETC	Excitation transcription coupling
GPCR	G-protein-coupled receptor
GPI	Glycophosphatidylinositol
I_{Ca}	Ca^{2+} current
I_{NCX}	NCX current
$InsP_3R$	Inositol triphosphate receptor
LTCC	L-type Ca^{2+} channel
MCU	Mitochondrial Ca^{2+} uniporter

MTPT	Mitochondrial permeability transition pore
NCX	Na^+-Ca^{2+} exchanger
PKA	Protein kinase A
RyR	Ryanodine receptor
SERCA	Sarco-endoplasmatic reticulum ATPase
SR	Sarcoplasmatic reticulum
TRP	Transient receptor potential
t-tubules	Transverse tubules

15.1 Introduction

The ubiquitous second messenger Ca^{2+} is the main key player during cardiac excitation-contraction coupling (ECC), and its main role is the translation of electrical activity into the mechanical contraction.

During an action potential (AP), Ca^{2+} enters the cell via voltage-gated Ca^{2+} channels and activates ryanodine receptors (RyRs) that are located in the membrane of the sarcoplasmic reticulum (SR) causing Ca^{2+} release from this Ca^{2+} store (Wier 2007; Bers 2008). This so-called Ca^{2+}-induced Ca^{2+} release (CICR) supplies enough Ca^{2+} to induce a contraction (Valdeolmillos et al. 1989). To restore a situation where another contraction is possible, Ca^{2+} needs to be removed from the cytosol. The major part of the cytosolic Ca^{2+} is pumped back into the stores via a Ca^{2+} pump located in the SR (SERCA), and another major removal mechanism is extrusion out of the cell via the Na^+-Ca^{2+} exchanger (NCX) in the plasma membrane. The relative contribution of the various removal pathways is species dependent, while in rat myocytes, about 7–9% of the removed Ca^{2+} is transported out of the cell via NCX and 87–92% pumped back into the SR; this distribution is significantly different in human myocytes with NCX contributing about 27% and SERCA 63%. The latter is further shifted in failing myocytes to 42% for NCX and 58% for SERCA contribution (Bers 2008). Furthermore, Ca^{2+} is transported into the mitochondria via uniporters. This fraction of Ca^{2+} removal contributes with less than 2% to the total Ca^{2+} removed during ECC, but it is crucial for the metabolic demand of the cell (Huser et al. 2000a; Boyman et al. 2014).

The unique morphology of ventricular cardiac myocytes is vital for the simultaneous activation of all release sites within the cell.

The transverse tubules (t-tubules) are invaginations of the plasma membrane that form postnatal by penetrating inward (Di Maio et al. 2007). A large number of proteins contributing to ECC are located at these invaginations (Brette and Orchard 2003). The vast majority of L-type Ca^{2+} channels (LTCCs), responsible for the I_{Ca} inducing Ca^{2+} release from the stores, are located in the t-tubules, and a loss of the latter causes alterations in Ca^{2+} handling and ECC (Lipp et al. 1996; Leach et al. 2005; Brette et al. 2006).

The SR as the major Ca^{2+} storing organelle is located close to the t-tubules where functional couplons (Moore et al. 1984), the so-called dyadic junctions, are the sites of CICR. These sites have about 10–25 LTCCs located in the plasma membrane and about 100–200 RyRs on the SR (Franzini-Armstrong et al. 1999; Bers 2008).

Ca^{2+} is a universal second messenger involved in various cellular functions. The above described function in ECC reflects one aspect which can be further modulated due to spatial and temporal variations of the Ca^{2+} signal. The unique structure of the cardiac plasma membrane with its invaginations contributes largely to the well-orchestrated ECC. The t-tubules represent about 30% of the total membrane (Orchard et al. 2009), and numerous currents involved in ECC are predominantly found at the t-tubules [including I_{Ca}, I_{NCX}, I_{Na}, I_{NKA} (Yang et al. 2002; Despa et al. 2003; Brette et al. 2004; Brette and Orchard 2006)]. The synchrony of Ca^{2+} release due to an electrical stimulus strongly depends on the architecture of the cardiac myocyte, and a loss of t-tubules leads to a desynchronization of the released Ca^{2+} throughout the cell (Lipp et al. 1996; Heinzel et al. 2002; Louch et al. 2004).

The global Ca^{2+} signal is the uniform increase of cytosolic Ca^{2+} and is composed of the sum of local release events (Ca^{2+} sparks) within a cell. The opening of the Ca^{2+} channels in the membrane causes influx and increase of Ca^{2+} in the dyadic cleft (space between SR and plasma membrane) which then activates the RyRs, releasing Ca^{2+} from the SR. This Ca^{2+} can then bind to troponin C and thus cause a contraction of the cell.

Figure 15.1 displays a schematic drawing of the spatial organization of Ca^{2+} release events from intracellular stores. Clusters of RyRs are the Ca^{2+} release sites of the SR that consist of a group of single channels located at the dyadic junction. Ca^{2+} release from a single RyR is called a quark, while the activity of a cluster of RyRs is referred to as a spark. If such a spark activates neighboring clusters via CICR, the Ca^{2+} signal is spreading throughout the cell forming a wave. Since cardiac myocytes are functionally connected via gap junctions at the intercalated discs, the signal can theoretically spread to the neighboring cells initiating intercellular waves (Berridge et al. 2000). However, experimental studies found that Ca^{2+} waves do not readily pass cell borders (Kaneko et al. 2000; Baader et al. 2002; Wasserstrom et al. 2010; Hammer et al. 2015).

In this chapter, we will highlight the importance of cardiac microdomains for SR-Ca^{2+} leak as well as the importance of local Ca^{2+} release for Ca^{2+} alternans. We will only briefly touch on the Ca^{2+} relevant microdomains found in cardiac myocytes as this will be the topic of another chapter in this book. Our focus will be set on the relevance of local Ca^{2+} handling for the initiation and maintenance of Ca^{2+} alternans.

15.2 Ca^{2+} Microdomains in Cardiac Myocytes

The cytosol contains several microdomains with confined, local Ca signaling. Up to 65% of total cell volume is taken up by the cytosol, whereas the remaining 35% are taken by mitochondria (30–35%), nuclei (~2%) and the SR with about 1–3.5% (Bers 2008). All these compartments contribute to the cellular Ca^{2+} signaling and depending on the relative location within the cells and their ability to take up and/or release Ca^{2+}.

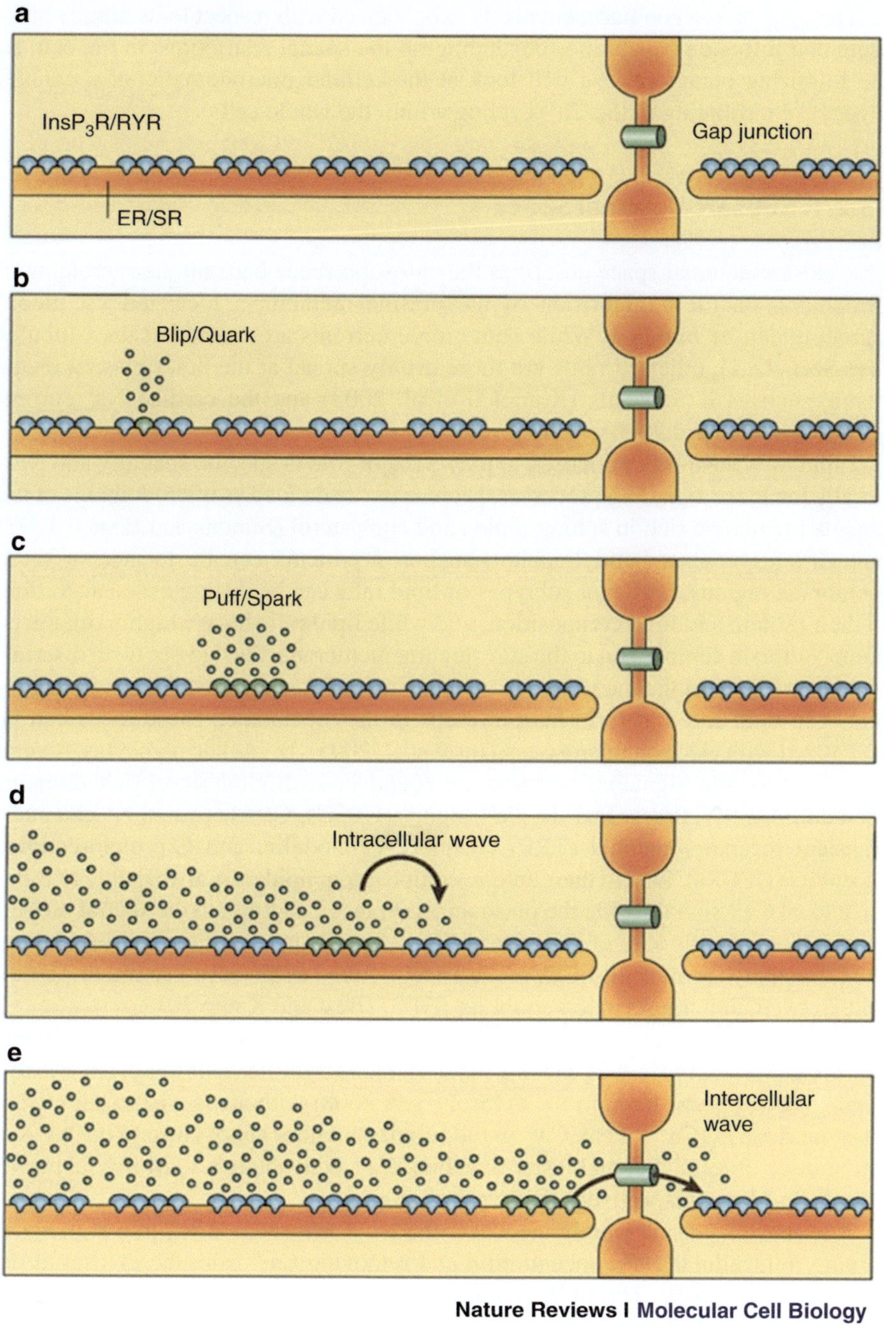

Fig. 15.1 RyRs and inositol triphosphate receptors ($InsP_3R$) are found in clusters on the SR of cardiac myocytes (**a**). The Ca^{2+} release from a single receptor is referred to as a quark (**b**) and the release from a cluster of RyRs is called a spark (**c**). A signal propagating from one cluster to the neighboring throughout the cell is an intracellular wave (**d**), whereas the propagation of such a wave via gap junctions to the neighboring cell is referred to as an intercellular wave (**e**). Image taken with permission from Berridge et al. (2000)

The cytosol as a compartment needs to be viewed with respect to its unique buffering and diffusion capabilities depending on the spatial restrictions in the cell. In the following paragraph, we will look at the cellular microdomains of a cardiac myocyte contributing to the Ca^{2+} cycling within the whole cell.

15.2.1 Subsarcolemmal Space

The subsarcolemmal space describes the cytosolic space beneath the sarcolemma. Depending on the composition of the plasma membrane, localized vs. global signaling can be initiated. While some ionic currents are located at the t-tubules (see Sect. 15.1), other currents are more evenly spread at the outer plasma membrane, such as K^+ currents (Komukai et al. 2002) and the cardiac Na^+ current (Brette and Orchard 2006).

Lipid rafts have been proposed to play a major role in specific spatially and temporally localized signaling processes. Lipid rafts are defined as microdomains of the plasma membrane rich in sphingolipids and cholesterol (Simons and Ikonen 1997) and GPI (glycophosphatidylinositol)-anchored proteins can be located at these membrane regions. Different subtypes of lipid rafts can be distinguished according to their protein and lipid composition, and while lipid rafts have a highly organized composition in comparison to the surrounding membrane, they do not form discriminable structures in the membrane. Caveolae instead contain caveolin to which proteins can bind and that form hairpin loops in the membrane. These loops can be visualized with electron microscopy (Insel et al. 2005). In cardiac myocytes, several ion channels and signaling proteins are found in such cholesterol-rich domains including Ca^{2+}-ATPases, $InsP_3R$, Ca^{2+} pumps, LTCCs, Ca^{2+}-activated K^+ channels, transient receptor potential (TRP) channels, calmodulin, and G-protein-coupled receptors (GPCRs). Due to their unique composition, lipid rafts are capable to influence local Ca^{2+} signaling via the plasma membrane in addition to the spatial properties on the cytosolic side of the latter. LTCCs that are localized to the t-tubules are predominantly involved in ECC, while LTCCs found at the caveolae rather participate in excitation transcription coupling (ETC) (Best and Kamp 2012).

If the cytosolic space is not confined by organelles in close proximity, Ca^{2+} entering the cell via LTCCs will cause a local increase right beneath the plasma membrane but will be buffered in the cytosol (Bers 2008) without causing a significant local increase in $[Ca^{2+}]_{i.}$ Peak Ca^{2+} in the submembranous space close to the NCX is about three times higher and faster than bulk Ca^{2+} with about 3.2 μM in <32 ms vs. 1.1 μM in 81 ms, respectively (Weber et al. 2002).

It is also the plasma membrane where the NCX is located, a key player in maintaining intracellular Na^+ concentration and removing Ca^{2+} from the cytosol at the end of a cardiac cycle. The NCX is found throughout the cell at the outer membrane, the t-tubules, and the intercalated disc, although the relative distribution has been under debate (Frank et al. 1992; Despa et al. 2003; Fowler et al. 2004; Jayasinghe et al. 2009). The contribution of the exchanger to the Ca^{2+} removal at the end of a cardiac cycle is species dependent and plays a more dominant role in large mammals and humans compared to smaller species such as small rodents. The NCX largely contributes to ECC because it transports Ca^{2+} out of the cell and thus ensures cytosolic Ca^{2+} baseline levels that are crucial for ECC (Bers 2002, 2008).

Another microdomain of the plasma membrane is found at the short ends of cardiac myocytes where the desmosomes are located. Specialized proteins form the desmosomal complexes that constitute a mechanical contact point between two adjacent cells (Kowalczyk and Green 2013). In direct neighborhood of these complexes, several ion channels are located and directly associate to the desmosomal protein complex (Noorman et al. 2013; Makara et al. 2014). They contribute to the intercellular communication by ion exchange and signal transduction. These complexes are directly modulated by Ca^{2+} and in turn affect the global Ca^{2+} signaling through their influence on the cellular excitability (Sato et al. 2009, 2011).

15.2.2 Dyadic Cleft

In order to provide localized and specific signal transduction in myocytes, the architecture of the myocyte plays a fundamental role. A close proximity to the SR enables the formation of so-called dyadic junctions where LTCCs in the plasma membrane directly influence the RyRs in the SR membrane. Cleft sizes are variable and can range from 9.3×10^3 to 3.98×10^6 nm^3 (Hayashi et al. 2009). The small and confined space in the dyadic junctions facilitates a rapid and strong increase in local Ca^{2+} concentration following the opening of LTCCs which in turn activate the RyRs in a cluster in the SR opposing the LTCCs. The narrow space prevents diffusion into the cytosol and thus the dissipation of the signal, culminating in peak Ca^{2+} values of 1.27 μM in the cleft vs. 0.56 μM in the bulk cytosol (Despa et al. 2014). The brief and very steep increase in local Ca^{2+} concentration in the cleft triggers the opening of clusters of RyRs to release a large amount of Ca^{2+} from the internal stores into the cytosol leading to a rapid, global increase of cytosolic Ca^{2+} necessary to initiate the contraction of the myocyte.

During diastole, leak from the SR can thus increase $[Ca^{2+}]_{cleft}$ from 100 nmol/L found in the bulk cytosol to 194 nmol/L depending on the leak rate (Despa et al. 2014).

15.2.3 Mitochondria

About 35% of the cell volume of a ventricular myocyte is taken by mitochondria. They supply up to 90% of the cellular ATP and are critical mediators in cell death through apoptosis and necrosis (Harris and Das 1991; Bers 2008). The energy production requires Ca^{2+} in order to activate key players in the Krebs cycle; however, excessive mitochondrial Ca^{2+} concentration has been associated with cellular dysfunction (Kwong and Molkentin 2015). The mitochondrial role in Ca^{2+} cycling has been controversial. Mitochondrial Ca^{2+} rises rapidly during SR-Ca^{2+} release but declines much slower compared to cytosolic dynamics, causing a beat-to-beat integration of the signal. The transient amplitude of the $[Ca^{2+}]_{mito}$ is small (~10 nM) compared to the cytosolic transient (>1 μM) (Andrienko et al. 2009). Under physiological conditions, mitochondrial Ca^{2+} signaling does not significantly alter cytosolic Ca^{2+} handling; hence, the mitochondria do not act as important dynamic Ca^{2+} buffers during ECC in ventricular cells (Boyman et al. 2014). This stands in contrast to pathophysiological conditions such as heart failure, where a leaky SR can cause mitochondrial Ca^{2+} overload leading to mitochondrial dysfunction and cell death (Santulli et al. 2015).

15.2.4 Nucleus

Nuclear Ca^{2+} signaling plays a key role during ETC in cardiac myocytes by inducing and modulating long-term adaptation in the nucleus, ultimately altering gene transcription (Bootman et al. 2009; Ljubojevic and Bers 2015). Ventricular myocytes are capable of decoding Ca^{2+} signals in order to regulate gene expression independent of the global changes in $[Ca^{2+}]_i$ during ECC. Several Ca^{2+}-dependent signaling pathways link extracellular information to cardiomyocyte reprogramming. Some of the transcription factors are influenced by cytosolic and nucleoplasmic Ca^{2+} (Molkentin et al. 1998; Passier et al. 2000), and others are strictly locally controlled by Ca^{2+} signaling at the nuclear envelope (Chawla et al. 1998; Bootman et al. 2009). Nuclear Ca^{2+} is influenced by and follows cytoplasmic Ca^{2+} as the nuclear envelope contains nuclear pore complexes allowing bidirectional, passive diffusion of ions such as Ca^{2+} (Kockskamper et al. 2008). However, nuclear Ca^{2+} can also be regulated independent of the cytosolic Ca^{2+} by direct stimulation of the perinuclear envelop, which acts as a Ca^{2+} store (Ljubojevic et al. 2011). Accordingly, nuclear Ca^{2+} follows cytosolic Ca^{2+} with some delay in a proportional manner that becomes disproportionate with increasing frequencies. Together with Ca^{2+} signaling independent of cytoplasmic Ca^{2+}, a long-term cellular response can be initiated and has been shown to start early during the remodeling process at the onset of hypertrophy and heart failure (Ljubojevic et al. 2014).

15.2.5 Sarcoplasmic/Endoplasmic Reticulum

The sarcoplasmic reticulum (SR) is the primary Ca^{2+} storing organelle in a cardiac myocyte. It holds enough Ca^{2+} to support a contraction and is consequently the main source of Ca^{2+} released during an ECC cycle (about 74% of the systolic Ca^{2+}) (Bers 2014). Ca^{2+} is released through the Ca^{2+}-sensitive RyR in the SR membrane which is found in highly organized arrays of >100 single channels at the dyadic junctions where they face the LTCCs in the plasma membrane (Moore et al. 1984; Franzini-Armstrong et al. 1999; Bers 2008). They not only serve as Ca^{2+} release channels but also as scaffolding proteins localizing regulatory proteins to the junctions on the luminal side (triadin, junctin, and calsequestrin) and on the cytosolic side (calmodulin, FKBP 12.6, protein kinase A (PKA), and phosphatases) (Zhang et al. 1997). Luminal SR-Ca^{2+} as well as cytosolic Ca^{2+} concentration influence the open probability of the channel (Sitsapesan and Williams 1994; Lukyanenko et al. 1996; Gyorke et al. 2002). If the cytosolic Ca^{2+} rises, the RyRs in a cluster open near simultaneously to release the Ca^{2+} stored in the SR. On the luminal side, Ca^{2+}, if rising above 4 mM, dissociates calsequestrin from the receptor and increases the open probability

(Beard et al. 2005). Calsequestrin can bind a substantial amount of Ca^{2+} and is the reason that only a fraction of the total intra-SR Ca^{2+} is actually free (Fryer and Stephenson 1996). The refilling of the SR after the release of its stored Ca^{2+} is achieved with a Ca^{2+} pump located in the SR membrane. The SERCA actively transports Ca^{2+} from the cytosol into the SR at the end of a contraction to restore diastolic conditions in the cell. Depending on the species, 70–90% of the Ca^{2+} released into the cytosol will be pumped back into the SR via the SERCA (Bers 2008). The remaining Ca^{2+} will be transported out of the cell mainly via the NCX and Ca^{2+} pumps ensuring that the diastolic Ca^{2+} levels are reached (Bers 2008, 2014).

These release and reuptake mechanisms guarantee a complete and effective ECC cycle following an excitatory stimulus. Small variations in either the release or the reuptake will cause a shift in the homeostasis and can be found early on during disease progression.

15.3 Triggered Ca^{2+} Release

The main pathway for Ca^{2+} entry from the extracellular space is via voltage-gated Ca^{2+} channels in the plasma membrane. The LTCC is activated by membrane depolarization and can be inactivated in a Ca^{2+}-dependent manner at the cytosolic side, limiting the amount of Ca^{2+} entering during an action potential (Sham 1997; Bers 2008). The LTCCs are located predominantly at the junctional cleft in close proximity of the RyRs in the SR (Balijepalli et al. 2006; Makarewich et al. 2012). The Ca^{2+} release through these channels furthers the inactivation of the LTCCs by rapidly raising the cleft Ca^{2+}. Indeed the Ca^{2+} released from the SR reduces the I_{Ca} by 50% (Sipido et al. 1995; Puglisi et al. 1999).

Theoretically, the opening of one L-type Ca^{2+} channel can trigger full release of the RyRs in a couplon; nevertheless, about 10–15 LTCCs are located in a couplon, creating a safety margin for proper and complete activation (Bers 2002). This so-called Ca^{2+}-induced Ca^{2+} release is the most widely accepted mechanism triggering release from the SR. Although several other mechanisms have been proposed, such as Ca^{2+} influx through the NCX, voltage-dependent activation of the RyRs or $InsP_3$-mediated Ca^{2+} release, these mechanisms most likely only serve as modulators or backup systems as they all share one major drawback of not being spatially focused on the RyR clusters (Bers 2002).

15.4 Spontaneous Ca^{2+} Release

Spontaneous Ca^{2+} release from the SR is defined as release events that do not require a trigger such as I_{Ca} to induce opening of the RyRs in a cluster. Spontaneous Ca^{2+} release through a group of RyRs (6–20 receptors in a cluster) is called a spark and can occur spontaneously at rest without trigger (Cheng et al. 1993; Bridge et al. 1999; Wier and Balke 1999). Sparks are stochastic events that happen rarely under physiological conditions but increase in number and size under diseased conditions (Shannon et al. 2003; Ai et al. 2005; Curran et al. 2010).

The opening of the RyRs is dependent on the $[Ca^{2+}]_i$ on luminal as well as the cytosolic side in such a manner that increased cytosolic Ca^{2+} will increase the open probability (Bers 2008) and may foster the posttranslational modification of the RyRs via phosphorylation through PKA or CaMKII (Ca^{2+}/calmodulin-dependent protein kinase II) (Valdivia et al. 1995; Wehrens et al. 2006). Both will ultimately lead to a shift of the single receptors toward a higher open probability and thus enable the moderately increased $[Ca^{2+}]_i$ to open the channels and release Ca^{2+} from the stores. The $[Ca^{2+}]_{SR}$ can influence the open probability through the regulators that are tightly connected to the RyRs on the luminal side such as calsequestrin, triadin, and junctin and will respond to increased SR-Ca^{2+} levels or SR-Ca^{2+} overload with the release of Ca^{2+} into the cytosol (Beard et al. 2005).

Under physiological conditions, this leak is not large enough to significantly increase cytosolic Ca^{2+} levels or cause further alterations in the cells' physiology (Cheng et al. 1993; Cheng and Lederer 2008). A single spark represents the opening of a part (~ 6–20) of the RyRs in a couplon and usually doesn't spread across several neighboring clusters due to the distance between the single clusters and the buffering capabilities of the cytosol. However, if the spark is getting larger or the global cytosolic Ca^{2+} level is increased, it is possible that the released Ca^{2+} can trigger release from the neighboring cluster and accumulate to a Ca^{2+} wave travelling through the cell and causing arrhythmogenic spontaneous contractions (see also Fig. 15.1).

15.5 Alternans

Cardiac alternans is defined as a beat-to-beat alternation in Ca^{2+}-transient (CaT) amplitude or action potential duration (APD) at the cellular level or T-wave alternans (TWA) at the organ level. Clinically, T-wave alternans is seen as important marker for the risk assessment for ventricular arrhythmia (Leach et al. 2005; Brette et al. 2006). At the cellular level, several aspects of ECC and SR-Ca^{2+} release are involved in alternans. Ca^{2+} alternans is often accompanied by changes in SR load, whereas the initiation of alternans can most likely be traced back to RyR refractoriness or availability (Wier 2007; Despa et al. 2003). With increasing pacing rate i, RyR refractoriness seems to be the first factor to initiate alternans, and both SR-Ca^{2+} load and APD alternans at higher rates can be the result (Louch et al. 2004).

15.5.1 APD Vs. Ca^{2+} Alternans

It has been proposed that the generation of APD alternans is dependent on the steepness of the restitution curve, i.e., the steeper the restitution curve is, the more likely it is for a short diastolic interval (DI) to evoke a shortened APD, which in turn is followed by a relatively longer DI, leading to a longer APD and so on (Nolasco and Dahlen 1968). APD alternans is usually accompanied by Ca^{2+} alternans, i.e., alternating Ca^{2+}-transient amplitudes. In cardiac myocytes, the shorter APD is generally accompanied by a shorter Ca^{2+} transient (Choi and Salama 2000; Pruvot et al. 2004), although the opposite has been described as well (Huser et al. 2000b; Blatter et al. 2003).

Ca^{2+}-transient alternans in turn does not necessarily depend on APD alternans and indeed Ca^{2+} alternans can be evoked in voltage-clamped myocytes (Chudin et al. 1999; Wan et al. 2005). The unstable Ca^{2+} behavior in myocytes can be understood by studying the cellular Ca^{2+} fluxes during a cardiac cycle. The Ca^{2+} entering the cytosol must be completely removed before the next beat is initiated. If the net fluxes are $\neq 0$, an instability evolves that can be measured as CaT alternans. In other words, the Ca^{2+} entering the cytosol through LTCCs and RyRs need to be removed by the two main removal pathways via SERCA into the SR and NCX out of the cell. If the equilibrium between these two pathways is shifted toward one or the other, the Ca^{2+} redistribution within the cell is disturbed. It has been suggested that at increased SR-Ca^{2+} values even small variations in $[Ca^{2+}]_{SR}$ can cause large variations in the amount of released Ca^{2+} and thus the CaT amplitude (Shannon et al. 2000; Xie et al. 2008).

15.5.2 SR-Ca^{2+} and Alternans

As implicated above, the $[Ca^{2+}]_{SR}$ plays a crucial role during Ca^{2+} alternans. Increased Ca^{2+} load in the SR affects the open probability of the SR in such a way that already small changes, even very localized, can cause the opening of RyRs. Such very local events can be detected as Ca^{2+} sparks, and this "leak" from the SR, if happening at a dyadic junction, can increase the cleft Ca^{2+} to levels that are significantly higher than the cytosolic Ca^{2+}, creating a diastolic gradient (Despa et al. 2014). High cleft Ca^{2+} can trigger CaMKII and calcineurin activation (Saucerman and Bers 2008; Dries et al. 2016), altering their targets such as ion channels and transporters and may eventually cause transcriptional adaptation processes (Molkentin 2000). Spontaneous Ca^{2+} leak from the SR will also likely affect the open probability and availability of the RyRs. And elevated $[Ca^{2+}]_{SR}$ will further increase the open probability of the channels, eventually releasing more Ca^{2+} into the cytosol.

For alternans to develop, an appropriate substrate and a trigger to induce alternans need to be available. Increased SR-Ca^{2+} and alterations in the SR-Ca^{2+} uptake and release machinery might set the basis for Ca^{2+} alternans. And in fact, it has been shown that ventricular myocytes that were more susceptible to alternans had reduced expression levels of SERCA2a and RyR (Huser et al. 2000b; Diaz et al. 2004; Wan et al. 2005). A trigger event, such as shortened DI or Ca^{2+} waves, might cause sustained Ca^{2+} alternans at constant pacing rates by initiating an AP before diastolic Ca^{2+} levels are reached. Under these circumstances, less Ca^{2+} is released into the cytosol. On the subsequent beat, less Ca^{2+} needs to be removed from the cytosol and transported into the SR within the same time frame, making more SR-Ca^{2+} available for the next beat. However, SR-Ca^{2+} content alone is not sufficient to cause Ca^{2+} alternans, and indeed, it has been shown that Ca^{2+} alternans does not require SR-Ca^{2+} alternans (Picht et al. 2006). The refilling rate, determined by the SERCA activity, seems to be a likely key factor for sustained Ca^{2+} alternans. Data from intact hearts suggest a prolonged recovery rate with increased Ca^{2+} alternans ratio (indicating the intensity of alternans; Fig. 15.2) (Hammer et al. 2015), and the inotropic effects of β-adrenergic stimulation, which has been proposed as a potential trigger for alternans (Merchant and Armoundas 2012), can act protective against Ca^{2+} alternans

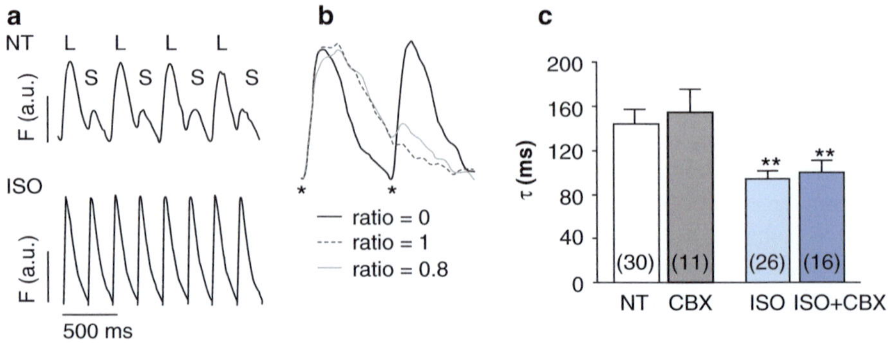

Fig. 15.2 In isolated cardiac myocytes, Ca^{2+} alternans ratio observed at a fixed basic cycle length of 200 ms can be completely abolished by β-adrenergic stimulation (**a**). The transients from single myocytes within intact tissue displayed prolonged recovery times with increasing alternans ratio (**b**), indicating that the protective effect of β-adrenergic stimulation might be due to enhanced SERCA function as observed in normally coupled tissue and tissue with reduced intercellular coupling. Taken with permission from Hammer et al. (2015)

(Huser et al. 2000b; Florea and Blatter 2012; Hammer et al. 2015). β-adrenergic stimulation will also affect the open probability of the RyRs and thus release from the SR. In intact hearts, the refractoriness of the RyRs is a determinant during the onset of frequency-induced alternans, and modulating the refractory period of the receptors will shift the threshold for alternans onset (Wang et al. 2014).

15.5.3 Subcellular-Cellular-Multicellular Organ

Ca^{2+} alternans can be initiated and maintained by the dependence of the APD on the preceding DI or a steep relationship of SR-Ca^{2+} release versus SR-Ca^{2+} load. CaT alternans can occur as concordant alternans describing a pattern where all cellular regions alternate with the same temporal sequence or as discordant alternans where subcellular regions alternate with different temporal sequences (Pastore et al. 1999; Qian et al. 2001; Sato et al. 2006). These patterns can be complex and develop dynamically within single cells of the intact tissue (Aistrup et al. 2009; Sato et al. 2013; Hammer et al. 2015). Spatial heterogeneities in the Ca^{2+} cycling machinery are likely to be the underlying mechanism of phase-mismatched CaT alternans (Cordeiro et al. 2007; Sato et al. 2013; Hammer et al. 2014). At a cellular level, the bidirectional coupling between Ca^{2+} and voltage seems likely to influence the patterns found, as Ca^{2+}-sensitive currents (i.e., NCX and LTCC) might be influenced in a feedback mechanism (Kanaporis and Blatter 2015). Although CaT alternans might occasionally be out of phase with the corresponding APD alternans (Hayashi et al. 2007; Armoundas 2009), it is more likely that the two are in phase and recent modeling studies (Sato et al. 2013) have found that APD and CaT alternans tend to move toward concordant alternans over time (concordant electromechanical coupling).

At a single-cell level, Ca^{2+} alternans is not dependent on APD alternans, but it can affect APD by influencing the two most prominent Ca^{2+}-sensitive currents in

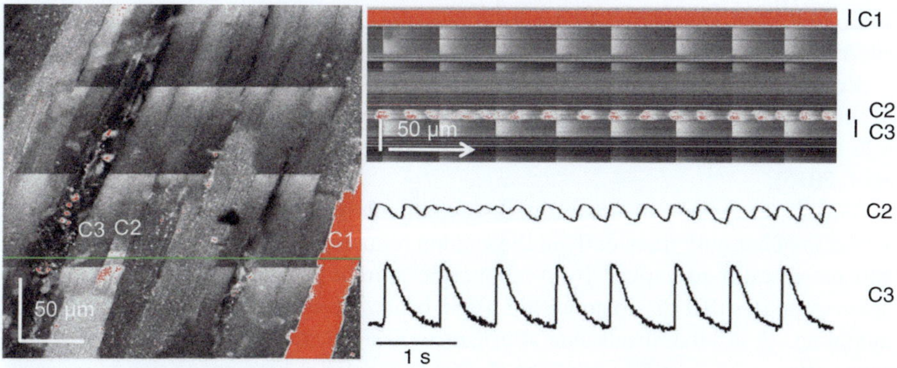

Fig. 15.3 Overloaded and spontaneously active cells within the intact tissue. The snapshot in (**a**) shows the 2D view of the surface of a heart loaded (bar graphs = 50 µm). The green line indicates the position of the line scan that is depicted in (**b**). Three cells are highlighted: C1 is an overloaded cell that no longer reacts to any stimulus; C2 is a spontaneously active cell with a high frequency of Ca^{2+} waves neighboring C3, a cell regularly exhibiting Ca transients in synchrony with the other cells in the field of view. The traces of the Ca^{2+} signals of the adjacent cells C2 and C3 confirm that the two cells seem to be disconnected from each other. Image taken with permission from Hammer et al. (2015)

myocytes (I_{Ca} and I_{NCX}) (Sato et al. 2006; Kanaporis and Blatter 2015). It is perceivable that the membrane voltage patterns of one cell might influence the voltage behavior of its neighboring cells. At the organ level, spatially discordant APD alternans is a proarrhythmic factor that can lead to wave break and reentry (Pastore et al. 1999; Myles et al. 2008). Electrotonic coupling between myocytes attenuates differences in individual cellular properties, and indeed it has been shown that Ca^{2+} waves do not readily pass the gap junction connecting the myocytes (Kaneko et al. 2000; Baader et al. 2002; Wasserstrom et al. 2010; Hammer et al. 2015). The Ca^{2+} signals of a single cell will remain within the borders of this cell, but the voltage signal might travel across several cells (see also Fig. 15.3). However, alterations in a confined region of the heart will not cause disturbances at the organ level. In other words, aberrant electrical properties of a single cell will not cause changes of the surrounding tissue as the signal would dissipate within its surrounding neighbors ("sink").

15.5.4 Gap Junctions and Alternans

Electrotonic coupling of myocytes allows the communication and interaction of neighboring myocytes, and it is crucial in maintaining cardiac function (Rohr 2004). Alterations in coupling efficacy will affect not only the conduction velocity of electrical signals necessary for the simultaneous excitation of the heart but also the ability of the myocardium to act as a current sink in case of a single aberrant cell. The latter is crucial for the synchronization of neighboring cells, and the loss of intercellular coupling efficacy will most likely lead to enhanced spatial gradients, giving rise to arrhythmogenic behavior. In fact it has been shown that a disruption of cellular communication by scarred tissue promotes the production of spatially discordant alternans and wave break (Ohara et al. 2001; Krogh-Madsen and Christini 2007).

Reducing the intercellular coupling by partially inhibiting the gap junctions, the occurrence and intensity of Ca^{2+} alternans are significantly increased, indicating the role of intercellular communication for the formation of Ca^{2+} alternans (Hammer et al. 2015).

The development of spatially discordant alternans (between adjacent cells) seems to occur only transiently at the sudden reduction of basic cycle length (BCL), and increased load evoked by β-adrenergic stimulation seems to enhance spatial heterogeneities that dissipate within a few beats. However it has been proposed by modeling studies that discordant alternans in tissue due to shortened BCL will only occur after more than 50 s (Sato et al. 2013). Regional heterogeneities of intercellular coupling might be crucial for the development of spatially discordant alternans. A source-sink mismatch caused by partial gap junctional coupling will increase the risk of premature ventricular contractions in intact tissue by promoting spatial heterogeneities (Myles et al. 2012).

15.5.5 NCX and Alternans

As mentioned above, aberrant Ca^{2+} cycling in cardiac myocytes, such as spontaneous Ca^{2+} release or cyclic fluctuations of CaT amplitude as seen during Ca^{2+} alternans, will affect and modulate Ca^{2+}-sensitive currents such as I_{NCX} in the plasma membrane. NCX activity has been shown to be involved in promoting triggered activity by spontaneous Ca^{2+} release in the form of Ca^{2+} waves (Fujiwara et al. 2008).

The released Ca^{2+} from the SR will influence NCX activity and drive its depolarizing effects and thereby modulating the APD. If the effect of Ca^{2+} alternans on the NCX outweighs the ones on LTCC, the APD will be in phase with the Ca^{2+} alternans. However, if the I_{Ca} is the predominant electrogenic current, APD and Ca^{2+} alternans will be discordant, where a large CaT will lead to a shortened APD (Wan et al. 2012). The latter scenario is less often observed, and the likelihood of I_{NCX} being the predominant electrogenic current influenced by the Ca^{2+} released from the SR is backed by the finding that myocytes that are more prone to Ca^{2+} alternans show an increased expression of NCX (Wan et al. 2005).

15.5.6 Mitochondria and Alternans

Mitochondrial Ca^{2+} signaling contributes about 1% to the global Ca^{2+} cycling during a cardiac cycle. It is thus perceivable that mitochondrial Ca^{2+} signaling does not significantly influence cytosolic Ca^{2+} (Boyman et al. 2014). In fact it has been shown that mitochondrial Ca is integrated over several Ca^{2+} transients (Andrienko et al. 2009). However, a leaky SR can induce mitochondrial Ca^{2+} overload and subsequent malfunction, which is a determinant during heart failure (Santulli et al. 2015). One potential pathway of mitochondrial involvement in aberrant Ca^{2+} cycling behavior is an accumulation of Ca^{2+} in the mitochondrial matrix by increased cytosolic Ca^{2+}

entering through the mitochondrial permeability transition pore (MTPT) or the mitochondrial Ca^{2+} uniporter (MCU), which will cause an increased metabolic output in response to increased demand (Lu et al. 2013; Kwong and Molkentin 2015). If the increased Ca^{2+} levels in the mitochondrial matrix persist, mitochondrial dysfunction and cell death will follow. Uncoupling of mitochondria has a facilitating effect on cardiac alternans and has been proposed to be carried by Ca^{2+} overload caused by a drop in ATP levels (Florea and Blatter 2010; Smith et al. 2013). Interestingly, there seems to be a tight connection between SR-Ca^{2+} signaling and mitochondrial function, as mitochondrial Ca^{2+} signaling is dependent on the relative location in the cell (Lu et al. 2013). While mitochondrial dysfunction has been proposed to promote alternans, recent evidence shows that upregulated SERCA function can mitigate the detrimental effects of metabolic dysfunction (Stary et al. 2016).

Conclusion

In cardiac myocytes, Ca^{2+} handling is tightly controlled, spatially and temporally. Microdomains provide the framework for spatially focused signaling that enables the cell to specifically control its functions and dissect, for example, ECC from ETC. Microdomains also allow the cell to react to quickly changing requirements such as increased metabolic demand.

Local Ca^{2+} signaling in the cleft is tightly controlled by LTCCs and RyRs and in turn is responsible for the modulation of both of them. Even small changes in either the RyRs or LTCCs influence the $[Ca^{2+}]_{cleft}$ and have significant effects on global Ca^{2+} handling. Small perturbations in this fine-tuned machinery can lead to the onset of alternans at the cellular level and might extend to the whole tissue. The gap junctional coupling between myocytes plays a crucial role for impulse propagation, and alterations in coupling efficacy will contribute to arrhythmogenesis.

Understanding the cellular mechanisms leading to cardiac alternans and the importance of localized signaling in cardiac myocytes during the early stages of alternans might help to elucidate potential targets for treating arrhythmogenic conditions.

Compliance with Ethical Standards

Conflict of Interest Statement K. P. Hammer has no conflict of interest. L. S. Maier receives funding from Gilead and Sanofi.

References

Ai X, Curran JW, Shannon TR et al (2005) Ca^{2+}/calmodulin-dependent protein kinase modulates cardiac ryanodine receptor phosphorylation and sarcoplasmic reticulum Ca^{2+} leak in heart failure. Circ Res 97:1314–1322. doi:10.1161/01.RES.0000194329.41863.89

Aistrup GL, Shiferaw Y, Kapur S et al (2009) Mechanisms underlying the formation and dynamics of subcellular calcium alternans in the intact rat heart. Circ Res 104:639–649. doi:10.1161/CIRCRESAHA.108.181909

Andrienko TN, Picht E, Bers DM (2009) Mitochondrial free calcium regulation during sarcoplasmic reticulum calcium release in rat cardiac myocytes. J Mol Cell Cardiol 46(6):1027–1036. doi:10.1016/j.yjmcc.2009.03.015

Armoundas AA (2009) Discordant calcium transient and action potential alternans in a canine left-ventricular myocyte. IEEE Trans Biomed Eng 56(9):2340–2344. doi:10.1109/TBME.2009.2023671
Baader AP, Buchler L, Bircher-Lehmann L et al (2002) Real time, confocal imaging of Ca^{2+} waves in arterially perfused rat hearts. Cardiovasc Res 53(1):105–115. doi:10.1016/S0008-6363(01)00423-0
Balijepalli RC, Foell JD, Hall DD et al (2006) Localization of cardiac L-type Ca^{2+} channels to a caveolar macromolecular signaling complex is required for β_2-adrenergic regulation. Proc Natl Acad Sci U S A 103(19):7500–7505. doi:10.1073/pnas.0503465103
Beard NA, Casarotto MG, Wei L et al (2005) Regulation of ryanodine receptors by calsequestrin: effect of high luminal Ca^{2+} and phosphorylation. Biophys J 88(5):3444–3454. doi:10.1529/biophysj.104.051441
Berridge MJ, Lipp P, Bootman MD (2000) The versatility and universality of calcium signalling. Nat Rev Mol Cell Biol 1:11–21
Bers DM (2002) Cardiac excitation-contraction coupling. Nature 415:198–205. doi:10.1038/415198a
Bers DM (2008) Excitation-contraction coupling and cardiac contractile force. Springer, Dordrecht
Bers DM (2014) Cardiac sarcoplasmic reticulum calcium leak: basis and roles in cardiac dysfunction. Annu Rev Physiol 76:107–127. doi:10.1146/annurev-physiol-020911-153308
Best JM, Kamp TJ (2012) Different subcellular populations of L-type Ca^{2+} channels exhibit unique regulation and functional roles in cardiomyocytes. J Mol Cell Cardiol 52(2):376–387. doi:10.1016/j.yjmcc.2011.08.014
Blatter LA, Kockskamper J, Sheehan KA et al (2003) Local calcium gradients during excitation-contraction coupling and alternans in atrial myocytes. J Physiol 546(Pt 1):19–31. doi:10.1113/jphysiol.2002.025239
Bootman MD, Fearnley C, Smyrnias I et al (2009) An update on nuclear calcium signalling. J Cell Sci 122:2337–2350. doi:10.1242/jcs.028100
Boyman L, Chikando AC, Williams GS et al (2014) Calcium movement in cardiac mitochondria. Biophys J 107(6):1289–1301. doi:10.1016/j.bpj.2014.07.045
Brette F, Orchard C (2003) T-tubule function in mammalian cardiac myocytes. Circ Res 92:1182–1192. doi:10.1161/01.RES.0000074908.17214.FD
Brette F, Orchard CH (2006) No apparent requirement for neuronal sodium channels in excitation-contraction coupling in rat ventricular myocytes. Circ Res 98:667–674. doi:10.1161/01.RES.0000209963.02720.70
Brette F, Salle L, Orchard CH (2004) Differential modulation of L-type Ca^{2+} current by SR Ca^{2+} release at the T-tubules and surface membrane of rat ventricular myocytes. Circ Res 95:e1–e7. doi:10.1161/01.RES.0000135547.53927.F6
Brette F, Salle L, Orchard CH (2006) Quantification of calcium entry at the T-tubules and surface membrane in rat ventricular myocytes. Biophys J 90:381–389
Bridge JH, Ershler PR, Cannell MB (1999) Properties of Ca^{2+} sparks evoked by action potentials in mouse ventricular myocytes. J Physiol 518(Pt 2):469–478. doi:10.1111/j.1469-7793.1999.0469p.x
Chawla S, Hardingham GE, Quinn DR et al (1998) CBP: a signal-regulated transcriptional coactivator controlled by nuclear calcium and CaM kinase IV. Science 281:1505–1509
Cheng H, Lederer WJ (2008) Calcium sparks. Physiol Rev 88(4):1491–1545. doi:10.1152/physrev.00030.2007
Cheng H, Lederer WJ, Cannell MB (1993) Calcium sparks: elementary events underlying excitation-contraction coupling in heart muscle. Science 262:740–744
Choi BR, Salama G (2000) Simultaneous maps of optical action potentials and calcium transients in guinea-pig hearts: mechanisms underlying concordant alternans. J Physiol 529(Pt 1):171–188. doi:10.1111/j.1469-7793.2000.00171.x
Chudin E, Goldhaber J, Garfinkel A et al (1999) Intracellular Ca^{2+} dynamics and the stability of ventricular tachycardia. Biophys J 77(6):2930–2941. doi:10.1016/S0006-3495(99)77126-2
Cordeiro JM, Malone JE, Di Diego JM et al (2007) Cellular and subcellular alternans in the canine left ventricle. Am J Physiol Heart Circ Physiol. doi:10.1152/ajpheart.00757.2007
Curran J, Brown KH, Santiago DJ et al (2010) Spontaneous Ca waves in ventricular myocytes from failing hearts depend on Ca^{2+}-calmodulin-dependent protein kinase II. J Mol Cell Cardiol 49(1):25–32. doi:10.1016/j.yjmcc.2010.03.013

Despa S, Brette F, Orchard CH et al (2003) Na/Ca exchange and Na/K-ATPase function are equally concentrated in transverse tubules of rat ventricular myocytes. Biophys J 85(5):3388–3396. doi:10.1016/S0006-3495(03)74758-4
Despa S, Shui B, Bossuyt J et al (2014) Junctional cleft $[Ca^{2+}]_i$ measurements using novel cleft-targeted Ca^{2+} sensors. Circ Res 115:339–347. doi:10.1161/CIRCRESAHA.115.303582
Di Maio A, Karko K, Snopko RM et al (2007) T-tubule formation in cardiacmyocytes: two possible mechanisms? J Muscle Res Cell Motil 28:231–241
Diaz ME, O'Neill SC, Eisner DA (2004) Sarcoplasmic reticulum calcium content fluctuation is the key to cardiac alternans. Circ Res 94:650–656. doi:10.1161/01.RES.0000119923.64774.72
Dries E, Santiago DJ, Johnson DM et al (2016) Calcium/calmodulin-dependent kinase II and nitric oxide synthase 1-dependent modulation of ryanodine receptors during beta-adrenergic stimulation is restricted to the dyadic cleft. J Physiol 594(20):5923–5939. doi:10.1113/JP271965
Florea SM, Blatter LA (2010) The role of mitochondria for the regulation of cardiac alternans. Front Physiol 1:141. doi:10.3389/fphys.2010.00141
Florea SM, Blatter LA (2012) Regulation of cardiac alternans by beta-adrenergic signaling pathways. Am J Physiol Heart Circ Physiol 303(8):H1047–H1056. doi:10.1152/ajpheart.00384.2012
Fowler MR, Dobson RS, Orchard CH et al (2004) Functional consequences of detubulation of isolated rat ventricular myocytes. Cardiovasc Res 62:529–537
Frank JS, Mottino G, Reid D et al (1992) Distribution of the Na^+-Ca^{2+} exchange protein in mammalian cardiac myocytes: an immunofluorescence and immunocolloidal gold-labeling study. J Cell Biol 117:337–345
Franzini-Armstrong C, Protasi F, Ramesh V (1999) Shape, size, and distribution of Ca^{2+} release units and couplons in skeletal and cardiac muscles. Biophys J 77(3):1528–1539. doi:10.1016/S0006-3495(99)77000-1
Fryer MW, Stephenson DG (1996) Total and sarcoplasmic reticulum calcium contents of skinned fibres from rat skeletal muscle. J Physiol 493(Pt 2):357–370
Fujiwara K, Tanaka H, Mani H et al (2008) Burst emergence of intracellular Ca^{2+} waves evokes arrhythmogenic oscillatory depolarization via the Na^+-Ca^{2+} exchanger: simultaneous confocal recording of membrane potential and intracellular Ca^{2+} in the heart. Circ Res 103:509–518. doi:10.1161/CIRCRESAHA.108.176677
Gyorke S, Gyorke I, Lukyanenko V et al (2002) Regulation of sarcoplasmic reticulum calcium release by luminal calcium in cardiac muscle. Front Biosci 7:d1454–d1463
Hammer KP, Hohendanner F, Blatter LA et al (2014) Variations in local calcium signaling in adjacent cardiac myocytes of the intact mouse heart detected with two-dimensional confocal microscopy. Front Physiol 5:517. doi:10.3389/fphys.2014.00517
Hammer KP, Ljubojevic S, Ripplinger CM et al (2015) Cardiac myocyte alternans in intact heart: influence of cell-cell coupling and beta-adrenergic stimulation. J Mol Cell Cardiol 84:1–9. doi:10.1016/j.yjmcc.2015.03.012
Harris DA, Das AM (1991) Control of mitochondrial ATP synthesis in the heart. Biochem J 280 (Pt 3):561–573
Hayashi H, Shiferaw Y, Sato D et al (2007) Dynamic origin of spatially discordant alternans in cardiac tissue. Biophys J 92(2):448–460. doi:10.1529/biophysj.106.091009
Hayashi T, Martone ME, Yu Z et al (2009) Three-dimensional electron microscopy reveals new details of membrane systems for Ca^{2+} signaling in the heart. J Cell Sci 122:1005–1013. doi:10.1242/jcs.028175
Heinzel FR, Bito V, Volders PG et al (2002) Spatial and temporal inhomogeneities during Ca^{2+} release from the sarcoplasmic reticulum in pig ventricular myocytes. Circ Res 91:1023–1030
Huser J, Blatter LA, Sheu SS (2000a) Mitochondrial calcium in heart cells: beat-to-beat oscillations or slow integration of cytosolic transients? J Bioenerg Biomembr 32:27–33
Huser J, Wang YG, Sheehan KA et al (2000b) Functional coupling between glycolysis and excitation-contraction coupling underlies alternans in cat heart cells. J Physiol 524(Pt 3): 795–806
Insel PA, Head BP, Patel HH et al (2005) Compartmentation of G-protein-coupled receptors and their signalling components in lipid rafts and caveolae. Biochem Soc Trans 33(5):1131–1134. doi:10.1042/BST20051131

Jayasinghe ID, Cannell MB, Soeller C (2009) Organization of ryanodine receptors, transverse tubules, and sodium-calcium exchanger in rat myocytes. Biophys J 97(10):2664–2673. doi:10.1016/j.bpj.2009.08.036

Kanaporis G, Blatter LA (2015) The mechanisms of calcium cycling and action potential dynamics in cardiac alternans. Circ Res 116:846–856. doi:10.1161/CIRCRESAHA.116.305404

Kaneko T, Tanaka H, Oyamada M et al (2000) Three distinct types of Ca^{2+} waves in Langendorff-perfused rat heart revealed by real-time confocal microscopy. Circ Res 86:1093–1099

Kockskamper J, Seidlmayer L, Walther S et al (2008) Endothelin-1 enhances nuclear Ca^{2+} transients in atrial myocytes through ins(1,4,5)P_3-dependent Ca^{2+} release from perinuclear Ca^{2+} stores. J Cell Sci 121:186–195. doi:10.1242/jcs.021386

Komukai K, Brette F, Yamanushi TT et al (2002) K^+ current distribution in rat sub-epicardial ventricular myocytes. Pflugers Arch 444(4):532–538. doi:10.1007/s00424-002-0851-8

Kowalczyk AP, Green KJ (2013) Structure, function, and regulation of desmosomes. Prog Mol Biol Transl Sci 116:95–118. doi:10.1016/B978-0-12-394311-8.00005-4

Krogh-Madsen T, Christini DJ (2007) Action potential duration dispersion and alternans in simulated heterogeneous cardiac tissue with a structural barrier. Biophys J 92(4):1138–1149. doi:10.1529/biophysj.106.090845

Kwong JQ, Molkentin JD (2015) Physiological and pathological roles of the mitochondrial permeability transition pore in the heart. Cell Metab 21(2):206–214. doi:10.1016/j.cmet.2014.12.001

Leach RN, Desai JC, Orchard CH (2005) Effect of cytoskeleton disruptors on L-type Ca channel distribution in rat ventricular myocytes. Cell Calcium 38:515–526

Lipp P, Huser J, Pott L et al (1996) Spatially non-uniform Ca^{2+} signals induced by the reduction of transverse tubules in citrate-loaded Guinea-pig ventricular myocytes in culture. J Physiol 497(Pt 3):589–597

Ljubojevic S, Bers DM (2015) Nuclear calcium in cardiac myocytes. J Cardiovasc Pharmacol 65(3):211–217. doi:10.1097/FJC.0000000000000174

Ljubojevic S, Walther S, Asgarzoei M et al (2011) In situ calibration of nucleoplasmic versus cytoplasmic Ca^{2+} concentration in adult cardiomyocytes. Biophys J 100(10):2356–2366. doi:10.1016/j.bpj.2011.03.060

Ljubojevic S, Radulovic S, Leitinger G et al (2014) Early remodeling of perinuclear Ca^{2+} stores and nucleoplasmic Ca^{2+} signaling during the development of hypertrophy and heart failure. Circulation 130:244–255. doi:10.1161/CIRCULATIONAHA.114.008927

Louch WE, Bito V, Heinzel FR et al (2004) Reduced synchrony of Ca^{2+} release with loss of T-tubules-a comparison to Ca^{2+} release in human failing cardiomyocytes. Cardiovasc Res 62(1):63–73. doi:10.1016/j.cardiores.2003.12.031

Lu X, Ginsburg KS, Kettlewell S et al (2013) Measuring local gradients of intramitochondrial [Ca^{2+}] in cardiac myocytes during sarcoplasmic reticulum Ca^{2+} release. Circ Res 112:424–431. doi:10.1161/CIRCRESAHA.111.300501

Lukyanenko V, Gyorke I, Gyorke S (1996) Regulation of calcium release by calcium inside the sarcoplasmic reticulum in ventricular myocytes. Pflugers Arch 432:1047–1054

Makara MA, Curran J, Little SC et al (2014) Ankyrin-G coordinates intercalated disc signaling platform to regulate cardiac excitability in vivo. Circ Res 115:929–938. doi:10.1161/CIRCRESAHA.115.305154

Makarewich CA, Correll RN, Gao H et al (2012) A caveolae-targeted L-type Ca^{2+} channel antagonist inhibits hypertrophic signaling without reducing cardiac contractility. Circ Res 110:669–674. doi:10.1161/CIRCRESAHA.111.264028

Merchant FM, Armoundas AA (2012) Role of substrate and triggers in the genesis of cardiac alternans, from the myocyte to the whole heart: implications for therapy. Circulation 125:539–549. doi:10.1161/CIRCULATIONAHA.111.033563

Molkentin JD (2000) Calcineurin and beyond: cardiac hypertrophic signaling. Circ Res 87:731–738

Molkentin JD, Lu JR, Antos CL et al (1998) A calcineurin-dependent transcriptional pathway for cardiac hypertrophy. Cell 93(2):215–228. doi:10.1016/S0092-8674(00)81573-1

Moore LE, Schmid A, Isenberg G (1984) Linear electrical properties of isolated cardiac cells. J Membr Biol 81:29–40

Myles RC, Burton FL, Cobbe SM et al (2008) The link between repolarisation alternans and ventricular arrhythmia: does the cellular phenomenon extend to the clinical problem? J Mol Cell Cardiol 45(1):1–10. doi:10.1016/j.yjmcc.2008.03.024

Myles RC, Wang L, Kang C et al (2012) Local beta-adrenergic stimulation overcomes source-sink mismatch to generate focal arrhythmia. Circ Res 110:1454–1464. doi:10.1161/CIRCRESAHA.111.262345

Nolasco JB, Dahlen RW (1968) A graphic method for the study of alternation in cardiac action potentials. J Appl Physiol 25:191–196

Noorman M, Hakim S, Kessler E et al (2013) Remodeling of the cardiac sodium channel, connexin43, and plakoglobin at the intercalated disk in patients with arrhythmogenic cardiomyopathy. Heart Rhythm 10(3):412–419. doi:10.1016/j.hrthm.2012.11.018

Ohara T, Ohara K, Cao JM et al (2001) Increased wave break during ventricular fibrillation in the epicardial border zone of hearts with healed myocardial infarction. Circulation 103:1465–1472

Orchard CH, Pasek M, Brette F (2009) The role of mammalian cardiac t-tubules in excitation-contraction coupling: experimental and computational approaches. Exp Physiol 94(5): 509–519. doi:10.1113/expphysiol.2008.043984

Passier R, Zeng H, Frey N et al (2000) CaM kinase signaling induces cardiac hypertrophy and activates the MEF2 transcription factor in vivo. J Clin Invest 105(10):1395–1406. doi:10.1172/JCI8551

Pastore JM, Girouard SD, Laurita KR et al (1999) Mechanism linking T-wave alternans to the genesis of cardiac fibrillation. Circulation 99:1385–1394

Picht E, DeSantiago J, Blatter LA et al (2006) Cardiac alternans do not rely on diastolic sarcoplasmic reticulum calcium content fluctuations. Circ Res 99:740–748. doi:10.1161/01.RES.0000244002.88813.91

Pruvot EJ, Katra RP, Rosenbaum DS et al (2004) Role of calcium cycling versus restitution in the mechanism of repolarization alternans. Circ Res 94:1083–1090. doi:10.1161/01.RES.0000125629.72053.95

Puglisi JL, Yuan W, Bassani JW et al (1999) Ca^{2+} influx through Ca^{2+} channels in rabbit ventricular myocytes during action potential clamp: influence of temperature. Circ Res 85:e7–e16

Qian YW, Clusin WT, Lin SF et al (2001) Spatial heterogeneity of calcium transient alternans during the early phase of myocardial ischemia in the blood-perfused rabbit heart. Circulation 104:2082–2087

Rohr S (2004) Role of gap junctions in the propagation of the cardiac action potential. Cardiovasc Res 62(2):309–322. doi:10.1016/j.cardiores.2003.11.035

Santulli G, Xie W, Reiken SR et al (2015) Mitochondrial calcium overload is a key determinant in heart failure. Proc Natl Acad Sci U S A 112(36):11389–11394. doi:10.1073/pnas.1513047112

Sato D, Shiferaw Y, Garfinkel A et al (2006) Spatially discordant alternans in cardiac tissue: role of calcium cycling. Circ Res 99:520–527. doi:10.1161/01.RES.0000240542.03986.e7

Sato PY, Musa H, Coombs W et al (2009) Loss of plakophilin-2 expression leads to decreased sodium current and slower conduction velocity in cultured cardiac myocytes. Circ Res 105:523–526. doi:10.1161/CIRCRESAHA.109.201418

Sato PY, Coombs W, Lin X et al (2011) Interactions between ankyrin-G, plakophilin-2, and connexin43 at the cardiac intercalated disc. Circ Res 109:193–201. doi:10.1161/CIRCRESAHA.111.247023

Sato D, Bers DM, Shiferaw Y (2013) Formation of spatially discordant alternans due to fluctuations and diffusion of calcium. PLoS One 8(12):e85365. doi:10.1371/journal.pone.0085365

Saucerman JJ, Bers DM (2008) Calmodulin mediates differential sensitivity of CaMKII and calcineurin to local Ca^{2+} in cardiac myocytes. Biophys J 95(10):4597–4612. doi:10.1529/biophysj.108.128728

Sham JS (1997) Ca2+ release-induced inactivation of Ca^{2+} current in rat ventricular myocytes: evidence for local Ca^{2+} signalling. J Physiol 500(Pt 2):285–295

Shannon TR, Ginsburg KS, Bers DM (2000) Potentiation of fractional sarcoplasmic reticulum calcium release by total and free intra-sarcoplasmic reticulum calcium concentration. Biophys J 78(1):334–343. doi:10.1016/S0006-3495(00)76596-9

Shannon TR, Pogwizd SM, Bers DM (2003) Elevated sarcoplasmic reticulum Ca^{2+} leak in intact ventricular myocytes from rabbits in heart failure. Circ Res 93:592–594. doi:10.1161/01.RES.0000093399.11734.B3

Simons K, Ikonen E (1997) Functional rafts in cell membranes. Nature 387:569–572. doi:10.1038/42408

Sipido KR, Callewaert G, Carmeliet E (1995) Inhibition and rapid recovery of Ca^{2+} current during Ca^{2+} release from sarcoplasmic reticulum in Guinea pig ventricular myocytes. Circ Res 76:102–109

Sitsapesan R, Williams AJ (1994) Regulation of the gating of the sheep cardiac sarcoplasmic reticulum Ca^{2+}-release channel by luminal Ca^{2+}. J Membr Biol 137:215–226

Smith RM, Visweswaran R, Talkachova I et al (2013) Uncoupling the mitochondria facilitates alternans formation in the isolated rabbit heart. Am J Physiol Heart Circ Physiol 305(1):H9–H18. doi:10.1152/ajpheart.00915.2012

Stary V, Puppala D, Scherrer-Crosbie M et al (2016) SERCA2a upregulation ameliorates cellular alternans induced by metabolic inhibition. J Appl Physiol 120(8):865–875. doi:10.1152/japplphysiol.00588.2015

Valdeolmillos M, O'Neill SC, Smith GL et al (1989) Calcium-induced calcium release activates contraction in intact cardiac cells. Pflugers Arch 413:676–678

Valdivia HH, Kaplan JH, Ellis-Davies GC et al (1995) Rapid adaptation of cardiac ryanodine receptors: modulation by Mg^{2+} and phosphorylation. Science 267:1997–2000

Wan X, Laurita KR, Pruvot EJ et al (2005) Molecular correlates of repolarization alternans in cardiac myocytes. J Mol Cell Cardiol 39(3):419–428. doi:10.1016/j.yjmcc.2005.06.004

Wan X, Cutler M, Song Z et al (2012) New experimental evidence for mechanism of arrhythmogenic membrane potential alternans based on balance of electrogenic I_{NCX}/I_{Ca} currents. Heart Rhythm 9(10):1698–1705. doi:10.1016/j.hrthm.2012.06.031

Wang L, Myles RC, De Jesus NM et al (2014) Optical mapping of sarcoplasmic reticulum Ca^{2+} in the intact heart: ryanodine receptor refractoriness during alternans and fibrillation. Circ Res 114:1410–1421. doi:10.1161/CIRCRESAHA.114.302505

Wasserstrom JA, Shiferaw Y, Chen W et al (2010) Variability in timing of spontaneous calcium release in the intact rat heart is determined by the time course of sarcoplasmic reticulum calcium load. Circ Res 107:1117–1126. doi:10.1161/CIRCRESAHA.110.229294

Weber CR, Piacentino V 3rd, Ginsburg KS et al (2002) Na^+-Ca^{2+} exchange current and submembrane [Ca^{2+}] during the cardiac action potential. Circ Res 90:182–189

Wehrens XH, Lehnart SE, Reiken S et al (2006) Ryanodine receptor/calcium release channel PKA phosphorylation: a critical mediator of heart failure progression. Proc Natl Acad Sci U S A 103(3):511–518. doi:10.1073/pnas.0510113103

Wier WG (2007) Gain and cardiac E-C coupling: revisited and revised. Circ Res 101:533–535. doi:10.1161/CIRCRESAHA.107.160929

Wier WG, Balke CW (1999) Ca^{2+} release mechanisms, Ca^{2+} sparks, and local control of excitation-contraction coupling in normal heart muscle. Circ Res 85:770–776

Xie LH, Sato D, Garfinkel A et al (2008) Intracellular Ca alternans: coordinated regulation by sarcoplasmic reticulum release, uptake, and leak. Biophys J 95(6):3100–3110. doi:10.1529/biophysj.108.130955

Yang Z, Pascarel C, Steele DS et al (2002) Na^+-Ca^{2+} exchange activity is localized in the T-tubules of rat ventricular myocytes. Circ Res 91:315–322

Zhang L, Kelley J, Schmeisser G et al (1997) Complex formation between junctin, triadin, calsequestrin, and the ryanodine receptor. Proteins of the cardiac junctional sarcoplasmic reticulum membrane. J Biol Chem 272:23389–23397

The Control of Sub-plasma Membrane Calcium Signalling by the Plasma Membrane Calcium ATPase Pump PMCA4

16

Nicholas Stafford, Ludwig Neyses, and Delvac Oceandy

Contents

N. Stafford • D. Oceandy (✉)
Division of Cardiovascular Sciences, The University of Manchester,
AV Hill Building, Oxford Road, Manchester M13 9PT, UK
e-mail: delvac.oceandy@manchester.ac.uk

L. Neyses
Division of Cardiovascular Sciences, The University of Manchester,
AV Hill Building, Oxford Road, Manchester M13 9PT, UK

University of Luxembourg, 2, Avenue de l'Universite, Esch-sur-Alzette, L-4365, Luxembourg

V. Nikolaev, M. Zaccolo (eds.), *Microdomains in the Cardiovascular System*,
Cardiac and Vascular Biology 3, DOI 10.1007/978-3-319-54579-0_16

Abstract
Within cardiomyocytes cytosolic calcium levels rise and fall by an order of magnitude in each cardiac cycle, yet amidst the noise of this "global" calcium, a separate pool of "local" calcium is able to act as a second messenger in a multitude of signalling networks. The cell is equipped to deal with this through utilising the calcium-binding messenger protein calmodulin which in turn activates calcium/calmodulin-dependent targets and through compartmentalisation. This allows decoding of the calcium signal within such subcellular microdomains as the mitochondrion, the nucleus, the sarcoplasmic reticulum and the plasma membrane. In recent years our group and others have identified isoform 4 of the plasma membrane calcium/calmodulin-dependent ATPase (PMCA4) as a major regulator of local subplasmalemmal calcium in a number of cardiovascular cell types including the cardiomyocyte. Here we review techniques developed for the study of calcium levels local to PMCA4, the protein interaction and signalling complexes formed and regulated by the pump and the physiological implications of these in the heart and vascular systems.

16.1 Introduction

Just a handful of second messengers regulate the multitude of active cellular signalling processes. In the cardiovascular system, these govern such crucial events as cardiac contraction and relaxation, physiological and pathological growth and vascular tone making it essential for pathways sharing common signalling molecules to be compartmentalised within specific subcellular microdomains.

Signalling microdomains are generated at a number of specialised sites in the plasma membrane including caveolae/lipid rafts, t-tubules and sarcomeres, as well as in proximity to specific organelles such as mitochondria and the nuclear envelope. These compartments are rich in scaffolding proteins, allowing the formation of macromolecular protein complexes which then utilise second messengers to transduce local signals. Examples of pathways exhibiting such spatial confinement include the localisation of β-adrenoceptors to caveolae, which allows for subsequent protein kinase A (PKA)-dependent phosphorylation of ion channels and myofilaments in order to regulate inotropic function (Harvey and Calaghan 2012), phosphodiesterase-mediated breakdown of the cyclic nucleotides cAMP and cGMP (Zaccolo 2006) and the confinement of nitric oxide derived from endothelial- or neuronal-NOS in order to exert differential downstream effects in terms of cGMP-mediated signalling or S-nitrosylation (Zhang and Casadei 2012; Hare and Stamler 2005; Tsui et al. 2011; Damy et al. 2004).

Given the large beat-to-beat fluctuations in intracellular Ca^{2+} concentration ($[Ca^{2+}]_i$) during excitation-contraction coupling (EC coupling), the need for local signalling in cardiomyocytes is no more apparent than that required for calcium-dependent pathways. Through its interaction with the calcium sensor calmodulin (CaM) and compartmentalisation, the tight regulation of calcium-dependent enzymes, ion channels and transcription factors governing a plethora of pathways is not only essential for normal cardiac function but also in determining disease progression (Frey et al. 2000). At the t-tubule, for instance, dyadic microdomains facilitate synchronised release of calcium from the sarcoplasmic reticulum (SR) during calcium-induced calcium release in order to optimise inotropy (Trafford et al. 2013), and the loss of these in the atria has been shown to be associated with persistent atrial fibrillation (Lenaerts et al. 2009). At the perinuclear compartment, Ca^{2+} release via inositol 1,4,5-trisphosphate receptors ($InsP_3Rs$) regulates the activity of the Ca^{2+}/CaM-dependent enzymes CaMKII and calcineurin, which in turn regulate nuclear transcription and hypertrophic signalling (Wu et al. 2006; Higazi et al. 2009; Nakayama et al. 2010). In contrast, when in proximity to mitochondria, calcium-induced calcineurin activation has been shown to regulate apoptosis, whilst at the SR activated calcineurin induces nuclear translocation of the transcription factor MEF2 (Heineke and Ritter 2012).

Over the past decade and a half, isoform 4 of the plasma membrane calcium ATPase pump (PMCA4) has emerged as a major player in the regulation of local calcium signalling at the subplasmalemmal domain through a number of protein-protein interactions and in a variety of cardiovascular cell types. This chapter will give an overview of PMCA4-mediated calcium signalling, the physiological implications of these pathways and detail novel techniques for the measurement of local calcium.

16.2 General Overview of the Plasma Membrane Calcium ATPases (PMCAs)

The presence of a calcium-pumping ATPase located in the plasma membrane was first identified in erythrocytes nearly half a century ago (Schatzmann 1966). The PMCA has since been classified as a member of the family of P-type transport ATPases due to a covalent intermediate state formed during its catalytic cycle upon phosphorylation of an aspartic acid residue that is highly conserved amongst the group (Pedersen and Carafoli 1987). This occurs following the binding of a single intracellular calcium ion with high affinity (0.2–0.5 μM under optimal conditions) and leads to conformational changes allowing transport of the ion through the plasma membrane and its subsequent extrusion to the extracellular space by lowering the pump affinity, thus promoting dissociation (Mangialavori et al. 2010; Brini and Carafoli 2009). The PMCA therefore has a low capacity for extrusion, removing only one calcium ion per ATP molecule hydrolysed, whereas related ATPases such as the sarcoendoplasmic reticulum calcium ATPase (SERCA) have a second calcium-binding site and exhibit a 2:1 stoichiometry (Guerini et al. 2000). In non-excitable cells the PMCA represents the main pathway for calcium extrusion; however, in excitable cells such as the cardiomyocyte, the presence of the higher-capacity sodium-calcium exchanger (NCX) means that only a small amount of calcium is removed through the PMCAs, which in terms of diastolic cytosolic Ca^{2+} clearance during EC coupling equates to less than 2% (Bers 2000). This led the research community to explore the possibility that, rather than being functionally redundant in these cells, the PMCAs may be involved in the regulation of local calcium signalling.

16.2.1 PMCA Isoforms and Alternative Splicing

There are four known PMCA isoforms (PMCA1–4), encoded by separate genes *ATP2B1–4* located at human chromosomal loci 12q21-q23, 3p25-p26, Xq28 and 1q25-q32, respectively (Strehler and Zacharias 2001). The gene products amongst these isoforms share approximately 75–85% identity, whilst around 85–90% of the primary sequence is conserved (Strehler 1991). In addition, a total of over 25 splice variants have been identified, which vary in calcium affinity and enzymatic activity. Many of these show cell-specific distributions, and this has led to the consensus that each isoform and variant may perform a unique function (Strehler and Zacharias 2001).

16.2.2 PMCA Expression

The developing mouse embryo displays evidence of PMCA1 expression at least as early as day 9.5 post coitum (pc) and displays abundant and ubiquitous expression thereafter. By day 12.5pc, all four isoforms are expressed, which in the case of PMCA2 is confined to the brain and PMCA3 to the nervous system, lung and skeletal muscle. PMCA4 expression is highest in the brain, bladder, heart and spinal cord although it is significantly less abundant than PMCA1 in all tissues (Zacharias and Kappen 1999). The early and widespread nature of PMCA1 expression has led to the opinion that this isoform is the major housekeeping isoform.

In the adult PMCA1 and PMCA4 show a largely ubiquitous pattern of expression. In human lung, liver, kidney, stomach and skeletal muscle, isoform 1 transcript levels have been found to make up roughly two-thirds of total PMCA abundance, with PMCA4 accounting for the remainder (Stauffer et al. 1993). In the heart, however, isoform 1 and 4 transcript levels are present in similar amounts (47% and 51% of total PMCA expression, respectively). Conversely, PMCA2 displays a highly specific expression pattern, being the most prominent in Purkinje and inner ear cells, as well as being abundant in lactating mammary glands (Strehler and Zacharias 2001). Isoform 3 displays the most restricted pattern of expression in the neonate and adult, exclusively neuronal in human.

As the name suggests, the PMCA is localised to the plasma membrane. In some cell types such as neurons and astrocytes, PMCA distribution appears uniform throughout the plasma membrane (Blaustein et al. 2002). In contrast, in cell types in which PMCAs play a role in active calcium transport such as renal distal convoluted tubule cells and osteoblasts, distribution appears targeted to specific sites where unidirectional Ca^{2+} transport is required (Magyar et al. 2002; van der Eerden et al. 2012). Interestingly, in cardiomyocytes, vascular endothelial and smooth muscle cells PMCA4 have been shown to be particularly enriched in caveolae (Fujimoto 1993) where it colocalises with caveolin-3 (Hammes et al. 1998). This finding led to the speculation that PMCA4 in particular may be involved in signal transduction pathways in the cardiovascular system.

16.2.3 PMCA Structure and Regulation of Function

The general structure of the PMCA is comparable to other members of the P-type ATPase family, having ten hydrophobic transmembrane (TM) domains and two major intracellular loops, flanked by cytosolic N- and C-terminals (Fig. 16.1) (Strehler and

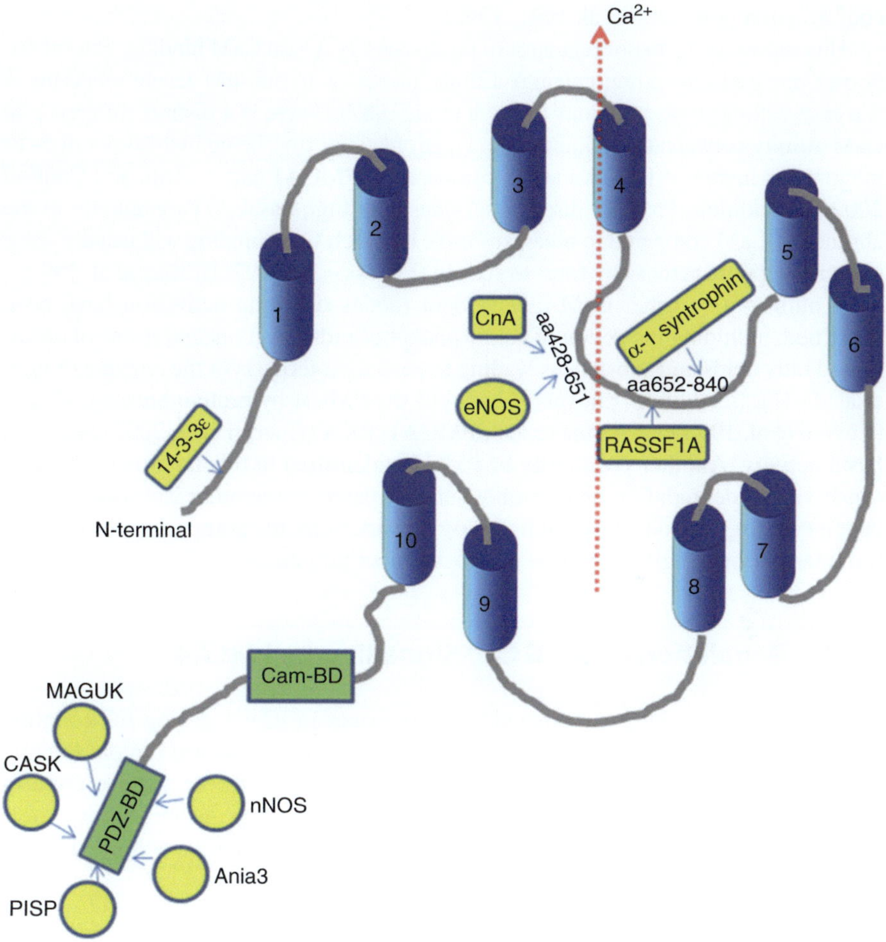

Fig. 16.1 PMCA4 structure and interaction partner binding domains. Schematic diagram illustrating general PMCA structure, consisting of ten transmembrane (TM) domains. The intracellular loop connecting TM 4 and 5 contains the catalytic core with phosphorylation and ATP binding sites. PMCA4 interaction partners are shown, with *arrows* detailing to which regions they bind. *CnA* calcineurin A, *eNOS* endothelial nitric oxide synthase, *RASSF1A* Ras-associated factor 1A, *CASK* calcium/calmodulin-dependent serine protein kinase, *MAGUK* membrane-associated guanylate kinase, *PISP* PMCA-interacting single-PDZ protein, *nNOS* neuronal nitric oxide synthase, *PDZ-BD* PDZ protein-binding domain, *CaM-BD* calmodulin-binding domain (adapted from Di Leva et al. 2008)

Zacharias 2001; Di Leva et al. 2008). The N-terminal comprises of 80–90 amino acids and contains the most variation amongst isoforms. The two intracellular loops span TM domains 2–3 and 4–5, the second of which contains conserved aspartate and lysine residues critical for catalytic phosphorylation and ATP binding, respectively; hence, this is the catalytic domain of the PMCA. The long carboxyl terminal contains the CaM-binding site which, in the absence of CaM, inhibits pump activity by interacting with a region on each of the intracellular loops to render the PMCA in a closed conformation (Falchetto et al. 1991, 1992).

This serves as the major regulator of pump activity. Upon CaM binding, autoinhibition is released and pump calcium affinity increases to sub-μM levels rendering it active at cellular concentrations (Enyedi et al. 1987). There is a distinct difference in CaM affinity between the ubiquitous and neuronal isoforms, being in the order of 5–10 magnitudes higher in PMCA2 and 3 compared to PMCA1 and 4 (Brini and Carafoli 2009). In addition, PMCA2 displays an unusually high-basal ATPase activity in the absence of CaM compared to other isoforms, in which CaM binding will usually serve to increase pump activity by four- to sixfold (Elwess et al. 1997; Hilfiker et al. 1993).

A number of further CaM-independent modes of pump activation have been identified, including the action of acidic phospholipids, low concentrations of unsaturated fatty acids and substances leading to partial proteolysis of the enzyme (Niggli et al. 1981). In addition, phosphorylation of the PMCA by protein kinase C (PKC) (Enyedi et al. 1996) and, in the case of PMCA1, PKA (Guerini et al. 2003) increases basal activity. Another peculiarity of PMCA1 identified in this latter study was its sensitivity to degradation by calpain when compared to the other isoforms. Hence, despite sharing a good degree of homology in structure, the pumps display substantial inter-isoform variation in certain biochemical properties.

16.3 Regulation of Local Ca^{2+} Signalling by PMCA4

In cardiomyocytes the noise created by global rises in $[Ca^{2+}]_i$ during EC coupling poses a significant challenge when attempting to accurately measure rapid fluctuations in specific microdomains. Over the past 25 years, however, the evolution of improved calcium indicators and arrival of novel-targeting techniques has now given the researcher a range of options when monitoring local Ca^{2+} not only in vitro but also in vivo.

16.3.1 Techniques for Monitoring Local Ca^{2+}

The advent of synthetic small molecule Ca^{2+} indicators with a wide dynamic range and the capability to rapidly respond to changes in calcium opened the way for the detection of localised calcium hotspots. Using the indicator fluo-3 and laser scanning confocal microscopy, Cheng and colleagues were able to detect local rises at the SR termed "Ca^{2+} sparks" which could be attributed to the opening of single RyRs, providing information on spontaneous and synchronised Ca^{2+} release events underpinning EC coupling (Cheng et al. 1993).

Targeting of small molecule indicators to specific sites proved problematic, however, limiting their use in measuring subcellular Ca^{2+}. In order to achieve this, genetically encoded calcium indicator proteins (GECIs) were developed which could be targeted to organ, tissue or cell-specific subcellular domains, organelles and proteins. The first generation of GECIs employed Forster resonance energy transfer (FRET)-based sensors with CaM and the M13 peptide from the myosin light chain kinase inserted between the donor and acceptor molecules. Tsien and colleagues dubbed these "cameleons" and were able to specifically monitor cytosolic, nuclear and ER calcium levels (Miyawaki et al. 1997). The dynamic range of cameleons was restricted, however, a problem which was overcome by replacing CaM with troponin C as the linker protein (Heim and Griesbeck 2004). Griesbeck and colleagues further developed these "TN-XL" indicators to increase signal strength and also enable in vivo mammalian delivery via viral injection (Mank et al. 2006, 2008), and the TN-XL indicator has since been coupled with specific calcium channels in order to measure nanodomain Ca^{2+} (Tay et al. 2012).

Shortly after the advent of FRET-based sensors, a novel class of circularly permuted (cp) GECIs was developed which fused together CaM with a single fluorophore. These cpGECIs were termed pericams (Nagai et al. 2001) or GCaMPs (Nakai et al. 2001) dependent on whether utilising enhanced yellow (EYFP) or green fluorescent protein (EGFP), respectively, and benefited from higher signal intensity than FRET-based probes.

16.3.2 The GCaMP Calcium Indicator

Nakai and colleagues developed the first generation of GCaMP sensors (GCaMP1) by connecting the N- and C- termini of cpEGFP to M13 and CaM, respectively (Nakai et al. 2001). Upon binding of Ca^{2+} to CaM, a conformational change in cpEGFP altered fluorescent signal intensity in line with $[Ca^{2+}]$. Through a number of amino acid changes to GCaMP1, the group was subsequently able to lower its dissociation constant from 235 nM to 146 nM and further increase the fluorescent signal intensity (Ohkura et al. 2005). When subjected to PCR-based random mutagenesis, further alterations to this GCaMP1.6 protein enabled the fluorescence to remain stable at 37 °C therefore making the indicator suitable for in vivo use (Tallini et al. 2006). By placing this GCaMP2 protein under the control of the α-myosin heavy chain promoter, Kotlikoff and colleagues were able to generate transgenic mice conditionally overexpressing GCaMP2 in cardiomyocytes and successfully image in vivo calcium transients. GCaMPs have now evolved to further increase sensitivity and kinetic properties, enabling their use in cell types displaying even the fastest calcium dynamics or exhibiting low peak calcium such as neurons (Chen et al. 2013).

The properties of GCaMP sensors have now enabled the development of a number of fusion proteins capable of being targeted to specific organelles and subcellular regions and hence monitoring local calcium. In the cardiovascular system, these include measurement of subplasmalemmal Ca^{2+} through fusion with α-subunits of the Na^+/K^+ ATPase, mitochondrial matrix Ca^{2+} upon addition of the mitochondrial targeting sequence of cytochrome oxidase and Ca^{2+} signalling around junctional dyadic clefts through fusion with triadin-1, junctin or the RyR-binding protein FKBP12.6 (Lee et al. 2006; Iguchi et al. 2012; Shang et al. 2014; Despa et al. 2014).

16.3.3 The PMCA4-GCaMP2 Fusion Protein

In recent years our group and others have identified a significant role for the PMCAs in regulating the activity of a number of key signalling molecules in a variety of cell types including those of the cardiovascular system (Oceandy et al. 2011). In particular, PMCA4 has been shown to functionally interact with proteins in the heart and vasculature capable of influencing cardiac contractility (Oceandy et al. 2007; Mohamed et al. 2011), hypertrophic remodelling (Mohamed et al. 2016; Wu et al. 2009) and vascular tone (Gros et al. 2003; Schuh et al. 2003a). Given the potential importance of PMCA4s' activity on these signalling pathways, we sought to develop new tools with which to monitor local Ca^{2+} in its vicinity.

We generated a PMCA4-GCaMP2 fusion molecule by cloning GCaMP2 cDNA to the N-terminal of PMCA4, which was subsequently subcloned into an adenoviral vector (Mohamed et al. 2013a). In order to characterise the function of this molecule and verify that the fusion protein could indeed report Ca^{2+} dynamics local to PMCA4, we also generated adenovirus-expressing cytoplasmic GCaMP2 (ad-GCaMP2) and an inactive mutant PMCA4-GCaMP2 construct (PMCA4^{mut}-GCaMP2). This contained an Asp^{672}Glu amino acid substitution in the nucleotide-binding region on PMCA4s' second intracellular loop, which has been shown to reduce the pumps' ability to transport Ca^{2+} by around 90% (Adamo et al. 1995). Expression of all three constructs was driven by a cytomegalovirus promoter.

Confocal microscopy confirmed that when transfected into neonatal and adult rat cardiomyocytes, the PMCA4-GCaMP2 and PMCA4^{mut}-GCaMP2 proteins were each targeted to the plasma membrane where they colocalised with caveolin-3, whilst ad-GCaMP2 showed staining throughout the cytoplasm (Mohamed et al. 2013a). We confirmed that PMCA4 ATPase activity was not affected by fusion to GCaMP2 compared to wild-type PMCA4 via a coupled enzyme assay and that only residual activity could be detected in the PMCA4^{mut}-GCaMP2 molecule on a par with ad-GCaMP2 and a LacZ construct.

All three constructs successfully recorded Ca^{2+} transients in electrically stimulated adult and neonatal cardiomyocytes, which were of significantly greater amplitude and decayed faster when detected by PMCA4-GCaMP2 compared to PMCA4^{mut}-GCaMP2. Application of caffeine to empty the SR evoked a large rise in signal detected by PMCA4-GCaMP2, whilst having little effect on the mutant. Similarly, isoprenaline stimulation increased the amplitude and decay rate of oscillations recorded by the ad-GCaMP2 and PMCA4-GCaMP2 constructs, whilst not altering the signal detected by PMCA4^{mut}-GCaMP2.

In order to study whether PMCA4 exclusively regulates local sub-membrane Ca^{2+}, we required a specific inhibitor. We therefore screened a library of medically optimised bioactive small molecules via a modified coupled enzyme assay (Mohamed et al. 2010) and identified aurintricarboxylic acid (ATA) as displaying potent-specific

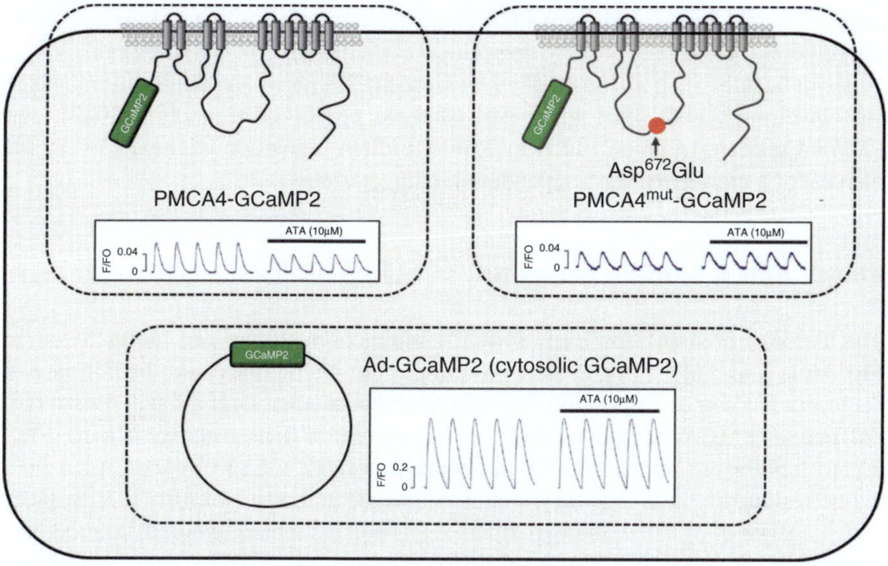

Fig. 16.2 Structure and characterisation of PMCA4-GCaMP2 fusion proteins. Cartoon detailing the structure and localisation of PMCA4-GCaMP2 fusion proteins. GCaMP2 was fused to the N-termini of PMCA4 (PMCA4-GCaMP2) and a largely inactive mutant PMCA4 (PMCA4^{mut}-GCaMP2) with an Asp^{672}Glu amino acid substitution in its catalytic domain, as well as being placed in an adenoviral vector (ad-GCaMP2) which diffused throughout the cytoplasm. Example traces of local calcium oscillations recorded by each GCaMP2 construct upon electrical stimulation in cardiomyocytes are shown before and after incubation with the specific PMCA4 inhibitor aurintricarboxylic acid (ATA)

inhibition (IC_{50} 150 nM) of PMCA4, whilst not affecting other related ATPases expressed in the cardiovascular system. Importantly, ATA treatment did not affect the signal detected by the cytoplasmic or mutant GCaMP2 but significantly reduced the amplitude and decay rate of systolic and caffeine-evoked oscillations detected by PMCA4-GCaMP2 to a level comparable to PMCA4^{mut}-GCaMP2. These findings are summarised in Fig. 16.2, along with the structures and subcellular targeting of the three GCaMP2 constructs. These results provided strong evidence that PMCA4 indeed regulates local sub-sarcolemmal Ca^{2+} but is not involved in bulk Ca^{2+} extrusion in cardiomyocytes. Furthermore, they demonstrate that the PMCA4-GCaMP2 fusion protein can be used as a tool to assess Ca^{2+} dynamics around PMCA4, which may prove useful when studying modulators of PMCA4 function in the future.

16.4 PMCA4 Transduction of the Subplasmalemmal Calcium Signal

Protein-protein interactions lie at the heart of signal transduction networks, tethering effector molecules to the same subcellular locale as their upstream regulators and compartmentalising second messenger signalling. A typical example of such a network would be the binding of PKA isoforms to the scaffolding A-kinase anchoring proteins (AKAPs) in order to direct local cAMP production (Kritzer et al. 2012). Other molecules such as focal adhesion kinase (FAK) have the ability to transduce signals through both scaffolding and catalytic properties (Franchini 2012), and PMCA4 appears to be one such molecule through regulation of local Ca^{2+} in the vicinity of a number of Ca^{2+}-dependent binding partners.

16.4.1 PMCA Binding Domains

The binding domains for each PMCA4 interaction partner are summarised in Fig. 16.1. The majority of PMCA protein-protein interactions have been found to occur via binding to a PDZ domain located at the terminal end of the carboxyl tail. All human PMCA isoforms adhere to the minimum consensus for class I PDZ ligand binding by terminating in the consensus motif ETSX. In PMCA4 a final valine residue replaces leucine which is conserved amongst isoforms 1–3 (Strehler and Zacharias 2001). Interacting partners at this domain include calcium/calmodulin-dependent serine protein kinase (CASK) (Schuh et al. 2003b), members of the membrane-associated guanylate kinase (MAGUK) family (Kim et al. 1998; DeMarco and Strehler 2001), homer protein Ania-3 (Sgambato-Faure et al. 2006), PMCA-interacting single-PDZ protein (PISP) (Goellner et al. 2003) and neuronal nitric oxide synthase (nNOS) (Schuh et al. 2001). Interactions also occur at the second intracellular loop between the PMCA and the Ca^{2+}/CaM-dependent phosphatase calcineurin (Buch et al. 2005), the tumour suppressor Ras-associated factor 1A (RASSF1A) (Armesilla et al. 2004), endothelial nitric oxide synthase (eNOS) (Holton et al. 2010) and α-1 syntrophin (Williams et al. 2006), whilst the N-terminal region has been shown to associate with isoform ε of 14–3-3 (Rimessi et al. 2005). The PMCA signalling complexes which have been functionally characterised will be discussed below.

16.4.2 The PMCA4-nNOS Complex

This interaction was first identified in HEK293 cells where PMCA4 was found to negatively regulate nNOS activity. Overexpression of wild-type PMCA4 decreased NO production, whilst mutant PMCA4 either lacking ATPase activity or the PDZ-binding domain did not regulate nNOS (Schuh et al. 2001). It was later shown that PMCA4 and nNOS form a macromolecular complex at the plasma membrane with α-1 syntrophin, through an interaction between its linker region and the second intracellular loop of PMCA4 (Williams et al. 2006), and the binding of the two molecules acted synergistically to further reduce nNOS activity. The PMCA4-nNOS complex has since been expanded to include the cardiac sodium channel $Na_v1.5$ and found to be of relevance in the regulation of vascular tone as well as cardiac rhythm, contractility, β-adrenergic responsiveness and remodelling (Gros et al. 2003; Schuh et al. 2003a; Ueda et al. 2008; Mohamed et al. 2009, 2011; Oceandy et al. 2007).

16.4.3 The PMCA4-Calcineurin Complex

An interaction between the catalytic domains of PMCA4 and the calcineurin A subunit was first identified in HEK cells (Buch et al. 2005). Overexpression of the PMCA resulted in calcineurin's recruitment from the cytoplasm to the plasma membrane and a reduction in calcineurin substrate nuclear factor of activated T-cells (NFAT) transcriptional activity. Calcineurin has since been noted to also interact with PMCA4 in cardiomyocytes thereby influencing cardiac hypertrophy (Wu et al. 2009), as well as endothelial and breast cancer cells (Holton et al. 2007, 2010), and the PC12 pheochromocytoma cell line where the interaction was shown to influence catecholamine secretion (Kosiorek et al. 2011).

16.4.4 Additional Characterised PMCA4 Interaction Complexes

Further examples of signalling pathways downstream of PMCA4 have also been identified. A region in the catalytic domain of PMCA4 was found to bind the tumour suppressor RASSF1A in HEK293 cells, thereby negatively regulating epidermal growth factor-dependent activation of the mitogen-activated protein kinase Erk and hence Ras-mediated signalling (Armesilla et al. 2004).

One further interaction partner at the catalytic domain of PMCA4 is eNOS. This interaction was found to be present in both human umbilical vascular endothelial cells (HUVECs) and human dermal microvascular endothelial cells (HDMECs) and to be common to isoforms 1, 2 and 4 of the PMCA (Holton et al. 2010). PMCA4 was found to negatively regulate eNOS through phosphorylation of the inhibitory Thr-495 residue and consequently reduce NO bioavailability and NO-dependent cGMP production.

In addition to these regulatory pathways, the calcium/calmodulin-dependent serine protein kinase (CASK) has been found to co-precipitate with PMCA4 via an interaction with its C-terminus in the brain and kidney, resulting in inhibition of T-element-dependent transcriptional activity (Schuh et al. 2003b).

16.5 Physiological Implications of PMCA4-Mediated Signalling in the Cardiovascular System

The previous section detailed a variety of signalling complexes regulated by PMCA4. Through the use of transgenic mouse models, in recent years the physiological consequences of PMCA4-mediated signal regulation in a number of cardiovascular cell types have begun to be elucidated.

16.5.1 PMCA4 Signal Regulation in Cardiomyocytes

The PMCA4-nNOS and PMCA4-calcineurin complexes have been shown to be of utmost importance in regulating myocyte function in both health and disease progression. Work in PMCA4 overexpressing and knockout mice has revealed that

PMCA4-mediated regulation of nNOS affects both basal and β-adrenergic-stimulated inotropic function. We found PMCA4 overexpression to attenuate the inotropic response to β-adrenergic stimulation to an extent comparable to that seen upon nNOS-specific inhibition, whilst this effect was absent in mice overexpressing a mutant form of PMCA4 unable to interact with nNOS (Oceandy et al. 2007). We have since further characterised this pathway, demonstrating that the interaction affects local cGMP and cAMP levels, ultimately modulating PKA activity at the SR to elevate systolic Ca^{2+} and enhance basal contractility, whilst blunting the adrenergic response (Mohamed et al. 2009, 2011). A further physiological role of the PMCA4-nNOS-α-1 syntrophin complex appears to be the regulation of the sodium channel $Na_v1.5$. A mutation in the syntrophin gene affecting PMCA4 binding was found to increase $Na_v1.5$ nitrosylation and dramatically prolong QT duration, potentially increasing susceptibility to sudden cardiac death (Ueda et al. 2008).

Calcineurin-NFAT signalling is one of the best characterised pathways in the development of pathological hypertrophy and remodelling (Molkentin et al. 1998). Hence, it was hypothesised that negative regulation of calcineurin by PMCA4 may be a determinant of disease progression (Buch et al. 2005), and indeed PMCA4 overexpression in cardiomyocytes was subsequently found to significantly reduce the extent of pathological remodelling during pressure overload and phenylephrine/angiotensin II-induced hypertrophy (Wu et al. 2009). Following transverse aortic constriction (TAC), PMCA4 transgenic mice displayed attenuated hypertrophy, reduced fibrosis and improved contractility compared to controls, and it was determined that this was the result of an increased interaction between PMCA4 and calcineurin and the subsequent reduction of NFAT transcriptional activity.

16.5.2 PMCA4 Signal Regulation in Cardiac Fibroblasts

Pathological cardiac hypertrophy is a complex process involving both the myocyte population and a number of other cell types including the cardiac fibroblast. These play a key role in the remodelling process, not only through regulating extracellular matrix composition but also through the secretion of signalling molecules which are able to act upon myocytes themselves (Souders et al. 2009).

We therefore decided to examine the role of PMCA4-mediated signalling in cardiac fibroblasts. We recently reported that cardiac fibroblasts lacking PMCA4 display a 25% higher basal calcium level compared to wild-type cells (Mohamed et al. 2016). Through the regulation of NF-κB and Pax2 transcription factors, this led to increased expression of Wnt pathway regulators including secreted frizzled-related protein 2 (sFRP2) and insulin-like growth factor-binding proteins 4 and 5 (IGFBP-4 and IGFBP-5). In vitro we determined that secreted sFRP2 acted in a paracrine fashion to prevent agonist-induced cardiomyocyte hypertrophy through reducing Wnt/β-catenin activation. In vivo, systemic and fibroblast-specific PMCA4 deletion attenuated the development of TAC-induced hypertrophy whilst preserving contractility and improving survival. As in the in vitro model, this protection was conferred via the paracrine secretion of sFRP2 from fibroblasts and a subsequent reduction in Wnt/β-catenin activation. Moreover, we demonstrated that pharmacological administration of a specific PMCA4 inhibitor was able to reverse established hypertrophy, providing proof of principle that PMCA4-mediated signalling in cardiac fibroblasts may be a target for an anti-hypertrophic therapy in the clinical setting.

16.5.3 PMCA4 Signal Regulation in Vascular Smooth Muscle Cells

A wealth of genome-wide association studies has found that polymorphisms in *ATP2B1*-encoding PMCA1 show a strong association with the development of hypertension (Little et al. 2016); however, studies in the vasculature have also found PMCA4-mediated signalling to influence vascular tone, as well as vascular smooth muscle cell (VSMC) proliferation.

In 2003, Schuh and colleagues demonstrated that mice overexpressing PMCA4 specifically in VSMCs under the control of the SM22α promoter displayed a 10 mmHg elevation in systolic blood pressure compared to controls (Schuh et al. 2003a). In addition, de-endothelialised aortas from PMCA4 transgenic mice exhibited increased maximal contraction to KCl. This response was found to be comparable to that seen in nNOS knockout arteries or wild-type arteries treated with a specific nNOS inhibitor. Furthermore PMCA4 was found to co-precipitate with nNOS in aortic lysate suggesting that these effects were due to PMCA4-nNOS-mediated signalling. Meanwhile Gros and colleagues demonstrated similarly elevated blood pressure, as well as a lowered set point of myogenic response and increased phenylephrine-induced constriction in mesenteric arteries isolated from SM22α-PMCA4 transgenic mice (Gros et al. 2003). This was again attributed to the interaction between PMCA4 and nNOS and found to be associated with reduced levels of cGMP in isolated PMCA4 transgenic VSMCs.

To follow up the finding, Afroze and co-workers have since gone on to define a role for PMCA4 in regulating VSMC cell cycle progression. VSMCs from PMCA4 knockout mice were found to exhibit significantly reduced G_1 to S phase cell cycle progression suggesting G_1 phase arrest (Afroze et al. 2014). G_1 arrest could be prevented by restoring PMCA4 function through electroporation of either wild-type PMCA4a, PMCA4b or PMCA4 lacking its PDZ domain but not an inactive mutant PMCA4 suggesting that the regulatory role of PMCA4 was not nNOS-dependent. Microarray analysis of G_1-arrested PMCA4KO VSMCs and electroporation-rescued cells revealed that a- and b-splice variants of PMCA4 differentially regulated cell cycle progression via independent pathways, by targeting expression of the anti-proliferative transcription factor AP-2β and p15, a cyclin-dependent kinase inhibitor.

16.5.4 PMCA4 Signal Regulation in Vascular Endothelial Cells

Whilst the physiological implications of PMCA4-eNOS-mediated signalling have yet to be characterised in vivo, PMCA4-mediated signal transduction in the endothelium has recently emerged to be of significance via its interaction with calcineurin.

Baggott and colleagues found PMCA4 to negatively regulate vascular endothelial growth factor (VEGF)-mediated calcineurin-NFAT induction in HUVECs, which in turn controlled expression of the proangiogenic protein RCAN1.4 (Baggott et al. 2014). Functionally, PMCA4-calcineurin signalling was found to inhibit migration of endothelial cells and VEGF-mediated angiogenesis in vitro and in vivo, whilst PMCA4-ablated mice displayed increased perfusion in the hind limb following femoral artery ligation suggesting that PMCA4 inhibition may be a novel target to treat post-ischaemic injury.

Conclusions

Subcellular compartmentalisation of local Ca^{2+} is essential in order to precisely direct signalling events governing an array of cellular processes. The development of a novel GECI fusion protein PMCA4-GCaMP2 has provided evidence that this Ca^{2+}-ATPase exclusively regulates local as opposed to global calcium in cardiomyocytes and a means with which to examine Ca^{2+} dynamics in its vicinity. Through direct interactions with a number of binding partners, PMCA4 is able to tether Ca^{2+}-dependent effector molecules to subplasmalemmal microdomains and orchestrate downstream signalling through their regulation. These interactions can intrinsically influence both cardiovascular performance in health and determine disease progression. PMCA4-mediated signalling in the cardiac myocyte and fibroblast can profoundly affect pathological growth and remodelling, whilst regulating basal myocyte contractility and adrenergic responsiveness. Meanwhile, cells of the vasculature signalling via PMCA4 is able to regulate vascular tone as well as angiogenic and proliferative processes. As we have demonstrated in the cardiac fibroblast (Mohamed et al. 2016), the inherent druggability of PMCA4 being an ATPase located at the plasma membrane may therefore lend itself to becoming a suitable therapeutic target in the fields of cardiac hypertrophy, blood pressure regulation and post-ischaemic injury. Furthermore, the ubiquitous nature of the pump may mean that PMCA4-mediated signalling could be targeted for therapeutics in wider fields outside of the cardiovascular system, such as contraception and infectious disease (Mohamed et al. 2013b), with roles having been identified in the regulation of sperm motility (Schuh et al. 2004) and malarial resistance (Timmann et al. 2012). Given that other isoforms of the PMCA share common interaction partners with PMCA4 including calcineurin (Holton et al. 2007), eNOS (Holton et al. 2010) and α1-syntrophin (Williams et al. 2006), future studies may find that PMCA-mediated control of local Ca^{2+} extends beyond PMCA4 to provide isoform-specific refinement of the signal.

Compliance with Ethical Standards

Conflict of Interest Statement The authors declare that they have no conflict of interest.

References

Adamo HP, Filoteo AG, Enyedi A, Penniston JT (1995) Mutants in the putative nucleotide-binding region of the plasma membrane Ca(2+)-pump. A reduction in activity due to slow dephosphorylation. J Biol Chem 270(50):30111–30114

Afroze T, Yang G, Khoshbin A, Tanwir M, Tabish T, Momen A, Husain M (2014) Calcium efflux activity of plasma membrane Ca2+ ATPase-4 (PMCA4) mediates cell cycle progression in vascular smooth muscle cells. J Biol Chem 289(10):7221–7231. doi:10.1074/jbc.M113.533638

Armesilla AL, Williams JC, Buch MH, Pickard A, Emerson M, Cartwright EJ, Oceandy D, Vos MD, Gillies S, Clark GJ, Neyses L (2004) Novel functional interaction between the plasma membrane Ca2+ pump 4b and the proapoptotic tumor suppressor Ras-associated factor 1 (RASSF1). J Biol Chem 279(30):31318–31328. doi:10.1074/jbc.M307557200

Baggott RR, Alfranca A, Lopez-Maderuelo D, Mohamed TM, Escolano A, Oller J, Ornes BC, Kurusamy S, Rowther FB, Brown JE, Oceandy D, Cartwright EJ, Wang W, Gomez-del Arco P, Martinez-Martinez S, Neyses L, Redondo JM, Armesilla AL (2014) Plasma membrane calcium ATPase isoform 4 inhibits vascular endothelial growth factor-mediated angiogenesis through interaction with calcineurin. Arterioscler Thromb Vasc Biol 34(10):2310–2320. doi:10.1161/atvbaha.114.304363

Bers DM (2000) Calcium fluxes involved in control of cardiac myocyte contraction. Circ Res 87(4):275–281

Blaustein MP, Juhaszova M, Golovina VA, Church PJ, Stanley EF (2002) Na/Ca exchanger and PMCA localization in neurons and astrocytes: functional implications. Ann N Y Acad Sci 976:356–366

Brini M, Carafoli E (2009) Calcium pumps in health and disease. Physiol Rev 89(4):1341–1378. doi:10.1152/physrev.00032.2008

Buch MH, Pickard A, Rodriguez A, Gillies S, Maass AH, Emerson M, Cartwright EJ, Williams JC, Oceandy D, Redondo JM, Neyses L, Armesilla AL (2005) The sarcolemmal calcium pump inhibits the calcineurin/nuclear factor of activated T-cell pathway via interaction with the calcineurin A catalytic subunit. J Biol Chem 280(33):29479–29487. doi:10.1074/jbc.M501326200

Chen TW, Wardill TJ, Sun Y, Pulver SR, Renninger SL, Baohan A, Schreiter ER, Kerr RA, Orger MB, Jayaraman V, Looger LL, Svoboda K, Kim DS (2013) Ultrasensitive fluorescent proteins for imaging neuronal activity. Nature 499(7458):295–300. doi:10.1038/nature12354

Cheng H, Lederer WJ, Cannell MB (1993) Calcium sparks: elementary events underlying excitation-contraction coupling in heart muscle. Science 262(5134):740–744

Damy T, Ratajczak P, Shah AM, Camors E, Marty I, Hasenfuss G, Marotte F, Samuel JL, Heymes C (2004) Increased neuronal nitric oxide synthase-derived NO production in the failing human heart. Lancet 363(9418):1365–1367. doi:10.1016/s0140-6736(04)16048-0

DeMarco SJ, Strehler EE (2001) Plasma membrane Ca2 + –ATPase isoforms 2b and 4b interact promiscuously and selectively with members of the membrane-associated guanylate kinase family of PDZ (PSD95/Dlg/ZO-1) domain-containing proteins. J Biol Chem 276(24):21594–21600. doi:10.1074/jbc.M101448200

Despa S, Shui B, Bossuyt J, Lang D, Kotlikoff MI, Bers DM (2014) Junctional cleft [Ca(2)(+)]i measurements using novel cleft-targeted Ca(2)(+) sensors. Circ Res 115(3):339–347. doi:10.1161/circresaha.115.303582

Di Leva F, Domi T, Fedrizzi L, Lim D, Carafoli E (2008) The plasma membrane Ca2+ ATPase of animal cells: structure, function and regulation. Arch Biochem Biophys 476(1):65–74. doi:10.1016/j.abb.2008.02.026

Elwess NL, Filoteo AG, Enyedi A, Penniston JT (1997) Plasma membrane Ca2+ pump isoforms 2a and 2b are unusually responsive to calmodulin and Ca2+. J Biol Chem 272(29):17981–17986

Enyedi A, Flura M, Sarkadi B, Gardos G, Carafoli E (1987) The maximal velocity and the calcium affinity of the red cell calcium pump may be regulated independently. J Biol Chem 262(13):6425–6430

Enyedi A, Verma AK, Filoteo AG, Penniston JT (1996) Protein kinase C activates the plasma membrane Ca2+ pump isoform 4b by phosphorylation of an inhibitory region downstream of the calmodulin-binding domain. J Biol Chem 271(50):32461–32467

Falchetto R, Vorherr T, Brunner J, Carafoli E (1991) The plasma membrane Ca2+ pump contains a site that interacts with its calmodulin-binding domain. J Biol Chem 266(5):2930–2936

Falchetto R, Vorherr T, Carafoli E (1992) The calmodulin-binding site of the plasma membrane Ca2+ pump interacts with the transduction domain of the enzyme. Protein Sci 1(12):1613–1621. doi:10.1002/pro.5560011209

Franchini KG (2012) Focal adhesion kinase—the basis of local hypertrophic signaling domains. J Mol Cell Cardiol 52(2):485–492. doi:10.1016/j.yjmcc.2011.06.021

Frey N, McKinsey TA, Olson EN (2000) Decoding calcium signals involved in cardiac growth and function. Nat Med 6(11):1221–1227. doi:10.1038/81321

Fujimoto T (1993) Calcium pump of the plasma membrane is localized in caveolae. J Cell Biol 120(5):1147–1157

Goellner GM, DeMarco SJ, Strehler EE (2003) Characterization of PISP, a novel single-PDZ protein that binds to all plasma membrane Ca2 + −ATPase b-splice variants. Ann N Y Acad Sci 986:461–471

Gros R, Afroze T, You XM, Kabir G, Van Wert R, Kalair W, Hoque AE, Mungrue IN, Husain M (2003) Plasma membrane calcium ATPase overexpression in arterial smooth muscle increases vasomotor responsiveness and blood pressure. Circ Res 93(7):614–621. doi:10.1161/01.res.0000092142.19896.d9

Guerini D, Zecca-Mazza A, Carafoli E (2000) Single amino acid mutations in transmembrane domain 5 confer to the plasma membrane Ca2+ pump properties typical of the Ca2+ pump of endo(sarco)plasmic reticulum. J Biol Chem 275(40):31361–31368. doi:10.1074/jbc.M003474200

Guerini D, Pan B, Carafoli E (2003) Expression, purification, and characterization of isoform 1 of the plasma membrane Ca2+ pump: focus on calpain sensitivity. J Biol Chem 278(40):38141–38148. doi:10.1074/jbc.M302400200

Hammes A, Oberdorf-Maass S, Rother T, Nething K, Gollnick F, Linz KW, Meyer R, Hu K, Han H, Gaudron P, Ertl G, Hoffmann S, Ganten U, Vetter R, Schuh K, Benkwitz C, Zimmer HG, Neyses L (1998) Overexpression of the sarcolemmal calcium pump in the myocardium of transgenic rats. Circ Res 83(9):877–888

Hare JM, Stamler JS (2005) NO/redox disequilibrium in the failing heart and cardiovascular system. J Clin Invest 115(3):509–517. doi:10.1172/jci24459

Harvey RD, Calaghan SC (2012) Caveolae create local signalling domains through their distinct protein content, lipid profile and morphology. J Mol Cell Cardiol 52(2):366–375. doi:10.1016/j.yjmcc.2011.07.007

Heim N, Griesbeck O (2004) Genetically encoded indicators of cellular calcium dynamics based on troponin C and green fluorescent protein. J Biol Chem 279(14):14280–14286. doi:10.1074/jbc.M312751200

Heineke J, Ritter O (2012) Cardiomyocyte calcineurin signaling in subcellular domains: from the sarcolemma to the nucleus and beyond. J Mol Cell Cardiol 52(1):62–73. doi:10.1016/j.yjmcc.2011.10.018

Higazi DR, Fearnley CJ, Drawnel FM, Talasila A, Corps EM, Ritter O, McDonald F, Mikoshiba K, Bootman MD, Roderick HL (2009) Endothelin-1-stimulated InsP3-induced Ca2+ release is a nexus for hypertrophic signaling in cardiac myocytes. Mol Cell 33(4):472–482. doi:10.1016/j.molcel.2009.02.005

Hilfiker H, Strehler-Page MA, Stauffer TP, Carafoli E, Strehler EE (1993) Structure of the gene encoding the human plasma membrane calcium pump isoform 1. J Biol Chem 268(26):19717–19725

Holton M, Yang D, Wang W, Mohamed TM, Neyses L, Armesilla AL (2007) The interaction between endogenous calcineurin and the plasma membrane calcium-dependent ATPase is isoform specific in breast cancer cells. FEBS Lett 581(21):4115–4119. doi:10.1016/j.febslet.2007.07.054

Holton M, Mohamed TM, Oceandy D, Wang W, Lamas S, Emerson M, Neyses L, Armesilla AL (2010) Endothelial nitric oxide synthase activity is inhibited by the plasma membrane calcium ATPase in human endothelial cells. Cardiovasc Res 87(3):440–448. doi:10.1093/cvr/cvq077

Iguchi M, Kato M, Nakai J, Takeda T, Matsumoto-Ida M, Kita T, Kimura T, Akao M (2012) Direct monitoring of mitochondrial calcium levels in cultured cardiac myocytes using a novel fluorescent indicator protein, GCaMP2-mt. Int J Cardiol 158(2):225–234. doi:10.1016/j.ijcard.2011.01.034

Kim E, DeMarco SJ, Marfatia SM, Chishti AH, Sheng M, Strehler EE (1998) Plasma membrane Ca2+ ATPase isoform 4b binds to membrane-associated guanylate kinase (MAGUK) proteins via their PDZ (PSD-95/Dlg/ZO-1) domains. J Biol Chem 273(3):1591–1595

Kosiorek M, Podszywalow-Bartnicka P, Zylinska L, Zablocki K, Pikula S (2011) Interaction of plasma membrane Ca(2+)-ATPase isoform 4 with calcineurin A: implications for catecholamine secretion by PC12 cells. Biochem Biophys Res Commun 411(2):235–240. doi:10.1016/j.bbrc.2011.06.098

Kritzer MD, Li J, Dodge-Kafka K, Kapiloff MS (2012) AKAPs: the architectural underpinnings of local cAMP signaling. J Mol Cell Cardiol 52(2):351–358. doi:10.1016/j.yjmcc.2011.05.002

Lee MY, Song H, Nakai J, Ohkura M, Kotlikoff MI, Kinsey SP, Golovina VA, Blaustein MP (2006) Local subplasma membrane Ca2+ signals detected by a tethered Ca2+ sensor. Proc Natl Acad Sci U S A 103(35):13232–13237. doi:10.1073/pnas.0605757103

Lenaerts I, Bito V, Heinzel FR, Driesen RB, Holemans P, D'Hooge J, Heidbuchel H, Sipido KR, Willems R (2009) Ultrastructural and functional remodeling of the coupling between Ca2+ influx and sarcoplasmic reticulum Ca2+ release in right atrial myocytes from experimental persistent atrial fibrillation. Circ Res 105(9):876–885. doi:10.1161/circresaha.109.206276

Little R, Cartwright EJ, Neyses L, Austin C (2016) Plasma membrane calcium ATPases (PMCAs) as potential targets for the treatment of essential hypertension. Pharmacol Ther 159:23–34. doi:10.1016/j.pharmthera.2016.01.013

Magyar CE, White KE, Rojas R, Apodaca G, Friedman PA (2002) Plasma membrane Ca2 + −ATPase and NCX1 Na+/Ca2+ exchanger expression in distal convoluted tubule cells. Am J Physiol Renal Physiol 283(1):F29–F40. doi:10.1152/ajprenal.00252.2000

Mangialavori I, Ferreira-Gomes M, Pignataro MF, Strehler EE, Rossi JP (2010) Determination of the dissociation constants for Ca2+ and calmodulin from the plasma membrane Ca2+ pump by a lipid probe that senses membrane domain changes. J Biol Chem 285(1):123–130. doi:10.1074/jbc.M109.076679

Mank M, Reiff DF, Heim N, Friedrich MW, Borst A, Griesbeck O (2006) A FRET-based calcium biosensor with fast signal kinetics and high fluorescence change. Biophys J 90(5):1790–1796. doi:10.1529/biophysj.105.073536

Mank M, Santos AF, Direnberger S, Mrsic-Flogel TD, Hofer SB, Stein V, Hendel T, Reiff DF, Levelt C, Borst A, Bonhoeffer T, Hubener M, Griesbeck O (2008) A genetically encoded calcium indicator for chronic in vivo two-photon imaging. Nat Methods 5(9):805–811. doi:10.1038/nmeth.1243

Miyawaki A, Llopis J, Heim R, McCaffery JM, Adams JA, Ikura M, Tsien RY (1997) Fluorescent indicators for Ca2+ based on green fluorescent proteins and calmodulin. Nature 388(6645):882–887. doi:10.1038/42264

Mohamed TM, Oceandy D, Prehar S, Alatwi N, Hegab Z, Baudoin FM, Pickard A, Zaki AO, Nadif R, Cartwright EJ, Neyses L (2009) Specific role of neuronal nitric-oxide synthase when tethered to the plasma membrane calcium pump in regulating the beta-adrenergic signal in the myocardium. J Biol Chem 284(18):12091–12098. doi:10.1074/jbc.M809112200

Mohamed TM, Baudoin-Stanley FM, Abou-Leisa R, Cartwright E, Neyses L, Oceandy D (2010) Measurement of plasma membrane calcium-calmodulin-dependent ATPase (PMCA) activity. Methods Mol Biol 637:333–342. doi:10.1007/978-1-60761-700-6_18

Mohamed TM, Oceandy D, Zi M, Prehar S, Alatwi N, Wang Y, Shaheen MA, Abou-Leisa R, Schelcher C, Hegab Z, Baudoin F, Emerson M, Mamas M, Di Benedetto G, Zaccolo M, Lei M, Cartwright EJ, Neyses L (2011) Plasma membrane calcium pump (PMCA4)-neuronal nitric-oxide synthase complex regulates cardiac contractility through modulation of a compartmentalized cyclic nucleotide microdomain. J Biol Chem 286(48):41520–41529. doi:10.1074/jbc.M111.290411

Mohamed TM, Abou-Leisa R, Baudoin F, Stafford N, Neyses L, Cartwright EJ, Oceandy D (2013a) Development and characterization of a novel fluorescent indicator protein PMCA4-GCaMP2 in cardiomyocytes. J Mol Cell Cardiol 63:57–68. doi:10.1016/j.yjmcc.2013.07.007

Mohamed TM, Zakeri SA, Baudoin F, Wolf M, Oceandy D, Cartwright EJ, Gul S, Neyses L (2013b) Optimisation and validation of a high throughput screening compatible assay to identify inhibitors of the plasma membrane calcium ATPase pump—a novel therapeutic target for contraception and malaria. J Pharm Pharm Sci 16(2):217–230

Mohamed TM, Abou-Leisa R, Stafford N, Maqsood A, Zi M, Prehar S, Baudoin-Stanley F, Wang X, Neyses L, Cartwright EJ, Oceandy D (2016) The plasma membrane calcium ATPase 4 signalling in cardiac fibroblasts mediates cardiomyocyte hypertrophy. Nat Commun 7:11074. doi:10.1038/ncomms11074

Molkentin JD, Lu JR, Antos CL, Markham B, Richardson J, Robbins J, Grant SR, Olson EN (1998) A calcineurin-dependent transcriptional pathway for cardiac hypertrophy. Cell 93(2):215–228

Nagai T, Sawano A, Park ES, Miyawaki A (2001) Circularly permuted green fluorescent proteins engineered to sense Ca2+. Proc Natl Acad Sci U S A 98(6):3197–3202. doi:10.1073/pnas.051636098

Nakai J, Ohkura M, Imoto K (2001) A high signal-to-noise Ca(2+) probe composed of a single green fluorescent protein. Nat Biotechnol 19(2):137–141. doi:10.1038/84397

Nakayama H, Bodi I, Maillet M, DeSantiago J, Domeier TL, Mikoshiba K, Lorenz JN, Blatter LA, Bers DM, Molkentin JD (2010) The IP3 receptor regulates cardiac hypertrophy in response to select stimuli. Circ Res 107(5):659–666. doi:10.1161/circresaha.110.220038

Niggli V, Adunyah ES, Carafoli E (1981) Acidic phospholipids, unsaturated fatty acids, and limited proteolysis mimic the effect of calmodulin on the purified erythrocyte Ca2+ – ATPase. J Biol Chem 256(16):8588–8592

Oceandy D, Cartwright EJ, Emerson M, Prehar S, Baudoin FM, Zi M, Alatwi N, Venetucci L, Schuh K, Williams JC, Armesilla AL, Neyses L (2007) Neuronal nitric oxide synthase signaling in the heart is regulated by the sarcolemmal calcium pump 4b. Circulation 115(4):483–492. doi:10.1161/circulationaha.106.643791

Oceandy D, Mohamed TM, Cartwright EJ, Neyses L (2011) Local signals with global impacts and clinical implications: lessons from the plasma membrane calcium pump (PMCA4). Biochim Biophys Acta 1813(5):974–978. doi:10.1016/j.bbamcr.2010.12.007

Ohkura M, Matsuzaki M, Kasai H, Imoto K, Nakai J (2005) Genetically encoded bright Ca2+ probe applicable for dynamic Ca2+ imaging of dendritic spines. Anal Chem 77(18):5861–5869. doi:10.1021/ac0506837

Pedersen PL, Carafoli E (1987) Ion motive ATPases. I. Ubiquity, properties, and significance to cell function. Trends Biochem Sci 12:146–150

Rimessi A, Coletto L, Pinton P, Rizzuto R, Brini M, Carafoli E (2005) Inhibitory interaction of the 14-3-3 protein with isoform 4 of the plasma membrane Ca(2+)-ATPase pump. J Biol Chem 280(44):37195–37203. doi:10.1074/jbc.M504921200

Schatzmann HJ (1966) ATP-dependent Ca++ – extrusion from human red cells. Experientia 22(6):364–365

Schuh K, Uldrijan S, Telkamp M, Rothlein N, Neyses L (2001) The plasma membrane calmodulin-dependent calcium pump: a major regulator of nitric oxide synthase I. J Cell Biol 155(2): 201–205. doi:10.1083/jcb.200104131

Schuh K, Quaschning T, Knauer S, Hu K, Kocak S, Roethlein N, Neyses L (2003a) Regulation of vascular tone in animals overexpressing the sarcolemmal calcium pump. J Biol Chem 278(42):41246–41252. doi:10.1074/jbc.M307606200

Schuh K, Uldrijan S, Gambaryan S, Roethlein N, Neyses L (2003b) Interaction of the plasma membrane Ca2+ pump 4b/CI with the Ca2+/calmodulin-dependent membrane-associated kinase CASK. J Biol Chem 278(11):9778–9783. doi:10.1074/jbc.M212507200

Schuh K, Cartwright EJ, Jankevics E, Bundschu K, Liebermann J, Williams JC, Armesilla AL, Emerson M, Oceandy D, Knobeloch KP, Neyses L (2004) Plasma membrane Ca2+ ATPase 4 is required for sperm motility and male fertility. J Biol Chem 279(27):28220–28226. doi:10.1074/jbc.M312599200

Sgambato-Faure V, Xiong Y, Berke JD, Hyman SE, Strehler EE (2006) The Homer-1 protein Ania-3 interacts with the plasma membrane calcium pump. Biochem Biophys Res Commun 343(2):630–637. doi:10.1016/j.bbrc.2006.03.020

Shang W, Lu F, Sun T, Xu J, Li LL, Wang Y, Wang G, Chen L, Wang X, Cannell MB, Wang SQ, Cheng H (2014) Imaging Ca2+ nanosparks in heart with a new targeted biosensor. Circ Res 114(3):412–420. doi:10.1161/circresaha.114.302938

Souders CA, Bowers SL, Baudino TA (2009) Cardiac fibroblast: the renaissance cell. Circ Res 105(12):1164–1176. doi:10.1161/circresaha.109.209809

Stauffer TP, Hilfiker H, Carafoli E, Strehler EE (1993) Quantitative analysis of alternative splicing options of human plasma membrane calcium pump genes. J Biol Chem 268(34):25993–26003

Strehler EE (1991) Recent advances in the molecular characterization of plasma membrane Ca2+ pumps. J Membr Biol 120(1):1–15

Strehler EE, Zacharias DA (2001) Role of alternative splicing in generating isoform diversity among plasma membrane calcium pumps. Physiol Rev 81(1):21–50

Tallini YN, Ohkura M, Choi BR, Ji G, Imoto K, Doran R, Lee J, Plan P, Wilson J, Xin HB, Sanbe A, Gulick J, Mathai J, Robbins J, Salama G, Nakai J, Kotlikoff MI (2006) Imaging cellular signals in the heart in vivo: cardiac expression of the high-signal Ca2+ indicator GCaMP2. Proc Natl Acad Sci U S A 103(12):4753–4758. doi:10.1073/pnas.0509378103

Tay LH, Dick IE, Yang W, Mank M, Griesbeck O, Yue DT (2012) Nanodomain Ca(2)(+) of Ca(2)(+) channels detected by a tethered genetically encoded Ca(2)(+) sensor. Nat Commun 3:778. doi:10.1038/ncomms1777

Timmann C, Thye T, Vens M, Evans J, May J, Ehmen C, Sievertsen J, Muntau B, Ruge G, Loag W, Ansong D, Antwi S, Asafo-Adjei E, Nguah SB, Kwakye KO, Akoto AO, Sylverken J, Brendel M, Schuldt K, Loley C, Franke A, Meyer CG, Agbenyega T, Ziegler A, Horstmann RD (2012) Genome-wide association study indicates two novel resistance loci for severe malaria. Nature 489(7416):443–446. doi:10.1038/nature11334

Trafford AW, Clarke JD, Richards MA, Eisner DA, Dibb KM (2013) Calcium signalling microdomains and the t-tubular system in atrial myocytes: potential roles in cardiac disease and arrhythmias. Cardiovasc Res 98(2):192–203. doi:10.1093/cvr/cvt018

Tsui AK, Marsden PA, Mazer CD, Adamson SL, Henkelman RM, Ho JJ, Wilson DF, Heximer SP, Connelly KA, Bolz SS, Lidington D, El-Beheiry MH, Dattani ND, Chen KM, Hare GM (2011) Priming of hypoxia-inducible factor by neuronal nitric oxide synthase is essential for adaptive responses to severe anemia. Proc Natl Acad Sci U S A 108(42):17544–17549. doi:10.1073/pnas.1114026108

Ueda K, Valdivia C, Medeiros-Domingo A, Tester DJ, Vatta M, Farrugia G, Ackerman MJ, Makielski JC (2008) Syntrophin mutation associated with long QT syndrome through activation of the nNOS-SCN5A macromolecular complex. Proc Natl Acad Sci U S A 105(27):9355–9360. doi:10.1073/pnas.0801294105

van der Eerden BC, Weissgerber P, Fratzl-Zelman N, Olausson J, Hoenderop JG, Schreuders-Koedam M, Eijken M, Roschger P, de Vries TJ, Chiba H, Klaushofer K, Flockerzi V, Bindels RJ, Freichel M, van Leeuwen JP (2012) The transient receptor potential channel TRPV6 is dynamically expressed in bone cells but is not crucial for bone mineralization in mice. J Cell Physiol 227(5):1951–1959. doi:10.1002/jcp.22923

Williams JC, Armesilla AL, Mohamed TM, Hagarty CL, McIntyre FH, Schomburg S, Zaki AO, Oceandy D, Cartwright EJ, Buch MH, Emerson M, Neyses L (2006) The sarcolemmal calcium pump, alpha-1 syntrophin, and neuronal nitric-oxide synthase are parts of a macromolecular protein complex. J Biol Chem 281(33):23341–23348. doi:10.1074/jbc.M513341200

Wu X, Zhang T, Bossuyt J, Li X, McKinsey TA, Dedman JR, Olson EN, Chen J, Brown JH, Bers DM (2006) Local InsP3-dependent perinuclear Ca2+ signaling in cardiac myocyte excitation-transcription coupling. J Clin Invest 116(3):675–682. doi:10.1172/jci27374

Wu X, Chang B, Blair NS, Sargent M, York AJ, Robbins J, Shull GE, Molkentin JD (2009) Plasma membrane Ca2 + −ATPase isoform 4 antagonizes cardiac hypertrophy in association with calcineurin inhibition in rodents. J Clin Invest 119(4):976–985. doi:10.1172/jci36693

Zaccolo M (2006) Phosphodiesterases and compartmentalized cAMP signalling in the heart. Eur J Cell Biol 85(7):693–697. doi:10.1016/j.ejcb.2006.01.002

Zacharias DA, Kappen C (1999) Developmental expression of the four plasma membrane calcium ATPase (Pmca) genes in the mouse. Biochim Biophys Acta 1428(2–3):397–405

Zhang YH, Casadei B (2012) Sub-cellular targeting of constitutive NOS in health and disease. J Mol Cell Cardiol 52(2):341–350. doi:10.1016/j.yjmcc.2011.09.006

Calcium Microdomains in Cardiac Cells

17

A.M. Gómez, T.R.R. Mesquita, J.J. Mercadier, J.L. Álvarez, and J.P. Benitah

Contents

Abstract

Calcium (Ca^{2+}) is a universal intracellular second messenger. In the heart, it plays a key role by activating contraction through the excitation-contraction coupling (***EC coupling***) mechanism. Although this is its key role in the heart, Ca^{2+} has other important functions, not only being involved in cell growth (in the heart named excitation-transcription coupling, ***ET coupling***) but also in mitochondrial

A.M. Gómez (✉) • T.R.R. Mesquita • J.P. Benitah
Inserm, UMR-S 1180, Univ. Paris-Sud, Université Paris-Saclay, F-92296 Châtenay-Malabry, France
e-mail: ana-maria.gomez@inserm.fr

J.J. Mercadier
Inserm, UMR-S 1180, Univ. Paris-Sud, Université Paris-Saclay, F-92296 Châtenay-Malabry, France

Univ. Paris Diderot, Sorbonne Paris Cité and Assistance Publique—Hôpitaux de Paris, F-75018 Paris, France

J.L. Álvarez
Inserm, UMR-S 1180, Univ. Paris-Sud, Université Paris-Saclay, F-92296 Châtenay-Malabry, France

Laboratorio de Electrofisiología, Instituto de Cardiología y Cirugía Cardiovascular, La Habana, Cuba

V. Nikolaev, M. Zaccolo (eds.), *Microdomains in the Cardiovascular System*, Cardiac and Vascular Biology 3, DOI 10.1007/978-3-319-54579-0_17

function (excitation-metabolism coupling, ***EM coupling***) and cell **death**. Moreover, as Ca^{2+} is electrically charged, its movement across membranes generates an electrical current, which is important in cardiomyocyte electrophysiology and, if disturbed, may be involved in **arrhythmias**. The cardiac myocyte may discriminate between Ca^{2+} signals by creating "spaces" where Ca^{2+} diffusion is limited, creating gradients of $[Ca^{2+}]_i$ at the micrometer scale, which are named microdomains. They are maintained by the cellular architecture and location of Ca^{2+}-handling proteins and buffers.

Ca^{2+} is highly compartmentalized by membranes in the cardiac myocyte. In the extracellular media, $[Ca^{2+}]$ is over 1 mM (around 1.5 mM), while in the cytoplasm, it is about 100 nM during cell rest (diastole). In the plasma membrane are located Ca^{2+} channels, both voltage dependent and independent, which let Ca^{2+} pass according to its concentration gradient when opened, the Na^+/Ca^{2+} exchanger (NCX) which lets Ca^{2+} go either way depending on Na^+ and Ca^{2+} concentration gradients and membrane potential, and the plasmalemmal Ca^{2+} pump (PMCA), which extrudes Ca^{2+} out of the cell. Cytosolic $[Ca^{2+}]_i$ increases at each heartbeat up to around 1 μM, activating contraction, but the sarcolemmal Ca^{2+} channels do not provide as much (see below). Inside the cell, there are some organelles where $[Ca^{2+}]$ can be also much higher than in the cytoplasm. The sarcoplasmic reticulum (SR) is the major Ca^{2+} reservoir in the cardiomyocyte. $[Ca^{2+}]$ inside can reach about 1 mM. Other organelles can also store Ca^{2+} such as the mitochondria and nuclear envelope (Bers 2008). Ca^{2+}-release channels (ryanodine receptors, RyR_2) are located on the SR membrane; they release Ca^{2+} when activated by Ca^{2+}. Other intracellular Ca^{2+}-release channels are expressed in cardiomyocytes such as the IP_3 receptors (IP_3R); although in adult cardiomyocytes, they are restricted mainly to the nuclear envelope, participating in activation of gene transcription. However, IP_3R may also be detected in the SR, mostly in pathological states (Harzheim et al. 2009; Nakayama et al. 2010; Wehrens et al. 2005). The NCX is also found in the mitochondria, together with the mitochondrial Ca^{2+} uniporter (MCU). The MCU lets Ca^{2+} enter into the mitochondria from the cytosol and is involved in its function (Finkel et al. 2015). Thus Ca^{2+} is highly compartmentalized in the cardiomyocyte, which is achieved by cell architecture. The location of Ca^{2+} channels and transporters in these compartments, together with Ca^{2+} buffers, creates Ca^{2+} microdomains, where $[Ca^{2+}]$ differs from the bulk cytosol during short periods of time, the size of these gradients extending in the μm range. In this chapter, we will focus on three mechanisms involving Ca^{2+} microdomains: EC coupling, ET coupling, and EM coupling.

17.1 EC Coupling

EC coupling is the process by which electrical excitation activates contraction. As a consequence of using tap water for cardiac contraction experiments, S. Ringer found that Ca^{2+} is needed for heart contraction (Ringer 1883). Lately, Fabiato and

Fabiato found that Ca^{2+} activates cardiac contraction by the mechanism of Ca^{2+}-induced Ca^{2+} release (CICR) (Fabiato and Fabiato 1975). The principle of this mechanism is as follows: in each heartbeat, membrane depolarization during an action potential (AP) activates the sarcolemmal Ca^{2+} channels (mainly L-type, LTCC), which let some Ca^{2+} enter into the cell. This produces only a slight increase in $[Ca^{2+}]_i$, which is not enough to activate contraction. However, it activates the Ca^{2+}-release channel, known as ryanodine receptor (type 2 in the heart, RyR_2). In fact, the number and conductance of RyR_2 channels are at least five times bigger than those of the LTCC (Bers 2008; Fill and Copello 2002), resulting in an amplification of the initial Ca^{2+} signal and providing enough Ca^{2+} to activate contraction. This amplification factor may be also referred to as "gain." Because LTCCs are scarce and have low conductance, their activation results in only slight increase of bulk cytoplasmic $[Ca^{2+}]_i$. If it were global cytosolic $[Ca^{2+}]_i$ that activates RyR_2, it would imply that RyR_2 needs to be highly Ca^{2+} sensitive. This would result in the problem of providing an all-or-none response and being highly unstable. On these considerations, a model was proposed by Stern in 1992 (Stern 1992) where the Ca^{2+} release is not controlled by global cytosolic $[Ca^{2+}]_i$ but by local $[Ca^{2+}]$ limited in time and space. In this model, the RyR_2 is less sensitive to Ca^{2+}, but by being located close to the Ca^{2+} entry channels, it can sense much higher $[Ca^{2+}]$ and be activated providing amplification and stability. This is the *local control theory* of EC coupling, which necessitates Ca^{2+} microdomains, where temporally and spatially $[Ca^{2+}]$ may show big differences with the bulk cytosol.

This local control theory is possible owing to the cell architecture, which locates the LTCC in close vicinity to RyR_2. In ventricular cardiomyocytes, the sarcolemma is invaginated in a network of transverse tubules (TT). In its surface are found LTCCs. The junctional SR is located in close vicinity to these TT (10–15 nm), making possible the local control of EC coupling. Thus, protein location and geometry support the Ca^{2+} microdomain at the dyad (Tanskanen et al. 2007). Figure 17.1 represents an optical section of a membrane-labeled mouse ventricular cardiomyocyte taken by two-photon microscopy. As it can be seen, besides the cell periphery, there are fluorescence dots deep inside the cell, which correspond to the TT. Owing to these TT, the LTCCs (voltage-dependent Ca^{2+} entry channels) bring Ca^{2+} close to the Ca^{2+}-release channels, the RyR_2, which may sense high $[Ca^{2+}]_i$ before it diffuses away. In non-ventricular cardiomyocytes, without or with few TT, Ca^{2+} release is initiated at the cell periphery by the LTCC and neighbor RyR_2 and then propagated deep into the cell from a Ca^{2+} release unit (formed by a cluster of RyR_2) to the next one in a fire-diffuse-fire way.

The local control theory of EC coupling was functionally validated with the visualization of Ca^{2+} sparks 1 year later (Cheng et al. 1993), as the elementary events whose temporal and spatial summation produce the global $[Ca^{2+}]_i$ transient. Ca^{2+} sparks are seen as local elevation in $[Ca^{2+}]_i$ which expands about 2 μm and lasts about 30 ms when visualized with confocal microscopy and Ca^{2+} fluorescence dyes (Cheng and Lederer 2008). Ca^{2+} sparks can occur spontaneously in a quiescent myocyte, stochastically. Their frequency in quiescent cells depends on the open probability of RyR_2 at diastolic $[Ca^{2+}]_i$. Although they do not need LTCC openings to occur, they may be influenced by some spontaneous LTCC

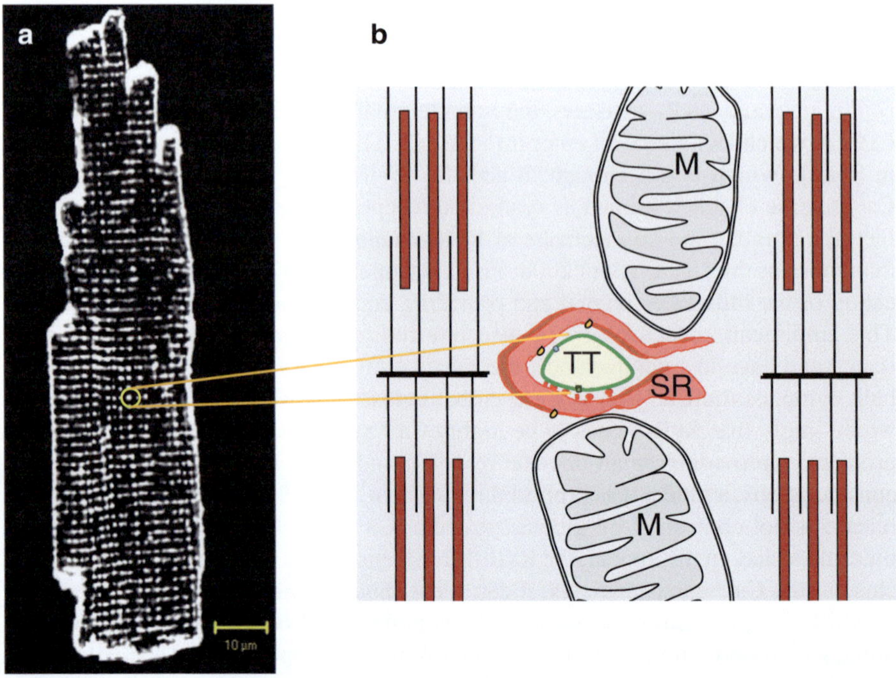

Fig. 17.1 Dyad structure. (**a**) two-photon microscopy image of an adult mouse ventricular cardiomyocyte labeled with the voltage-sensitive dye di-4-ANEPPS. (**b**) Cartoon of the structure of a cardiac dyad with transverse tubule (TT), junctional sarcoplasmic reticulum (SR), and mitochondria (M)

openings. Even if the LTCC open probability at resting potential is very low, it is not null, and the single-channel current is high at negative potentials, due to the high electrochemical gradient. Thus, the relationship of the Ca^{2+} spark probability normalized to the LTCC current (I_{CaL}) as a function of voltage resembles the shape of the Nernst-Planck equation, the Ca^{2+} spark probability being highest at more negative potentials, where the LTCC unitary current is bigger (Santana et al. 1996). Supporting this finding, in conditions where LTCC expression is augmented, the frequency of spontaneous Ca^{2+} sparks increases (Benitah et al. 2001). However, the main factor determining spontaneous Ca^{2+} sparks is the open probability of the RyR_2, which depends on $[Ca^{2+}]$ on both sides of the SR membrane as the RyR_2 is sensitive to both cytosolic and luminal Ca^{2+} (Gyorke and Terentyev 2008; Pessah et al. 1985). Ca^{2+} sparks constitute a Ca^{2+} microdomain, as they are $[Ca^{2+}]_i$ gradients confined within a volume with µm dimension. Their maximum amplitude, which corresponds to the maximum $[Ca^{2+}]$, is determined by the number of RyR_2 in a cluster, also known as calcium release unit (CRU), the single-channel current (which also depends on the amount of Ca^{2+}

stored in the SR), and the coupled gating within individual RyR_2. The maximum amplitude thus corresponds to the release site and is located on the Z lines, where the RyR_2 aligns (Cheng et al. 1996; Gómez et al. 1997). Ca^{2+} spark duration depends mainly on diffusion, although Ca^{2+} reuptake by the SR Ca^{2+}-ATPase (SERCA) also contributes (Gómez et al. 1996). Spontaneous Ca^{2+} sparks (during diastolic periods) are very rare in the working heart and appear in conditions of Ca^{2+} overload. During the normal cardiomyocyte cycle, Ca^{2+} sparks are the elementary events constituting a $[Ca^{2+}]_i$ transient, and in this way, the number of evoked Ca^{2+} sparks in a twitch and their amplitude grades the global $[Ca^{2+}]_i$ transient and thus the force of contraction. For example, under β-adrenergic stimulation, phosphorylation of LTCC increases the Ca^{2+} entry, evoking more Ca^{2+} sparks. Individual Ca^{2+} sparks are also more prone to fire, as RyR_2 phosphorylation increases their response to Ca^{2+} (Valdivia et al. 1995). Moreover, phospholamban phosphorylation releases its inhibitory action on SERCA, thus increasing Ca^{2+} stored in the SR, which both favors RyR_2 openings and enhances Ca^{2+} spark amplitude (Bers 2008). As explained earlier, the local control theory of EC coupling requires initial $[Ca^{2+}]_i$ gradients, which activate Ca^{2+} sparks. These local elevations in $[Ca^{2+}]_i$ produced by Ca^{2+} entry through LTCC have been named Ca^{2+} sparklets when visualized with confocal microscopy (Wang et al. 2001). The ability of voltage-independent Ca^{2+} permeating channels evoking Ca^{2+} sparks has not been studied, although it is thought that some transient receptor (TRP) channels or other store-operated channels (see below) may be involved in arrhythmia. Ca^{2+} sparks have their counterpart in the luminal side of the junctional SR because each Ca^{2+} spark produces a local depletion on $[Ca^{2+}]_{SR}$, which is also a Ca^{2+} microdomain, as the local $[Ca^{2+}]$ is different than the global, in this case lower. These local depletions in Ca^{2+} were evidenced by Brochet et al. in 2005 and named Ca^{2+} blinks (Brochet et al. 2005). Imaging Ca^{2+} microdomains at the dyadic space has evolved along with technical developments. They can be visualized by confocal microscopy and appropriate Ca^{2+} dyes (Sham 1997), which can be targeted to the dyadic space (Shang et al. 2014), using two-photon microscopy (Awasthi et al. 2016) or by electrophysiological measurements of I_{CaL} and NCX currents (Acsai et al. 2011).

Ca^{2+} sparks modulation. As building blocks of the global $[Ca^{2+}]_i$ transient that activates contraction, Ca^{2+} sparks have a key role in cardiac physiology, and their modulation has an important pathophysiological impact (Eisner et al. 2004). As they are evoked by I_{CaL} (Santana et al. 1996), any modulation of LTCC would affect Ca^{2+} sparks, and as they are modulated by $[Ca^{2+}]_{SR}$, any factor altering the amount of Ca^{2+} stored in the SR would also modulate Ca^{2+} sparks frequency and characteristics. We will mention here some factors that modulate this Ca^{2+} microdomain in the dyad by modulating RyR_2 function.

The RyR_2, at the Ca^{2+} spark origin, is a protein that can be modulated by many factors both at the cytosolic and luminal (intra SR) sides. It has a large cytosolic regulatory region which binds smaller proteins which can modulate its function.

Among these proteins, one can find not only the FK506-binding proteins (FKBP12 and 12.6, also known as calstabin 1 and 2), sorcin, and calmodulin but also kinases (PKA, CaMKII) and phosphatases (PP1 and PP2A) (O'Brien et al. 2015). Other proteins such as junctin, triadin, and calsequestrin bind directly or indirectly at the luminal side (O'Brien et al. 2015). The association of the RyR_2 to ancillary proteins has caused it to be called a macromolecular complex (Marx et al. 2000; O'Brien et al. 2015; Valdivia 1998; Wehrens et al. 2005). RyR_2 can also be modulated at least by phosphorylation, nitrosylation, and oxidation. The latter posttranslational modifications have been found in some cardiac diseases such as heart failure (HF) and diabetic cardiomyopathy, and it has been suggested that they may contribute to the pathogenesis of the disease. The location in the dyad of enzymes such as kinases and microscopic gradients of its soluble modulators is able to regulate this Ca^{2+} microdomain. Moreover, a number of mutations in the RyR_2 are related to arrhythmic diseases.

FKBP12 was shown to stabilize the RyR_1 (skeletal muscle type) in closed states and promote openings of full conductance (Brillantes et al. 1994). Another FKBP with close molecular weight to FKBP12 (FKBP12.6) was found to bind RyR_2 in dog heart (Lam et al. 1995), and it was thought to be specific for cardiac tissue (Timerman et al. 1996). Lately it was found that this specificity was species dependent, and today it is accepted that both FKBP12 and 12.6 bind to RyR_2 (Jeyakumar et al. 2001). It was first proposed that FKBP12.6 stabilizes the RyR_2 channel in its closed state and, when open, promotes coupled gating (Marx et al. 2000).This would limit Ca^{2+} leak during diastole and improve $[Ca^{2+}]_i$ transients. Thus, the FKBP12.6 binding to RyR_2 could be used as a therapeutic target for contractile dysfunction in HF and to prevent arrhythmias (see below). FK506 or rapamycin unbinds FKBP12/12.6 from the RyR_2 producing long openings and long-lasting Ca^{2+} sparks (Xiao et al. 1997), and the opposite was found with FKBP12.6 overexpression (Gellen et al. 2008; Gómez et al. 2004; Loughrey et al. 2004). Thus, these functional data in cardiomyocytes supported the first findings in single-channel data by the Marks' group. However, a study on single RyR_2 current has challenged this assumption (Galfre et al. 2012). This study showed that while FKBP12 reduces the open probability of the skeletal isoform of the RyR (RyR_1), it activates the cardiac isoform (RyR_2). FKBP12.6 effects on decreasing RyR_2 open probability would not be a direct effect, but the consequence of a competition with FKBP12, while not having major effect by itself (Galfre et al. 2012).

FKBP12.6 was proposed to unbind from the PKA-phosphorylated RyR_2 (Marx et al. 2000). This has been widely challenged (Guo et al. 2010; Jiang et al. 2002; Stange et al. 2003; Xiao et al. 2004). The depression in FKBP12.6 content in the failing heart (Stange et al. 2003) has been confirmed by other groups (Huang et al. 2006; Ono et al. 2000; Yano et al. 2000). This depression could be involved in the pathology of HF and arrhythmias (Gómez et al. 2009; Yano et al. 2006). In fact, FKBP12.6-KO mice show ventricular arrhythmia (Wehrens et al. 2004). Although some authors failed to see the link between FKBP12.6 and arrhythmia (Xiao et al. 2007), others found that FKBP12.6 overexpression protects the mice from stress-induced arrhythmias (Gellen et al. 2008; Vinet et al. 2012) and from maladaptive remodeling observed after Thoracic Aortic Constriction (Gellen et al. 2008; Vinet et al. 2012). On the other hand, the same mice failed to show positive effects in maladaptive HF remodeling in the myocardial infarction model (Bito et al. 2013). Thus, there is still more information needed, and the validity of stabilizing FKBP12.6 binding to the RyR2 as a therapeutic target is under study (Kushnir and Marks 2012).

Sorcin is a protein that can bind to RyR_2 (Meyers et al. 1995a; Zamparelli et al. 2000). Its action on RyR_2 could be similar to FKBP12.6 by depressing its open probability (Lokuta et al. 1997) and depressing the Ca^{2+} sparks occurrence (Farrell et al. 2003), but in a dynamic way as its binding depends on $[Ca^{2+}]$ in the dyad space (Meyers et al. 1995b). Sorcin also binds to LTCC (Meyers et al. 1998) reducing Ca^{2+} current (Farrell et al. 2003), so its potential to be used as therapeutic target has not been so much addressed. Nevertheless, Suarez et al. showed that sorcin overexpression protects from diabetic cardiomyopathy (Suarez et al. 2004).

Calmodulin (CaM) is a small ubiquitous protein that binds Ca^{2+}, acting as a Ca^{2+} sensor. It has different actions when it is not bound to Ca^{2+} (apo-CaM) than when it is bound. CaM may be bound to channels modulating them or be free in the microdomains, as in the dyad, where it senses Ca^{2+} variations, and the Ca^{2+}-CaM complex activates kinases (as Ca^{2+}-CaM Kinase, CaMK) and phosphatases (as calcineurin, Cn) (Saucerman and Bers 2012). In the dyad, CaM binds and modulates LTCCs and RyR_2. During EC coupling, Ca^{2+}-CaM inactivates the **LTCC** and modulates the I_{CaL} window current, limiting Ca^{2+} entry when dyadic $[Ca^{2+}]$ is high (Fernández-Velasco et al. 2011; Sham 1997). CaM may serve as a sensor for the cross talk between Ca^{2+} released by the RyR_2 and the LTCC, whose inactivation and window current depend on Ca^{2+} (Adachi-Akahane et al. 1996; Delgado et al. 1999; Fernández-Velasco et al. 2011). A number of excellent reviews discuss in detail the effects of CaM on LTCC (Ben-Johny and Yue 2014; Benitah et al. 2010). The CaM effects on LTCC with regard to the nuclear Ca^{2+} microdomain will be mentioned below. CaM binds to RyR_2 and inhibits it, stabilizing the channel in the closed state (Balshaw et al. 2001). Hypertrophic stimuli have been shown to unbind CaM from the RyR_2 in neonatal cardiomyocytes (Gangopadhyay and Ikemoto 2011). However, the best established effects of CaM on RyR_2 are through its activation of CaMKII and subsequent phosphorylation of the channel.

Phosphorylation. RyR_2 has many residues that can be phosphorylated. The main consensus residues are S2830 and S2808 (the amino acid number refers to the human sequence) as targets for PKA and S2814 for CaMKII (Huke and Bers 2008), although S2808 can also be phosphorylated by CaMKII (Rodriguez et al. 2003) and protein kinase G (Ho et al. 2016). Much has been discussed over the last decade about RyR_2 phosphorylation, mostly regarding pathology. In 1995, Valdivia et al. showed that **PKA** RyR_2 phosphorylation enhances its Ca^{2+} sensitivity and accelerates its adaptation (depression of RyR_2 response to a maintained Ca^{2+} stimulus, in the ms range) (Valdivia et al. 1995). Later on, a PKA hyperphosphorylation was pointed out as a major pathophysiological mechanism involved in HF and related to RyR_2 depletion of FKBP12.6 (Marx et al. 2000; Reiken et al. 2003). This finding was challenged by other groups (Benkusky et al. 2007; Houser 2014; Li et al. 2002; Xiao et al. 2004; Zhang et al. 2012), pointing to CaMKII as the main kinase involved in Ca^{2+} sparks' modulation (Guo et al. 2006) and in pathology (Ai et al. 2005; Valdivia 2012). Some authors have shown that RyR_2 is highly phosphorylated at S2808 in normal conditions, at 75% (what would make three out of four monomers phosphorylated). Some authors have shown that RyR_2 open probability is low and with normal gating at this basal high level of phosphorylation. When the channel is fully phosphorylated, it shows long openings, and when it is dephosphorylated, it increases its open probability (Carter et al. 2006). These data are consistent with findings from Terentyev et al. who showed that phosphatases activity enhanced Ca^{2+} sparks frequency. PKA-RyR_2 phosphorylation has been also shown to modulate luminal Ca^{2+} sensitivity (Ullrich et al. 2012).

Today most data point to RyR_2 phosphorylation at the **CaMKII** site as being involved in pathology, both HF and arrhythmias. In vitro analyses have shown that RyR_2 phosphorylation by CaMKII enhances Ca^{2+} sparks occurrence (Erickson et al. 2013) and that RyR_2 phosphorylation at the CaMKII site is augmented in disease conditions, being involved in the pathophysiology of HF and arrhythmias (Ai et al. 2005; Sag et al. 2011; Valdivia 2012). In fact, the enhanced activity of the CaMKII-phosphorylated RyR_2 induces Ca^{2+} leak from the SR that would contribute to SR Ca^{2+} depletion. If less Ca^{2+} is stored at the SR, less Ca^{2+} is available to be released at each twitch, which produces weaker $[Ca^{2+}]_i$ transients and contractions in HF. Nevertheless, it is worth to keep in mind that HF is a complex disease, where many factors are altered, each with diverse consequences in EC coupling. Thus, incriminating the RyR_2 alteration as the unique defect in HF *cannot* be exact. In fact, some elegant experiments from the Eisner group showed that in healthy cardiomyocytes, enhancing or decreasing RyR_2 activity only produces transitory effects (Trafford et al. 2000).

Enhanced RyR_2 phosphorylation in HF could imply an increase in the Ca^{2+} sparks frequency. However, the RyR_2 is also sensitive to luminal (intra SR) Ca^{2+}, and its activity is strongly depressed when the SR Ca^{2+} content is decreased (Gyorke and Gyorke 1998), a situation that is associated with HF as a consequence of the reduction in SERCA expression and function and the enhancement of NCX expression (Egger et al. 2001; Zima et al. 2014). In fact, Ca^{2+} sparks occurrence shows a steep relationship with SR Ca^{2+} content (Bers 2014). Moreover other changes in the RyR_2 environment, such as a low ATP level, might contribute to depress RyR_2 activity. In fact, we found that evoked Ca^{2+} sparks were depressed in experimental HF (Gómez et al. 1997, 2001), and spontaneous Ca^{2+} sparks have been shown to be less frequent in human HF (Lindner et al. 2002). In a recent study, we showed that intact resting cardiomyocytes from failing mice present less Ca^{2+} sparks occurrence than control ones. However, the enhanced RyR_2 activity was unmasked in permeabilized cells where the cytosolic environment was equivalent in both cell types as well as the SR Ca^{2+} load (Ruiz-Hurtado et al. 2015). Besides, changes in the amount of dyads also affect Ca^{2+} signaling in this microdomain, resulting in altered EC coupling as it was first proposed (Benitah et al. 2002; Gómez et al. 1997, 2001) and then evidenced by the TT remodeling (Crocini et al. 2014; Pinali et al. 2013; Shah et al. 2014).

CaMKII phosphorylation of RyR_2 has also been involved in arrhythmia. Enhancing spontaneous Ca^{2+} release activates the NCX to extrude Ca^{2+} out of the cell, which is exchanged by Na^+. Because of the NCX stoichiometry, net inward (depolarizing current) is generated by NCX when working in forward mode (extruding Ca^{2+}) that can produce delayed afterdepolarizations (DADs) and if threshold for an AP is reached, a triggered activity (Bers 2014). One Ca^{2+} spark is not enough to induce any detectable alteration in membrane voltage, but when RyR_2 are more prone to open, there is also more probability that one Ca^{2+} spark activates neighboring Ca^{2+} release units, initiating a propagated Ca^{2+} wave (Cheng et al. 1996), which is big enough to produce DADs. Besides RyR_2 phosphorylation, this mechanism has been involved in arrhythmic genetic diseases that affect RyR_2 function, by either mutation in the RyR_2 itself or in its modulatory proteins (calsequestrin, triadin, CaM). These diseases are named catecholaminergic polymorphic ventricular tachycardia (CPVT) due to the role of the sympathetic nervous system in their triggering. In these conditions either by gain or loss of RyR_2 function, the probability of Ca^{2+} waves is enhanced (Fernandez-Velasco et al. 2009; Watanabe et al. 2009; Zhao et al. 2015). Mutations or alteration in macromolecular complexes of other channels, able to modify the dyadic Ca^{2+} microdomain, are also involved in cardiac arrhythmias (Abriel et al. 2015).

Oxidation, nitrosylation, and O-glcNAcylation (Erickson et al. 2013) as well as **pH** (Xu et al. 1996) are able to modulate RyR_2, among others. Moreover, there are gradients, making, for example, the redox potential in the dyad different than the bulk cytosol (Hanna et al. 2014), thus affecting Ca^{2+} microdomain at the dyad (Zhang et al. 2013). RyR_2 oxidation enhances RyR_2 activity and may be involved in HF (Terentyev et al. 2008) and ischemia-reperfusion damage, together with low pH (Calderon-Sanchez et al. 2011; Overend et al. 2001). RyR_2 oxidation can also be transiently produced by mechanical stress resulting in enhanced Ca^{2+} sparks occurrence (Prosser et al. 2011). Mechanical stress may also activate stretch-activated channels modifying Ca^{2+} microdomains (see below). Oxidation also interferes with RyR_2 nitrosylation, as it has been shown that inhibition of xanthine oxidase restores RyR_2 nitrosylation and rescues HF (Gonzalez et al. 2010). Other authors have shown that RyR_2 nitrosylation inhibits its activity and that RyR_2 is hyponitrosylated in HF (Gonzalez et al. 2010), while there is a benefic action of nNOS overexpression (Loyer et al. 2008). Recently, Becerra et al. have shown that preventing RyR_2 S-nitrosylation during reperfusion is involved in arrhythmias (Becerra et al. 2016).

Luminal RyR2-interacting proteins. As mentioned just above, mutations in calsequestrin and triadin are causative of CPVT. Calsequestrin is able to chelate Ca^{2+} in the SR, and triadin and junctin interact with both calsequestrin and RyR_2 modulating its luminal Ca^{2+} responsiveness (Chopra and Knollmann 2013; Gyorke and Terentyev 2008).

Stretch-activated channels (SACs). SACs are currently defined as nonselective and voltage-independent channels (Sasaki et al. 1992; Zhang et al. 2000). As the result of EC coupling, the heartbeat maintains a pulsatile physiological level of cellular tension; therefore, SACs may represent a cellular control of beat-to-beat tension via mechano-electric feedback in order to maintain the cardiac output (Lab

1996). Although the molecular identities of cardiac SACs are not fully characterized, there are several potential candidates and signaling pathways involved in the mechano-control of the Ca^{2+}-signaling network. In cardiac cells, SACs have been located at TT (Huang et al. 2009), caveolae (Kohl et al. 2003), and intercalated discs (Iwata et al. 2003). Thus, SACs may modulate functional units of EC coupling and Ca^{2+} microdomains in the cardiac dyad and non-dyad junctions.

According to electrophysiology recordings, SACs are permeant to Na^+, Ca^{2+}, and K^+. Thus, activation of SACs may cause an increase $[Na^+]_i$ and $[Ca^{2+}]_I$, although the increase in $[Ca^{2+}]_i$ may be indirect, as SACs have greater permeability for Na^+ than Ca^{2+}. Hence, they could indirectly increase $[Ca^{2+}]_i$ via the reverse-mode NCX (Alvarez et al. 1999; Baartscheer et al. 2003; Isenberg et al. 2003; Kamkin et al. 2003).

Beat-to-beat mechano-transduction pathway is essential to the maintenance of membrane integrity as well as TT network in cardiomyocytes. Iribe et al. (2009) have shown that physiological axial stretch (~8% of the cellular diastolic length), applied using carbon fibers, leads to a transient increase in the Ca^{2+} spark rate (~30% within 5 s), followed by a return to background levels within 1 min of maintained diastolic distension in whole-cell-stretch ventricular myocytes. Subsequently, the authors demonstrated that microtubules are located close to SR and TT membranes (to within 7 and 13 nm, respectively). Thereby, mechanical deformation of this functional complex causes an increase in the sensitivity of RyR_2 (Iribe et al. 2009). Similarly, Prosser et al. (2011) demonstrated that a stepwise stretch of single cardiomyocytes within the physiologic sarcomere range induces a rapid (~10 s) and reversible increase in the rate of Ca^{2+} spark occurrence (~30%). Importantly, Prosser et al. (2011) added valuable information regarding the contribution ROS generated by stretch-activated NOX2 in the sensitization of RyR_2, a mechanism named X-ROS signaling. Interestingly, although diastolic stretch of healthy cardiomyocytes induces a stretch-dependent "tuning" of RyR_2 and consequent Ca^{2+} spark burst, in mice cardiomyocytes from Duchenne muscular dystrophy, well known for exacerbated ROS generation, identical stretch was able to trigger arrhythmogenic Ca^{2+} waves. Consistently, the authors demonstrated that X-ROS signaling enhances RyR_2 Ca^{2+} sensitivity by recruiting a larger number of active individual CRU in the junctional SR and may underlie the increase in Ca^{2+} spark frequency.

Reactive nitrogen species have also been recognized as modulators of Ca^{2+} microdomains in ventricular myocytes (Petroff et al. 2001; Xu et al. 1998). Petroff et al. (2001) demonstrated that mechanical stretch of isolated cardiomyocytes, embedded in agarose-filled polyethylene tubing, directly modulates endogenous nitric oxide (NO) production (11% of increase) followed by increased $[Ca^{2+}]_i$ transient amplitude and myocyte length-dependent increase of Ca^{2+} sparks frequency (twofold and fourfold increase with 7% and 12% of stretch, respectively), being fully reversible when sarcomere stretch was removed. Moreover, these effects were fully abolished in cardiomyocytes derived from eNOS-deficient mice, suggesting that NO produced by the Pi3K/Akt/eNOS axis acts as a second messenger involved in the enhancement of RyR_2 activity.

17.2 ET Coupling

Excitation-transcription (ET) coupling refers to the process whereby plasma membrane depolarization or receptors stimulation lead to gene activation or inactivation, in which Ca^{2+} is a central player by directly or indirectly stimulating transcriptional factors or the chromatin structure. Mainly, Ca^{2+} acts in conjunction with CaM, activating either specific kinases or the phosphatases. Among the kinases one can find CaMKII, PKC, and the mitogen-activated protein kinases (MAPK), which phosphorylate transcription factors such as cyclic AMP response element-binding protein (CREB), histone deacetylase (HDAC), or nuclear factor kappa-light-chain-enhancer of activated B cells (NFκB), inducing their export from the nucleus. Ca^{2+} also activates the phosphatase calcineurin (Cn) (protein phosphatase 2B) leading to dephosphorylation of members of the nuclear factor of activated T-cells (NFAT) transcription factor family, the myocyte enhancer factor-2 (MEF2) within the cytosol, which then translocate to the nucleus to regulate genes involved in cardiac hypertrophy (Heineke and Molkentin 2006). In addition, Ca^{2+} can also influence transcription directly through the transcriptional repressor, the downstream regulatory element agonist modulator (DREAM) (Naranjo and Mellstrom 2012). A cross talk between these direct and indirect pathways might also exist, since in cardiomyocytes Ca^{2+} activation of CaMKIIδ induced DREAM translocation to the nucleus (Ronkainen et al. 2011).

The Ca^{2+}-induced transcriptional responses are not all-or-none events but are shaped by the spatial, temporal, and amplitude characteristics of specific Ca^{2+} signaling, which are particularly important in the heart, taking into account that Ca^{2+} signals occur rhythmically in each heartbeat. Two nonexclusive mechanisms might be put forward in order to meet these needs. The first is the difference of sensitivity to Ca^{2+} of different targets. For example, Cn has a particularly high affinity for Ca^{2+}-CaM and is at least one order of magnitude more sensitive than CaMKII. As a consequence, Cn responds to sustained, low-amplitude Ca^{2+} rises, whereas CaMK-dependent gene expression is used preferentially in response to transient, high-amplitude Ca^{2+} increases. The second mechanism is the creation of high Ca^{2+} microdomain that is insulated from the Ca^{2+} fluctuation occurring at each heart contraction. Indeed, the control of the activity of specific transcriptional networks by Ca^{2+} is regulated by cytosolic and nuclear mechanisms. Thus, the Ca^{2+} entry site or its intracellular source makes the Ca^{2+} ions face different microdomains that are composed of specific sets of proteins and determines the biological outcome of the Ca^{2+} signal by inducing temporal and spatial changes in specific nuclear interactomes (Holton et al. 2010; Ong and Ambudkar 2011).

The LTCCs are the main source of extracellular calcium in the adult heart (Benitah et al. 2010) providing a critical link between cellular excitability and gene regulation (Dominguez-Rodriguez et al. 2012). In several forms of heart disease, increases in LTCC activity are a proximal trigger that activates pathological signaling cascades, including mitogen-activated protein kinases, PKC, and Cn (Vlahos et al. 2003). It has been shown (Chen et al. 2011) that increasing I_{CaL} is sufficient to activate the Cn-NFAT and CaMKII-HDAC signaling pathways, involving both cytosolic and SR/ER-nuclear envelop Ca^{2+} pools. There are two functionally distinct pools of CaM in the LTCC microdomain (Saucerman and Bers 2012). Firstly, there is CaM that is associated with the $Ca_V1.2$ C-terminus (the main pore-forming subunit of LTCC) as a virtual subunit and is the sensor for Ca^{2+}-dependent inactivation. Secondly, there is a local pool of signaling CaM that, when bound to Ca^{2+}, can be used to locally activate Ca^{2+}-CaM-dependent effectors such as Cn and CaMKII. As for CaM, there is evidence that both Cn (Tandan et al. 2009) and CaMKII (Hudmon et al. 2005) directly interact with N-terminal and C-terminal intracellular domains of cardiac Cav1.2 which appeared to be essential to activate the NFAT (Nichols et al. 2010; Tandan et al. 2009) and the CREB (Wheeler et al. 2008) pathways, respectively. Of interest this might be limited to distinct subpopulations of LTCCs that have been identified in cardiac myocytes (Best and Kamp 2012). Hence, it has been suggested that LTCCs housed in caveolae, away from EC coupling proteins, are a local microdomain for Ca^{2+} activation of Cn-NFAT signaling (Makarewich et al. 2012). Indeed, Cn has been demonstrated to associate with Cav1.2, AKAP5, and Cav-3 in ventricular myocytes (Nichols et al. 2010). On the other hand, there is evidence that Ca^{2+} influx through Cav1.2 triggers translocation of CaM from the cytosol to the nucleus. At rest, most intracellular CaM appears highly localized to the Z-line (or transverse tubule), and when $[Ca^{2+}]$ is elevated (from 100 to 500 nM), there is a significant translocation of CaM into the nucleus (Bossuyt and Bers 2013; Wu and Bers 2007). Moreover, it has been proposed that Ca^{2+}-CaM shuttling is controlled by the conformational rearrangement of the Cav1.2 C-terminal tail that provides the precise targeting of the Ca^{2+} signal transduction (Morad and Soldatov 2005) participating to a periodic CREB-dependent activity localized in spatially segregated nuclear microdomains and mediated in part by the Cav1.2-initiated CaMKII-insensitive mechanisms (Kobrinsky et al. 2011).

Besides Cav1.2, but more controversial, Cn-NFAT and CaMKII activation may take place in signaling microdomains also housing T-type Ca^{2+} channels (Chiang et al. 2009; Horiba et al. 2008) or transient receptor potential (TRP) channels (Bush et al. 2006; Eder and Molkentin 2011; Gao et al. 2012; Kuwahara et al. 2006; Makarewich et al. 2014; Nakayama et al. 2006; Onohara et al. 2006; Poteser et al. 2011; Wu et al. 2010). Interestingly, TRPC3 has been suggested to govern gene transcription by mechanisms involving a linkage to voltage-dependent Ca^{2+} entry (Onohara et al. 2006). Indeed, the organization of TRPC3/4/6 channels along with LTCC in caveolae membrane microdomains influences their ability to orchestrate Cn-NFAT signaling in conjunction with Na^+ entry via NHE-1 and induce Ca^{2+} entry via the NCX (Makarewich et al. 2014). Moreover, STIM1-dependent intracellular Ca^{2+} microdomains in the heart have been linked to higher Ca^{2+} influx associated with increased NFAT and CaMKII activity (Correll et al. 2015; Luo et al. 2012; Parks et al. 2016).

Although the source or microdomain of Ca^{2+} that regulates calcineurin-NFAT in the heart has not been fully determined, store-operated Ca^{2+} entry (SOCE) has been suggested as a potential source, able to create specialized pools of Ca^{2+} that ultimately affect Ca^{2+}-sensitive signaling proteins such as Cn (Hunton et al. 2004). Mechanistically, it was demonstrated that TRPC channels are engaged with Cn-NFAT signaling in the heart, and overexpression of TRPC3 (Nakayama et al. 2006) or upregulation of TRPC6 by pressure overload as well as in failing human hearts (Kuwahara et al. 2006) induces a greater Ca^{2+} entry and, consequently, pathologic cardiac hypertrophy via Ca^{2+}-activated Cn-NFAT signaling. Although TRPC channels do not represent the main SR Ca^{2+} refilling mechanism in healthy hearts, overexpression of TRPC3 or C6 leads to SR Ca^{2+} overload and causes a robust Ca^{2+} spark activity, which was partially inhibited by CaMKII inhibition (Makarewich et al. 2014). Moreover, the authors have demonstrated that TRPC 3/4/6 were found at caveolar domains, where they co-localize with LTCC and caveolin-3. Therefore, the molecular arrangement involving TRPC and LTCC channels at caveolar microdomains represents a restricted diffusional space of lipid rafts, which may have implications for local regulation of Ca^{2+}-sensitive proteins and signaling pathways resident in the caveolae.

Another Ca^{2+} microdomain that likely regulates the ET coupling mainly linked to activation of membrane receptors is at the nuclear envelope through a complex involving the **IP3R**, CaMKII, and HDAC5 and has been recently reviewed (Bers 2013; Ibarra et al. 2014; Ljubojevic and Bers 2015).

17.3 EM Coupling

The main sites for ATP production in the cardiac cell are the mitochondria. A picture has emerged over the years that mitochondria are not only simple "ATP factories" but instead that they constitute complex centers of control inside the cell. According to cell's conditions, in mitochondria the right chemicals to efficiently produce ATP are selected, and complex mechanisms such as fusion, division, and even apoptosis are controlled and performed. It is thus evident that mitochondrial function requires the integration of signals about nutrient availability, redox status, and overall cell health status. The precise knowledge of signaling mechanisms in mitochondria is fundamental to understand mitochondrial function and dysfunction in cardiac diseases in order to design therapies to minimize mitochondrial and cell damage.

The cardiac cycle is extremely energy demanding, and the main ATP hydrolysers are the Ca^{2+}-dependent myofilament ATPase and ion transport ATPases (e.g., the SR Ca-ATPase and the Na/K-ATPase). During each cardiac cycle, large amounts of Na^+ ions enter the cardiac cell, mainly, via the Na^+-Ca^{2+} exchanger (NCX; coupled to Ca^{2+} extrusion) and the sarcolemmal voltage-gated Na^+ channels. The resulting intracellular Na^+ accumulation requires an intense Na/K-ATPase activity to maintain the cell's thermodynamics steady state. Therefore, most of the ATP consumption could be regulated by Ca^{2+} or involved in Ca^{2+} transport mechanisms.

Mitochondria exhibit a high permeability to Ca^{2+} (Gunter and Pfeiffer 1990), and fundamental enzymes in the tricarboxylic acid cycle are known to be Ca^{2+} sensitive. It is possible then to consider that metabolic signaling by Ca^{2+} may help to explain the control of mitochondrial oxidative phosphorylation process. Indeed, there is evidence suggesting that intramitochondrial Ca^{2+} concentration ($[Ca^{2+}]_{Mit}$) might regulate various sites of the electron transport chain and the mitochondrial ATP synthase (Balaban et al. 2003; Cali et al. 2012; Denton and McCormack 1990; Territo et al. 2000). Thus, mitochondrial Ca^{2+} is important to provide the necessary feedback between energy supply and demand. In fact, when cardiac workload is increased due to augmented Ca^{2+} transients (amplitude or frequency), changes in [NADH] (abruptly decreased due to ATP consumption) match the time course of the slow rise in $[Ca^{2+}]_{Mit}$ (Brandes and Bers 1997, 2002). This finding suggests that $[Ca^{2+}]_{Mit}$-dependent activation of dehydrogenases could play an important role in energy supply as a function of demand during (i.e., inotropic mechanisms). Whether variations in $[Ca^{2+}]_{Mit}$ are fast in a beat-to-beat fashion and are energetically feasible (favorable) has been a matter of debate (Drago et al. 2012; Huser et al. 2000; O'Rourke and Blatter 2009; Williams et al. 2013). In any case, due to their ability to sense the cytosolic Ca^{2+} fluctuations and by balancing their rates of Ca^{2+} uptake and release, mitochondria are far from being passive Ca^{2+} sinks. Thus mitochondria cooperate in the control of changes in intracellular Ca^{2+} concentration to fine-tune the Ca^{2+}-regulated processes (Brini and Carafoli 2000). This is especially important to adequately handle opposite processes such as energy production or apoptotic cell death (Cali et al. 2012; Williams et al. 2013). The Ca^{2+} uptake system of mitochondria has a low affinity. However, the presence of well-defined, localized Ca^{2+} microdomains allows the mitochondria to rapidly and efficiently sense variations in $[Ca^{2+}]$ in these regions. Such Ca^{2+} microdomains are achieved by a close proximity of mitochondrial Ca^{2+} uptake sites to specific structures like the IP_3 receptors and RyR_2 in the endo-(sarco)plasmic reticulum (Rizzuto et al. 1993) and voltage-gated Ca^{2+} channels at the plasma membrane (Cali et al. 2012; Hoth et al. 2000; Malli et al. 2003); for review, see (Bers 2014; Rizzuto and Pozzan 2006).

Control of $[Ca^{2+}]_{Mit}$ in cardiomyocytes. At the resting state, cytosolic Ca^{2+} ($[Ca^{2+}]_{in}$) concentration is around 100 nM. This concentration is achieved through an active Ca^{2+} extrusion via the Ca^{2+} ATPases at the sarcolemmal and sarcoplasmic reticulum membranes and the NCX exchanger (Bers 2008). During a normal cardiac cycle, two main mechanisms increase intracellular Ca^{2+}: activation of the LTCC (mainly distributed at the T-tubules) and the Ca^{2+}-induced Ca^{2+}-release mechanism via the RyR_2 at the SR. Other mechanisms, such as activation of TRP channels or store-operated calcium entry mechanisms, can also contribute to increase $[Ca^{2+}]_{in}$ (see above). Within 500 ms, cytosolic Ca^{2+} is "pumped back" by the SERCA and extruded by the NCX (Cheng and Lederer 2008), but the increases in $[Ca^{2+}]_{in}$ can, in turn, raise microdomain $[Ca^{2+}]$ large enough (estimated in tens of micromoles) to induce Ca^{2+} uptake by the mitochondria (Rizzuto and Pozzan 2006). As seen in the Fig. 17.1 scheme, mitochondria are close enough to the Ca^{2+} dyad to sense a Ca^{2+} spark. In fact, in myotubes, miniature $[Ca^{2+}]_{Mit}$ transients have been recorded, which correspond to Ca^{2+} sparks, although they last longer (Pacher et al. 2002). Even if the mitochondrial Ca^{2+} uptake mechanisms exhibit a surprisingly low affinity ($K_D \approx 10$–20 mM), the close proximity with microdomains makes the mitochondria to undergo rapid changes in $[Ca^{2+}]_{Mit}$, comparable to those of $[Ca^{2+}]_i$ upon SR Ca^{2+} release (Maack et al. 2006; Robert et al. 2001; Sharma et al. 2000; Trollinger et al. 1997). However, the exposure time to high local $[Ca^{2+}]$ in microdomains during a heartbeat could be very short [≈ 10 ms in the heart; (Cheng and Lederer 2008)], and the Ca^{2+} fluxes due to NCX and Ca-ATPases are significantly larger than mitochondrial Ca^{2+} uptake suggesting that mitochondrial fluxes play a minor role in $[Ca^{2+}]_i$ dynamics (Andrienko et al. 2009; Bassani et al. 1994; Williams et al. 2013). Nonetheless, the small mitochondrial Ca^{2+} fluxes can play a critical role in regulating mitochondrial function. Yet, a substantial Ca^{2+} load in mitochondria can be achieved by increasing the frequency or the amplitude of Ca^{2+} transients during cardiac activity or when $[Ca^{2+}]_i$ levels are chronically elevated (Bers 2008).

Several important mechanisms are involved in mitochondrial Ca^{2+} load: the mitochondrial Ca^{2+} uniporter (MCU) and the Na^+-Ca^{2+} antiporter (one of the two Ca^{2+} extrusion system) which have been molecularly identified (Baughman et al. 2011; De Stefani et al. 2011, 2016; Palty et al. 2010). The MCU is a multiprotein complex composed of MCU and MCUb (as channel pore-forming subunits), the "gatekeepers" MICU1 and MICU2 proteins and EMRE, and a Ca^{2+} sensor and channel regulator that keep attached MICU1 and MICU2 (De Stefani et al. 2016; Sancak et al. 2013; Vais et al. 2016). The MCU, proposed to be a gated and highly selective ion channel (≈ 6–7 pS conductance at 105 mM $[Ca^{2+}]$; (De Stefani et al. 2011; De Stefani et al. 2016; Kirichok et al. 2004; Saris and Allshire 1989), and the so-called rapid mode (RaM) of uptake activated at lower Ca^{2+} concentrations (Bazil and Dash 2011; Sparagna et al. 1995) are responsible for mitochondrial Ca^{2+} uptake. Ca^{2+} efflux is accomplished by a Na^+-dependent efflux (associated with the activity of the $3Na^+/Ca^{2+}$ exchanger) and a Na^+-independent Ca^{2+} efflux mechanism associated to the $2H^+/Ca^{2+}$ antiporter (of still debated nature), whose action is coupled to the proton-motive force developed by the respiratory chain (Carafoli et al. 1974).

During permeation, Ca^{2+} must overcome two barriers with different ion permeabilities: the outer (OMM) and the inner (IMM) mitochondrial membranes. The OMM has been considered loosely permeable to Ca^{2+}; however, recent evidence suggests that its permeability is regulated by the voltage-dependent anion-selective channels VDAC (Cali et al. 2012; Szabadkai et al. 2006). The IMM is the location where the oxidative phosphorylation occurs and is an impermeable barrier to the free diffusion of ions and metabolites where the proteins responsible for Ca^{2+} transport (influx and efflux from the matrix) are expressed. Mitochondrial Ca^{2+} uptake is an electrogenic process that occurs down a large electrochemical gradient set up by proton extrusion (Nicholls and Crompton 1980). The translocation of H^+ from the matrix to the intermembrane space coupled to electron fluxes between the complexes of the respiratory chain generates a ≈ −180 mV membrane potential at the IMM (inside-vs.-cytosolic potential); diastolic $[Ca^{2+}]_{Mit}$ is similar to $[Ca^{2+}]_i$ (Mitchell 1961) which is the driving force for Ca^{2+} entry. Taking into account that free $[Ca^{2+}]_i$ is around 100 nM, for an equilibrium $[Ca^{2+}]_{Mit}$ should be at 0.1 M Ca^{2+}, but this condition is never reached since Ca^{2+} uptake is matched by Ca^{2+} efflux via distinct pathways.

We already mentioned that an increased mitochondrial Ca^{2+} load can be generated by increasing either the frequency or amplitude of Ca^{2+} transients. Furthermore, Ca^{2+} extrusion is up a large electrochemical driving force for Ca^{2+} entry (−180 mV inside-vs.-cytosolic potential; diastolic $[Ca^{2+}]_{Mit}$ is similar to $[Ca^{2+}]_i$), and it is thus very expensive energetically. Thus, even if beat-to-beat changes in $[Ca]_{Mit}$ are small, cumulative changes in $[Ca]_{Mit}$ over several or tens of seconds may cause regulatory changes in ATP production. Nonetheless, Ca^{2+} uptake via the MCU can be dramatically increased when $[Ca^{2+}]_i$ levels (and hence, $[Ca^{2+}]$ in the associated microdomains) are chronically elevated to greater than 1 μM ($K_D \approx 30$ μM), although this activation seems to take time at high $[Ca^{2+}]_i$. This implies that Ca^{2+} uptake, via MCU during normal excitation-contraction coupling, allows more Ca^{2+} to be used by the contractile machinery. Instead, during cellular Ca^{2+} overload, mitochondria are able to accumulate large amounts of Ca^{2+} to protect the cytosol from Ca^{2+} overload and may influence $[Ca^{2+}]_{in}$ dynamics (Bers 2008; Williams et al. 2013). However, the idea of beat-to-beat dynamic changes of $[Ca]_{Mit}$ is far from being definitively overruled. In a series of elegant experiments, it has been demonstrated that direct activation of I_{CaL} can increase mitochondrial calcium uptake (Viola et al. 2013). As a result, mitochondrial NADH production, oxygen consumption, and reactive oxygen species production are also increased. The existence of Ca^{2+} microdomains between LTCC and mitochondria could be sufficient to explain these findings. But the interplay between LTCCs and mitochondria appears to be more complex than the simple existence of variations of $[Ca^{2+}]$ in a restricted Ca^{2+} microdomain. Conformational changes in the Ca^{2+} channel during activation and inactivation seem to regulate the mitochondrial membrane potential and ATP/ADP trafficking via cytoskeletal proteins. It was proposed that the Ca^{2+} channel can regulate mitochondrial function through cytoskeletal proteins as a result of transmission of movement from the β_2 subunit of the channel and that an association between cytoskeletal proteins and the mitochondrial voltage-dependent anion channel (VDAC) may play a role in this response (Viola et al. 2007, 2009, 2013; Viola and Hool 2013, 2014). Moreover, since ROS and Ca^{2+} participate as partners in pathological remodeling and in the development of cardiac hypertrophy, such a coordinated interaction between L-type Ca^{2+} channels and mitochondria could be crucial in hypertrophic and apoptotic signaling pathways. An increase in cytoplasmic calcium can potentiate cellular oxidative stress via effects on mitochondrial metabolism (Viola et al. 2007; Viola and Hool 2011).

17.3.1 Cardiomyocyte Death by Necrosis and Apoptosis

Mitochondrial Ca^{2+} overload can be involved in both necrosis and apoptosis. Under pathological conditions such as ischemia-reperfusion, Ca^{2+} overload of mitochondria may limit myocyte contracture and activation of calpains, thus preventing cell damage and arrhythmias. However, at the same time, Ca^{2+} uptake can depolarize mitochondria, and ATP production can be decreased since proton influx is coupled to ATP synthesis (see Bers 2008). When $[Ca]_{Mit}$ is extremely high, the mitochondrial permeability transition pore (MPTP) is activated, allowing molecules of up to ~1500 Da to flow across the mitochondrial membrane. The MPTP is a multimeric-protein complex whose exact molecular properties are yet to be determined although evidence exists that MPTP forms from F-ATP synthase [see for review (Bernardi and Di Lisa 2015)]. The voltage-dependent anion channel (VDAC), the adenine nucleotide translocase (ANT), and the cyclophilin D (Baines 2009; Bernardi and Forte 2007) modify the MPTP sensitivity to Ca^{2+}, regardless of whether they take directly part in PTP formation. Increases in mitochondrial Ca^{2+} enhance MPTP opening due to the presence of a Ca^{2+}-binding site, which is competitively inhibited by other ions such as Mg^{2+}, Sr^{2+}, and Mn^{2+} (Nicolli et al. 1996). In addition to the Ca^{2+} increase in the mitochondrial matrix, other factors such as pH, adenine nucleotides, free radicals, and mitochondrial membrane potential modulate the opening of the MPTP. Mitochondrial Ca^{2+} overload and increases in ROS in the mitochondrial matrix are the "point of no return" that causes permeabilization of the inner mitochondrial membrane, decrease in the proton electrochemical gradient, ATP depletion, increased ROS production, organelle swelling, and hypercontracture of the cardiomyocyte. These events cause the release of cytochrome C and culminate in apoptotic cell death (Cali et al. 2012). Ca^{2+} overload during ischemia-reperfusion may result in calpains activation that may cleave essential contractile proteins such as TnI provoking the long-term contractile dysfunction characteristic of myocardial stunning (Barta et al. 2005; Bolli and Marban 1999); see (Bers 2008).

In conclusion, perhaps, no other organ uses calcium as heart does, participating, among others, to excitability, contractility, metabolism, and gene transcription. To achieve these multiple and varied functions, a fine control of Ca^{2+} homeostasis is mandatory, and its signals are decoded thanks to the existence of Ca^{2+} microdomains in different compartments. These Ca^{2+} microdomains are adapted to the development of highly specialized and sophisticated cellular response not only according to their location in the cell but also according to the time constant of their involvement. For instance, the fight-or-flight response involves very rapid alteration in EC coupling microdomains, without affecting much ET coupling microdomains, the latter being involved only when stress conditions are sustained. EM coupling microdomains are probably secondarily involved as a response to the initial increase in EC coupling for an acute adaptation to unusual stress conditions and, if the stress conditions persist, as a chronic adaptation to optimize the ATP supply of the cell. This necessitates a very narrow and subtle coupling between EC, EM, and ET microdomains and signaling both on a location and time basis.

Acknowledgments Funding was provided by the National Agency of Research (ANR-13-BSV1-0023, ANR-15-CE14-0005) Inserm, and University Paris-Sud. We thank the program Jean d'Alembert (Université Paris-Sud, Université Paris-Saclay) for providing an invited scientist fellowship to JLA. We are in debt to Françoise Boussac for administrative assistance.

Compliance with Ethical Standards

Conflict of Interest Statement The authors declare that they have no conflict of interest.

References

Abriel H, Rougier JS, Jalife J (2015) Ion channel macromolecular complexes in cardiomyocytes: roles in sudden cardiac death. Circ Res 116:1971–1988

Acsai K, Antoons G, Livshitz L et al (2011) Microdomain [Ca(2)(+)] near ryanodine receptors as reported by L-type Ca(2)(+) and Na+/Ca(2)(+) exchange currents. J Physiol 589:2569–2583

Adachi-Akahane S, Cleemann L, Morad M (1996) Cross-signaling between L-type Ca2+ channels and ryanodine receptors in rat ventricular myocytes. J Gen Physiol 108:435–454

Ai X, Curran JW, Shannon TR et al (2005) Ca2+/calmodulin-dependent protein kinase modulates cardiac ryanodine receptor phosphorylation and sarcoplasmic reticulum Ca2+ leak in heart failure. Circ Res 97:1314–1322

Alvarez BV, Perez NG, Ennis IL et al (1999) Mechanisms underlying the increase in force and Ca(2+) transient that follow stretch of cardiac muscle: a possible explanation of the Anrep effect. Circ Res 85:716–722

Andrienko TN, Picht E, Bers DM (2009) Mitochondrial free calcium regulation during sarcoplasmic reticulum calcium release in rat cardiac myocytes. J Mol Cell Cardiol 46:1027–1036

Awasthi S, Izu LT, Mao Z et al (2016) Multimodal SHG-2PF imaging of microdomain Ca2 + –contraction coupling in live cardiac myocytes. Circ Res 118:e19–e28

Baartscheer A, Schumacher CA, van Borren MM et al (2003) Increased Na+/H + –exchange activity is the cause of increased [Na+]i and underlies disturbed calcium handling in the rabbit pressure and volume overload heart failure model. Cardiovasc Res 57:1015–1024

Baines CP (2009) The molecular composition of the mitochondrial permeability transition pore. J Mol Cell Cardiol 46:850–857

Balaban RS, Bose S, French SA, Territo PR (2003) Role of calcium in metabolic signaling between cardiac sarcoplasmic reticulum and mitochondria in vitro. Am J Physiol Cell Physiol 284:C285–C293

Balshaw DM, Xu L, Yamaguchi N et al (2001) Calmodulin binding and inhibition of cardiac muscle calcium release channel (ryanodine receptor). J Biol Chem 276:20144–20153

Barta J, Toth A, Edes I et al (2005) Calpain-1-sensitive myofibrillar proteins of the human myocardium. Mol Cell Biochem 278:1–8

Bassani JW, Bassani RA, Bers DM (1994) Relaxation in rabbit and rat cardiac cells: species-dependent differences in cellular mechanisms. J Physiol 476:279–293

Baughman JM, Perocchi F, Girgis HS et al (2011) Integrative genomics identifies MCU as an essential component of the mitochondrial calcium uniporter. Nature 476:341–345

Bazil JN, Dash RK (2011) A minimal model for the mitochondrial rapid mode of Ca(2) + uptake mechanism. PLoS One 6:e21324

Becerra R, Roman B, Di Carlo MN et al (2016) Reversible redox modifications of ryanodine receptor ameliorate ventricular arrhythmias in the ischemic-reperfused heart. Am J Physiol Heart Circ Physiol 311:H713–H724

Benitah JP, Perrier E, Gómez AM, Vassort G (2001) Effects of aldosterone on transient outward K+ current density in rat ventricular myocytes. J Physiol Lond 537:151–160

Benitah JP, Kerfant BG, Vassort G et al (2002) Altered communication between L-type calcium channels and ryanodine receptors in heart failure. Front Biosci 7:E263–E275

Benitah JP, Alvarez JL, Gómez AM (2010) L-type Ca(2+) current in ventricular cardiomyocytes. J Mol Cell Cardiol 48:26–36

Ben-Johny M, Yue DT (2014) Calmodulin regulation (calmodulation) of voltage-gated calcium channels. J Gen Physiol 143:679–692

Benkusky NA, Weber CS, Scherman JA et al (2007) Intact beta-adrenergic response and unmodified progression toward heart failure in mice with genetic ablation of a major protein kinase a phosphorylation site in the cardiac ryanodine receptor. Circ Res 101:819–829

Bernardi P, Di Lisa F (2015) The mitochondrial permeability transition pore: molecular nature and role as a target in cardioprotection. J Mol Cell Cardiol 78:100–106

Bernardi P, Forte M (2007) The mitochondrial permeability transition pore. Novartis Found Symp 287:157–164; discussion 164–159

Bers DM (2008) Calcium cycling and signaling in cardiac myocytes. Annu Rev Physiol 70:23–49
Bers DM (2013) Membrane receptor neighborhoods: snuggling up to the nucleus. Circ Res 112:224–226
Bers DM (2014) Cardiac sarcoplasmic reticulum calcium leak: basis and roles in cardiac dysfunction. Annu Rev Physiol 76:107–127
Best JM, Kamp TJ (2012) Different subcellular populations of L-type Ca2+ channels exhibit unique regulation and functional roles in cardiomyocytes. J Mol Cell Cardiol 52:376–387
Bito V, Biesmans L, Gellen B et al (2013) FKBP12.6 overexpression does not protect against remodelling after myocardial infarction. Exp Physiol 98:134–148
Bolli R, Marban E (1999) Molecular and cellular mechanisms of myocardial stunning. Physiol Rev 79:609–634
Bossuyt J, Bers DM (2013) Visualizing CaMKII and CaM activity: a paradigm of compartmentalized signaling. J Mol Med (Berl) 91:907–916
Brandes R, Bers DM (1997) Intracellular Ca2+ increases the mitochondrial NADH concentration during elevated work in intact cardiac muscle. Circ Res 80:82–87
Brandes R, Bers DM (2002) Simultaneous measurements of mitochondrial NADH and Ca(2+) during increased work in intact rat heart trabeculae. Biophys J 83:587–604
Brillantes AB, Ondrias K, Scott A et al (1994) Stabilization of calcium release channel (ryanodine receptor) function by FK506-binding protein. Cell 77:513–523
Brini M, Carafoli E (2000) Calcium signalling: a historical account, recent developments and future perspectives. Cell Mol Life Sci 57:354–370
Brochet DX, Yang D, Di Maio A et al (2005) Ca2+ blinks: rapid nanoscopic store calcium signaling. Proc Natl Acad Sci U S A 102:3099–3104
Bush EW, Hood DB, Papst PJ et al (2006) Canonical transient receptor potential channels promote cardiomyocyte hypertrophy through activation of calcineurin signaling. J Biol Chem 281:33487–33496
Calderon-Sanchez EM, Ruiz-Hurtado G, Smani T et al (2011) Cardioprotective action of urocortin in postconditioning involves recovery of intracellular calcium handling. Cell Calcium 50:84–90
Cali T, Ottolini D, Brini M (2012) Mitochondrial Ca(2+) as a key regulator of mitochondrial activities. Adv Exp Med Biol 942:53–73
Carafoli E, Tiozzo R, Lugli G et al (1974) The release of calcium from heart mitochondria by sodium. J Mol Cell Cardiol 6:361–371
Carter S, Colyer J, Sitsapesan R (2006) Maximum phosphorylation of the cardiac ryanodine receptor at serine-2809 by protein kinase a produces unique modifications to channel gating and conductance not observed at lower levels of phosphorylation. Circ Res 98:1506–1513
Chen X, Nakayama H, Zhang X et al (2011) Calcium influx through Cav1.2 is a proximal signal for pathological cardiomyocyte hypertrophy. J Mol Cell Cardiol 50:460–470
Cheng H, Lederer WJ (2008) Calcium sparks. Physiol Rev 88:1491–1545
Cheng H, Lederer WJ, Cannell MB (1993) Calcium sparks: elementary events underlying excitation-contraction coupling in heart muscle. Science 262:740–744
Cheng H, Lederer MR, Lederer WJ, Cannell MB (1996) Calcium sparks and [Ca2+]i waves in cardiac myocytes. Am J Phys 270:C148–C159
Chiang CS, Huang CH, Chieng H et al (2009) The Ca(v)3.2 T-type Ca(2+) channel is required for pressure overload-induced cardiac hypertrophy in mice. Circ Res 104:522–530
Chopra N, Knollmann BC (2013) Triadin regulates cardiac muscle couplon structure and microdomain Ca(2+) signalling: a path towards ventricular arrhythmias. Cardiovasc Res 98:187–191
Correll RN, Goonasekera SA, van Berlo JH et al (2015) STIM1 elevation in the heart results in aberrant Ca(2)(+) handling and cardiomyopathy. J Mol Cell Cardiol 87:38–47
Crocini C, Coppini R, Ferrantini C et al (2014) Defects in T-tubular electrical activity underlie local alterations of calcium release in heart failure. Proc Natl Acad Sci U S A 111:15196–15201
De Stefani D, Raffaello A, Teardo E et al (2011) A forty-kilodalton protein of the inner membrane is the mitochondrial calcium uniporter. Nature 476:336–340
De Stefani D, Rizzuto R, Pozzan T (2016) Enjoy the trip: calcium in mitochondria back and forth. Annu Rev Biochem 85:161–192

Delgado C, Artiles A, Gómez AM, Vassort G (1999) Frequency-dependent increase in cardiac Ca2+ current is due to reduced Ca2+ release by the sarcoplasmic reticulum. J Mol Cell Cardiol 31:1783–1793
Denton RM, McCormack JG (1990) Ca2+ as a second messenger within mitochondria of the heart and other tissues. Annu Rev Physiol 52:451–466
Dominguez-Rodriguez A, Ruiz-Hurtado G, Benitah JP, Gómez AM (2012) The other side of cardiac Ca(2+) signaling: transcriptional control. Front Physiol 3:452
Drago I, De Stefani D, Rizzuto R, Pozzan T (2012) Mitochondrial Ca2+ uptake contributes to buffering cytoplasmic Ca2+ peaks in cardiomyocytes. Proc Natl Acad Sci U S A 109:12986–12991
Eder P, Molkentin JD (2011) TRPC channels as effectors of cardiac hypertrophy. Circ Res 108:265–272
Egger M, Gómez AM, Schwaller B et al (2001) Na+/Ca2+ exchange activity in the post-myocardial infarction rat model. Biophys J 80:2710
Eisner DA, Diaz ME, O'Neill SC, Trafford AW (2004) Physiological and pathological modulation of ryanodine receptor function in cardiac muscle. Cell Calcium 35:583–589
Erickson JR, Pereira L, Wang L et al (2013) Diabetic hyperglycaemia activates CaMKII and arrhythmias by O-linked glycosylation. Nature 502:372–376
Fabiato A, Fabiato F (1975) Contractions induced by a calcium-triggered release of calcium from the sarcoplasmic reticulum of single skinned cardiac cells. J Physiol 249:469–495
Farrell EF, Antaramián A, Rueda A et al (2003) Sorcin inhibits calcium release and modulates excitation-contraction coupling in the heart. J Biol Chem 278:34660–34666
Fernandez-Velasco M, Rueda A, Rizzi N et al (2009) Increased Ca2+ sensitivity of the ryanodine receptor mutant RyR2R4496C underlies catecholaminergic polymorphic ventricular tachycardia. Circ Res 104:201–209
Fernández-Velasco M, Ruiz-Hurtado G, Rueda A et al (2011) RyRCa2+ leak limits cardiac Ca2+ window current overcoming the tonic effect of calmodulinin mice. PLoS One 6:e20863
Fill M, Copello JA (2002) Ryanodine receptor calcium release channels. Physiol Rev 82:893–922
Finkel T, Menazza S, Holmstrom KM et al (2015) The ins and outs of mitochondrial calcium. Circ Res 116:1810–1819
Galfre E, Pitt SJ, Venturi E et al (2012) FKBP12 activates the cardiac ryanodine receptor Ca2 + – release channel and is antagonised by FKBP12.6. PLoS One 7:e31956
Gangopadhyay JP, Ikemoto N (2011) Aberrant interaction of calmodulin with the ryanodine receptor develops hypertrophy in the neonatal cardiomyocyte. Biochem J 438:379–387
Gao H, Wang F, Wang W et al (2012) Ca(2+) influx through L-type Ca(2+) channels and transient receptor potential channels activates pathological hypertrophy signaling. J Mol Cell Cardiol 53:657–667
Gellen B, Fernandez-Velasco M, Briec F et al (2008) Conditional FKBP12.6 overexpression in mouse cardiac myocytes prevents triggered ventricular tachycardia through specific alterations in excitation-contraction coupling. Circulation 117:1778–1786
Gómez AM, Cheng HP, Lederer WJ, Bers DM (1996) Ca2+ diffusion and sarcoplasmic reticulum transport both contribute to Ca2+ (i) decline during Ca2+ sparks in rat ventricular myocytes. J Physiol Lond 496:575–581
Gómez AM, Valdivia HH, Cheng H et al (1997) Defective excitation-contraction coupling in experimental cardiac hypertrophy and heart failure. Science 276:800–806
Gómez AM, Guatimosim S, Dilly KW, Vassort G, Lederer WJ (2001) Heart failure after myocardial infarction: altered excitation-contraction coupling. Circulation 104:688–693
Gómez AM, Schuster I, Fauconnier J et al (2004) FKBP12.6 overexpression decreases Ca2+ spark amplitude but enhances [Ca2+](i) transient in rat cardiac myocytes. Am J Phys Heart Circ Phys 287:H1987–H1993
Gómez AM, Rueda A, Sainte-Marie Y et al (2009) Mineralocorticoid modulation of cardiac ryanodine receptor activity is associated with downregulation of FK506-binding proteins. Circulation 119:2179–U2189
Gonzalez DR, Treuer AV, Castellanos J et al (2010) Impaired S-nitrosylation of the ryanodine receptor caused by xanthine oxidase activity contributes to calcium leak in heart failure. J Biol Chem 285:28938–28945

Gunter TE, Pfeiffer DR (1990) Mechanisms by which mitochondria transport calcium. Am J Phys 258:C755–C786

Guo T, Zhang T, Mestril R, Bers DM (2006) Ca2+/calmodulin-dependent protein kinase II phosphorylation of ryanodine receptor does affect calcium sparks in mouse ventricular myocytes. Circ Res 99:398–406

Guo T, Cornea RL, Huke S et al (2010) Kinetics of FKBP12.6 binding to ryanodine receptors in permeabilized cardiac myocytes and effects on Ca sparks. Circ Res 106:1743–1752

Gyorke I, Gyorke S (1998) Regulation of the cardiac ryanodine receptor channel by luminal Ca2+ involves luminal Ca2+ sensing sites. Biophys J 75:2801–2810

Gyorke S, Terentyev D (2008) Modulation of ryanodine receptor by luminal calcium and accessory proteins in health and cardiac disease. Cardiovasc Res 77:245–255

Hanna AD, Lam A, Thekkedam C et al (2014) Cardiac ryanodine receptor activation by a high Ca(2)(+) store load is reversed in a reducing cytoplasmic redox environment. J Cell Sci 127:4531–4541

Harzheim D, Movassagh M, Foo RS et al (2009) Increased InsP3Rs in the junctional sarcoplasmic reticulum augment Ca2+ transients and arrhythmias associated with cardiac hypertrophy. Proc Natl Acad Sci U S A 106:11406–11411

Heineke J, Molkentin JD (2006) Regulation of cardiac hypertrophy by intracellular signalling pathways. Nat Rev Mol Cell Biol 7:589–600

Ho HT, Belevych AE, Liu B et al (2016) Muscarinic stimulation facilitates sarcoplasmic reticulum Ca release by modulating ryanodine receptor 2 phosphorylation through protein kinase G and Ca/calmodulin-dependent protein kinase II. Hypertension 68(5):1171–1178

Holton ML, Wang W, Emerson M et al (2010) Plasma membrane calcium ATPase proteins as novel regulators of signal transduction pathways. World J Biol Chem 1:201–208

Horiba M, Muto T, Ueda N et al (2008) T-type Ca2+ channel blockers prevent cardiac cell hypertrophy through an inhibition of calcineurin-NFAT3 activation as well as L-type Ca2+ channel blockers. Life Sci 82:554–560

Hoth M, Button DC, Lewis RS (2000) Mitochondrial control of calcium-channel gating: a mechanism for sustained signaling and transcriptional activation in T lymphocytes. Proc Natl Acad Sci U S A 97:10607–10612

Houser SR (2014) Role of RyR2 phosphorylation in heart failure and arrhythmias: protein kinase A-mediated hyperphosphorylation of the ryanodine receptor at serine 2808 does not alter cardiac contractility or cause heart failure and arrhythmias. Circ Res 114:1320–1327; discussion 1327

Huang F, Shan J, Reiken S et al (2006) Analysis of calstabin2 (FKBP12.6)-ryanodine receptor interactions: rescue of heart failure by calstabin2 in mice. Proc Natl Acad Sci U S A 103:3456–3461

Huang H, Wang W, Liu P et al (2009) TRPC1 expression and distribution in rat hearts. Eur J Histochem: EJH 53:e26

Hudmon A, Schulman H, Kim J et al (2005) CaMKII tethers to L-type Ca2+ channels, establishing a local and dedicated integrator of Ca2+ signals for facilitation. J Cell Biol 171:537–547

Huke S, Bers DM (2008) Ryanodine receptor phosphorylation at serine 2030, 2808 and 2814 in rat cardiomyocytes. Biochem Biophys Res Commun 376:80–85

Hunton DL, Zou L, Pang Y, Marchase RB (2004) Adult rat cardiomyocytes exhibit capacitative calcium entry. Am J Physiol Heart Circ Physiol 286:H1124–H1132

Huser J, Blatter LA, Sheu SS (2000) Mitochondrial calcium in heart cells: beat-to-beat oscillations or slow integration of cytosolic transients? J Bioenerg Biomembr 32:27–33

Ibarra C, Vicencio JM, Varas-Godoy M et al (2014) An integrated mechanism of cardiomyocyte nuclear Ca(2+) signaling. J Mol Cell Cardiol 75:40–48

Iribe G, Ward CW, Camelliti P et al (2009) Axial stretch of rat single ventricular cardiomyocytes causes an acute and transient increase in Ca2+ spark rate. Circ Res 104:787–795

Isenberg G, Kazanski V, Kondratev D et al (2003) Differential effects of stretch and compression on membrane currents and [Na+]c in ventricular myocytes. Prog Biophys Mol Biol 82:43–56

Iwata Y, Katanosaka Y, Arai Y et al (2003) A novel mechanism of myocyte degeneration involving the Ca2 + −permeable growth factor-regulated channel. J Cell Biol 161:957–967

Jeyakumar LH, Ballester L, Cheng DS et al (2001) FKBP binding characteristics of cardiac microsomes from diverse vertebrates. Biochem Biophys Res Commun 281:979–986

Jiang MT, Lokuta AJ, Farrell EF et al (2002) Abnormal Ca2+ release, but normal ryanodine receptors, in canine and human heart failure. Circ Res 91:1015–1022
Kamkin A, Kiseleva I, Isenberg G (2003) Ion selectivity of stretch-activated cation currents in mouse ventricular myocytes. Pflugers Arch 446:220–231
Kirichok Y, Krapivinsky G, Clapham DE (2004) The mitochondrial calcium uniporter is a highly selective ion channel. Nature 427:360–364
Kobrinsky E, Duong SQ, Sheydina A, Soldatov NM (2011) Microdomain organization and frequency-dependence of CREB-dependent transcriptional signaling in heart cells. FASEB J 25:1544–1555
Kohl P, Cooper PJ, Holloway H (2003) Effects of acute ventricular volume manipulation on in situ cardiomyocyte cell membrane configuration. Prog Biophys Mol Biol 82:221–227
Kushnir A, Marks AR (2012) Ryanodine receptor patents. Recent Pat Biotechnol 6:157–166
Kuwahara K, Wang Y, McAnally J et al (2006) TRPC6 fulfills a calcineurin signaling circuit during pathologic cardiac remodeling. J Clin Invest 116:3114–3126
Lab MJ (1996) Mechanoelectric feedback (transduction) in heart: concepts and implications. Cardiovasc Res 32:3–14
Lam E, Martin MM, Timerman AP et al (1995) A novel FK506 binding protein can mediate the immunosuppressive effects of FK506 and is associated with the cardiac ryanodine receptor. J Biol Chem 270:26511–26522
Li Y, Kranias EG, Mignery GA, Bers DM (2002) Protein kinase a phosphorylation of the ryanodine receptor does not affect calcium sparks in mouse ventricular myocytes. Circ Res 90:309–316
Lindner M, Brandt MC, Sauer H et al (2002) Calcium sparks in human ventricular cardiomyocytes from patients with terminal heart failure. Cell Calcium 31:175–182
Ljubojevic S, Bers DM (2015) Nuclear calcium in cardiac myocytes. J Cardiovasc Pharmacol 65:211–217
Lokuta AJ, Meyers MB, Sander PR et al (1997) Modulation of cardiac ryanodine receptors by sorcin. J Biol Chem 272:25333–25338
Loughrey CM, Seidler T, Miller SL et al (2004) Over-expression of FK506-binding protein FKBP12.6 alters excitation-contraction coupling in adult rabbit cardiomyocytes. J Physiol 556(Pt 3):919–934
Loyer X, Gómez AM, Milliez P et al (2008) Cardiomyocyte overexpression of neuronal nitric oxide synthase delays transition toward heart failure in response to pressure overload by preserving calcium cycling. Circulation 117:3187–3198
Luo X, Hojayev B, Jiang N et al (2012) STIM1-dependent store-operated Ca(2) entry is required for pathological cardiac hypertrophy. J Mol Cell Cardiol 52:136–147
Maack C, Cortassa S, Aon MA et al (2006) Elevated cytosolic Na + decreases mitochondrial Ca2+ uptake during excitation-contraction coupling and impairs energetic adaptation in cardiac myocytes. Circ Res 99:172–182
Makarewich CA, Correll RN, Gao H et al (2012) A caveolae-targeted L-type Ca(2) + channel antagonist inhibits hypertrophic signaling without reducing cardiac contractility. Circ Res 110:669–674
Makarewich CA, Zhang H, Davis J et al (2014) Transient receptor potential channels contribute to pathological structural and functional remodeling after myocardial infarction. Circ Res 115:567–580
Malli R, Frieden M, Osibow K, Graier WF (2003) Mitochondria efficiently buffer subplasmalemmal Ca2+ elevation during agonist stimulation. J Biol Chem 278:10807–10815
Marx SO, Reiken S, Hisamatsu Y et al (2000) PKA phosphorylation dissociates FKBP12.6 from the calcium release channel (ryanodine receptor): defective regulation in failing hearts. Cell 101:365–376
Meyers MB, Pickel VM, Sheu SS et al (1995a) Association of sorcin with the cardiac ryanodine receptor. J Biol Chem 270:26411–26418
Meyers MB, Zamparelli C, Verzili D et al (1995b) Calcium-dependent translocation of sorcin to membranes: functional relevance in contractile tissue. FEBS Lett 357:230–234

Meyers MB, Puri TS, Chien AJ et al (1998) Sorcin associates with the pore-forming subunit of voltage-dependent L-type Ca2+ channels. J Biol Chem 273:18930–18935
Mitchell P (1961) Coupling of phosphorylation to electron and hydrogen transfer by a chemi-osmotic type of mechanism. Nature 191:144–148
Morad M, Soldatov N (2005) Calcium channel inactivation: possible role in signal transduction and Ca2+ signaling. Cell Calcium 38:223–231
Nakayama H, Wilkin BJ, Bodi I, Molkentin JD (2006) Calcineurin-dependent cardiomyopathy is activated by TRPC in the adult mouse heart. FASEB J 20:1660–1670
Nakayama H, Bodi I, Maillet M et al (2010) The IP3 receptor regulates cardiac hypertrophy in response to select stimuli. Circ Res 107:659–666
Naranjo JR, Mellstrom B (2012) Ca2 + −dependent transcriptional control of Ca2+ homeostasis. J Biol Chem 287:31674–31680
Nicholls DG, Crompton M (1980) Mitochondrial calcium transport. FEBS Lett 111:261–268
Nichols CB, Rossow CF, Navedo MF et al (2010) Sympathetic stimulation of adult cardiomyocytes requires association of AKAP5 with a subpopulation of L-type calcium channels. Circ Res 107:747–756
Nicolli A, Basso E, Petronilli V et al (1996) Interactions of cyclophilin with the mitochondrial inner membrane and regulation of the permeability transition pore, and cyclosporin A-sensitive channel. J Biol Chem 271:2185–2192
O'Brien F, Venturi E, Sitsapesan R (2015) The ryanodine receptor provides high throughput Ca2 + −release but is precisely regulated by networks of associated proteins: a focus on proteins relevant to phosphorylation. Biochem Soc Trans 43:426–433
Ong HL, Ambudkar IS (2011) The dynamic complexity of the TRPC1 channelosome. Channels (Austin) 5:424–431
Ono K, Yano M, Ohkusa T et al (2000) Altered interaction of FKBP12.6 with ryanodine receptor as a cause of abnormal Ca(2+) release in heart failure. Cardiovasc Res 48:323–331
Onohara N, Nishida M, Inoue R et al (2006) TRPC3 and TRPC6 are essential for angiotensin II-induced cardiac hypertrophy. EMBO J 25:5305–5316
O'Rourke B, Blatter LA (2009) Mitochondrial Ca2+ uptake: tortoise or hare? J Mol Cell Cardiol 46:767–774
Overend CL, Eisner DA, O'Neill SC (2001) Altered cardiac sarcoplasmic reticulum function of intact myocytes of rat ventricle during metabolic inhibition. Circ Res 88:181–187
Pacher P, Thomas AP, Hajnoczky G (2002) Ca2+ marks: miniature calcium signals in single mitochondria driven by ryanodine receptors. Proc Natl Acad Sci U S A 99:2380–2385
Palty R, Silverman WF, Hershfinkel M et al (2010) NCLX is an essential component of mitochondrial Na+/Ca2+ exchange. Proc Natl Acad Sci U S A 107:436–441
Parks C, Alam MA, Sullivan R, Mancarella S (2016) STIM1-dependent Ca(2+) microdomains are required for myofilament remodeling and signaling in the heart. Sci Rep 6:25372
Pessah IN, Waterhouse AL, Casida JE (1985) The calcium-ryanodine receptor complex of skeletal and cardiac muscle. Biochem Biophys Res Commun 128:449–456
Petroff MG, Kim SH, Pepe S et al (2001) Endogenous nitric oxide mechanisms mediate the stretch dependence of Ca2+ release in cardiomyocytes. Nat Cell Biol 3:867–873
Pinali C, Bennett H, Davenport JB et al (2013) Three-dimensional reconstruction of cardiac sarcoplasmic reticulum reveals a continuous network linking transverse-tubules: this organization is perturbed in heart failure. Circ Res 113:1219–1230
Poteser M, Schleifer H, Lichtenegger M et al (2011) PKC-dependent coupling of calcium permeation through transient receptor potential canonical 3 (TRPC3) to calcineurin signaling in HL-1 myocytes. Proc Natl Acad Sci U S A 108:10556–10561
Prosser BL, Ward CW, Lederer WJ (2011) X-ROS signaling: rapid mechano-chemo transduction in heart. Science 333:1440–1445
Reiken S, Gaburjakova M, Guatimosim S et al (2003) Protein kinase a phosphorylation of the cardiac calcium release channel (ryanodine receptor) in normal and failing hearts - role of phosphatases and response to isoproterenol. J Biol Chem 278:444–453

Ringer S (1883) A further contribution regarding the influence of the different constituents of the blood on the contraction of the heart. J Physiol 4(29–42):23

Rizzuto R, Pozzan T (2006) Microdomains of intracellular Ca2+: molecular determinants and functional consequences. Physiol Rev 86:369–408

Rizzuto R, Brini M, Murgia M, Pozzan T (1993) Microdomains with high Ca2+ close to IP3-sensitive channels that are sensed by neighboring mitochondria. Science 262:744–747

Robert V, Gurlini P, Tosello V et al (2001) Beat-to-beat oscillations of mitochondrial [Ca2+] in cardiac cells. EMBO J 20:4998–5007

Rodriguez P, Bhogal MS, Colyer J (2003) Stoichiometric phosphorylation of cardiac ryanodine receptor on serine 2809 by calmodulin-dependent kinase II and protein kinase a. J Biol Chem 278:38593–38600

Ronkainen JJ, Hanninen SL, Korhonen T et al (2011) Ca2 + −calmodulin-dependent protein kinase II represses cardiac transcription of the L-type calcium channel alpha(1C)-subunit gene (Cacna1c) by DREAM translocation. J Physiol 589:2669–2686

Ruiz-Hurtado G, Li L, Fernández-Velasco M et al (2015) Reconciling depressed Ca2+ sparks occurrence with enhanced RyR2 activity in failing mice cardiomyocytes. J Gen Physiol 146(4):295–306

Sag CM, Kohler AC, Anderson ME et al (2011) CaMKII-dependent SR Ca leak contributes to doxorubicin-induced impaired Ca handling in isolated cardiac myocytes. J Mol Cell Cardiol 51:749–759

Sancak Y, Markhard AL, Kitami T et al (2013) EMRE is an essential component of the mitochondrial calcium uniporter complex. Science 342:1379–1382

Santana LF, Cheng H, Gómez AM et al (1996) Relation between the sarcolemmal Ca2+ current and Ca2+ sparks and local control theories for cardiac excitation-contraction coupling. Circ Res 78:166–171

Saris NE, Allshire A (1989) Calcium ion transport in mitochondria. Methods Enzymol 174:68–85

Sasaki N, Mitsuiye T, Noma A (1992) Effects of mechanical stretch on membrane currents of single ventricular myocytes of Guinea-pig heart. Jpn J Physiol 42:957–970

Saucerman JJ, Bers DM (2012) Calmodulin binding proteins provide domains of local Ca2+ signaling in cardiac myocytes. J Mol Cell Cardiol 52:312–316

Shah SJ, Aistrup GL, Gupta DK et al (2014) Ultrastructural and cellular basis for the development of abnormal myocardial mechanics during the transition from hypertension to heart failure. Am J Physiol Heart Circ Physiol 306:H88–100

Sham JS (1997) Ca2+ release-induced inactivation of Ca2+ current in rat ventricular myocytes: evidence for local Ca2+ signalling. J Physiol 500(Pt 2):285–295

Shang W, Lu F, Sun T et al (2014) Imaging Ca2+ nanosparks in heart with a new targeted biosensor. Circ Res 114:412–420

Sharma VK, Ramesh V, Franzini-Armstrong C, Sheu SS (2000) Transport of Ca2+ from sarcoplasmic reticulum to mitochondria in rat ventricular myocytes. J Bioenerg Biomembr 32:97–104

Sparagna GC, Gunter KK, Sheu SS, Gunter TE (1995) Mitochondrial calcium uptake from physiological-type pulses of calcium. A description of the rapid uptake mode. J Biol Chem 270:27510–27515

Stange M, Xu L, Balshaw D et al (2003) Characterization of recombinant skeletal muscle (Ser-2843) and cardiac muscle (Ser-2809) ryanodine receptor phosphorylation mutants. J Biol Chem 278:51693–51702

Stern MD (1992) Theory of excitation-contraction coupling in cardiac muscle. Biophys J 63:497–517

Suarez J, Belke DD, Gloss B et al (2004) In vivo adenoviral transfer of sorcin reverses the cardiac contractile abnormalities of diabetic cardiomyopathy. Am J Physiol Heart Circ Physiol 286:H68–H75

Szabadkai G, Bianchi K, Varnai P et al (2006) Chaperone-mediated coupling of endoplasmic reticulum and mitochondrial Ca2+ channels. J Cell Biol 175:901–911

Tandan S, Wang Y, Wang TT et al (2009) Physical and functional interaction between calcineurin and the cardiac L-type Ca2+ channel. Circ Res 105:51–60
Tanskanen AJ, Greenstein JL, Chen A et al (2007) Protein geometry and placement in the cardiac dyad influence macroscopic properties of calcium-induced calcium release. Biophys J 92:3379–3396
Terentyev D, Gyorke I, Belevych AE et al (2008) Redox modification of ryanodine receptors contributes to sarcoplasmic reticulum Ca2+ leak in chronic heart failure. Circ Res 103:1466–1472
Territo PR, Mootha VK, French SA, Balaban RS (2000) Ca(2+) activation of heart mitochondrial oxidative phosphorylation: role of the F(0)/F(1)-ATPase. Am J Physiol Cell Physiol 278:C423–C435
Timerman AP, Onoue H, Xin HB et al (1996) Selective binding of FKBP12.6 by the cardiac ryanodine receptor. J Biol Chem 271:20385–20391
Trafford AW, Diaz ME, Sibbring GC, Eisner DA (2000) Modulation of CICR has no maintained effect on systolic Ca2+: simultaneous measurements of sarcoplasmic reticulum and sarcolemmal Ca2+ fluxes in rat ventricular myocytes. J Physiol 522(Pt 2):259–270
Trollinger DR, Cascio WE, Lemasters JJ (1997) Selective loading of Rhod 2 into mitochondria shows mitochondrial Ca2+ transients during the contractile cycle in adult rabbit cardiac myocytes. Biochem Biophys Res Commun 236:738–742
Ullrich ND, Valdivia HH, Niggli E (2012) PKA phosphorylation of cardiac ryanodine receptor modulates SR luminal Ca2+ sensitivity. J Mol Cell Cardiol 53:33–42
Vais H, Mallilankaraman K, Mak DO et al (2016) EMRE is a matrix Ca(2+) sensor that governs gatekeeping of the mitochondrial Ca(2+) uniporter. Cell Rep 14:403–410
Valdivia HH (1998) Modulation of intracellular Ca2+ levels in the heart by sorcin and FKBP12, two accessory proteins of ryanodine receptors. Trends Pharmacol Sci 19:479–482
Valdivia HH (2012) Ryanodine receptor phosphorylation and heart failure: phasing out S2808 and "criminalizing" S2814. Circ Res 110:1398–1402
Valdivia HH, Kaplan JH, Ellis-Davies GC, Lederer WJ (1995) Rapid adaptation of cardiac ryanodine receptors: modulation by Mg2+ and phosphorylation. Science 267:1997–2000
Vinet L, Pezet M, Bito V et al (2012) Cardiac FKBP12.6 overexpression protects against triggered ventricular tachycardia in pressure overloaded mouse hearts. Basic Res Cardiol 107:246
Viola HM, Hool LC (2011) Targeting calcium and the mitochondria in prevention of pathology in the heart. Curr Drug Targets 12:748–760
Viola HM, Hool LC (2013) Role of the cytoskeleton in communication between L-type Ca(2+) channels and mitochondria. Clin Exp Pharmacol Physiol 40:295–304
Viola HM, Hool LC (2014) How does calcium regulate mitochondrial energetics in the heart?—new insights. Heart Lung Circ 23:602–609
Viola HM, Arthur PG, Hool LC (2007) Transient exposure to hydrogen peroxide causes an increase in mitochondria-derived superoxide as a result of sustained alteration in L-type Ca2+ channel function in the absence of apoptosis in ventricular myocytes. Circ Res 100:1036–1044
Viola HM, Arthur PG, Hool LC (2009) Evidence for regulation of mitochondrial function by the L-type Ca2+ channel in ventricular myocytes. J Mol Cell Cardiol 46:1016–1026
Viola HM, Davies SM, Filipovska A, Hool LC (2013) L-type Ca(2+) channel contributes to alterations in mitochondrial calcium handling in the mdx ventricular myocyte. Am J Physiol Heart Circ Physiol 304:H767–H775
Vlahos CJ, McDowell SA, Clerk A (2003) Kinases as therapeutic targets for heart failure. Nat Rev Drug Discov 2:99–113
Wang SQ, Song LS, Lakatta EG, Cheng H (2001) Ca2+ signalling between single L-type Ca2+ channels and ryanodine receptors in heart cells. Nature 410:592–596
Watanabe H, Chopra N, Laver D et al (2009) Flecainide prevents catecholaminergic polymorphic ventricular tachycardia in mice and humans. Nat Med 15:380–383
Wehrens XH, Lehnart SE, Reiken SR et al (2004) Protection from cardiac arrhythmia through ryanodine receptor-stabilizing protein calstabin2. Science 304:292–296
Wehrens XH, Lehnart SE, Marks AR (2005) Intracellular calcium release and cardiac disease. Annu Rev Physiol 67:69–98

Wheeler DG, Barrett CF, Groth RD et al (2008) CaMKII locally encodes L-type channel activity to signal to nuclear CREB in excitation-transcription coupling. J Cell Biol 183:849–863
Williams GS, Boyman L, Chikando AC et al (2013) Mitochondrial calcium uptake. Proc Natl Acad Sci U S A 110:10479–10486
Wu X, Bers DM (2007) Free and bound intracellular calmodulin measurements in cardiac myocytes. Cell Calcium 41:353–364
Wu X, Eder P, Chang B, Molkentin JD (2010) TRPC channels are necessary mediators of pathologic cardiac hypertrophy. Proc Natl Acad Sci U S A 107:7000–7005
Xiao RP, Valdivia HH, Bogdanov K et al (1997) The immunophilin FK506-binding protein modulates Ca2+ release channel closure in rat heart. J Physiol Lond 500:343–354
Xiao B, Sutherland C, Walsh MP, Chen SR (2004) Protein kinase a phosphorylation at serine-2808 of the cardiac Ca2 + –release channel (ryanodine receptor) does not dissociate 12.6-kDa FK506-binding protein (FKBP12.6). Circ Res 94:487–495
Xiao J, Tian X, Jones PP et al (2007) Removal of FKBP12.6 does not alter the conductance and activation of the cardiac ryanodine receptor or the susceptibility to stress-induced ventricular arrhythmias. J Biol Chem 282:34828–34838
Xu L, Mann G, Meissner G (1996) Regulation of cardiac Ca2+ release channel (ryanodine receptor) by Ca2+, H+, Mg2+, and adenine nucleotides under normal and simulated ischemic conditions. Circ Res 79:1100–1109
Xu L, Eu JP, Meissner G, Stamler JS (1998) Activation of the cardiac calcium release channel (ryanodine receptor) by poly-S-nitrosylation. Science 279:234–237
Yano M, Ono K, Ohkusa T et al (2000) Altered stoichiometry of FKBP12.6 versus ryanodine receptor as a cause of abnormal Ca(2+) leak through ryanodine receptor in heart failure. Circulation 102:2131–2136
Yano M, Yamamoto T, Ikeda Y, Matsuzaki M (2006) Mechanisms of disease: ryanodine receptor defects in heart failure and fatal arrhythmia. Nat Clin Pract Cardiovasc Med 3:43–52
Zamparelli C, Ilari A, Verzili D et al (2000) Structure-function relationships in sorcin, a member of the penta EF-hand family. Interaction of sorcin fragments with the ryanodine receptor and an *Escherichia coli* model system. Biochemistry 39:658–666
Zhang YH, Youm JB, Sung HK et al (2000) Stretch-activated and background non-selective cation channels in rat atrial myocytes. J Physiol 523(Pt 3):607–619
Zhang H, Makarewich CA, Kubo H et al (2012) Hyperphosphorylation of the cardiac ryanodine receptor at serine 2808 is not involved in cardiac dysfunction after myocardial infarction. Circ Res 110:831–840
Zhang HL, Gómez AM, Wang XH et al (2013) ROS regulation of microdomain Ca2+ signalling at the dyads. Cardiovasc Res 98:248–258
Zhao YT, Valdivia CR, Gurrola GB et al (2015) Arrhythmogenesis in a catecholaminergic polymorphic ventricular tachycardia mutation that depresses ryanodine receptor function. Proc Natl Acad Sci U S A 112:E1669–E1677
Zima AV, Bovo E, Mazurek SR et al (2014) Ca handling during excitation-contraction coupling in heart failure. Pflugers Arch 466:1129–1137

Correction to: Microdomains in the Cardiovascular System

Viacheslav Nikolaev and Manuela Zaccolo

Correction to:
V. Nikolaev, M. Zaccolo (eds.), *Microdomains in the Cardiovascular System*, Cardiac and Vascular Biology, DOI 10.1007/978-3-319-54579-0

The original version of this volume was revised as it was originally published unnumbered. The revised version now has been numbered; numbering is done following the order of appearance in the book series Cardiac and Vascular Biology.

The updated online version of the book can be found at 10.1007/978-3-319-54579-0

V. Nikolaev, M. Zaccolo (eds.), *Microdomains in the Cardiovascular System*, Cardiac and Vascular Biology 3, https://doi.org/10.1007/978-3-319-54579-0_18

MIX
Papier aus verantwortungsvollen Quellen
Paper from responsible sources
FSC® C105338

If you have any concerns about our products,
you can contact us on
ProductSafety@springernature.com

In case Publisher is established outside the EU,
the EU authorized representative is:
Springer Nature Customer Service Center GmbH
Europaplatz 3, 69115 Heidelberg, Germany

Printed by Libri Plureos GmbH
in Hamburg, Germany